Lecture Notes in Economics and Mathematical Systems

481

Springer-Verlag Berlin Heidelberg GmbH

Van Hien Nguyen   Jean-Jacques Strodiot
Patricia Tossings

Optimization

Proceedings of the 9th Belgian-French-
German Conference on Optimization
Namur, September 7-11, 1998

 Springer

Editors:

Prof. Van Hien Nguyen
Prof. Jean-Jacques Strodiot

University of Namur
Department of Mathematics
Rue de Bruxelles 61
5000 Namur, Belgium

Dr. Patricia Tossings
University of Liège
Institut de Mathématiques (B37)
Grande Traverse 12
4000 Liège, Belgium

Library of Congress Cataloging-in-Publication Data

French-German Conference on Optimization (9th : 1996 : Namur, Belgium)
   Optimization : proceedings of the 9th Belgian-French-German Conference on
Optimization, Namur, September 7-11, 1998 / Van Hien Nguyen, Jean-Jacques Strodiot,
Patricia Tossings (eds.).
      p. cm. -- (Lecture notes in economics and mathematical systems, ISSN 0075-8442 ; 481)
   Includes bibliographical references.
   ISBN 3540669051 (softcover : alk. paper)
   1. Mathematical optimization--Congresses. I. Nguyen, Van Hien, 1943- II. Strodiot,
Jean-Jacques, 1944- III. Tossings, Patricia, 1964- IV. Title. V. Series.

QA402.5 .F736 1996b
519.3--dc21

                                                                                99-059997

ISBN 978-3-540-66905-0   ISBN 978-3-642-57014-8 (eBook)
DOI 10.1007/978-3-642-57014-8

Springer-Verlag is a company in the specialist publishing group BertelsmannSpringer
© Springer-Verlag Berlin Heidelberg 2000
Originally published by Spinger-Verlag Berlin Heidelberg New York in 2000

Preface

The 9th Belgian–French–German Conference on Optimization has been held in Namur (Belgium) on September 7–11, 1998. This volume is a collection of papers presented at this Conference.

Originally, this Conference was a French–German Conference but this year, in accordance with the organizers' wishes, a third country, Belgium, has joined the founding members of the Conference. Hence the name: Belgian–French–German Conference on Optimization.

Since the very beginning, the purpose of these Conferences has been to bring together researchers working in the area of Optimization and particularly to encourage young researchers to present their work. Most of the participants come from the organizing countries. However the general tendancy is to invite outside researchers to attend the meeting. So this year, among the 101 participants at this Conference, twenty researchers came from other countries.

The general theme of the Conference is everything that concerns the area of Optimization without specification of particular topics. So theoretical aspects of Optimization, in addition to applications and algorithms of Optimization, will be developed. However, and this point was very important for the organizers, the Conference must retain its convivial character. No more than two parallel sessions are organized. This would allow useful contacts between researchers to be promoted. The editors express their sincere thanks to all those who took part in this Conference. Their invaluable discussions have made this volume possible.

The Scientific Committee of the Conference consisted of D. Azé (Perpignan), V. Blondel (Liège), J.C. Gilbert (Inria–Rocquencourt), J. Gwinner (München), J. Jahn (Erlangen), F. Jarre (Würsburg), C. Malivert (Limoges), C. Michelot (Dijon), Y. Nesterov (Louvain–la–Neuve), V.H. Nguyen (Namur), D. Noll (Toulouse), P. Recht (Dormund), W. Römisch (Berlin), A. Seeger (Avignon), J.J. Strodiot (Namur), R. Tichatschke (Trier), Ph. Toint (Namur), P. Tossings (Liège).

The organizers of the Conference greatfully acknowledge the financial support granted by the *Belgian Fonds National de la Recherche Scientifique*, the *Communauté Française de Belgique*, the *European Commission DG XII*, the bank *BACOB* and the three Belgian universities: *Facultés Universitaires N.D. de la Paix, Namur, Facultés Universitaires Saint Louis, Brussels* and *Université de Liège, Liège*.

The editors are indebted to the referees for their helpful comments and suggestions. Finally, they want to express their special thanks to Springer–Verlag for having offered to publish their volume and for their constant encouragement and cooperation.

V.H. Nguyen   J.J. Strodiot   P. Tossings

Contents

Linearization Method for Solving Equilibrium Programming Problems

Anatoly Antipin

Computing Center Russian Academy of Sciences,
Vavilov str., 40, Moscow 117967, Russia

Abstract. In this paper, we study the canonical form of the equilibrium programming problem. For solving it, we consider the linearization method and we prove its convergence.

1   Statement of problem

Let us consider the problem of computing a fixed point of the extreme constrained mapping [1],[2]

$$find\ v^* \in \Omega_0\ such\ that\ v^* \in \mathrm{Argmin}\{\Phi(v^*, w) \mid g(w) \leq 0,\ w \in \Omega_0\}. \tag{1}$$

Here the function $\Phi(v, w)$ is defined on the product space $R^n \times R^n$ and $\Omega_0 \subset R^n$ is a convex closed set. It is supposed that $\Phi(v, w)$ is convex in $w \in \Omega_0$ for any $v \in \Omega_0$. The vector-valued function $g(w)$ has the dimension m. Every component of this function is convex. The variable $v \in \Omega_0$ is a parameter and $w \in \Omega_0$ is the optimization variable. We also assume that the extreme (marginal) mapping $w(v) \equiv \mathrm{Argmin}\{\Phi(v, w) \mid g(w) \leq 0,\ w \in \Omega_0\}$ is defined for all $v \in \Omega_0$ and the solution set $\Omega^* = \{v^* \in \Omega_0 \mid v^* \in w(v^*)\}$ of the initial problem is nonempty. According to Kakutani's fixed point theorem the latter assertion follows from the continuity of $\Phi(v, w)$ and the convexity of $\Phi(v, w)$ in w for any $v \in \Omega_0$, where Ω_0 is compact. In this case $w(v)$ is an upper semi-continuous mapping that maps each point of the convex, compact set Ω_0 into a closed convex subset of Ω_0 [3]. According to (1), any fixed point satisfies the inequality

$$\Phi(v^*, v^*) \leq \Phi(v^*, w)\ \forall w \in \Omega, \tag{2}$$

where $\Omega = \{w \mid g(w) \leq 0,\ w \in \Omega_0\}$ is an admissible set of (1). Let us introduce the function $\Psi(v, w) = \Phi(v, w) - \Phi(v, v)$ and using it, present (2) as

$$\Psi(v^*, w) \geq 0\ \forall w \in \Omega. \tag{3}$$

So this inequality is a consequence of (1). But if this inequality is considered as primary, then it is known as Ky Fan's inequality [3] since it is proved in [15] that the solution of (3) exists and is the vector v^*. In this case, the existence of the fixed point of (1) results from (3). The problem (1) can be regarded as a scalar convolution of diverse game problems, which describe the situation of reconciliation of contradictory interests and factors for many agents. For instance, we show, that a n−person game with Nash equilibrium can be scalarized and presented as an equilibrium programming problem (1). Really, let $f_i(x_i, x_{-i})$ be the payoff function of the i-th player, $i \in I$. This function depends on both its own strategy $x_i \in X_i$, and the strategies $x_{-i} = (x_j)_{j \in I \setminus i}$ of all other players. Each $f_i(x_i, x_{-i})$ is convex in its own variable and each $g_i(x_i)$ is a convex function. The equilibrium solution of a n−person game is the solution of the system of extreme inclusions [16]

$$x_i^* \in \mathrm{Argmin}\{f_i(x_i, x_{-i}^*) \mid g_i(x_i) \leq 0, \ x_i \in X_i\}. \tag{4}$$

In particular, if $i = 1, 2$, and $f_1(x_1, x_2) + f_2(x_2, x_1) = 0$, we obtain a saddle programming problem [4]. Let us introduce a normalized function of the type

$$\Phi(v, w) = \sum_{i=1}^{n} f_i(x_i, x_{-i}),$$

where $v = (x_{-i})$, $w = (x_i)$, $g(w) = (g_i(x_i))$, $i = 1, 2, ..., n$ and $\Omega_0 = X_1 \times X_2 \times ... \times X_n$, therewith $(v, w) = (x_i, x_{-i}) \in \Omega_0$. With the help of this convolution, the problem (4) can be written in the form (1). Many inverse optimization problems [5] can be introduced also alike (1). Really, we shall consider an inverse convex programming problem in the form

$$x^* \in \mathrm{Argmin}\{\langle \lambda^*, f(x) \rangle \mid g(x) \leq 0, \ x \in Q\}, \ G(x^*) \leq d. \tag{5}$$

In this problem, one must choose nonnegative coefficients of linear convolutions $\lambda = \lambda^*$ so that the optimal solution $x = x^*$ corresponding to these weights belongs to the preassigned convex set. In particular, this set may contain only one point. It is supposed that all functions in this problem are convex. The system (5) can be recorded as a two-person game with the Nash equilibrium [1]

$$x^* \in \mathrm{Argmin}\{\langle \lambda^*, f(x) \rangle \mid g(x) \leq 0, \ x \in Q\},$$
$$\lambda^* \in \mathrm{Argmin}\{\langle \lambda, G(x^*) - d \rangle \mid \lambda \geq 0\}. \tag{6}$$

Problem (6) in turn, as it was shown above, can be written in the form of computing a fixed point for the extreme map (1). For other examples of problems, which can be considered as (1), see in [6]. The adduced examples show, that the area of possible applications of equilibrium problems is extensive, therefore the development of methods for solving equilibrium problems is presented with sizeable interest. In the present paper, the linearization method is considered and justified for solving equilibrium problems.

2 Splitting of functions

It is known that the linear space of square matrices consists of the linear subspaces of symmetric and anti-symmetric matrices. Every square matrix can be represented as the sum of projections onto these subspaces. Drawing an analogy between square matrices and the objective functions $\Phi(v,w)$ of problem (1) we choose two linear subspaces in the linear space of these functions. These subspaces are characterized by the following properties:

$$\Phi(w,v) - \Phi(v,w) = 0 \ \forall w \in \Omega_0, \ \forall v \in \Omega_0, \tag{7}$$

$$\Phi(w,v) + \Phi(v,w) = 0 \ \forall w \in \Omega_0, \ \forall v \in \Omega_0. \tag{8}$$

The functions of the first class are called symmetric; those of the second class, anti-symmetric. If these functions are defined on a square grid, we have the conventional classes of symmetric and anti-symmetric matrices. Recall that a pair of points with coordinates w, v and v, w is situated symmetrically concerning the diagonal of the square $\Omega_0 \times \Omega_0$, i.e. with respect to the linear manifold $v = w$. This allows us to introduce the concept of a transposed function [4]. If we assign the value of $\Phi(w,v)$ calculated at the point w, v to every point with coordinates v, w, we obtain the transposed function $\Phi^T(v,w) = \Phi(w,v)$. In terms of this function, conditions (7) and (8) look like

$$\Phi(v,w) = \Phi^T(v,w), \ \Phi(v,w) = -\Phi^T(v,w).$$

Using the obvious relations $\Phi(v,w) = (\Phi^T(v,w))^T$, $(\Phi_1(v,w) + \Phi_2(v,w))^T = \Phi_1^T(v,w) + \Phi_2^T(v,w)$, we can readily verify that any real function $\Phi(v,w)$ can be represented as the sum

$$\Phi(v,w) = S(v,w) + K(v,w), \tag{9}$$

where $S(v,w)$ and $K(v,w)$ are symmetric and anti-symmetric functions, respectively. This expansion is unique, and

$$S(v,w) = \frac{1}{2}\left(\Phi(v,w) + \Phi^T(v,w)\right), \ K(v,w) = \frac{1}{2}\left(\Phi(v,w) - \Phi^T(v,w)\right). \tag{10}$$

The classes of symmetric and anti-symmetric functions are subsets of larger functional classes, namely, the classes of pseudo-symmetric and skew-symmetric functions. Now we give the following definitions.

Definition 1. A differentiable function $\Phi(v,w)$ from $\mathbb{R}^n \times \mathbb{R}^n$ in $\mathbb{R}^1$ is called pseudo-symmetric on $\Omega_0 \times \Omega_0$, if there exists a differentiable function $P(v)$ such that

$$\nabla P(v) = 2\nabla_w \Phi(v,w)|_{w=v} \ \forall v \in \Omega_0, \tag{11}$$

where $\nabla P(v)$ is the gradient of $P(v)$ and $\nabla_w \Phi(v,w)$ is the partial gradient of the function $\Phi(v,w)$ in w. The function $P(v)$ is called the potential of the operator $\nabla_w \Phi(v,w)|_{w=v}$.

Below, we will point out that any symmetric function is pseudo-symmetric.

Definition 2. A function $\Phi(v, w)$ from $I\!\!R^n \times I\!\!R^n$ to $I\!\!R^1$ is called skew-symmetric onto $\Omega_0 \times \Omega_0$, if it satisfies the inequality [11]

$$\Phi(w, w) - \Phi(w, v) - \Phi(v, w) + \Phi(v, v) \geq 0 \ \forall w \in \Omega_0, \ \forall v \in \Omega_0. \tag{12}$$

It will be shown below that every anti-symmetric function is skew-symmetric. Since any sum of pseudo-symmetric and skew-symmetric functions can be considered as the example of decomposition (9), then it is possible to consider that any objective function of problem (1) can be splitted to a sum of pseudo-symmetric and skew-symmetric functions, certainly, not in a unique way. So, expansion (9), where functions $S(v, w)$ and $K(v, w)$ are pseudo-symmetric and skew-symmetric is justified. From (9), we have

$$\nabla_w \Phi(v, w)|_{v=w} = \nabla_w S(v, w)|_{v=w} + \nabla_w K(v, w)|_{v=w}. \tag{13}$$

Using condition of pseudo-symmetry (2), it is possible to rewrite the equality (13) in the form

$$\nabla_w \Phi(v, v) = \frac{1}{2} \nabla P(v) + \nabla_w K(v, v). \tag{14}$$

Recall that the fixed point of (1) satisfies inequality (2). If $K(v, w)$ is differentiable in $w \in \Omega_0$ for any $v \in \Omega_0$, then from (2) and (14), the necessary condition is

$$\langle \frac{1}{2} \nabla P(v^*) + \nabla_w K(v^*, v^*), w - v^* \rangle \geq 0 \ \forall w \in \Omega. \tag{15}$$

If the function $P(w) + \Phi(v^*, w)$ is convex, then this condition is sufficient and $v^* \in \Omega^*$ is the solution of the problem

$$v^* \in \text{Argmin}\{R(v^*, w) = P(w) + K(v^*, w) \mid g(w) \leq 0, \ w \in \Omega_0\}. \tag{16}$$

The representation of the objective function of problem (1) in the form $R(v, w) = P(w) + K(v, w)$ will be called **canonical.** Here $P(w)$ is any real function and $K(v, w)$ any skew-symmetric one. In particular, if $S(v, w)$ is a symmetric function, then $P(v) = S(v, v)$. It is important to underline that the **canonical** representation always exists for any objective function $\Phi(v, w)$ of problem (1). The function $R(v, w)$ is in turn skew-symmetric and consequently, $\Phi(v, w)$ can be considered as skew-symmetric. In the case when $P(w)$ is a potential for the operator $\nabla_w S(v, v)$, the condition (15) is only necessary. This condition is sufficient for (1) when $\Phi(v, w)$ is convex in w for any v. In the next sections we shall assume that the objective function of problems (1) is given in the canonical form. Note that if in this representation the component $K(v, w)$ is away, we deal with a nonlinear programming problem.

3   Pseudo-symmetric functions

In the next two sections, we consider properties of pseudo-symmetric and skew-symmetric functions. By definition, the pseudo-symmetric function meets condition (2). The latter means that there exists a function $P(w)$ such that its gradient coincides with the operator $2\nabla_w\Phi(v,w)|_{v=w}$. If the function $P(w)$ is twice continuously differentiable, then the Lagrange formula follows from (2)

$$P(v+h) = P(v) + 2\int_0^1 \langle \nabla_w\Phi(v+th, v+th), h\rangle dt. \tag{17}$$

On the contrary, if the Jacobi matrix $\nabla F(v)$ for the operator $F(v) = \nabla_w\Phi(v,v)$ is symmetric for all $v \in \Omega_0$, then (3) holds and in this case, the operator $\nabla_w\Phi(v,v)$ is a potential [7]. So, if the objective function in (1) satisfies (2) or (3), then the equilibrium problem is said to be potential. The set of all pseudo-symmetric functions is by itself a linear space. The pseudo-symmetric functions include all symmetric functions (7). Indeed, if $\Phi(v,w)$ is a differentiable function, then by differentiating identity (7) in w, we obtain

$$\nabla_v\Phi(w,v) = \nabla_w\Phi(v,w) \; \forall w \in \Omega_0, \; \forall v \in \Omega_0. \tag{18}$$

Let us assume that $w = v$ in (4). Then we have

$$\nabla_v\Phi(v,v) = \nabla_w\Phi(v,v) \; \forall v \in \Omega_0. \tag{19}$$

Thus, we can formulate the following property

Property 1. The contractions of partial derivatives of symmetric functions onto the diagonal of the square $\Omega_0 \times \Omega_0$ are identical.

By definition of the differentiability of the function $\Phi(v,w)$ we get [8]

$$\Phi(v+h, w+k) = \Phi(v,w) + \langle\nabla_v\Phi(v,w), h\rangle +$$
$$+\langle\nabla_w\Phi(v,w), k\rangle + \omega(v,w,h,k), \tag{20}$$

where $\omega(v,w,h,k)/(|h|^2 + |k|^2)^{1/2} \to 0$ as $|h|^2 + |k|^2 \to 0$. Let $w = v$ and $h = k$. Then with regard to (5), we get from (6) that

$$\Phi(v+h, v+h) = \Phi(v,v) + 2\langle\nabla_w\Phi(v,v), h\rangle + \omega(v,h), \tag{21}$$

where $\omega(v,h)/|h| \to 0$ as $|h| \to 0$. Since formula (7) is a particular case of (6), it means that the contraction of the gradient $\nabla_w\Phi(v,w)$ on the diagonal of the square $\Omega_0 \times \Omega_0$ is the gradient $\nabla\Phi(v,v)$ of function $\Phi(v,v)$, i.e.

$$2\nabla_w\Phi(v,w)|_{v=w} = \nabla\Phi(v,v) \; \forall v \in \Omega_0. \tag{22}$$

Thus, we can prove

Property 2. If $\Phi(v, w)$ is a symmetric function, then the operator $\nabla_w \Phi(v, v)$ is a potential and coincides with the contraction of the gradient of the function $\Phi(v, w)$ on the diagonal of square, i.e. $\nabla_w \Phi(v, w)|_{w=v} = \nabla \Phi(v, v) = \nabla P(v)$.

The concept of potentiality in the scientific literature has been considered rather for a long time. Apparently, one of the first article, where the potential was used for the substantiation of asymptotic stability for a gradient method to solve the n–person game, was the publication [9]. In one of the recent papers [10], on an example of n–person game, the concept of potential game was introduced as follows. If for a n–person game with Nash equilibrium

$$x_i^* \in \mathrm{Argmin}\{f_i(x_i, x_{-i}^*) \mid x_i \in X_i\},$$

where $f_i(x_i, x_{-i})$ is a payoff function of the i-th player, $i \in I = \{1, 2, ..., n\}$, $x_i \in X_i = (x_i)_{i \in I}$, $x_{-i} = X_{-i} = (x_j)_{j \in I \setminus i}$, $(x_i, x_{-i}) = x \in X = X_1 \times X_2, ..., \times X_n$, there is a function $P(x_1, x_2, ..., x_n)$ such that

$$\frac{\partial P(x_1, x_2, ...x_n)}{\partial x_i} = \frac{\partial f_i(x_1, x_2, ..., x_n)}{\partial x_i}, \quad i \in I, \tag{23}$$

then the game is called potential. In other words, the partial derivatives of payoff functions in the own variables of the players constitute the gradient of some function $P(x_1, x_2, ..., x_n)$ which is called a potential. We shall write the right-hand side of (9) in the form

$$\frac{\partial f_i(x_1, x_2, ...x_n)}{\partial x_i} = \frac{\partial f_i(x_i, x_{-i})}{\partial x_i}, \quad i \in I, \tag{24}$$

and introduce a normalized function for the considered game

$$\Phi(v, w) = \sum_{i=1}^{n} f_i(x_i, x_{-i}),$$

where $v = (x_{-i})$, $w = (x_i)$, $i = 1, ..., n$ and $(v, w) = (x_i, x_{-i}) \in \Omega_0 \times \Omega_0$. Since the function $\Phi(v, w)$ is separable in w and the set X has a block structure, then

$$\nabla_w \Phi(v, w) |_{v=w} = \left(\frac{\partial f_i(x_i, x_{-i})}{\partial x_i} \right) \quad i \in I, \tag{25}$$

Comparing (9), (10) and (11), we have

$$\nabla P(v) = \nabla_w \Phi(v, w) |_{v=w}.$$

Thus, potential games in the sense of (9) are potential games in the sense of (2). If the objective function of problem (1) is subordinated to the condition (2), then the equilibrium problem can be considered as an optimization problem. Really, from (2) we have

$$\langle \nabla_w \Phi(v^*, v^*), w - v^* \rangle \geq 0 \; \forall w \in \Omega. \tag{26}$$

By virtue of (2) and from (12), we get

$$\langle \nabla P(v^*), w - v^* \rangle \geq 0 \ \forall w \in \Omega. \tag{27}$$

If the operator $\nabla P(v)$ is monotone, then $P(v)$ is a convex function over Ω and $v^* \in \Omega^*$ is its optimal solution. In this case, an equilibrium potential problem (1) can be replaced by an optimization problem of the function $P(v)$ over Ω. However, we shall mark that the function $\Phi(v^*, w)$, generally speaking, is not convex and condition (12) is only a necessary one. But, if $\Phi(v^*, w)$ is a convex function in $w \in \Omega_0$, then v^* is an equilibrium solution of (1). Now we formulate the notion of sharpness for the potential equilibrium solution. Whereas in the potential case the equilibrium solutions simultaneously are minimum points, it is enough to re-state fitting definition for sharp minima [19],[5]. A pseudo-symmetric equilibrium problem is said to have the $1 + \nu$−order sharpness for equilibrium solution $v^* \in \Omega^*$ if condition

$$P(w) - P(v^*) \geq \gamma_1 |w - v^*|^{1+\nu} \ \forall w \in \Omega \tag{28}$$

holds, where $\gamma_1 \geq 0$ is constant. If $\nu = 0$, the equilibrium is called sharp; if $\nu = 1$, it is called a quadratic equilibrium.

4   Skew-symmetric functions

By definition, a skew-symmetric function satisfies inequality (1) but if this inequality holds with respect to the solution $v^* \in \Omega^*$ of

$$\Phi(w, w) - \Phi(w, v^*) - \Phi(v^*, w) + \Phi(v^*, v^*) \geq 0 \ \forall w \in \Omega_0, \tag{29}$$

then the function $\Phi(v, w)$ is called skew-symmetric concerning an equilibrium point. The class of skew-symmetric functions is non-empty, as it includes in itself all anti-symmetric functions (8). Put $v = w$ in (8), then $\Phi(v, v) + \Phi(v, v) = 0$, that is, the anti-symmetric function is identically equal to zero on the diagonal of square $\Omega_0 \times \Omega_0$. If the anti-symmetric function is convex in w, then it follows from (8) it is concave in v, that is, in this case, $\Phi(v, w)$ is a saddle point function. To illustrate condition (1), consider the normalized function $\Phi(v, w)$ of the saddle-point problem, which can be obtained from (4) with $i = 1, 2$. In this case, $\Phi(v, w)$ satisfies the relations [11]

$$\Phi(v, v) = 0, \ \Phi(v, w) + \Phi(w, v) = 0 \ \forall w \in \Omega_0, v \in \Omega_0.$$

Note that the authors of [13] earlier attempted to extend these conditions to non-saddle-point problems. The skew-symmetric functions have properties, which can be considered as analogs of monotonicity conditions of a gradient for convex functions.

Property 3. If the function $\Phi(v, w)$ is differentiable, skew-symmetric and convex in w, then the contraction of the partial gradient $\nabla \Phi_w(v, v)$ on the diagonal of the square $\Omega_0 \times \Omega_0$ is monotone i.e.

$$\langle \nabla_w \Phi(v+h, v+h) - \nabla_w \Phi(v, v), h \rangle \geq 0 \ \forall v \in \Omega_0, \ \forall v+h \in \Omega_0. \tag{30}$$

This inequality follows from (1) if we apply the condition of convexity for $\Phi(v, w)$ in w

$$\langle \nabla f(x), y - x \rangle \leq f(y) - f(x) \leq \langle \nabla f(y), y - x \rangle \tag{31}$$

for all x and y in some set. In particular, if $\Phi(v, w)$ is a normalized function of the saddle-point problem, it follows from (1) that $(-\nabla_x L(x, y), \nabla_y L(x, y))^T$ is a monotone operator, where $L(x, y) = f_1(x, y) = -f_2(x, y)$ from (4). This fact was established in [20]. From above, it follows that the skew-symmetric equilibrium problems largely inherit properties of saddle-point problems. In connection with this, it is important to present the skew-symmetric functions in a more investigated form.

Definition 3. The function $\Phi(v, w)$ from $\mathbb{R}^n \times \mathbb{R}^n$ in $\mathbb{R}^1$ is called bi-differentiable at point $v, v \in \Omega_0 \times \Omega_0$, if there exists a quadratic matrix $D(v, v)$ such that

$$\{\Phi(v+h, v+k) - \Phi(v+h, v)\} - \{\Phi(v, v+k) - \Phi(v, v)\}$$
$$= \langle D(v, v)h, k \rangle + \omega(v, h, k), \tag{32}$$

where $\omega(v, h, k)/|h||k| \to 0$ as $|h|, |k| \to 0$ for all $h \in \mathbb{R}^n$, $k \in \mathbb{R}^n$. The bilinear function $\langle D(v, v)h, k \rangle$ is called the bi-linear differential of the function $\Phi(v, v)$ at the point $v, v \in \Omega_0 \times \Omega_0$.

The function $\Phi(v, w)$ is called bi-differentiable on the diagonal of the square $\Omega_0 \times \Omega_0$ if it is differentiable at all points of this set. It is not hard to see that if the function $\Phi(v, w) = \varphi(v) + \varphi(w)$ is separable with respect to their variables, then bi-differential of such function is equal to zero,i.e. $D(v, v) = 0$, $(d_{ij}(v, v) = 0 \ \forall i, j \in N)$. The converse statement is true as well. Thus, the matrix $D(v, v)$, the bi-differential of the function $\Phi(v, w)$, characterizes a degree of nonseparability or nonseparation of variables and allows us to estimate a measure of functional dependency on one variable when changing the other. The introduced differential has a simple geometric sense and enables to estimate deviation in v, namely, $\{\Phi(v+h, v+k) - \Phi(v+h, v)\}$ and $\{\Phi(v, v+k) - \Phi(v, v)\}$ under transition from a point v, v to the point $v+h, v+k$. A bilinear differential represents a saddle tangent surface and describes some singularities of the behaviour of double increment of $\Phi(v, w)$ with respect to this differential. Now does the bilinear differential introduced above correspond with the classical one in the case of differentiability of $\Phi(v, w)$? We show that if $\Phi(v, w)$ is a twice continuously differentiable function, then the matrix $D(v, v)$ coincides with the contraction of the mixed derivative matrix $\frac{\partial^2 \Phi(v, w)}{\partial w, \partial v} |_{v=w}$ on the main diagonal of the square for the second differential. So, if the function $f(x)$ is twice continuously differentiable, then Taylor's formula [8] gives

$$F(x + y) - f(x) = \langle \nabla f(x), y \rangle + \frac{1}{2} \langle \nabla^2 f(x + \vartheta y)y, y \rangle, \qquad (33)$$

where $0 \leq \vartheta \leq 1$. Using this formula, it is possible to get the expansion for the first three terms $\Phi(v + \varepsilon h, v + \varepsilon k), \Phi(v + \varepsilon h, v)$ and $\Phi(v, v + \varepsilon k)$ from (32), assuming that the increments h and k have the form εh and εk, where $\varepsilon > 0$. Then, after substitution of expansions in (32) and mutual reductions of terms with different signs, we obtain

$$\frac{1}{2} \varepsilon^2 \left\langle \left\{ \nabla^2_{vv} \Phi(v + \vartheta_1 \varepsilon h, v + \vartheta_1 \varepsilon k) - \nabla^2_{vv} \Phi(v + \vartheta_2 \varepsilon h, v) \right\} h, h \right\rangle +$$

$$+ \varepsilon^2 \langle \nabla^2_{wv} \Phi(v + \vartheta_1 \varepsilon h, v + \vartheta_1 \varepsilon k)h, k \rangle +$$

$$+ \frac{1}{2} \varepsilon^2 \left\langle \left\{ \nabla^2_{ww} \Phi(v + \vartheta_1 \varepsilon h, v + \vartheta_1 \varepsilon k) - \nabla^2_{ww} \Phi(v, v + \vartheta_3 \varepsilon k) \right\} k, k \right\rangle =$$

$$= \varepsilon^2 \langle D(v, v)h, k \rangle + \omega(v, \varepsilon h, \varepsilon k). \quad (34)$$

Taking into account continuity of the operators $\nabla^2_{vv} \Phi(v, w)$, $\nabla^2_{ww} \Phi(v, w)$ and $\frac{\omega(v, \varepsilon h, \varepsilon k)}{\varepsilon^2} \to 0$, when $\varepsilon \to 0$, $\varepsilon > 0$, we have $\langle \nabla^2_{wv} \Phi(v, v)h, k \rangle = \langle D(v, v)h, k \rangle \ \forall h, k \in \mathbb{R}^n$. So

$$\nabla^2_{wv} \Phi(v, v) = D(v, v) \ \forall v \in \Omega_0. \qquad (35)$$

Thus, we can formulate the following statement.

Property 4. If the objective function $\Phi(v, w)$ is twice continuously differentiable, then the matrix $D(v, v)$, the bilinear differential, coincides with the contraction of the matrix of mixed derivatives of the second differential on the main diagonal.

In section 3, we have shown that a symmetric function in the sense of (7) is pseudo-symmetric. We shall prove here that for symmetric functions, the matrix $D(v, v)$ in the expansion (32) is symmetric for all $v \in \Omega_0$. Really, let $\Phi(v, w)$ be a symmetric function and consider once again expansion (34)

$$\Phi(v + \varepsilon h, v + \varepsilon k) - \Phi(v + \varepsilon h, v) - \Phi(v, v + \varepsilon k) + \Phi(v, v) =$$

$$= \varepsilon^2 \langle D(v, v)h, k \rangle + \omega(v, \varepsilon h, \varepsilon k) \qquad (36)$$

for all $v \in \Omega_0$, $v + \varepsilon h \in \Omega_0$, $v + \varepsilon k \in \Omega_0$. As (36) is true for any pairs h, k, we can interchange the places of these variables in (36) to get

$$\Phi(v + \varepsilon k, v + \varepsilon h) - \Phi(v, v + \varepsilon h) - \Phi(v + \varepsilon k, v) + \Phi(v, v) =$$

$$\varepsilon^2 \langle D(v, v)k, h \rangle + \omega(v, \varepsilon k, \varepsilon h). \qquad (37)$$

By symmetry of functions, the left-hand sides of (36) and (37) are equal. Therefore, their right-hand sides are equal as well and

$$\langle D(v,v)h, k \rangle = \langle D^T(v,v)h, k \rangle + (\omega(v, \varepsilon h, \varepsilon k) - \omega(v, \varepsilon k, \varepsilon h))/\varepsilon^2.$$

So, we get as $\varepsilon \to 0$,

$$D(v,v) = D^T(v,v) \; \forall v \in \Omega_0.$$

We shall prove that the matrix $D(v,v)$ from (32) is anti-symmetric for any anti-symmetric function (8). Really, assuming the satisfiability of conditions (8), we add inequalities (36) and (37). By virtue of the anti-symmetry condition we obtain

$$0 = \langle (D(v,v) + D^T(v,v))h, k \rangle + (\omega(v, \varepsilon h, \varepsilon k) + \omega(v, \varepsilon k, \varepsilon h))/\varepsilon^2.$$

So, as $\varepsilon \to 0$, we have

$$D(v,v) = -D^T(v,v) \; \forall v \in \Omega_0.$$

From the last equality, in particular, it follows that $\langle D(v,v)h, h \rangle = 0 \forall h \in I\!R^n$. Using the bilinear differential, we can now introduce the class of bi-convex functions

Definition 4. If in (32), $\omega(v, h, k) \geq 0$, then the function $\Phi(v, w)$ is called bi-convex on $\Omega_0 \times \Omega_0$. This function is subject to the condition

$$\{\Phi(v + h, v + k) - \Phi(v + h, v)\} - \{\Phi(v, v + k) - \Phi(v, v)\} \geq \langle D(v,v)h, k \rangle \tag{38}$$

for $\forall v \in \Omega_0$ and $h, k \in R^n$. If $D(v,v) \geq 0$, then we call this function positive bi-convex.

We explore the mutual relations between classes of skew-symmetric and bi-convex functions.

Lemma 1. *Classes of skew-symmetric and positive bi-convex functions coincide.*

Really, let $\Phi(v, w)$ be some skew-symmetric function in the sense of (1). Assuming that the increment in (32) looks like εh and $k = h$, we have

$$\langle D(v,v)h, h \rangle + \omega(v, \varepsilon h, \varepsilon h)/\varepsilon^2 \geq 0 \; \forall h \in R^n.$$

So, as $\varepsilon \to 0$, it follows $\langle D(v,v)h, h \rangle \geq 0 \; \forall h \in I\!R^n$ and $v \in \Omega_0$. On the contrary, if the value $\langle D(v,v)h, h \rangle \geq 0 \; \forall h \in I\!R^n$ and $v \in \Omega_0$ in (32), then the fulfilment of (1) is obvious. The lemma is proved. In the general case, if the matrix $D(v,v) \neq 0$, $v \in \Omega_0$, then it has the representation $D(v,v) = S(v,v) + K(v,v)$, where $S(v,v) = \frac{1}{2}\left(D(v,v) + D^T(v,v)\right)$,

$K(v,v) = \frac{1}{2}\left(D(v,v) - D^T(v,v)\right)$, where $S(v,v)$, $K(v,v)$ are symmetric and anti-symmetric matrices i.e. $\langle K(v,v)h, h \rangle = 0 \forall h \in \mathbb{R}^n$ and $v \in \Omega_0$. We notice that condition (38), in the case of the convexity of function $\Phi(v,w)$ in $w \in \Omega_0$, implies the bi-monotonicity of the contraction of the partial gradient $\nabla_w \Phi(v,w)|_{v=w}$ in w on the main diagonal of the square $\Omega_0 \times \Omega_0$. Indeed, let the function $\Phi(v,w)$ be convex in w. Then using the system of inequalities of convexity (1) we have, from (38), that

$$\langle \nabla_w \Phi(v+h, v+h) - \nabla_w \Phi(v,v), h \rangle \geq \langle D(v,v)h, h \rangle \ \forall v \in \Omega_0, \ v+h \in \Omega_0. \tag{39}$$

If the condition $\langle D(v,v)h, h \rangle \geq 0 \ \forall h \in \mathbb{R}^n$ is accomplished, then monotonicity of contraction for gradients (30) follows from (39). We put one more useful inequality, which allows to estimate the growth rate of function $\Phi(v,w)$ in a neighbourhood of a point $v, w \in \Omega_0 \times \Omega_0$

$$|\{\Phi(w+h, v+k) - \Phi(w+h, v)\} - \{\Phi(w, v+k) - \Phi(w,v)\}| \leq L\,|h|\,|k| \tag{40}$$

for all w, $w+h \in \Omega_0$, v, $v+k \in \Omega_0$, where L is a constant. A class of functions satisfying condition (13) is non-empty [20]. It was determined earlier that the symmetric functions have the potential properties. However some of them are also skew-symmetric. Indeed, consider a subset of functions subject to the condition: $\Phi(v,w) \leq \sqrt{\Phi(w,w)\Phi(v,v)}\forall v, w \in \Omega_0 \times \Omega_0$. Let us write out an expression similar to the left-hand side of (1). Using (7) and the condition introduced, we rewrite this expression to obtain $\Phi(w,w) - \Phi(w,v) - \Phi(v,w) + \Phi(v,v) = \Phi(w,w) - 2\Phi(w,v) + \Phi(v,v) \geq \Phi(w,w) - 2\sqrt{\Phi(w,w)\Phi(v,v)} + \Phi(v,v) = (\sqrt{\Phi(w,w)} - \sqrt{\Phi(v,v)})^2 \geq 0 \ \forall v, w \in \Omega_0$. From here and from (30), it follows that if $\Phi(v,w)$ is convex in w for any $v \in \Omega_0$, then $\nabla_w \Phi(v,v)$ is a monotone operator. Now we examine the properties of the equilibrium solution of (1) in the case when the objective function is skew-symmetric. Let $v = v^*$ be in Ω^*, then taking into account (2), we get from (29) that

$$\Phi(w,w) \geq \Phi(w, v^*) \ \forall w \in \Omega. \tag{41}$$

Thus, any equilibrium solution of a skew-symmetric problem (1) satisfies condition (15). Inequalities (2) and (15) are of fundamental importance for proving the convergence of methods to equilibrium solutions of skew-symmetric problems [1],[2]. Using a function $\Psi(v,w)$ from (3) we present inequalities (2) and (15) as the following system

$$\Psi(w, v^*) \leq \Psi(v^*, v^*) \leq \Psi(v^*, w) \ \forall w \in \Omega. \tag{42}$$

Here the right-hand inequality is known as Ky Fan's inequality (3) [3]. It follows from (19) that v^*, v^* is a saddle point of $\Psi(v,w)$ on $\Omega \times \Omega$, and $\Psi(v^*, v^*) = 0$. Note that the function $\Psi(v,w)$ in general is never convex-concave, even if $\Phi(v,w)$ is convex-concave. Inequality (19) shows us that the skew-symmetric equilibrium problems inherit properties of saddle-point problems. There are classes of skew-symmetric functions $\Phi(v, w)$ and accordingly equilibrium problems answering to these functions, whose solutions satisfy inequalities more rigid, than (19), namely

$$\Psi(w, v^*) \leq -\gamma_2 |w - v^*|^{1+\nu} \ \forall w \in \Omega, \tag{43}$$

and (or)

$$\gamma_2 |w - v^*|^{1+\nu} \leq \Psi(v^*, w) \ \forall w \in \Omega, \tag{44}$$

where $v^* \in \Omega^*$ is solution of problem, $\gamma_2 \geq 0$ and $\nu \in [0, \infty]$ are parameters. We rewrite (20) as

$$\Phi(w, w) - \Phi(w, v^*) \geq \gamma_2 |w - v^*|^{1+\nu} \ \forall w \in \Omega. \tag{45}$$

The obtained inequality will be called condition of sharpness for skew-symmetric equilibrium. If $\gamma_2 > 0$, then with $\nu = 0$ we have a sharp equilibrium, and with $\nu = 1$ a quadratic equilibrium. If $\gamma_2 = 0$, we get (15) (see [20]).

5   Linearization method

In this section, we investigate the approach to solving equilibrium problems in canonical form (16) based on the idea of linearization. We recall that the variational inequality (15) is equivalent to the operator equation

$$v^* = \pi_\Omega(v^* - \alpha \nabla_w R(v^*, v^*)), \tag{46}$$

where $\pi_\Omega(...)$ is the projection of some vector on a feasible set Ω, $\alpha > 0$ is a parameter like the step length, $\nabla_w R(v, v)$ is the partial gradient of $R(v, w)$ in w. The operator $\pi_\Omega(...)$ is not the unique way to take into account the functional constraints of (16). Linear convolution in the form of Lagrange function is another approach to this problem. Following this approach, we write out a Lagrange function for this problem (16)

$$L(v^*, w, y) = R(v^*, w) + \langle y, g(w) \rangle \ \forall w \in \Omega_0, \ y \geq 0,$$

where v^* is a fixed vector (solution of the problem). A saddle point v^*, p^* of the Lagrange function $L(v^*, w, y)$ satisfies the system of inequalities

$$L(v^*, v^*, y) \leq L(v^*, v^*, p^*) \leq L(v^*, w, p^*) \ \forall w \in \Omega_0, \ y \geq 0.$$

In case of differentiability, this system can be expressed as variational inequalities

$$\langle \nabla_w R(v^*, v^*) + \nabla g^T(v^*) p^*, w - v^* \rangle \geq 0 \ \forall w \in \Omega_0, \tag{47}$$

$$\langle y - p^*, g(v^*) \rangle \leq 0 \ \forall y \geq 0. \tag{48}$$

The system of inequalities (47), (19) is equivalent to (15), and if $R(v, w)$ is convex in $w \in \Omega_0$, (15) is equivalent to (2). Going back to equation (46), a geometrical meaning of it is very simple: a step from a point v^* along the antigradient takes us back to the point v^* after a projection operation, i.e. v^* is a fixed point or an equilibrium point. As the right-hand side of equation (46) represents itself the nonexpanding operator, it is possible to expect that a gradient projection method

$$v^{n+1} = \pi_\Omega(v^n - \alpha \nabla_w R(v^n, v^n)). \tag{49}$$

will be converging to an equilibrium solution of the initial problem. From the numerical point of view, the difficulties are connected with the projection operation. From experience, it is known that this operation is justified when Ω is a simple set: positive orthant, parallelepiped, ball. If the feasible set is complicated, for instance $\Omega = \{w \mid g(w) \leq 0, \ w \in \Omega_0\}$, then the projection operation becomes laborious and in this case, it is reasonable to approximate the set Ω by more simple sets as for example, polyhedral sets. The linearization method (including continuous version) with reference to optimization problems is investigated rather in detail in the papers [3], [10], [11]. In the present article, a controlled linearization method is applied to equilibrium programming problems (1). The idea is simple: split in Taylor's series the vector-function $g(w)$ and take only the linear term. In terms of this expansion, we write out the linearized constraints as

$$\Theta_n = \{w \mid \nabla g(v^n)(w - v^n) + g(v^n) \leq 0, \ w \in \Omega_0\},$$

where $\nabla g(v)$ is the matrix whose every line is the gradient-vector of the appropriate functional constraint. Then we project the vector of motion direction on these constraints at each iteration. Process (49) in this case takes the form

$$v^{n+1} = \pi_{\Theta_n}(v^n - \alpha \nabla_w R(v^n, v^n)),$$

$$\Theta_n = \{w \mid \nabla g(v^n)(w - v^n) + g(v^n) \leq 0, \ w \in \Omega_0\}. \tag{50}$$

Here Θ_n (accurate to Ω_0) is a polyhedron, which can be expressed as the intersection of the following half-spaces

$$\nabla g_i(v^n)(w - v^n) + g_i(v^n) \leq 0, \ i = 1, 2, .., m.$$

This set is created at each point of the sequence v^n and the projection of the gradient-step $v^n - \alpha \nabla_w R(v^n, v^n)$ is unique. The polyhedral family, with the iterations, will approximate the feasible set in a neighbourhood of the solution more and more exactly. In the papers [3],[10],[11] it is shown that the explained approach is enough effective for solving optimization problems, and also equilibrium problems of type (1) with pseudo-symmetric objective function. In this case, the equilibrium problem is equivalent to some optimization problem. In general, the method (50) does not converge to an equilibrium solution and, in this case, it is possible to achieve success, if one takes advantage of the idea of control of the method with the help of feedbacks. With reference to the differential gradient methods, this approach has been developed in [4] for solving equilibrium problems. Using the idea of control by the method (50), by means of discrepancy of the equation (46), we come to a process of this kind

$$\bar{u}^n = \pi_{\Theta_n}(v^n - \alpha\nabla_w R(v^n, v^n)),$$

$$v^{n+1} = \pi_{\Theta_n}(v^n - \alpha\nabla_w R(\bar{u}^n, \bar{u}^n)),$$

$$\Theta_n = \{w \mid \nabla g(v^n)(w - v^n) + g(v^n) \le 0, \ w \in \Omega_0\}. \tag{51}$$

Here the first half-step is treated as the prediction. As a result of this half-step, there is the point $\bar{u}^n$, in which is calculated the prediction direction of the development of the system. Then realization of the second half-step completes the iteration. A differential analog of this process has the following form [9]

$$\frac{dv}{dt} + v = \pi_{\Theta(\cdot)}(v - \alpha\nabla_w R(\bar{u}, \bar{u})), \qquad v(t_0) = v^0,$$

$$\bar{u} = \pi_{\Theta(\cdot)}(v - \alpha\nabla_w R(v, v)),$$

$$\Theta(v(t)) = \{w \mid \nabla g(v)(w - v) + g(v) \le 0, \ w \in \Omega_0\}.$$

The projection operator in process (51) represents problem of minimization for a quadratic function over the set Θ_n. The latter means that this process can be recorded in the form

$$\bar{u}^n = \text{argmin}\{\frac{1}{2}|w - (v^n - \alpha\nabla_w R(v^n, v^n))|^2 \mid \nabla g(v^n)(w - v^n) + g(v^n) \le 0,$$

$$w \in \Omega_0\}, \tag{52}$$

$$v^{n+1} = \text{argmin}\{\frac{1}{2}|w - (v^n - \alpha\nabla_w R(\bar{u}^n, \bar{u}^n))|^2 \mid \nabla g(v^n)(w - v^n) + g(v^n) \le 0,$$

$$w \in \Omega_0\}. \tag{53}$$

The minimum of a strongly convex quadratic function exists on a polyhedral set and is always unique. The existence of this minimum follows from Slater's condition. Really, let vector $v_c \in \Omega_0$ be such that

$$g_i(v_c) < 0, \ i = 1, 2, ..., m. \tag{54}$$

Then, from convexity of $g(v)$, we have

$$\nabla g(v^n)(v_c - v^n) + g(v^n) \leq g(v_c) < 0, \tag{55}$$

i.e. the Slater condition holds for any linearized problem (51). The uniqueness of the minimum follows from the fact that a projection operation is a strongly convex minimization problem. Lagrange functions for auxiliary problems (52) and (53) look like

$$L_1(w, y) = \frac{1}{2}|w - (v^n - \alpha \nabla_w R(v^n, v^n))|^2 + \alpha \langle y, \nabla g(v^n)(w - v^n) + g(v^n) \rangle$$

for all $w \in \Omega_0$, $y \geq 0$, and

$$L_2(w, y) = \frac{1}{2}|w - (v^n - \alpha \nabla_w R(\bar{u}^n, \bar{u}^n))|^2 + \alpha \langle y, \nabla g(v^n)(w - v^n) + g(v^n) \rangle$$

for all $w \in \Omega_0$, $y \geq 0$. The saddle point $\bar{u}^n, \bar{p}^n$ of the function $L_1(w, y)$ satisfies the system of inequalities

$$L_1(\bar{u}^n, y) \leq L_1(\bar{u}^n, \bar{p}^n) \leq L_1(w, \bar{p}^n) \ \forall w \in \Omega_0, y \geq 0. \tag{56}$$

The saddle point v^{n+1}, p^{n+1} of function $L_2(y, w)$ fulfils a similar system of inequalities

$$L_2(v^{n+1}, y) \leq L_2(v^{n+1}, p^{n+1}) \leq L_2(w, p^{n+1}) \ \forall w \in \Omega_0, y \geq 0. \tag{57}$$

We recall that the points $\bar{u}^n$ and v^{n+1} are the minima of quadratic problems (52) and (53), and $\bar{p}^n$ and p^{n+1} are their dual solutions. The right-hand side and left-hand side inequalities of (56) can be presented as variational inequalities

$$\langle \bar{u}^n - v^n + \alpha(\nabla_w R(v^n, v^n) + \nabla g^T(v^n)\bar{p}^n), w - \bar{u}^n \rangle \geq 0 \ \forall w \in \Omega_0, \tag{58}$$

$$\langle \bar{p}^n - y, \nabla g(v^n)(\bar{u}^n - v^n) + g(v^n) \rangle \geq 0 \ \forall y \geq 0. \tag{59}$$

Since the inequality (59) is determined on the positive orthant, it is equivalent to a system of relations

$$\nabla g(v^n)(\bar{u}^n - v^n) + g(v^n) \leq 0, \tag{60}$$

$$\langle \nabla g(v^n)(\bar{u}^n - v^n) + g(v^n), \bar{p}^n \rangle = 0. \tag{61}$$

Similarly, the system of inequalities (57) can be also presented under the form of variational inequalities

$$\langle v^{n+1} - v^n + \alpha(\nabla_w R(\bar{u}^n, \bar{u}^n) + \nabla g^T(v^n)p^{n+1}), w - v^{n+1} \rangle \geq 0 \ \forall w \in \Omega_0, \tag{62}$$

$$\langle p^{n+1} - y, \nabla g(v^n)(v^{n+1} - v^n) + g(v^n) \rangle \geq 0 \ \forall y \geq 0. \tag{63}$$

Analogously to (60)and (61), we have

$$\nabla g(v^n)(v^{n+1} - v^n) + g(v^n) \leq 0, \tag{64}$$

$$\langle \nabla g(v^n)(v^{n+1} - v^n) + g(v^n), p^{n+1} \rangle = 0. \tag{65}$$

To make an estimate necessary hereinafter to prove the convergence theorem of the method, we take advantage of the Lipschitz inequality

$$g(w) - g(v) - \nabla g(v)(w - v) \leq \frac{L}{2}|w - v|^2 \ \forall w \in \Omega_0, \ v \in \Omega_0, \tag{66}$$

where L is a vectorial constant with components $L_i, i = 1, 2, ..., m$. Each $i-th$ component is a Lipschitz constant for the $i-th$ functional constraint. If we put $w = v^{n+1}$ and $v = v^n$ in inequality (66) and take into account (64), we have

$$g(v^{n+1}) \leq \frac{L}{2}|v^{n+1} - v^n|^2. \tag{67}$$

A similar reasoning concerning inequalities (66) and (60) yields

$$g(\bar{u}^n) \leq \frac{L}{2}|\bar{u}^n - v^n|^2. \tag{68}$$

We prove boundedness of trajectory $\bar{p}^n$ for all n under the condition that v^n is bounded. Really, let $w = v_c$ be in the right inequality of system (56), where $g(v_c) < 0$ (Slater condition), then

$$\frac{1}{2}|\bar{u}^n - (v^n + \alpha \nabla_w R(v^n, v^n))|^2 + \alpha\langle \bar{p}^n, \nabla g(v^n)(\bar{u} - v^n) + g(v^n) \rangle \leq$$

$$\leq \frac{1}{2}|v_c - (v^n - \alpha \nabla_w R(v^n, v^n))|^2 + \alpha\langle \bar{p}^n, \nabla g(v^n)(v_c - v^n) + g(v^n) \rangle.$$

Using the estimate $\langle \bar{p}^n, \nabla g(v^n)(v_c - v^n) + g(v^n) \rangle \leq \langle \bar{p}^n, g(v_c) \rangle$, which follows from (55), we get

$$0 < -\langle \bar{p}^n, g(v_c) \rangle < \frac{1}{2}|v_c - v^n + \alpha \nabla_w R(v^v, v^v))|^2 \leq Const.$$

As $g(v_c) < 0$, the fulfilment of this inequality is possible only in the case

$$\bar{p}^n \leq C \ \forall n \geq n_0, \tag{69}$$

where C is a vectorial constant. We consider the Lipschitz condition for the operator $\nabla_w R(v, v)$, which will be used to prove the convergence of the method

$$|\nabla_w R(v + h, v + h) - \nabla_w R(v, v)| \leq L_0|h| \tag{70}$$

for all v, $v + h \in \Omega_0$ from some set, where L_0 is a constant. Now we will derive an estimate of the deviation vectors v^{n+1} from $\bar{u}^n$ in (51). For this purpose let us put $y = p^{n+1}$ in (59) and $y = \bar{p}^n$ in (63) to get

$$\langle \bar{p}^n - p^{n+1}, \nabla g(v^n)(\bar{u}^n - v^n) + g(v^n) \rangle \geq 0,$$

$$\langle p^{n+1} - \bar{p}^n, \nabla g(v^n)(v^{n+1} - v^n) + g(v^n) \rangle \geq 0.$$

We add both inequalities

$$\langle \bar{p}^n - p^{n+1}, \nabla g(v^n)(\bar{u}^n - v^{n+1}) \rangle \geq 0. \tag{71}$$

Now let us put $w = v^{n+1}$ in (58) and $w = \bar{u}^n$ in (62). Then

$$\langle \bar{u}^n - v^n + \alpha(\nabla_w R(v^n, v^n) + \nabla g^T(v^n)\bar{p}^n), v^{n+1} - \bar{u}^n \rangle \geq 0,$$

$$\langle v^{n+1} - v^n + \alpha(\nabla_w R(\bar{u}^n, \bar{u}^n) + \nabla g^T(v^n)p^{n+1}), \bar{u}^n - v^{n+1} \rangle \geq 0.$$

Adding both inequalities, we have

$$\langle \bar{u}^n - v^{n+1} + \alpha(\nabla_w R(v^n, v^n) - \nabla_w R(\bar{u}^n, \bar{u}^n)) +$$

$$+ \alpha(\nabla g^T(v^n)\bar{p}^n - \nabla g^T(v^n)p^{n+1}), v^{n+1} - \bar{u}^n \rangle \geq 0.$$

So we get

$$-|\bar{u}^n - v^{n+1}|^2 + \alpha|\nabla_w R(v^n, v^n) - \nabla_w R(\bar{u}^n, \bar{u}^n)||v^{n+1} - \bar{u}^n| +$$

$$+ \alpha \langle \bar{p}^n - p^{n+1}, \nabla g(v^n)(v^{n+1} - \bar{u}^n) \rangle \geq 0.$$

Taking into account (70) and (71) from the last inequality, we have

$$|\bar{u}^n - v^{n+1}| \leq \alpha L_0 |v^n - \bar{u}^n|. \tag{72}$$

To prove the convergence of method (51), we use the property of convexity for function $\Phi(v, w)$ in w. In this case, the operator $\nabla R_w(v, v)$ satisfies the inequality (30). If we put in this inequality $v + h = w$ and $v = v^*$, then we have

$$\langle \nabla_w R(w, w) - \nabla_w R(v^*, v^*), w - v^* \rangle \geq 0 \ \forall w \in \Omega_0. \tag{73}$$

Comparing (73) and (47), we obtain

$$\langle \nabla_w R(w, w), w - v^* \rangle + \langle p^*, \nabla g(v^*)(w - v^*) \rangle \geq 0 \ \forall w \in \Omega_0.$$

Taking into account the convexity of each component of the operator $g(w)$, we have

$$\langle \nabla_w R(w, w), w - v^* \rangle \geq -\langle p^*, g(w) \rangle \ \forall w \in \Omega_0. \tag{74}$$

The condition (74) can be considered as the gradient analog for (15). The inequality (74) is a consequence of the monotonicity condition. However, as it will follow from the theorem below, this inequality will guarantee the convergence of the considered method without the monotonicity condition. Therefore it is possible to premise this inequality to prove theorem that does not use a monotonicity condition at all. We show now that the process (51) converges monotonically under the norm to one of equilibrium solutions (16).

Theorem 1. *Suppose the objective function of problem (1) is written in the canonical form $R(v, w) = P(w) + K(v, w)$, where $K(v, w)$ is skew-symmetric function. Suppose also $R(v, w)$ and $g(w)$ are convex in w for any v and differentiable functions whose gradients satisfy Lipschitz's condition (70) and (66) with constants L_0 and L (the latter is a vectorial constant). Suppose that Slater's condition (54) holds and that $\Omega_0 \subseteq R^n$ is a convex, closed, bounded set. Then the sequence v^n generated by the method (51) with parameter $0 < \alpha < \bar{\alpha}$*

$$where \quad \bar{\alpha} = \min \left\{ \frac{1}{4} \frac{1}{\langle C, L \rangle}, \frac{1}{\sqrt{4L_0^2 + \langle p^*, L \rangle} + \langle p^*, L \rangle} \right\}$$

converges monotonically under the norm of the space to the solution of problem (16).

Proof. If we put $w = v^* \in \Omega_0^*$ in (62), then we have

$$\langle v^{n+1} - v^n + \alpha(\nabla_w R(\bar{u}^n, \bar{u}^n) + \nabla g^T(v^n) p^{n+1}), v^* - v^{n+1} \rangle \geq 0.$$
(75)

Using (70) and (72), we can write that

$$\langle \nabla_w R(\bar{u}^n, \bar{u}^n), v^* - v^{n+1} \rangle = \langle \nabla_w R(\bar{u}^n, \bar{u}^n), v^* - \bar{u}^n \rangle + \langle \nabla_w R(\bar{u}^n, \bar{u}^n), \bar{u}^n - v^{n+1} \rangle$$

$$= \langle \nabla_w R(\bar{u}^n, \bar{u}^n), v^* - \bar{u}^n \rangle - \langle \nabla_w R(v^n, v^n) - \nabla_w R(\bar{u}^n, \bar{u}^n), \bar{u}^n - v^{n+1} \rangle +$$

$$+ \langle \nabla_w R(v^n, v^n), \bar{u}^n - v^{n+1} \rangle \leq \langle \nabla_w R(\bar{u}^n, \bar{u}^n), v^* - \bar{u}^n \rangle +$$

$$+ \alpha L_0^2 |v^n - \bar{u}^n|^2 + \langle \nabla_w R(v^n, v^n), \bar{u}^n - v^{n+1} \rangle.$$

In view of the estimation obtained, we can rewrite (75) as

$$\langle v^{n+1} - v^n, v^* - v^{n+1} \rangle + \alpha \langle \nabla_w R(\bar{u}^n, \bar{u}^n), v^* - \bar{u}^n \rangle + (\alpha L_0)^2 |v^n - \bar{u}^n|^2 +$$

$$+ \alpha \langle \nabla_w R(v^n, v^n), \bar{u}^n - v^{n+1} \rangle + \alpha \langle \nabla g^T(v^n) p^{n+1}, v^* - v^{n+1} \rangle \geq 0$$
(76)

Let us put $w = v^{n+1}$ in (58). Then

$$\langle \bar{u}^n - v^n, v^{n+1} - \bar{u}^n \rangle + \alpha \langle \nabla_w R(v^n, v^n), v^{n+1} - \bar{u}^n \rangle +$$

$$\alpha \langle \nabla g^T(v^n) \bar{p}^n, v^{n+1} - \bar{u}^n \rangle \geq 0.$$
(77)

Adding (76) and (77), we obtain

$$\langle v^{n+1} - v^n, v^* - v^{n+1} \rangle + \langle \bar{u}^n - v^n, v^{n+1} - \bar{u}^n \rangle + \alpha \langle \nabla_w R(\bar{u}^n, \bar{u}^n), v^* - \bar{u}^n \rangle +$$

$$+(\alpha L_0)^2 |v^n - \bar{u}^n|^2 + \alpha \langle \nabla g^T(v^n) p^{n+1}, v^* - v^{n+1} \rangle +$$
$$\alpha \langle \nabla g^T(v^n) \bar{p}^n, v^{n+1} - \bar{u}^n \rangle \geq 0. \tag{78}$$

Using convexity (1) and condition (65), the second to last term from (78) can be expressed as

$$\langle \nabla g^T(v^n) p^{n+1}, v^* - v^{n+1} \rangle = \langle p^{n+1}, \nabla g(v^n)(v^* - v^n) \rangle -$$

$$-\langle p^{n+1}, \nabla g(v^n)(v^{n+1} - v^n) \rangle \leq \langle p^{n+1}, g(v^*) - g(v^n) \rangle + \langle p^{n+1}, g(v^n) \rangle \leq$$
$$\leq \langle p^{n+1}, g(v^*) \rangle \leq 0.$$

Moreover, using convexity (1) and conditions (61), (67), we can evaluate the last term from (78) as

$$\langle \nabla g^T(v^n) \bar{p}^n, v^{n+1} - \bar{u}^n \rangle = \langle \bar{p}^n, \nabla g(v^n)(v^{n+1} - v^n) \rangle + \langle \bar{p}^n, \nabla g(v^n)(v^n - \bar{u}^n) \rangle \leq$$

$$\leq \langle \bar{p}^n, g(v^{n+1}) - g(v^n) \rangle + \langle \bar{p}^n, g(v^n) \rangle \leq \frac{1}{2} \langle \bar{p}^n, L \rangle |v^{n+1} - v^n|^2.$$

So (78) becomes

$$\langle v^{n+1} - v^n, v^* - v^{n+1} \rangle + \langle \bar{u}^n - v^n, v^{n+1} - \bar{u}^n \rangle + \alpha \langle \nabla_w R(\bar{u}^n, \bar{u}^n), v^* - \bar{u}^n \rangle +$$

$$+(\alpha L_0)^2 |v^n - \bar{u}^n|^2 + \frac{1}{2} \alpha \langle \bar{p}^n, L \rangle |v^{n+1} - v^n|^2 \geq 0. \tag{79}$$

Using (74) and (68), we can estimate the third term from (79) by

$$\langle \nabla_w R(\bar{u}^n, \bar{u}^n), \bar{u}^n - v^* \rangle \geq -\frac{1}{2} \langle p^*, L \rangle |\bar{u}^n - v^n|^2 \tag{80}$$

and so (79) can be expressed as

$$\langle v^{n+1} - v^n, v^{n+1} - v^* \rangle + \langle \bar{u}^n - v^n, \bar{u}^n - v^{n+1} \rangle -$$

$$-\alpha(\alpha L_0^2 + \frac{1}{2} \langle p^*, L \rangle) |v^n - \bar{u}^n|^2 - \frac{1}{2} \alpha \langle \bar{p}^n, L \rangle |v^{n+1} - v^n|^2 \leq 0.$$

Using the identity

$$|x_1 - x_2|^2 = |x_1 - x_3|^2 + 2\langle x_1 - x_3, x_3 - x_2 \rangle + |x_3 - x_2|^2, \tag{81}$$

we can reorganize the first two scalar products from the last inequality as

$$|v^{n+1} - v^*|^2 + |v^{n+1} - \bar{u}^n|^2 + |\bar{u}^n - v^n|^2 -$$

$$-\alpha(2\alpha L_0^2 + \langle p^*, L\rangle)|v^n - \bar{u}^n|^2 - \alpha\langle C, L\rangle|v^{n+1} - v^n|^2 \le |v^n - v^*|^2,$$

where $0 \le \bar{p}^n \le C$, C is a vectorial constant from (69). Further, using the estimation

$$\frac{1}{2}|x_1 - x_2|^2 \le |x_1 - x_3|^2 + |x_3 - x_2|^2, \tag{82}$$

we can transform the sum of the second and third terms of the latter inequality as follows

$$|v^{n+1} - v^*|^2 + \frac{1}{2}|v^{n+1} - \bar{u}^n|^2 + \frac{1}{4}|v^{n+1} - v^n|^2 +$$

$$+(\frac{1}{2} - \alpha(2\alpha L_0^2 + \langle p^*, L\rangle))|v^n - \bar{u}^n|^2 - \alpha\langle C, L\rangle|v^{n+1} - v^n|^2 \le |v^n - v^*|^2. \tag{83}$$

Further

$$|v^{n+1} - v^*|^2 + d_1|\bar{u}^n - v^n|^2 + d_2|v^{n+1} - v^n|^2 \le |v^n - v^*|^2,$$

where $d_1 = \frac{1}{2} - \alpha(2\alpha L_0^2 + \langle p^*, L\rangle)$ and $d_2 = \frac{1}{4} - \alpha\langle C, L\rangle > 0$, as $\alpha < \bar{\alpha}$. Adding these inequalities from $n = 1$ up to $n = N$, we obtain

$$|v^N + 1 - v^*|^2 + d_1\sum_{n=1}^{n+N}|v^{n+1} - v^n|^2 + d_2\sum_{n=1}^{n=N}|v^n - \bar{u}^n|^2 \le |v_0 - v^*|^2. \tag{84}$$

Then, boundedness of the term $|v^n - v^*|^2$, monotone decrease of the sequence v^n and convergence of the series $\sum_{n=1}^{n=N}|v^{n+1} - v^n|^2 < \infty$, $\sum_{n=1}^{n=N}|v^n - \bar{u}^n|^2 < \infty$ follow easily. Let a sequence $n_i \to \infty$ be such that $v_{n_i} \to v'$, $v_{n_i+1} - v^{n_i} \to 0$, $\bar{u}^{n_i} - v^{n_i} \to 0$, $\bar{p}^{n_i} \to p'$, (recall that $\bar{p}^n$ is limited). Passing to the limit on this sequence in inequalities (58), (59), we obtain (47),(19). Whereas the value $|v^n - v^*|^2$ monotonically decreases, the sequence v^n has a unique limit point v', which is the solution of systems (47),(19), and, therefore, solution of the variational inequality (15)

$$\langle \nabla_w R(v', v'), w - v'\rangle \ge 0 \; \forall w \in \Omega. \tag{85}$$

Thus, it is shown that any limit point of sequence v^n is a fixed point of the equilibrium problem (16), i.e. solution of the variational inequality (15). If the function $R(v, w) = P(w) + K(v, w)$ is convex in w, then we have from (1)

$$R(v', w) - R(v', v') \ge \langle \nabla_w R(v', v'), w - v'\rangle \; \forall w \in \Omega. \tag{86}$$

With the inequality (85), we get

$$R(v', v') \le R(v', w) \; \forall w \in \Omega.$$

The inequality obtained, obviously, means that $v' = v^* \in \Omega_0^*$ is the solution of problem (16). Since function $\Phi(v, w)$ is convex in $w \in \Omega_0$ for any $v \in \Omega_0$, then inequality (85) means that $v' = v^* \in \Omega_0^*$, i.e. that the resulting point is the equilibrium solution of problem (1). The theorem is proved. $\square$

6   Convergence with rate of geometrical progression

In this section we consider equilibrium problems for which the solutions satisfy the additional requirement of a sharp equilibrium. We say that the function $R(w, w)$ has the quadratic-order sharpness of equilibrium, if the following inequality holds

$$R(w, w) - R(w, v^*) \geq \gamma |w - v^*|^2 \ \forall w \in \Omega, \tag{87}$$

where $\gamma > 0$ is a constant. Inequality (87) allows us to strengthen essentially estimate (80) from the previous theorem. We consider left-hand side of this estimate and we transform it taking into account the convexity of the function $R(v, w)$ in w

$$\langle \nabla_w R(\bar{u}^n, \bar{u}^n), \bar{u}^n - v^* \rangle \geq R(\bar{u}^n, \bar{u}^n) - R(\bar{u}^n, v^*).$$

So, considering (87), we deduce that

$$\langle \nabla_w R(\bar{u}^n, \bar{u}^n), \bar{u}^n - v^* \rangle \geq \gamma |\bar{u}^n - v^*|^2. \tag{88}$$

If each component $P(w)$ and $K(v^*, w)$ in the expansion $R(v^*, w) = P(w) + K(v^*, w)$ satisfies the conditions of sharpness (28) and (22) in the equilibrium point with the sharp-order $\nu = 1$ and with constants γ_1 and γ_2, then the inequality (88) will hold with a constant $\gamma = \gamma_1 + \gamma_2$. Now we present the theorem on the rate of convergence of the gradient process for finding the equilibrium solution with the quadratic-order of sharpness.

Theorem 2. *If in addition to conditions of theorem 1, the equilibrium solution of (16) satisfies condition (87), then the sequence v^n, generated by the process (51) with parameter $0 < \alpha < \min \left\{ \frac{1}{4} \frac{1}{\langle C, L \rangle}, \frac{1}{2L_0} \right\}$, converges to the solution of problem (16) with geometric progression rate, i.e.*

$$|v^{n+1} - v^*|^2 \leq q(\alpha)^{n+1} |v^0 - v^*|^2 \ at \ n \to \infty,$$

where $q(\alpha) = \left(1 + 4(\alpha\gamma)^2/d_2 - 2\alpha\gamma \right) < 1$, $d_2 = 1 + 2\alpha\gamma - 2(\alpha L)^2$.

Proof. In view to estimate (88), we write out inequality (79) as

$$\langle v^{n+1} - v^n, v^* - v^{n+1} \rangle + \langle \bar{u}^n - v^n, v^{n+1} - \bar{u}^n \rangle - \alpha\gamma |\bar{u}^n - v^*|^2 +$$

$$+ (\alpha L_0)^2 |v^n - \bar{u}^n|^2 + \frac{1}{2}\alpha \langle C, L \rangle |v^{n+1} - v^n|^2 \geq 0, \tag{89}$$

where C is a constant from (69). Then using the identity (81), we can transform (89) as

$$|v^{n+1} - v^*|^2 + |v^{n+1} - \bar{u}^n|^2 + |\bar{u}^n - v^n|^2 + 2\gamma\alpha |\bar{u}^n - v^*|^2 -$$

$$2(\alpha L_0)^2|v^n - \bar{u}^n|^2 - \alpha\langle C, L\rangle|v^{n+1} - v^n|^2 \le |v^n - v^*|^2. \qquad (90)$$

Repeating the reasoning (82) and (83), we can change (90) as follows

$$|v^{n+1} - v^*|^2 + \frac{1}{2}|v^{n+1} - \bar{u}^n|^2 + \frac{1}{2}|\bar{u}^n - v^n|^2 + \frac{1}{4}|v^{n+1} - v^n|^2 +$$

$$+2\alpha\gamma|\bar{u}^n - v^*|^2 - 2(\alpha L_0)^2|v^n - \bar{u}^n|^2 - \alpha\langle C, L\rangle|v^{n+1} - v^n|^2 \le |v^n - v^*|^2. \qquad (91)$$

In this inequality we can transform the fifth term like this

$$|\bar{u}^n - v^*|^2 = |\bar{u}^n - v^n|^2 + 2\langle\bar{u}^n - v^n, v^n - v^*\rangle + |v^n - v^*|^2.$$

Then it is possible to present (91) under the form

$$|v^{n+1} - v^*|^2 + d_1|v^{n+1} - v^n|^2 + d_2|\bar{u}^n - v^n|^2 +$$

$$+4\alpha\gamma\langle\bar{u}^n - v^n, v^n - v^*\rangle \le (1 - 2\alpha\gamma)|v^n - v^*|^2,$$

where $d_1 = \frac{1}{4} - \alpha\langle C, L\rangle \ge 0$ and $d_2 = \frac{1}{2} - 2(\alpha L_0)^2 \ge 0$, as $\alpha < \{\frac{1}{4\langle C,L\rangle}, \frac{1}{2L_0}\}$. From the third and fourth terms, we can choose the full quadrate:

$$|v^{n+1} - v^*|^2 + d_1|v^{n+1} - v^n|^2 + \left|\sqrt{d_2}(\bar{u}^n - v^n) + \frac{2\alpha\gamma}{\sqrt{d_2}}(v^n - v^*)\right|^2 -$$

$$-\frac{4(\alpha\gamma)^2}{d_2}|v^n - v^*|^2 \le (1 - 2\alpha\gamma)|v^n - v^*|^2.$$

So

$$|v^{n+1} - v^*|^2 \le \left(1 + 4(\alpha\gamma)^2/d_2 - 2\alpha\gamma\right)|v^n - v^*|^2.$$

As $\alpha < 1/(\{\frac{1}{4\langle C,L\rangle}, \frac{1}{2L_0}\})$, size $Q(\alpha) = 1 + 4(\alpha\gamma)^2/d_2 - 2\alpha\gamma = 1 + 2\alpha\gamma\left(\frac{2\alpha\gamma}{d_2} - 1\right) < 1$. Here $\frac{2\alpha\gamma}{d_2} - 1 < 0$. Thus, $|v^{n+1} - v^*|^2 \le q(\alpha)|v^n - v^*|^2$. Hence,

$$|v^{n+1} - v^*|^2 \le q(\alpha)^{n+1}|v^0 - v^*|^2.$$

The denominator of the progression $q(\alpha)$ depends on the parameter α. By minimizing it on the segment $(0, 1/(\{\frac{1}{4\langle C,L\rangle}, \frac{1}{2L_0}\}))$, it is possible to select the best value for the denominator of the progression. The theorem is proved. □

Thus, we have established the following convergence result for the linearization method as applied for computing the fixed point of an equilibrium programming problem: in the convex case, the method monotonically converges to one of the fixed points of problem (16) and to the solution of problem (1) if $\Phi(v, w)$ is convex in $w \in \Omega$ for any $v \in \Omega$. The method converges to a quadratic equilibrium at the rate of a geometric progression.

References

1. Antipin, A.S. (1992) Controlled proximal differential systems for solving saddle problems. Differentsial, nye Uravneniya **28**, 11, 1846–1861 (English Transl.: Differential Equations **28**, 11, 1498–1510)
2. Antipin, A.S. (1993) Inverse problems of nonlinear programming. Nonlinear Dynamic Systems: Qualitative Analysis and Control. The Collection of Transactions of the Institute for Systems Analysis, Moscow) **2**, 5–32 (English transl.:(1996) Computational Mathematics and Modelling Plenum Publish. Corp. N.Y. **7**, 3, 263–287)
3. Antipin, A.S. (1994) Linearization method. Nonlinear Dynamic Systems: Qualitative Analysis and Control. The Collection of Transactions of the Institute for Systems Analysis **2**, 4–20 (English transl.: (1997) Computational Mathematics and Modelling. Plenum Publish. Corp., New York **8**, 1, 1–15)
4. Antipin, A.S. (1995) On differential prediction-type gradient methods for computing fixed points of extremal mappings. Differentsial,nye Uravneniya **31**, 11, 1786–1795 (English Transl.: Differential Equations **31**, 11, 1754–1763)
5. Antipin, A.S. (1995) On convergence rate estimation for gradient projection method. Izvest. VUZov. Matematika **6**, 16–24 (English transl.: Russian Mathematics (Iz.VUZ) **39**, 6, 14–22)
6. Antipin, A.S. (1995) The Convergence of proximal methods to fixed points of extremal mappings and estimates of their rate of convergence. Zhurnal vychisl. mat. i mat. fiz **35**, 5, 688–704 (English transl.: Comp.Maths.Math.Phys. **35**, 5, 539-551)
7. Antipin, A.S. (1997) Equilibrium programming: proximal methods. Zhurnal vychisl. matem. i matem. fiz. **37**, 11, 1327–1339 (English Transl.: Comp. Maths.Math.Phys. **37**, 11, 1285–1296)
8. Antipin, A.S. (1997) Equilibrium programming: gradient-type methods. Avtomatika i telemehanika **8**, 125–137 (English Transl.: Automation and Control **8**, 1–12)
9. Antipin, A.S. (1998) The differential linearization method in equilibrium programming. Differential Equations **34**, 11, 1445–1458 (English transl.: Differential Equations **34**, 11, 1–13)
10. Antipin, A.S., Nedich, A. (1996) Continuous linearization method of the second-order for convex programming problems. Vestnik Moskovsk. universiteta. Ser.15. Vychisl.mat.i kibern **2**, 3–12 (English transl.: Comput.Maths. and Cybernetics **2**, 1-9)
11. Amochkina, T., Antipin, A.S., Vasil'ev, F.P. (1997) Continuous linearization method with a variable metric for problems in convex programming. Zurnal vychisl.mat. i mat. fiz. **37**, 12, 1459–1466 (English transl.: Comp.Maths.Math.Phys. **37**, 12, 1415–1421)
12. Aubin, J.-P., Frankowska, H. (1990) Set Valued Analysis. Birkhäuser, Basel
13. Belen'ky, V.Z., Volkonsky, V.A. (1974) Iterative Methods at Game Theory and Programming. Nauka, Moscow (in Russian)
14. Blum, E., Oettli, W. (1993) From optimization and variational inequalities to equilibrium problems. The Mathematics Student **63**, 1-4, 1–23
15. Fan, Ky (1972) A minimax inequality and applications. Inequalities III. Acad. Press, 103–113

16. Garcia, C.B., Zangwill, W.I. (1981) Pathways to Solutions, Fixed Points, and Equilibria. Prentice-Hall, Englewood Cliffs, NJ
17. Monderer, D., Shapley, L.S. (1996) Potential games. Games and economic behavior **14**, 124–143
18. Ortega, J., Rheinbold, W. (1970) Iterative Solution of Nonlinear Equations in Several Variables. Academic Press, New York
19. Polyak, B.T. (1987) Introduction to Optimization. Optimization Software, New York, New York
20. Rockafellar, R. (1970) Convex Analysis. Princeton Univ.Press, NJ
21. Rosen, J.B. (1965) Existence and uniqueness of equilibrium points for concave n-person games. Econometrica **33**, 3, 520–534
22. Vasil'ev, F.P. (1988) Numerical Methods for Extremum Value Problems. Nauka, Moscow

The heavy ball with friction dynamical system for convex constrained minimization problems[*]

H. Attouch[1] and F. Alvarez[2]

[1] Département des Sciences Mathématiques, Université Montpellier II
    34095 Montpellier, France ; email : attouch@math.univ-montp2.fr
[2] Universidad de Chile, Santiago, Chile; email: falvarez@dim.uchile.cl.

Abstract. The "heavy ball with friction" dynamical system

$$\ddot{u} + \gamma\dot{u} + \nabla\Phi(u) = 0$$

is a non-linear oscillator with damping ($\gamma > 0$). In [2], Alvarez proved that when H is a real Hilbert space and $\Phi : H \to \mathbb{R}$ is a smooth convex function whose minimal value is achieved, then each trajectory $t \to u(t)$ of this system weakly converges towards a minimizer of Φ.

We prove a similar result in the convex constrained case by considering the corresponding gradient-projection dynamical system

$$\ddot{u} + \gamma\dot{u} + u - \mathrm{proj}_C(u - \mu\nabla\Phi(u)) = 0,$$

where C is a closed convex subset of H. This result holds when H is a possibly infinite dimensional space, and extends, by using different technics, previous results by Antipin [1].

Key words : The heavy ball with friction dynamical system, dissipative dynamical systems, asymptotical behavior, steepest descent, constrained convex minimization, projection method, fixed point of a contraction

1 Introduction

Throughout the paper, H is a real Hilbert space, $\langle .,. \rangle$ denotes the associated scalar product and $|.|$ stands for the corresponding norm, $|v|^2 = \langle v, v \rangle$ for $v \in H$. Let us consider a function $\Phi : H \to \mathbb{R}$ that is assumed to be convex and continuously differentiable, which is the objective function, and a closed convex nonempty subset C of H, which is the set of constraints.

We are interested in time second-order differential systems whose trajectories asymptotically converge as $t \to +\infty$ towards minimizers of the convex constrained problem

$$(\mathcal{P}) \qquad\qquad \min\{\Phi(v) : v \in C\}.$$

[*] Partially supported by Comisión Nacional de Investigación Científica y Tecnológica de Chile under Fondecyt grant 1990884

When $C = H$, that is in the unconstrained case, and for Φ smooth possibly non convex, the asymptotical behavior of the non-linear oscillator with damping $(\gamma > 0)$ system

(HBF) $$\ddot{u} + \gamma\dot{u} + \nabla\Phi(u) = 0,$$

also called the "heavy ball with friction" system, has been considered by several authors: Antipin [1], Attouch-Goudou-Redont [4], Alvarez [2], Haraux-Jendoubi [10]. In the convex unconstrained case, a general asymptotical result has been obtained by Alvarez [2] who proved that, when the set ArgminΦ of the minimizers of Φ is nonempty, then every trajectory of the (HBF) system weakly converges to a minimizer of Φ.

The Alvarez result presents striking similarities (the use of the Opial lemma, weak convergence of the trajectories) with the Bruck theorem [8]. This last paper deals, as an important particular case, with the asymptotic behavior (as $t \to +\infty$) of the trajectories of the generalized steepest descent equation

(SD) $$\dot{u} + \partial\varphi(u) \ni 0,$$

where φ is a convex function, which may now take the value $+\infty$ and is only assumed to be lower semicontinuous.

A natural extension of the Bruck theorem to the second-order setting would consist in considering

$$\ddot{u} + \gamma\dot{u} + \partial\varphi(u) \ni 0.$$

Indeed, this is a quite interesting system from a mechanical point of view which modelizes shocks, as soon as φ is not continuous. But, in this situation, $\dot{u}$ may be discontinuous and $\ddot{u}$ has to be interpreted as a measure, which makes this system quite involved and not easy to deal with from a numerical point of view.

In this paper, we follow a *different approach*. In order to study the problem $(\mathcal{P})$ we need only to consider functions φ of the following form

$$\varphi = \Phi + \delta_C,$$

where δ_C is the indicator of C, i.e. $\delta_C(v) = 0$ for $v \in C$ and $+\infty$ elsewhere. In that case, it can be shown, see Brezis [6], that the generalized steepest descent equation can be equivalently interpreted as a gradient-projection method

$$\dot{u}(t) = \text{proj}_{T_C(u)}(-\nabla\Phi(u)),$$

where $T_C(u)$ is the tangent cone to C at u. This suggests that looking at various formulations of the *gradient-projection* method may be more tractable from a numerical point of view, and well fitted to the dynamical system approach.

As noticed in Antipin [1], one can reformulate the optimality condition for $(\mathcal{P})$

$$\nabla \Phi(u) + N_C(u) \ni 0,$$

where $N_C(u)$ is the outwards normal cone to C at u, as

$$u - \text{proj}_C(u - \mu \nabla \Phi(u)) = 0,$$

where $\mu > 0$ is a parameter. By taking $\mu > 0$ adequately chosen, one can prove that the operator $Au = u - \text{proj}_C(u - \mu \nabla \Phi(u))$ is a nice maximal monotone operator (indeed it is a cocoercive maximal monotone operator). We thus consider the associated second order differential system

$$\ddot{u} + \gamma \dot{u} + u - \text{proj}_C(u - \mu \nabla \Phi(u)) = 0$$

and prove in theorem 1 that each trajectory of this differential system weakly converges as $t \to +\infty$ to an element of the set $A^{-1}(0)$, which is indeed a global solution of the minimization problem $(\mathcal{P})$.

As basic ingredients for the proof of theorem 1, we use the cocoerciveness of the operator A and the Opial lemma. The proof readily extends (theorem 2) to the case $A = I - T$ where T is a general contraction.

2   A gradient-projection formulation of problem $(\mathcal{P})$

Let us now make the following assumptions on Φ :

(H_ϕ) $\Phi : H \to I\!\!R$ is a convex, continuously differentiable function whose gradient $\nabla \Phi$ is Lipschitz continuous, namely there exists a constant $L \geq 0$ such that for all $v, w \in H$

$$|\nabla \phi(v) - \nabla \phi(w)| \leq L|v - w|.$$

On the other hand, C is a nonempty, closed convex subset of H.

We consider the convex constrained minimization problem

$$(\mathcal{P}) \qquad\qquad \min\{\Phi(v) : v \in C\}$$

and assume that the set $S := \{v \in C : \Phi(v) = \inf_C \Phi\}$ of the optimal solutions (also called minimizers) of $(\mathcal{P})$ is nonempty. Clearly S is a nonempty closed convex subset of H.

We use standard notions of convex analysis and formulate $(\mathcal{P})$ as

$$\min\{\Phi(v) + \delta_C(v) : v \in H\}, \tag{1}$$

where δ_C is the indicator function of C, $\delta_C(v) = 0$ when $v \in C$ and $+\infty$ elsewhere. The optimality condition for (1) reads as follows

$$\nabla\Phi(u) + N_C(u) \ni 0, \tag{2}$$

where $N_C(u)$ is the outwards normal cone to C at u and is equal to the subdifferential of the indicator function of C at u.

Equivalently, given a parameter $\mu > 0$ (which for the moment is arbitrary and whose value will be precised later on), one can reformulate (2) as

$$-\mu\nabla\Phi(u) \in N_C(u),$$

or equivalently,

$$u - \mu\nabla\Phi(u) \in u + N_C(u). \tag{3}$$

As a basic tool, we use the classical relation

$$(I + N_C)^{-1} = \mathrm{proj}_C$$

to obtain as an equivalent optimality condition for $(\mathcal{P})$

$$u = \mathrm{proj}_C(u - \mu\nabla\Phi(u)). \tag{4}$$

Note that because of the convexity assumption on Φ, and the positivity of μ, (4) is a necessary and sufficient condition of optimality for $(\mathcal{P})$. So, we may work either with (4) or with the problem $(\mathcal{P})$. It is worthwhile to introduce the operator $T_\mu : H \to H$ defined by

$$T_\mu(v) := \mathrm{proj}_C(v - \mu\nabla\Phi(v)).$$

Let us summarize in the following statement the properties of T_μ.

Lemma 1. *Let us assume that $0 < \mu \leq \frac{2}{L}$. Then T_μ is a contraction from H into H. More precisely,*

$$\forall v, w \in H \quad |T_\mu v - T_\mu w| \leq |v - w|^2 - \frac{\mu}{2}(2 - \mu L)|\nabla\phi(v) - \nabla\phi(w)|^2.$$

Proof of lemma 1. Since proj_C is a contraction, we just need to prove that, for $0 < \mu \leq \frac{2}{L}$, the operator $v \mapsto v - \mu\nabla\phi(v)$ is a contraction. Given arbitrary $v, w \in H$

$$|(v - \mu\nabla\phi(v)) - (w - \mu\nabla\phi(w))|^2 = |(v - w) - \mu(\nabla\phi(v) - \nabla\phi(w))|^2$$
$$= |v - w|^2 - 2\mu\langle\nabla\phi(v) - \nabla\phi(w), v - w\rangle + \mu^2|\nabla\phi(v) - \nabla\phi(w)|^2.$$

In order to prove that $v \mapsto v - \mu\nabla\phi(v)$ is a contraction, we need equivalently to show that

$$-2\mu\langle\nabla\phi(v) - \nabla\phi(w), v - w\rangle + \mu^2|\nabla\phi(v) - \nabla\phi(w)|^2 \leq 0,$$

that is,

$$\langle \nabla\phi(v) - \nabla\phi(w), v - w \rangle \geq \frac{\mu}{2}|\nabla\phi(v) - \nabla\phi(w)|^2.$$

Indeed, this last property, called firmly non expansiveness of $\nabla\phi$ or *cocoer-civeness* of $\nabla\phi$, is equivalent to the $\frac{2}{\mu}$-Lipschitz property of $\nabla\Phi$, see lemma 2 below. So, for $0 < \mu \leq \frac{2}{L}$, we obtain that T_μ is a contraction. □

Lemma 2 (Baillon-Haddad [5]). *For an operator* $\nabla\Phi$ *(the gradient of a convex function), the following properties (i) and (ii) are equivalent*

(i) $\forall v, w \in H$, $|\nabla\phi(v) - \nabla\phi(w)| \leq L|v - w|$,

(ii) $\forall v, w \in H$, $\langle \nabla\phi(v) - \nabla\phi(w), v - w \rangle \geq \frac{1}{L}|\nabla\phi(v) - \nabla\phi(w)|^2$.

One may notice that (ii) can be viewed as a dual property of (i). Indeed, property (ii) is equivalent to

$$\langle (\nabla\phi)^{-1}(v) - (\nabla\phi)^{-1}(w), v - w \rangle \geq \frac{1}{L}|v - w|^2.$$

Since $(\nabla\phi)^{-1} = \partial\phi^*$, the above lemma expresses the Lipschitz continuity of $\nabla\phi$ in terms of the strong coerciveness of $\partial\phi^*$.

Let us return to the optimality condition (4) which can be expressed as

$$A_\mu u = 0 \qquad\qquad (5)$$

where $A_\mu v := v - T_\mu v$. The properties of the operator $A_\mu = I - T_\mu$ follow from the general properties of operators of the form $I - T$ where T is a contraction, as stated in the following lemma:

Lemma 3. *Let* $T : H \to H$ *be a contraction. Then the operator* $A = I - T$ *is a maximal monotone operator which is* $\frac{1}{2}$*-cocoercive, that is*

$$\forall v, w \in H, \quad \langle Av - Aw, v - w \rangle \geq \frac{1}{2}|Av - Aw|^2.$$

Proof of lemma 3. A direct computation yields

$\langle Av - Aw, v - w \rangle - \frac{1}{2}|Av - Aw|^2$
$= |v - w|^2 - \langle Tv - Tw, v - w \rangle - \frac{1}{2}(|v - w|^2 + |Tv - Tw|^2 - 2\langle v - w, Tv - Tw \rangle) = \frac{1}{2}(|v - w|^2 - |Tv - Tw|^2) \geq 0,$

where this last inequality expresses that T is a contraction. Thus, A is a monotone operator which is everywhere defined, which implies that A is a maximal monotone operator, see [7]. □

Remark : Following Bruck [8], $A = I - T$ is a demipositive maximal monotone operator. This class of operators is interesting because it is for operators of this class, which contains subdifferentials of closed convex functions and operators $A = I - T$ where T is a contraction, that Bruck [8] proved the weak asymptotical convergence of the trajectories of the associated first order differential system. In [2], Alvarez extended the Bruck result to the second order differential system for $A = \nabla\Phi$. Here, we are doing the parallel extension by considering operators $A = I - T$.

All these considerations lead us to consider the time second order differential system

$$\ddot{u} + \gamma\dot{u} + u - \text{proj}_C(u - \mu\nabla\Phi(u)) = 0$$

with $0 < \mu \leq \frac{2}{L}$ and $\gamma > 0$.

3   The main result

We can now state the following

Theorem 1. *Let $\Phi : H \to I\!R$ be a convex, continuously differentiable function, whose gradient $\nabla\Phi$ is Lipschitz continuous with Lipschitz constant L. Let C be a closed convex nonempty subset of H, and let us assume that $S := \text{Argmin}_C\Phi$, the set of minimizers of*

$$(\mathcal{P}) \qquad\qquad \min\{\Phi(v) : v \in C\}$$

is nonempty.

Let us consider the second order differential system with Cauchy data u_0, v_0:

$$(E_{\gamma,\mu}; u_0, v_0) \quad \begin{cases} \ddot{u} + \gamma\dot{u} + u - \text{proj}_C(u - \mu\nabla\Phi(u)) = 0 \\ u(0) = u_0, \dot{u}(0) = v_0 \end{cases}$$

Let us assume that $\gamma > \sqrt{2}$ and $0 < \mu \leq \frac{2}{L}$. Then for each u_0 and v_0 belonging to H, there exists a unique solution $u \in C^2([0, +\infty[; H)$ of $(E_{\gamma,\mu}; u_0, v_0)$ which satisfies :

(i) $\dot{u} \in L^2(0, +\infty; H)$, $\ddot{u} \in L^2(0, +\infty; H)$

(ii) $\lim_{t \to +\infty} \dot{u}(t) = \lim_{t \to +\infty} \ddot{u}(t) = 0$

(iii) there exists $\bar{u} \in S := \text{Argmin}_C\Phi$ such that $u(t) \rightharpoonup \bar{u}$ weakly in H as $t \to +\infty$.

Proof of theorem 1. The equation $(E_{\gamma,\mu}; u_0, v_0)$ can equivalently be written as

$$\ddot{u} + \gamma\dot{u} + Au = 0 \qquad\qquad (6)$$

where $A = I - T$ and $Tv = \text{proj}_C(v - \mu\nabla\Phi(v))$.

It follows from lemma 1 that, for $0 < \mu \leq \frac{2}{L}$, the operator T is a contraction and from lemma 3 that $A = I - T$ is a $\frac{1}{2}$-cocoercive operator, that is, for any $v, w \in H$

$$\langle Av - Aw, v - w \rangle \geq \frac{1}{2}|Av - Aw|^2. \tag{7}$$

Indeed, these are the only ingredients that we need for the proof of theorem 1, in addition to the fact that $S = \text{Argmin}_C \Phi = \{v \in H : Av = 0\} \neq \emptyset$.

Clearly, since A is Lipschitz continuous, by the Cauchy-Lipschitz theorem there exists a unique solution to $(E_{\gamma,\mu}; u_0, v_0)$, with $u \in C^2([0, +\infty[; H)$. Take now $z \in S = A^{-1}(0)$ and define the auxiliary function $h(t) := \frac{1}{2}|u(t) - z|^2$. We have

$$\dot{h}(t) = \langle u(t) - z, \dot{u}(t) \rangle$$
$$\ddot{h}(t) = |\dot{u}(t)|^2 + \langle u(t) - z, \ddot{u}(t) \rangle.$$

It follows that

$$\ddot{h} + \gamma \dot{h} = |\dot{u}|^2 + \langle u - z, \ddot{u} + \gamma \dot{u} \rangle,$$

which, by using (6) yields

$$\ddot{h} + \gamma \dot{h} + \langle Au, u - z \rangle = |\dot{u}|^2.$$

Since $Az = 0$, by using the $\frac{1}{2}$-cocoerciveness of A as stated in (7), we obtain

$$\ddot{h} + \gamma \dot{h} + \frac{1}{2}|Au|^2 \leq |\dot{u}|^2,$$

that is, by using (6) again,

$$\ddot{h} + \gamma \dot{h} + \frac{1}{2}|\ddot{u} + \gamma \dot{u}|^2 \leq |\dot{u}|^2. \tag{8}$$

This is the basic inequality from which we are going to derive various estimations on u. Let us rewrite (8) as

$$\ddot{h} + \gamma \dot{h} + \frac{\gamma}{2}\frac{d}{dt}|\dot{u}(t)|^2 + (\frac{\gamma^2}{2} - 1)|\dot{u}|^2 + \frac{1}{2}|\ddot{u}|^2 \leq 0. \tag{9}$$

Since $\gamma > \sqrt{2}$, we have

$$\ddot{h} + \gamma \dot{h} + \frac{\gamma}{2}\frac{d}{dt}|\dot{u}(t)|^2 \leq 0,$$

which implies that the function $t \to \dot{h} + \gamma h + \frac{\gamma}{2}|\dot{u}(t)|^2$ is decreasing. Thus, for any $t \geq 0$

$$\dot{h}(t) + \gamma h(t) + \frac{\gamma}{2}|\dot{u}(t)|^2 \leq \dot{h}(0) + \gamma h(0) + \frac{\gamma}{2}|v_0|^2$$
$$\leq |u_0 - z||v_0| + \frac{\gamma}{2}|u_0 - z|^2 + \frac{\gamma}{2}|v_0|^2 := C_z \tag{10}$$

which implies

$$\dot{h}(t) + \gamma h(t) \le C_z. \tag{11}$$

After integration of (11) we obtain

$$h(t) \le \frac{1}{2}|u_0 - z|^2 + \frac{1}{\gamma}C_z,$$

which implies that each trajectory of the differential system (6) is bounded. We need now to obtain estimations on $\dot{u}$ and $\ddot{u}$.

Let us integrate (9) from 0 to t to obtain

$$\dot{h}(t) + \gamma h(t) + \frac{\gamma}{2}|\dot{u}(t)|^2 + (\frac{\gamma^2}{2} - 1)\int_0^t |\dot{u}(s)|^2 ds + \frac{1}{2}\int_0^t |\ddot{u}(s)|^2 ds \le C_z. \tag{12}$$

From (12) we derive that

$$\dot{h}(t) + \frac{\gamma}{2}|\dot{u}(t)|^2 \le C_z$$

(notice that all the other terms are nonnegative) which by definition of $h(.)$ yields

$$\langle u(t) - z, \dot{u}(t) \rangle + \frac{\gamma}{2}|\dot{u}(t)|^2 \le C_z. \tag{13}$$

Since $|u(t)|$ remains bounded on $[0, +\infty[$, the inequality (13) clearly implies that $|\dot{u}(t)|$ also remains bounded on $[0, +\infty[$:

$$\sup_{t \in [0,+\infty[} |\dot{u}(t)| < +\infty. \tag{14}$$

Using that $u(.)$ and $\dot{u}(.)$ remain bounded on $[0, +\infty[$ we deduce that $\dot{h}(t) = \langle u(t) - z, \dot{u}(t) \rangle$ is also bounded on $[0, +\infty[$.
This information and (12) immediately yield that

$$\int_0^{+\infty} |\dot{u}(s)|^2 ds < +\infty \tag{15}$$

and

$$\int_0^{+\infty} |\ddot{u}(s)|^2 ds < +\infty. \tag{16}$$

Let us now return to (6). Since u and $\dot{u}$ are bounded on $[0, +\infty[$ and A is Lipschitz continuous we deduce from (6) that

$$\sup_{t \in [0,+\infty[} |\ddot{u}(t)| < +\infty. \tag{17}$$

We now observe that the function $g(t) = \dot{u}(t)$ satisfies both

$$g \in L^2(0, +\infty; H) \quad \text{and} \quad \dot{g} \in L^\infty(0, +\infty; H).$$

According to a classical result, these two properties imply that

$$\lim_{t \to +\infty} \dot{u}(t) = 0. \tag{18}$$

Let us now prove that $\lim_{t \to +\infty} \ddot{u}(t) = 0$. To that end, let us write the equation (6) at the points t and $t + \varepsilon$, make the difference and divide by ε. We obtain that $u_\varepsilon(t) := \frac{1}{\varepsilon}(\dot{u}(t + \varepsilon) - \dot{u}(t))$ satisfies

$$\dot{u}_\varepsilon(t) + \gamma u_\varepsilon(t) = f_\varepsilon(t) \tag{19}$$

where $f_\varepsilon(t) = -\frac{1}{\varepsilon}(Au(t + \varepsilon) - Au(t))$.

We know that A is a 2-Lipschitz continuous operator on H. Hence,

$$|f_\varepsilon(t)| \le \frac{2}{\varepsilon}|u(t + \varepsilon) - u(t)| \le 2 \sup_{s \in [t,+\infty[} |\dot{u}(s)|.$$

Let us write $g(t) := 2\sup_{s \in [t,+\infty[} |\dot{u}(s)|$.
We notice that, since $\lim_{t \to +\infty} \dot{u}(t) = 0$, we have that $\lim_{t \to +\infty} g(t) = 0$.
Let us integrate (19) to obtain

$$|u_\varepsilon(t)| \le e^{-\gamma t}|u_\varepsilon(0)| + e^{-\gamma t} \int_0^t e^{\gamma s} g(s) ds,$$

which easily implies

$$\lim_{t \to +\infty} \left(\sup_{\varepsilon > 0} |u_\varepsilon(t)| \right) = 0.$$

Since, for all $t \ge 0$ the following inequality holds

$$|\ddot{u}(t)| \le \sup_{\varepsilon > 0} |u_\varepsilon(t)|$$

we conclude that

$$\lim_{t \to +\infty} \ddot{u}(t) = 0. \tag{20}$$

We are now in position to apply Opial's lemma, which we recall for the convenience of the reader.

Lemma 4 (Opial). *Let H be a Hilbert space and $u : [0, +\infty[\to H$ be a function such that there exists a non void set $S \subset H$ which verifies :*

(a) *$\forall t_n \to +\infty$ with $u(t_n) \rightharpoonup \bar{u}$ weakly in H, we have $\bar{u} \in S$.*

(b) *$\forall z \in S$, $\lim_{t \to +\infty} |u(t) - z|$ exists.*

Then, $u(t)$ weakly converges as $t \to +\infty$ to an element $\bar{u}$ of S.

End of the proof of theorem 1. Let us apply the Opial lemma with $S = A^{-1}(0) = \text{Argmin}_C \Phi$.

(a) By equation (6) we have

$$-\ddot{u}(t) - \gamma \dot{u}(t) = Au(t). \tag{21}$$

We know, by (18) and (20) that the left member of (21) strongly converges to zero as $t \to +\infty$. Let $u(t_n) \rightharpoonup \bar{u}$ weakly in H. Since the maximal monotone operator A is closed in $H \times H$ for the topology $w - H \times s - H$, see [6] for example, we obtain $A\bar{u} = 0$, i.e. $\bar{u} \in A^{-1}(0) = S$.

(b) Let us return to (9) which implies, as we have already noticed, that the function $\psi(t) := \dot{h}(t) + \gamma h(t) + \frac{\gamma}{2}|\dot{u}(t)|^2$ is decreasing. Hence $\lim_{t \to +\infty} \psi(t)$ exists. On the other hand, by (18) we know that $\lim_{t \to +\infty} \dot{u}(t) = 0$. Since $\dot{h}(t) = \langle u(t) - z, \dot{u}(t) \rangle$ and $u(t)$ is bounded, we also have $\lim_{t \to +\infty} \dot{h}(t) = 0$. By combining all these results we obtain

$$\lim_{t \to +\infty} \psi(t) = \gamma \lim_{t \to +\infty} h(t) \quad \text{exists.}$$

Hence, for any $z \in S$, $\lim_{t \to +\infty} |u(t) - z|$ exists.

So, the two assumptions of Opial lemma are satisfied, which completes the proof of theorem 1.   □

Remarks : 1) Clearly, if, by an a priori estimate, we know that the trajectory of the differential system $(E_{\gamma,\mu}; u_0, v_0)$ remains in a ball $\mathbb{B}(0, R)$, $0 \le R < \infty$ we need only to assume $\nabla \phi$ to be Lipschitz continuous on bounded sets.

2) All the proof of theorem 1 works without any modification for a general operator $A = I - T$ where T is just assumed to be a contraction on H. So, it is worthwhile to state it independently.

Theorem 2. *Let $T : H \to H$ be a contraction, i.e.*

$$|Tv - Tw| \le |v - w| \quad \text{for all } v, w \in H.$$

Let us consider the second order differential system with Cauchy data u_0, v_0:

$$(E_\gamma; u_0, v_0) \quad \begin{cases} \ddot{u} + \gamma \dot{u} + u - Tu = 0 \\ u(0) = u_0, \dot{u}(0) = v_0 \end{cases}$$

Let us assume that $\gamma > \sqrt{2}$. Then for each u_0 and v_0 belonging to H, there exists a unique solution $u \in C^2([0, +\infty[; H)$ of $(E_\gamma; u_0, v_0)$ which satisfies :

(i) $\dot{u} \in L^2(0, +\infty; H)$, $\ddot{u} \in L^2(0, +\infty; H)$

(ii) $\lim_{t \to +\infty} \dot{u}(t) = \lim_{t \to +\infty} \ddot{u}(t) = 0$

(iii) there exists $\bar{u} \in \text{Fix} \, T := \{v \in H : Tv = v\}$ such that $u(t) \rightharpoonup \bar{u}$ weakly in H as $t \to +\infty$.

Remark : It is an interesting question, for numerical purposes, to consider discretized versions, explicit or implicit, of theorems 1 and 2. This is naturally suggested by previous results obtained by Alvarez [2] in the unconstrained case.

References

1. Antipin, A.S. (1994) Minimization of convex functions on convex sets by means of differential equations. Differential Equations **30**, n. 9, 1365–1375
2. Alvarez, F. (1998) On the minimizing property of a second order dissipative system in Hilbert spaces, prepublication 1998/05, Département des Sciences Mathématiques de l'Université Montpellier II. To appear in SIAM J. Control and Optimization
3. Attouch, H., Cominetti, R. (1996) A dynamical approach to convex minimization coupling approximation with the steepest descent method. Journal of Differential Equations **128**, 519–540
4. Attouch, H., X. Goudou, X., Redont, P. (1998) The heavy ball with friction method. I. The continuous dynamical system, prepublication 1998/11, Département des Sciences Mathématiques, Université Montpellier II
5. Baillon, J.B., Haddad, G. (1977) Quelques propriétés des opérateurs angle-bornés et n-cycliquement monotones. Israel J. Math. **26**, 137–150
6. Brezis, H. (1973) Opérateurs maximaux monotones et semi-groupes de contractions dans les espaces de Hilbert. Mathematics Studies **5**, North Holland, Amsterdam
7. Brezis, H. (1978) Asymptotic behaviour of some evolution systems : Nonlinear Evolution Equations, Academic Press, New York
8. Bruck, R.E. (1975) Asymptotic convergence of nonlinear contraction semi-groups in Hilbert space. Journal of Functional Analysis **18**, 15–26
9. Haraux, A. (1991) Systèmes dynamiques dissipatifs et applications, in R.M.A. **17**, Masson, Paris
10. Haraux, A., Jendoubi, M.A. (1998) Convergence of solutions of second order gradient like systems with analytic nonlinearities. Journal of Differential Equations **144**, n. 2, 313–320
11. Lemaire, B. (1996) Stability of the iteration method for non expansive mappings. Serdica Math. J. **22**, 331–340

Active Set Strategy for Constrained Optimal Control Problems: the Finite Dimensional Case

Maïtine Bergounioux[1] and Karl Kunisch[2]

[1] UMR-CNRS 6628, Université d'Orléans, U.F.R. Sciences, B.P. 6759, F-45067 Orléans Cedex 2, France
[2] Institut für Mathematik, Universität Graz, A-8010 Graz, Austria.

Abstract. We consider control constrained and state constrained optimal control problems governed by elliptic partial differential equations, once they have been discretized. We propose and analyze an algorithm for efficient solution of these finite dimensional problems. It is based on an active set strategy involving primal as well as dual variables and is suggested by a generalized Moreau-Yosida regularization of the control (or state) constraint. Sufficient conditions for convergence in finitely many iterations are given. At last we present numerical examples and discuss the role of the strict complementarity condition.

Key words: Moreau-Yosida approximation, augmented Lagrangian, primal-dual method, optimal control, active sets.

1 Introduction

This paper is devoted to the numerical treatment of the following optimal control problem

$$(\mathcal{P}) \quad \begin{cases} \text{minimize} \quad J(y,u) = \dfrac{1}{2} \displaystyle\int_\Omega (y-z_d)^2 dx + \dfrac{\alpha}{2} \displaystyle\int_\Omega (u-u_d)^2 dx \\ \text{subject to } -\Delta y = u \text{ in } \Omega, \ y = 0 \text{ on } \Gamma, \\ \text{and} \qquad (y,u) \in K \times U_{ad} \,, \end{cases}$$

where

- Ω is a bounded subset of $\mathbb{R}^2$ with (smooth) boundary Γ,
- $\alpha > 0, z_d, u_d \in L^2(\Omega)$,
-

$$K = \left\{ \, y \in L^2(\Omega) \mid y \le \varphi \text{ a.e.} \right\},$$

$$U_{ad} = \left\{ \, u \in L^2(\Omega) \mid u \le b \text{ a.e.} \ \right\},$$

with $\varphi \in L^2(\Omega)$, $b \in L^2(\Omega)$.

We assume that the feasible domain is non empty, that is, there exists at least one pair $(y,u) \in H_o^1(\Omega) \times L^2(\Omega)$ such that $-\Delta y = u$, $u \le b$ and $y \le \varphi$. It is standard to argue the existence of a unique solution $(y^*, u^*) \in \mathcal{W} \times L^2(\Omega)$ to $(\mathcal{P})$ (we have set $\mathcal{W} = H^2(\Omega) \cap H^1_o(\Omega)$).

The analysis and the numerical treatment of $(\mathcal{P})$ is not easy because the constraints on both the control and the state appear simultaneously. Therefore, we shall consider the cases where these constraints appear separately. Moreover let us underline that these two cases are quite different from the theoretical and numerical points of view and the methods we are going to present here behave quite differently for each of them.

1.1   Control constraints

In this case $K = L^2(\Omega)$ and $(\mathcal{P})$ becomes:

$(\mathcal{P}^c)$
$$\begin{cases} \text{minimize} \quad J(y, u) = \frac{1}{2} \int_\Omega (y - z_d)^2 dx + \frac{\alpha}{2} \int_\Omega (u - u_d)^2 dx \\ \text{subject to } -\Delta y = u \text{ in } \Omega, \ y = 0 \text{ on } \Gamma, \\ \text{and} \qquad u \leq b \text{ a.e. in } \Omega. \end{cases}$$

Associated to u^* we define the active set $\mathcal{A}^* = \{ x \mid u^*(x) = b(x) \text{ a.e}\}$, and the inactive set $\mathcal{I}^* = \Omega \setminus \mathcal{A}^*$.

The optimal solution (y^*, u^*) is characterized by the existence of $(p^*, \lambda^*) \in \mathcal{W} \times L^2(\Omega)$ such that $(y^*, u^*, p^*, \lambda^*)$ satisfies the following optimality system

$(\mathcal{S}^c)$
$$\begin{cases} -\Delta y^* = u^* \text{ in } \Omega, \ y^* \in H^1_o(\Omega) \\ -\Delta p^* = -(y^* - z_d) \text{ in } \Omega, \ p^* \in H^1_o(\Omega), \\ u^* = u_d + \frac{1}{\alpha}(p^* - \lambda^*) \text{ in } \Omega, u^* \leq b, \\ \lambda^* \in \partial I_{U_{ad}}(u^*). \end{cases}$$

This result is wellknown (see for instance [1]). The differential inclusion appearing as the last condition in $(\mathcal{S}^c)$ is not amenable for numerical realization and we therefore replace it by

$$\lambda^* = c\left(u^* + \frac{\lambda^*}{c} - \Pi_{U_{ad}}(u^* + \frac{\lambda^*}{c})\right) = c\max(0, u^* + \frac{\lambda^*}{c} - b), \quad (1)$$

for any $c > 0$. Here $\Pi_{U_{ad}}$ denotes the Hilbert space projection of $L^2(\Omega)$ onto U_{ad} and max stands for the pointwise maximum as x varies in Ω.

1.2   State constraints

Now we set $U_{ad} = L^2(\Omega)$ and $(\mathcal{P})$ becomes:

$(\mathcal{P}^s)$
$$\begin{cases} \text{minimize} \quad J(y, u) = \frac{1}{2} \int_\Omega (y - z_d)^2 dx + \frac{\alpha}{2} \int_\Omega (u - u_d)^2 dx \\ \text{subject to } -\Delta y = u \text{ in } \Omega, \ y = 0 \text{ on } \Gamma, \\ \text{and} \qquad y \leq \varphi \text{ a.e. in } \Omega. \end{cases}$$

In this very case we assume $u_d \in H^2(\Omega)$ and $\varphi \in H^4(\Omega)$. A first order optimality condition which characterize the optimal solution is given next : there exists $p^* \in L^2(\Omega)$ and $\lambda^* \in \mathcal{M}(\Omega)$ such that

$$(\mathcal{S}^s) \quad \begin{cases} -\Delta y^* = u^* \text{ in } \Omega, \ y^* \in H_o^1(\Omega) \\ (p^*, -\Delta y)_\Omega + \langle \lambda^*, y \rangle_{\mathcal{C}^*, \mathcal{C}} = (z_d - y^*, y)_\Omega \quad \text{for all } y \in \mathcal{W}, \\ \langle \lambda^*, y - y^* \rangle_{\mathcal{C}^*, \mathcal{C}} \leq 0 \text{ for all } y \in \mathcal{C}_o(\Omega), \ y^* \leq \varphi, \\ p^* = \alpha \left(u^* - u_d \right). \end{cases}$$

Here $\mathcal{M}(\Omega)$ is the space of real regular Borel measures on Ω, $(\cdot, \cdot)_\Omega$ denotes the $L^2(\Omega)$-inner product on Ω and $\langle \cdot, \cdot \rangle_{\mathcal{C}^*, \mathcal{C}}$ stands for the duality pairing between $\mathcal{M}(\Omega)$ and $\mathcal{C}_o(\Omega)$. The proof can be found in [3].

Note the distinct difference with respect to regularity properties for solutions to $(\mathcal{S}^c)$ and $(\mathcal{S}^s)$. The adjoint state p^* is an $H^4(\Omega)$-function if $z_d \in H^2(\Omega)$ for $(\mathcal{P}^c)$ whereas it is only in $L^2(\Omega)$ for $(\mathcal{P}^s)$.

Optimal control problems are infinite dimensional problems which require discretization before they can be solved numerically. Once discretized they can, in principle, be treated as generic minimization problems. We present here a primal-dual method to solve the discretized problems. Anyway, though the discretized structures of $(\mathcal{P}_s)$ and $(\mathcal{P}_c)$ are similar we observe that the algorithm behaves differently. This comes from the fact that the method is based on a infinite dimensional analysis and depends on the regularity of the different variables, especially the Lagrange multiplier associated to the control constraint (respectiveley the state constraint) which is an element of $L^2(\Omega)$ in case of control constraints and only a measure otherwise.
In the next section we detail the discretized problems and the (discrete) algorithm. Then we give a convergence analysis for each case. We present numerical results and comment on the behavior of the method.

2   The discrete Algorithm

This section is devoted to the description of the algorithm which solves a discretized version of $(\mathcal{P})$. The separate treatment of the two cases is motivated by the fact that the analytical properties of their solutions and subsequently the behavior of the numerical algorithms differ significantly.

For this purpose assume that Ω is endowed with a grid and let Ω_h denote the set of interior grid points. The cardinality of Ω_h is denoted by N and we set $\Omega_N = \{ 1, \dots, N \}$. Let $z_{d,h}, u_{d,h}, b_h$ and φ_h stand for the grid functions corresponding to z_d, u_d, b and φ. Here we assume that these functions are well defined in the pointwise sense on Ω. Further let A_h denote a symmetric positive definite $N \times N$ matrix representing the discretization of $-\Delta$.

2.1 Control constraints

We first consider the discretized control constrained problem

$$(P_h^c) \qquad \begin{cases} \text{minimize } J_h(y_h, u_h), \\ \text{subject to } A_h y_h = u_h \text{ and } u_h \leq b_h , \end{cases}$$

where

$$J_h(y_h, u_h) = \frac{1}{2}(y_h - z_{d,h})^\top M_{1,h}(y_h - z_{d,h}) + \frac{\alpha}{2}(u_h - u_{d,h})^\top M_{2,h}(u_h - u_{d,h}) ,$$

$M_{1,h} \in \mathbb{R}^N \times \mathbb{R}^N$ is positive semi definite and $M_{2,h} \in \mathbb{R}^N \times \mathbb{R}^N$ is positive definite. Before we describe the algorithm we need to introduce some additional notation:

- For $v \in \mathbb{R}^N, w \in \mathbb{R}^N$ we denote by $\{ i \mid v \leq w \}$ the set of all indices such that the i-th coordinate of $v - w$ is non positive.
- For $v \in \mathbb{R}^N$, and M a $N \times N$ real, symmetric, positive matrix, we set $\|v\|_M^2 = v^\top M v$;

  $\| \cdot \|$ denotes the usual euclidian norm in $\mathbb{R}^N$: $\|v\|^2 = \sum_{i=1}^N v_i^2$.

- $(\cdot, \cdot)$ denotes the usual scalar product of $\mathbb{R}^N$:

$$\text{for } v \in \mathbb{R}^N, w \in \mathbb{R}^N, \ (v, w) = \sum_{i=1}^N v_i w_i .$$

In addition, if $\mathcal{I} \subset \Omega_N$, we set $(v, w)_{\mathcal{I}} = \sum_{i \in \mathcal{I}} v_i w_i$ and $\|v\|_{\mathcal{I}}^2 = (v, v)_{\mathcal{I}}$.

It is simple to derive the optimality system for (P_h^c). We point out that the complementarity condition has the form

$$\lambda_h^* = c \max\left(0, u_h^* + \frac{\lambda_h^*}{c} - b_h\right), \tag{2}$$

for $c > 0$, where the max is understood componentwise. This is the discretized version of (1). The essential ingredient of the Algorithm to solve (P_h^c) is a primal-dual active set strategy that is motivated by (2). Moreover we drop the subscripts h.

To describe the key step of this algorithm let us assume that (u_{n-1}, λ_{n-1}) is available from the previous iteration level. Then (2) suggests to define

$$\mathcal{A}_n = \left\{ i \mid u_{n-1} + \frac{\lambda_{n-1}}{c} > b \right\}, \tag{3}$$

We refer to $\mathcal{A}_n$ as the active set at the n-th iteration level and define $\mathcal{I}_n$ as the complement, which is formally described by $\mathcal{I}_n = \Omega_N \backslash \mathcal{A}_n$. We arrive at the following algorithm.

Algorithm (A^c)

1. Initialization: choose $y_o, u_o, \lambda_o \in \mathbb{R}^N$ and set $n = 1$.
2. Determine the subset of active/inactive indices:

$$\mathcal{A}_n = \left\{ i \mid u_{n-1} + \frac{\lambda_{n-1}}{c} > b \right\}, \ \mathcal{I}_n = \left\{ i \mid u_{n-1} + \frac{\lambda_{n-1}}{c} \leq b \right\}.$$

3. If $n \geq 2$ and $\mathcal{A}_n = \mathcal{A}_{n-1}$ then STOP; otherwise goto 4.
4. Find (y_n, p_n) such that

$$Ay_n = \begin{cases} b & \text{in } \mathcal{A}_n \\ u_d + \dfrac{1}{\alpha} M_2^{-1} p_n & \text{in } \mathcal{I}_n \end{cases}$$

$$Ap_n = M_1(z_d - y_n),$$

and set

$$u_n = \begin{cases} b & \text{in } \mathcal{A}_n \\ u_d + \dfrac{1}{\alpha} M_2^{-1} p_n & \text{in } \mathcal{I}_n. \end{cases}$$

5. Set $\lambda_n = p_n - \alpha M_2(u_n - u_d)$, $n = n + 1$ and goto 2.

In step 4 "in $\mathcal{A}_n$" stands for "indices that are contained in $\mathcal{A}_n$", and analogously for $\mathcal{I}_n$. Moreover, we say that equality is satisfied on $\mathcal{A}_n$ or $\mathcal{I}_n$ if equality holds for all coordinates in $\mathcal{A}_n$ respectively $\mathcal{I}_n$.

We may use different initialization schemes. The one that was used most frequently is the following one:

$$\begin{cases} Ay_o = u_o \ , \ u_o = b \ , \\ Ap_o = M_1(z_d - y_o) \ , \\ \lambda_o = \max(0, -\alpha M_2(b - u_d) + p_o) \ . \end{cases} \tag{4}$$

This choice of initialization has the property of feasibility.

2.2   State constraints

We suppose that the discretization of $(\mathcal{P}^s)$ can be expressed as

$$(P_h^s) \qquad \begin{cases} \text{minimize } J_h(y_h, u_h), \\ \text{subject to } A_h y_h = u_h \text{ and } y_h \leq \varphi_h \ . \end{cases}$$

Throughout it is assumed that there exists at least one feasible pair for (P_h^s). Then there exists a unique solution (y_h^*, u_h^*) to (P_h^s) which is characterized by the existence of $\lambda_h^* \in \mathbb{R}^N$, $p_h^* \in \mathbb{R}^N$ satisfying the first order optimality system

$$(S^s) \qquad \begin{cases} A_h y_h^* = u_h^* \ , \\ A_h p_h^* + \lambda_h^* = -M_{1,h}(y_h^* - z_{d,h}) \ , \\ y_h^* \leq \varphi_h \ , \\ (\lambda_h^*)^\top (y_h - y_h^*) \leq 0 \text{ for all } y_h \leq \varphi_h \ , \\ p_h^* = \alpha \, M_{2,h}(u_h^* - u_{d,h}) \ . \end{cases}$$

The basis for the algorithm that we shall propose is the observation that the relation

$$(\lambda_h^*)^\top (y_h - y_h^*) \leq 0 \text{ for all } y_h \leq \varphi_h ,$$

is equivalent to

$$\lambda_h^* = \max \left(0, \lambda_h^* + c(y_h^* - \varphi_h)\right) \text{ for each } c > 0 , \qquad (5)$$

where the max-operation is understood coordinatewise. The active and in-active sets corresponding to the solution (y_h^*, u_h^*) to (P_h^s) are then defined by

$$\mathcal{A}^* = \{ \, i \mid y_h^* = \varphi_h \, \} \text{ and } \mathcal{I}^* = \{ \, i \mid y_h^* < \varphi_h \, \} .$$

The complete algorithm is specified next. Once again, we drop the subscripts h.

Algorithm (A^s)

1. Initialization : choose y_o, u_o and $\lambda_o \in \mathbb{R}^N$ and set $n = 1$.
2. Determine the active and inactive sets

$$\mathcal{A}_n = \{ \, i \mid y_{n-1} + \frac{\lambda_{n-1}}{c} > \varphi \, \} , \ \mathcal{I}_n = \{ \, i \mid y_{n-1} + \frac{\lambda_{n-1}}{c} \leq \varphi \, \} .$$

3. If $n \geq 2$ and $\mathcal{A}_n = \mathcal{A}_{n-1}$ then STOP, else:
4. Find $(y_n, p_n, \lambda_n) \in \mathbb{R}^N \times \mathbb{R}^N \times \mathbb{R}^N$ such that

$$\begin{cases} Ay_n = \dfrac{1}{\alpha} M_2^{-1} p_n + u_d , \\ Ap_n + \lambda_n = M_1 (z_d - y_n) , \\ y_n = \varphi \text{ on } \mathcal{A}_n, \ \lambda_n = 0 \text{ on } \mathcal{I}_n , \end{cases}$$

and set

$$u_n = \frac{1}{\alpha} M_2^{-1} p_n + u_d .$$

5. Update $n = n + 1$ and goto 2.

The existence of a triple (y_n, p_n, λ_n) satisfying the linear system of step 4 follows from the fact that this system constitutes the optimality system for the auxiliary equality constrained problem

$$(\mathcal{P}_{aux}) \qquad \begin{cases} \text{minimize } J(y, u) \\ \text{subject to } Ay = u \text{ and } y = \varphi \text{ on } \mathcal{A}_n . \end{cases}$$

We have tested the Algorithm with several different initialization schemes. The one that we used most frequently consists in solving the unconstrained problem or equivalently solving the following system:

$$\begin{cases} Ay_o = \dfrac{1}{\alpha} M_2^{-1} p_o + u_d , \\ Ap_o = M_1 (z_d - y_o) , \\ u_o = \dfrac{1}{\alpha} M_2^{-1} p_o + u_d, \ \lambda_o = 0 . \end{cases} \qquad (6)$$

3   Analysis of the Algorithm

The first result of this section justifies the stopping rule of step 3 of Algorithms (A^c) and (A^s).

Theorem 1. *If $A_n = A_{n+1}$ for some $n \in \mathbb{N} - \{0\}$ then Algorithm (A^c) (respectively (A^s)) stops and the last iterate satisfies the (discrete) optimality system. Thus the last iterate provides a solution to the (discretized) optimal control problem.*

Proof - See [1,3].   □

The following Lemma will be useful in the sequel:

Lemma 1. *For all $n \in \mathbb{N} - \{0\}$ and (y, u) satisfying $Ay = u$ we have*

$$J(y_n, u_n) - J(y, u) =$$

$$-\frac{1}{2}\|y - y_n\|_{M_1}^2 - \frac{\alpha}{2}\|u - u_n\|_{M_2}^2 + \begin{cases} (\lambda_n, u - u_n) & , \ case\ [A^c] \\ (\lambda_n, y - y_n) & , \ case\ [A^s] \end{cases} \tag{7}$$

where (y_n, u_n) is given either by A^c (case $[A^c]$) or A^s (case $[A^s]$).

Proof - Using $\|a\|^2 - \|b\|^2 = -\|a - b\|^2 + 2(a - b, a)$ we find that

$$J(y_n, u_n) - J(y, u) =$$

$$-\frac{1}{2}\|y-y_n\|_{M_1}^2 - \frac{\alpha}{2}\|u-u_n\|_{M_2}^2 + (y_n-y)^\top M_1(y_n-z_d) + \alpha(u_n-u)^\top M_2(u_n-u_d).$$

Let us detail the two situations: Step 4 of Algorithm A^c implies
• Case $[A^c]$:

$$(y_n - y)^\top M_1(y_n - z_d) + \alpha(u_n - u)^\top M_2(u_n - u_d)$$

$$= -(y_n - y)^\top A p_n + (u_n - u)^\top (p_n - \lambda_n)$$

$$= -[A(y_n - y)]^\top p_n + (u_n - u)^\top (p_n - \lambda_n)$$

$$= (u_n - u)^\top (-p_n + p_n - \lambda_n) = -(u_n - u)^\top \lambda_n.$$

• Case $[A^s]$

$$(y_n - y)^\top M_1(y_n - z_d) + \alpha(u_n - u)^\top M_2(u_n - u_d)$$

$$= (y_n - y)^\top (-A p_n - \lambda_n) + \alpha(y_n - y)^\top A M_2(u_n - u_d)$$

$$= (y_n - y)^\top (-A p_n - \lambda_n + A p_n) = -(y_n - y)^\top \lambda_n.$$

                                                                            □

Now we detail the convergence analysis for the two cases separately.

3.1   Control constraints

The convergence analysis of the Algorithm is based on the decrease of appropriately chosen merit functions. For that purpose we define the following augmented Lagrangian functions on $\mathbb{R}^N \times \mathbb{R}^N \times \mathbb{R}^N$ by

$$L_c(y, u, \lambda) = \frac{1}{2}\|y - y_d\|^2_{M_1} + \frac{\alpha}{2}\|u - u_d\|^2_{M_2} + (\lambda, \hat{g}_c(u, \lambda)) + \frac{c}{2}\|\hat{g}_c(u, \lambda)\|^2 ,$$

and

$$\hat{L}_c(y, u, \lambda) = L_c(y, u, \lambda^+) ,$$

with $\hat{g}_c(u, \lambda) = (\max(u_1 - b_1, -\lambda_1/c), \ldots, \max(u_N - b_N, -\lambda_N/c))^\top$ and $\lambda^+ = \max(\lambda, 0)$.

In addition we assume that M_2 **is a diagonal matrix** and we denote

$$\underline{m_2} = \min_i(M_2)_{i,i}, \quad \overline{m_2} = \max_i(M_2)_{i,i} \text{ and } \kappa = \frac{\|A^{-1}M_1^{1/2}\|^2}{\underline{m_2}^2} ,$$

where $\|\cdot\|$ denotes any (subordinated to $\mathbb{R}^N$) matrix-norm.

Theorem 2. *If $\mathcal{A}_n \neq \mathcal{A}_{n-1}$ and if*

$$\overline{m_2}(\alpha + \gamma) \leq c \leq \alpha \underline{m_2} + \frac{\alpha^2}{\kappa} - \frac{\underline{m_2}^2 \alpha^2}{\overline{m_2}\gamma} . \tag{8}$$

holds for some $\gamma > 0$, then $\hat{L}_c(y_n, u_n, \lambda_n) \leq \hat{L}_c(y_{n-1}, u_{n-1}, \lambda_{n-1})$. In addition, if the second inequality of (8) is strict then either $\hat{L}_c(y_n, u_n, \lambda_n) < \hat{L}_c(y_{n-1}, u_{n-1}, \lambda_{n-1})$ or the Algorithm stops at the solution to (P^c).

Proof - We have assumed that M_2 is a diagonal matrix. This implies in particular that $\lambda_n = 0$ on $\mathcal{I}_n$.
Let us define

$$\mathcal{S}_{n-1} = \{ i \in \mathcal{A}_{n-1} \mid \lambda_{n-1} \leq 0 \} \text{ and } \mathcal{T}_{n-1} = \{ i \in \mathcal{I}_{n-1} \mid u_{n-1} > b \} .$$

These two sets can be paraphrased by calling $\mathcal{S}_{n-1}$ the set of elements that the active set strategy predicts to be active at level $n-1$ but the Lagrange multiplier indicates that they should be inactive, and by calling $\mathcal{T}_{n-1}$ the set of elements that was predicted to be inactive but the $n-1$ st iteration level corrects it to be active. We note that

$$\Omega_N = (\mathcal{I}_{n-1}\backslash\mathcal{T}_{n-1}) \cup \mathcal{T}_{n-1} \cup \mathcal{S}_{n-1} \cup (\mathcal{A}_{n-1}\backslash\mathcal{S}_{n-1}) \tag{9}$$

defines a decomposition of Ω_N in mutually disjoint sets. Moreover we have the following relation between these sets at each level n:

$$\mathcal{I}_n = (\mathcal{I}_{n-1}\backslash\mathcal{T}_{n-1}) \cup \mathcal{S}_{n-1} , \quad \mathcal{A}_n = (\mathcal{A}_{n-1}\backslash\mathcal{S}_{n-1}) \cup \mathcal{T}_{n-1} . \tag{10}$$

Table 1 below depicts the signs of primal and dual variables for two consecutive iteration levels.

Table 1.

	λ_{n-1}	λ_n	u_{n-1}	u_n
$\mathcal{T}_{n-1} = \mathcal{I}_{n-1} \cap \mathcal{A}_n$	0		$> b$	$= b$
$\mathcal{S}_{n-1} = \mathcal{A}_{n-1} \cap \mathcal{I}_n$	≤ 0	0	$= b$	
$\mathcal{I}_{n-1} \backslash \mathcal{T}_{n-1} \ (\subset \mathcal{I}_n)$	0	0	$\leq b$	
$\mathcal{A}_{n-1} \backslash \mathcal{S}_{n-1} \ (\subset \mathcal{A}_n)$	> 0		$= b$	$= b$

A short computation gives

$$(\lambda, \hat{g}_c(u,\lambda)) + \frac{c}{2}\|\hat{g}_c(u,\lambda)\|^2 = \left(\frac{1}{\sqrt{c}}\lambda, \sqrt{c}\hat{g}_c(u,\lambda)\right) + \frac{1}{2}\left(\sqrt{c}\hat{g}_c(u,\lambda), \sqrt{c}\,\hat{g}_c(u,\lambda)\right)$$

$$= \frac{1}{2}\left\|\sqrt{c}\max(g(u), -\frac{\lambda}{c}) + \frac{1}{\sqrt{c}}\lambda\right\|^2 - \frac{1}{2c}\|\lambda\|^2$$

$$\frac{1}{2c}\|\max(cg(u)+\lambda, 0)\|^2 - \frac{1}{2c}\|\lambda\|^2 ,$$

where we have set $g(u) = u - b$. Therefore for all (y, u, λ) we find

$$L_c(y, u, \lambda) = J(y, u) + \frac{1}{2c}\|\max(c\,g(u) + \lambda, 0)\|^2 - \frac{1}{2c}\|\lambda\|^2 . \tag{11}$$

By assumption $\mathcal{A}_n \neq \mathcal{A}_{n-1}$ and hence $\mathcal{S}_{n-1} \cup \mathcal{T}_{n-1} \neq \emptyset$. Using (11) we obtain

$$\hat{L}_c(y_n, u_n, \lambda_n) - \hat{L}_c(y_{n-1}, u_{n-1}, \lambda_{n-1}) = J(y_n, u_n) - J(y_{n-1}, u_{n-1}) +$$
$$\frac{1}{2c}\left[\|\max(cg(u_n)+\lambda_n^+, 0)\|^2 - \|\lambda_n^+\|^2 - \|\max(cg(u_{n-1})+\lambda_{n-1}^+, 0)\|^2 + \|\lambda_{n-1}^+\|^2\right]$$

and by relation (7) of Lemma 1

$$\hat{L}_c(y_n, u_n, \lambda_n) - \hat{L}_c(y_{n-1}, u_{n-1}, \lambda_{n-1})$$
$$= -\frac{1}{2}\|y_{n-1} - y_n\|_{M_1}^2 - \frac{\alpha}{2}\|u_{n-1} - u_n\|_{M_2}^2 + (u_{n-1} - u_n, \lambda_n)_{\mathcal{T}_{n-1}} +$$
$$\frac{1}{2c}\left[\|\max(cg(u_n)+\lambda_n^+, 0)\|^2 - \|\lambda_n^+\|^2 - \|\max(cg(u_{n-1})+\lambda_{n-1}^+, 0)\|^2 + \|\lambda_{n-1}^+\|^2\right]. \tag{12}$$

It will be convenient to introduce, for $i = 1 \ldots N$

$$d_i = |\max(c\,g(u_n)_i + \lambda_{n,i}^+, 0)|^2 - |\lambda_{n,i}^+|^2 - |\max(c\,g(u_{n-1})_i + \lambda_{n-1,i}^+, 0)|^2 + |\lambda_{n-1,i}^+|^2,$$

so that

$$\hat{L}_c(y_n, u_n, \lambda_n) - \hat{L}_c(y_{n-1}, u_{n-1}, \lambda_{n-1})$$

$$= -\frac{1}{2}\|y_{n-1} - y_n\|_{M_1}^2 - \frac{\alpha}{2}\|u_{n-1} - u_n\|_{M_2}^2 + (u_{n-1} - u_n, \lambda_n)_{\mathcal{T}_{n-1}} + \frac{1}{2c}\sum_{i=1}^{N} d_i ;$$

($u_{n,i}$ stands for the ith component of u_n). Let us estimate d_i on the four distinct subsets of Ω_N according to (9).

• On $\mathcal{I}_{n-1}\backslash\mathcal{T}_{n-1}$ we have $\lambda_n = \lambda_{n-1} = 0$, $u_{n-1} \le b$ $(g(u_{n-1}) \le 0)$ and

$$d_i = |\max(c\,g(u_n)_i, 0)|^2 - |\max(c\,g(u_{n-1})_i, 0)|^2 \le c^2|u_{n,i} - u_{n-1,i}|^2 \ .$$

Moreover as $\lambda_n = p_n - \alpha M_2(u_n - u_d) = 0$ and $\lambda_{n-1} = p_{n-1} - \alpha M_2(u_{n-1} - u_d) = 0$ we have $u_{n,i} - u_{n-1,i} = [M_2^{-1}(p_n - p_{n-1})]_i/\alpha$ so that

$$\|u_n - u_{n-1}\|_{\mathcal{I}_{n-1}\backslash\mathcal{T}_{n-1}} \le \frac{1}{\alpha m_2}\|p_n - p_{n-1}\|_{\mathcal{I}_{n-1}\backslash\mathcal{T}_{n-1}} \ .$$

• On $\mathcal{S}_{n-1}$, $\lambda_n = 0$, $\lambda_{n-1} \le 0$, $g(u_{n-1}) = 0$, so that $d_i = |\max(c\,g(u_n)_i, 0)|^2$. Here we used the definition of $\hat{L}_c$ which involves λ^+ rather than λ. To estimate d_i in detail we consider first the case where $u_n \ge b$. Since $i \in \mathcal{S}_{n-1} \subset \mathcal{I}_n$ we obtain $\lambda_n = p_n - \alpha[M_2(u_n - u_d)] = 0$ and hence $u_{n,i} = ([M_2^{-1}p_n]_i/\alpha) + u_{d,i}$. Moreover, $\lambda_{n-1} = p_{n-1} - \alpha[M_2(u_{n-1} - u_d)] \le 0$ so that $u_{d,i} - b_i \le -[M_2^{-1}p_{n-1}]_i/\alpha$ where we used $u_{n-1,i} = b_i$. Since by assumption $u_n \ge b$ these estimates imply

$$|u_{n,i} - u_{n-1,i}| = u_{n,i} - b_i = \frac{[M_2^{-1}p_n]_i}{\alpha} + u_{d,i} - b_i$$

$$\le \frac{[M_2^{-1}p_n]_i}{\alpha} - \frac{[M_2^{-1}p_{n-1}]_i}{\alpha} = \frac{1}{\alpha}|[M_2^{-1}(p_n - p_{n-1})]_i| \le \frac{1}{\alpha m_2}|(p_n - p_{n-1})_i| \ .$$

In addition it is clear that on the set $\mathcal{I}_n$:

$$d_i = |\max(c\,g(u_n)_i, 0)|^2 \le c^2|u_{n,i} - u_{n-1,i}|^2 \ .$$

In the second case, $u_n < b$ so that $\max(c\,g(u_n)_i, 0) = 0$ and $d_i = 0$.
Finally we have a precise estimate on the whole set $\mathcal{I}_n$. Let us denote

$$\mathcal{I}_n^* = \mathcal{I}_{n-1}\backslash\mathcal{T}_{n-1} \cup \{i \in \mathcal{S}_{n-1} \mid u_n \ge b\} \ = \mathcal{I}_n\backslash\{i \in \mathcal{S}_{n-1} \mid u_n < b\} \ ;$$

then

$$\sum_{i\in\mathcal{I}_n} d_i = \sum_{i\in\mathcal{I}_n^*} d_i = c^2\|\max(g(u_n), 0)\|_{\mathcal{I}_n^*}^2 \le c^2\ \|u_n - u_{n-1}\|_{\mathcal{I}_n^*}^2 \ . \tag{13}$$

We note that we have proved in addition that

$$\|u_n - u_{n-1}\|_{\mathcal{I}_n^*} \le \frac{1}{\alpha m_2}\|p_n - p_{n-1}\|_{\mathcal{I}_n^*} \ . \tag{14}$$

• On $\mathcal{A}_{n-1}\backslash\mathcal{S}_{n-1}$, we have $g(u_{n-1}) = g(u_n) = 0$, $\lambda_{n-1} > 0$ and hence

$$d_i = |\max(\lambda_{n,i}^+, 0)|^2 - |\lambda_{n,i}^+|^2 \le 0 \ . \tag{15}$$

- On $\mathcal{T}_{n-1}$ we have $\lambda_{n-1} = 0$, $g(u_n) = 0$, $g(u_{n-1}) > 0$ and thus

$$d_i = -c^2|g(u_{n-1})_i|^2 = -c^2|u_{n,i} - u_{n-1,i}|^2 \ . \tag{16}$$

Next we estimate the term $(\lambda_n, u_{n-1} - u_n)_{\mathcal{T}_{n-1}}$ in (12):

$$(\lambda_n, u_{n-1} - u_n)_{\mathcal{T}_{n-1}} = (\lambda_n - \lambda_{n-1}, u_{n-1} - u_n)_{\mathcal{T}_{n-1}}$$

$$= (p_n - p_{n-1}, u_{n-1} - u_n)_{\mathcal{T}_{n-1}} + \alpha\|u_n - u_{n-1}\|^2_{\mathcal{T}_{n-1}, M_2} \ .$$

and therefore

$$(\lambda_n, u_{n-1} - u_n)_{\mathcal{T}_{n-1}} \leq$$
$$\|p_n - p_{n-1}\|_{\mathcal{T}_{n-1}}\|u_n - u_{n-1}\|_{\mathcal{T}_{n-1}} + \alpha\|u_n - u_{n-1}\|^2_{\mathcal{T}_{n-1}, M_2} \ . \tag{17}$$

Inserting (13)-(17) into (12) we find

$$\hat{L}_c(y_n, u_n, \lambda_n) - \hat{L}_c(y_{n-1}, u_{n-1}, \lambda_{n-1}) \leq$$

$$-\frac{1}{2}\|y_{n-1} - y_n\|^2_{M_1} - \frac{\alpha}{2}\|u_{n-1} - u_n\|^2_{\Omega_N \setminus \mathcal{T}_{n-1}, M_2} + \frac{\alpha}{2}\|u_n - u_{n-1}\|^2_{\mathcal{T}_{n-1}, M_2}$$

$$+\|p_n - p_{n-1}\| \|u_n - u_{n-1}\|_{\mathcal{T}_{n-1}} + \frac{c}{2}\|u_{n-1} - u_n\|^2_{\mathcal{I}_n^*} - \frac{c}{2}\|u_{n-1} - u_n\|^2_{\mathcal{T}_{n-1}} \leq$$

$$-\frac{1}{2}\|y_{n-1} - y_n\|^2_{M_1}$$

$$-\frac{\alpha m_2}{2}\|u_{n-1} - u_n\|^2_{\mathcal{I}_n^*} - \frac{\alpha m_2}{2}\|u_{n-1} - u_n\|^2_{\mathcal{I}_n \setminus \mathcal{I}_n^*} + \frac{\alpha \overline{m_2}}{2}\|u_{n-1} - u_n\|^2_{\mathcal{T}_{n-1}}$$

$$+\|p_n - p_{n-1}\|\|u_n - u_{n-1}\|_{\mathcal{T}_{n-1}} + \frac{c}{2}\|u_{n-1} - u_n\|^2_{\mathcal{I}_n^*} - \frac{c}{2}\|u_{n-1} - u_n\|^2_{\mathcal{T}_{n-1}} \leq$$

$$-\frac{1}{2}\|y_{n-1} - y_n\|^2_{M_1} - \frac{(\alpha m_2 - c)}{2}\|u_{n-1} - u_n\|^2_{\mathcal{I}_n^*}$$

$$-\frac{(c - \alpha\overline{m_2})}{2}\|u_{n-1} - u_n\|^2_{\mathcal{T}_{n-1}} + \|p_n - p_{n-1}\| \|u_n - u_{n-1}\|_{\mathcal{T}_{n-1}} \ .$$

Using $2ab \leq (a^2/\varrho) + \varrho b^2$ for every $\varrho > 0$ and the above relation , we get for $c \geq \alpha\underline{m_2}$

$$\hat{L}_c(y_n, u_n, \lambda_n) - \hat{L}_c(y_{n-1}, u_{n-1}, \lambda_{n-1}) \leq$$
$$-\frac{1}{2}\|y_{n-1} - y_n\|^2_{M_1} + \frac{(c - \alpha\underline{m_2})}{2}\|u_{n-1} - u_n\|^2_{\mathcal{I}_n^*}$$
$$-\frac{(c - \varrho - \alpha\overline{m_2})}{2}\|u_{n-1} - u_n\|^2_{\mathcal{T}_{n-1}} + \frac{\|p_n - p_{n-1}\|^2}{2\varrho} \ .$$

Using relation (14) and

$$\|p_n - p_{n-1}\|^2 = \|A^{-1}M_1(y_n - y_{n-1})\|^2 \leq \|A^{-1}M_1^{1/2}\|^2 \|y_n - y_{n-1}\|^2_{M_1} \ ,$$

we have

$$\|u_{n-1} - u_n\|^2_{\mathcal{I}_n^*} \leq \frac{1}{\alpha^2\underline{m_2}^2}\|A^{-1}M_1^{1/2}\|^2 \|y_n - y_{n-1}\|^2_{M_1} \ .$$

Setting $\kappa = \left[\|A^{-1}M_1^{1/2}\|/\underline{m_2}\right]^2$ we finally obtain

$$\hat{L}_c(y_n, u_n, \lambda_n) - \hat{L}_c(y_{n-1}, u_{n-1}, \lambda_{n-1}) \le$$

$$-\frac{1}{2}\left[1 - (c - \alpha \underline{m_2})\frac{\kappa}{\alpha^2} - \frac{\underline{m_2}^2\kappa}{\varrho}\right]\|y_{n-1} - y_n\|_{M_1}^2$$
$$-\frac{(c - \alpha \overline{m_2} - \varrho)}{2}\|u_{n-1} - u_n\|_{\mathcal{I}_{n-1}}^2.$$

Setting $\varrho = \gamma \overline{m_2}$ then $\hat{L}_c(y_n, u_n, \lambda_n) \le \hat{L}_c(y_{n-1}, u_{n-1}, \lambda_{n-1})$ provided that

$$\left[1 - (c - \alpha \underline{m_2})\frac{\kappa}{\alpha^2} - \frac{\underline{m_2}^2\kappa}{\gamma \overline{m_2}}\right] \ge 0 \quad \text{and} \quad -\alpha \overline{m_2} + c - \gamma \overline{m_2} \ge 0.$$

Therefore it is sufficient to find $\gamma > 0$ such that

$$\overline{m_2}(\alpha + \gamma) \le c \le \alpha \underline{m_2} + \frac{\alpha^2}{\kappa} - \frac{\underline{m_2}^2\alpha^2}{\overline{m_2}\gamma}.$$

If the second inequality is strict then $\hat{L}_c(y_n, u_n, \lambda_n) < \hat{L}_c(y_{n-1}, u_{n-1}, \lambda_{n-1})$ except if $y_n = y_{n-1}$. In this latter case $u_n = u_{n-1}$ as well and the Algorithm stops at the solution to (P^c).   □

Remark 1. Note that for $\gamma = \alpha \underline{m_2}/\overline{m_2}$ condition (8) is satisfied with the second inequality strict if

$$\alpha\left(\overline{m_2} + \underline{m_2}\right) \le c < \frac{\alpha^2}{\kappa}.$$

Therefore, one can choose $c = \alpha\left(\overline{m_2} + \underline{m_2}\right)$ for any $\alpha > \kappa(\overline{m_2} + \underline{m_2})$.

Remark 2. The same analysis and a similar proof can be performed in the infinite dimensional case. We refer to [1].

Corollary 1. *If (8) with the second inequality strict holds for some $\gamma > 0$, then the Algorithm converges in finitely many steps to the solution to (P^c).*

Proof - First we observe that if the Algorithm stops in Step 3 then the current iterate gives the unique solution. Then with Theorem 2 we have $\hat{L}_c(y_n, y_n, \lambda_n) < \hat{L}_c(y_{n-1}, y_{n-1}, \lambda_{n-1})$ or $(y_n, u_n) = (y_{n-1}, u_{n-1})$.
If $(y_n, u_n) = (y_{n-1}, u_{n-1})$ then $\mathcal{A}_{n+1} = \mathcal{A}_n$ and the Algorithm stops at the solution. The case $\hat{L}_c(y_n, y_n, \lambda_n) < \hat{L}_c(y_{n-1}, y_{n-1}, \lambda_{n-1})$ cannot occur for infinitely many n since there are only finitely many different combinations of active index sets. In fact, assume that there exists $p < n$ such that $\mathcal{A}_n = \mathcal{A}_p$ and $\mathcal{I}_n = \mathcal{I}_p$. Since (y_n, u_n) is a solution of the optimality system of Step 4 if and only if (y_n, y_n) is the unique solution of

$$\min\{\, J(y, u) \mid A y = u, \ u = b \ \text{in } \mathcal{A}_n \,\},$$

it follows that $y_n = y_p$, $u_n = u_p$ and $\lambda_n = \lambda_p$.
This contradicts $\hat{L}_c(y_n, u_n, \lambda_n) < \hat{L}_c(y_p, u_p, \lambda_p)$ and ends the proof.   □

3.2   State constraints

As in the previous section, it will be useful to consider the following diagram
(Table 2) which allows to follow the active and inactive sets from one iteration
to the next: We denote

$$S_{n-1} = \{\, i \in \mathcal{A}_{n-1} \mid \lambda_{n-1} \leq 0 \,\} \text{ and } \mathcal{T}_{n-1} = \{\, i \in \mathcal{I}_{n-1} \mid y_{n-1} > \varphi \,\} .$$

Table 2.

	λ_{n-1}	λ_n	y_{n-1}	y_n
$\mathcal{T}_{n-1} = \mathcal{I}_{n-1} \cap \mathcal{A}_n$	0		$> \varphi$	$= \varphi$
$\mathcal{S}_{n-1} = \mathcal{A}_{n-1} \cap \mathcal{I}_n$	≤ 0	0	$= \varphi$	
$\mathcal{I}_{n-1} \backslash \mathcal{T}_{n-1} \ (\subset \mathcal{I}_n)$	0	0	$\leq \varphi$	
$\mathcal{A}_{n-1} \backslash \mathcal{S}_{n-1} \ (\subset \mathcal{A}_n)$	> 0		$= \varphi$	$= \varphi$

Let us note that $\mathcal{T}_{n-1} = \emptyset$ if y_{n-1} is feasible. We shall require the following
assumption.

$$(\mathcal{H}) \quad \begin{cases} \text{There exists } \bar{n} \geq 2 \text{ such that} \\ \text{for all } n \geq \bar{n} \text{ there exists } \beta_n < 1 \text{ such that} \\ 2\,(\lambda_n, y_{n-1} - y_n)_{\mathcal{T}_{n-1}} \leq \beta_n \left(\|y_{n-1} - y_n\|^2_{M_1} + \alpha \|u_{n-1} - u_n\|^2_{M_2} \right) . \end{cases}$$

$(\mathcal{H})$ is a rather implicit assumption which is difficult to check. Note that
$(\mathcal{H})$ is clearly satisfied if there exists $\bar{n} \geq 2$ such that y_n is feasible for all
$n \geq \bar{n} - 1$. We have observed for many numerical examples that $\mathcal{T}_{n-1}$ was
always empty and consequently $(\mathcal{H})$ was satisfied. $(\mathcal{H})$ can also be used as a
posteriori criterion that can be checked numerically to explain the behavior
of the Algorithm.

Theorem 3. *If $(\mathcal{H})$ holds then the Algorithm converges in finitely many
steps to a solution to (P^s).*

Proof - From Lemma 1 with $(y, u) = (y_{n-1}, u_{n-1})$ we find

$$J(y_n, u_n) - J(y_{n-1}, u_{n-1}) =$$

$$-\frac{1}{2}\|y_{n-1} - y_n\|^2_{M_1} - \frac{\alpha}{2}\|u_{n-1} - u_n\|^2_{M_2} + (\lambda_n, y_{n-1} - y_n)_{\mathcal{T}_{n-1}} ,$$

and by $(\mathcal{H})$

$$J(y_n, u_n) - J(y_{n-1}, u_{n-1}) \leq -\frac{1 - \beta_n}{2} \left(\|y_{n-1} - y_n\|^2_{M_1} + \alpha \|u_{n-1} - u_n\|^2_{M_2} \right) ,$$

for all $n \geq \bar{n}$. Thus the sequence $J(y_n, u_n)$ is strictly decreasing for $n \geq \bar{n}$ as
long as $u_n \neq u_{n-1}$. Since there are only finitely combinations of active sets

there must exist an index n^* such that $u_{n^*-1} = u_{n^*}$. For this index we have $y_{n^*-1} = y_{n^*}$, $p_{n^*-1} = p_{n^*}$ and consequently $\mathcal{A}_{n^*-1} = \mathcal{A}_{n^*}$. By Theorem 1 the Algorithm stops, $(y_{n^*}, u_{n^*}, p_{n^*}, \lambda_{n^*})$ satisfies the optimality system and (y_{n^*}, u_{n^*}) is a solution to $(\mathcal{P}_h^s)$. $\square$

For many numerical examples we observed that the iterates are feasible after a certain number of iterations. In these cases $(\mathcal{H})$ is satisfied and the active sets are decreasing. For one dimensional problems feasibility criteria are given in [3].

4   Numerical experiments

It has been verified in Theorem 1 that if the Algorithm terminates in step 3. then the exact solutions to (P_h^c) respectively (P_h^s) are obtained. In our numerical tests we found no example for which (A^c) or (A^s) do not terminate in 3. as long as $(\mathcal{P}^c)$ respectively $(\mathcal{P}^s)$ satisfy the strict complementarity condition. The case of lack of strict complementarity at the solution will be studied in Section 4.3. Both (A^c) and (A^s) have the property that from one iteration to the next many coordinates of the discretized control respectively state vector can move from $\mathcal{A}_{n-1}$ to $\mathcal{I}_n$ and vice versa. This correction process is especially efficient for control constrained problems where it changes the (discrete) interior of $\mathcal{A}_{n-1}$ or $\mathcal{I}_{n-1}$ from one iteration to the next. For state constrained problems changes from active to inactive sets and vice versa occur primarily along the boundary between the current active and inactive sets. This is due to the fact that λ_h for (A^s) is the discretization of the measure $\lambda^* \in \mathcal{M}(\Omega)$ whose singular part is concentrated along the boundary of the active set [3]. For the iterates u_n of (A^c) it is typically the case that they are all infeasible except the last one, i.e. (A^c) stops at the first feasible iterate (except for u_o). On the other hand the iterates for (A^s) are mostly feasible and the active sets $\mathcal{A}_n$ typically approximate $\mathcal{A}^*$ from outside. This approximation is typically monotone w.r.t. the cardinality of the active set but non monotone in the setwise sense. Finally we remark that for (A^c) also the infinite dimensional version of the algorithm can be successfully analyzed, [1].

The algorithm has been tested in one and two-dimensional cases. We present here two-dimensional examples with $\Omega =]0,1[\times]0,1[$. The discretization was chosen as the canonical five-point star finite-difference approximation of the Laplacian. The numerical integration rule is chosen such that $M_1 = M_2 = I_N$ (identity matrix). In addition we can assertain that, for the state-constraints case, assumption $(\mathcal{H})$ was satisfied for all examples that we tested and more precisely the set $\mathcal{T}_n$ was always empty. The fact that $(\mathcal{H})$ is satisfied and that $\mathcal{T}_n$ is empty does not, however, imply that the iterates are feasible. This is, in general, not the case.

4.1   Presentation of examples

Example 1 : Control constraints. We set

$$z_d(x_1, x_2) = \sin{(2\pi x_1)}\ \sin{(2\pi x_2)}\ \exp(2x_1)/6\ ,\ b \equiv 0\ .$$

Several tests for different values for α, c and u_d were performed. We present two of them ($N = 50$).

Table 3. Example 1.a : $u_d \equiv 0$, $\alpha = 10^{-2}$, $c = 10^{-1}$

Iteration	$\max(u_n - b)$	size of $\mathcal{A}_n$	$J(y_n, u_n)$	$\hat{L}_c(y_n, u_n, \lambda_n)$
1	4.8708e-02	1250	4.190703e-02	4.190785e-02
2	5.8230e-05	1331	4.190712e-02	4.190712e-02
3	0.0000e+00	1332	4.190712e-02	4.190712e-02
4	0.0000e+00	1332	4.190712e-02	4.190712e-02

Let us give plots of the optimal control and state for Example 1.a.

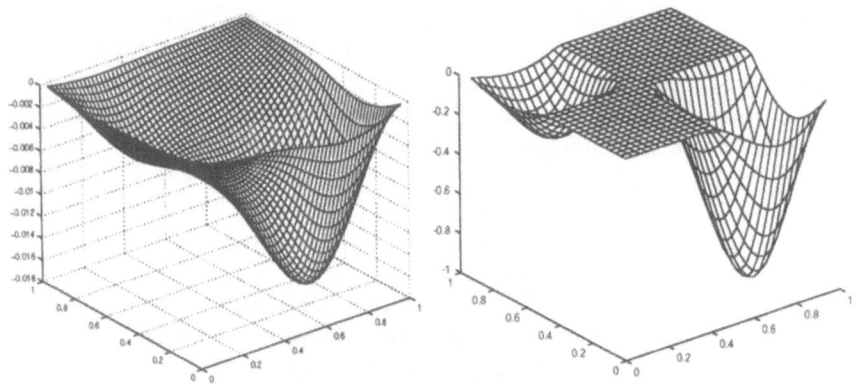

Fig. 1. Optimal State and Optimal control

We present in Table 4 a second example where (8) may be not fulfilled (it was for Table 3) because α is too small; in addition u_d has been chosen infeasible.

Though the size of the set $\mathcal{A}_n$, in the sense of number of grid points in $\mathcal{A}_n$ is increasing, the sequence $\mathcal{A}_n$ does not increase monotonically. More precisely points in $\mathcal{A}_n$ at iteration n may not belong to $\mathcal{A}_{n+1}$ at iteration $n + 1$.

We observe in our computations that the algorithm stops as soon as an iterate is feasible. So the sequence of iterates is not feasible until it reaches the solution. We could say that we have an "outer" method. We must also

Table 4. Example 1.b. : $u_d \equiv 1$, $\alpha = 10^{-6}$, $c = 10^{-2}$

Iteration	$\max(u_n - b)$	size of $\mathcal{A}_n$	$J(y_n, u_n)$	$\hat{L}_c(y_n, u_n, \lambda_n)$
1	5.0986e+02	1250	1.734351e-02	9.858325e+00
2	4.4728e+02	1487	2.089663e-02	7.688683e+00
3	3.6796e+02	1677	2.375001e-02	5.612075e+00
4	5.8313e+02	1831	2.603213e-02	4.526200e+00
5	6.7329e+02	1944	2.782111e-02	3.657995e+00
6	5.3724e+02	2039	2.911665e-02	2.402021e+00
7	3.6175e+02	2098	2.981378e-02	1.191161e+00
8	1.5071e+02	2146	3.011540e-02	3.678089e-01
9	6.5928e+01	2178	3.018832e-02	7.796022e-02
10	2.3420e+01	2196	3.019715e-02	3.344241e-02
11	3.4889e+00	2208	3.019762e-02	3.022994e-02
12	0.0000e+00	2210	3.019762e-02	3.019762e-02
13	0.0000e+00	2210	3.019762e-02	3.019762e-02

underline that differently from classical primal active set methods, the primal-dual method that we propose can move many points from the active set to the inactive set and vice versa from one iteration to the next.

Example 2 : State constraints. Here the problem data are given by

$$z_d(x, y) = \sin(4\pi x y) + 1.5 , \quad \varphi \equiv 1 , \quad u_d \equiv 0 .$$

The algorithm needs 23 iterations for a mesh size equal to 1/50 before it is terminated by the stopping criterion of step 3 at the solution to the discretized problem. $(\mathcal{H})$ is satisfied with $\bar{n} = 1$ (since $\mathcal{T}_n$ is always empty). The behavior of the convergence process is illustrated in Table 5. The optimal control and state are shown in Figure 2.

One can find further examples in [1,3,2].

4.2 Sensitivity with respect to parameters

Let us discuss at first the sensitivity of the convergence with respect to the parameters initialization and mesh size h and underline the different behavior of the method with respect to the type of constraints. We observe that the Algorithm is independent of the choice of c as long as $c > 0$.

Initialization In case of control constraints, the best initialization process was given by (4). Any other initialization process gives convergence with more iterations.

Table 5. Convergence process behavior

Iteration	$\max(u_n - b)$	size of $\mathcal{A}_n$	$J(y_n, u_n)$
1	0.000000e+00	1698	1.056996e+00
2	2.774069e-03	1320	8.983295e-01
3	2.190955e-03	1178	8.224091e-01
4	4.423866e-03	1026	7.859403e-01
5	1.356316e-03	912	7.708878e-01
6	5.099188e-03	782	7.649270e-01
7	2.162679e-03	706	7.625555e-01
8	2.418681e-03	664	7.616485e-01
9	3.385683e-04	631	7.611555e-01
10	1.612269e-03	625	7.610258e-01
11	2.841366e-04	619	7.609570e-01
12	8.572115e-04	615	7.609079e-01
13	3.201381e-04	611	7.608762e-01
14	6.004241e-04	613	7.607811e-01
15	4.977181e-04	604	7.607106e-01
16	3.451955e-05	594	7.606518e-01
17	2.887315e-04	582	7.605909e-01
18	0.000000e+00	572	7.605682e-01
19	0.000000e+00	560	7.605395e-01
20	6.917486e-05	552	7.605293e-01
21	0.000000e+00	543	7.605258e-01
22	0.000000e+00	539	7.605254e-01
23	0.000000e+00	539	7.605254e-01

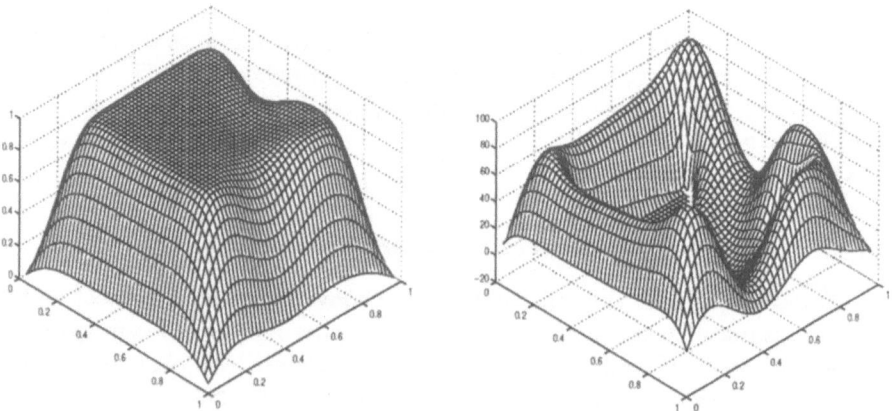

Fig. 2. Optimal State and Optimal control for $\alpha = 10^{-4}$

In case of state constraints, one possibility is given by initializing the algorithm with a control u so that the associated state y is feasible. However, numerical experiments show that a better choice is given by choosing y_o as the unconstrained minimum of J (initialization (6)). In case that the solution of (P^s) happens to be unconstrained the method stops after one iteration.

Dependence on the mesh-size   The algorithm applied to control constrained problems is independent of the mesh-size. The tests below (Tables 6 & 7) have been performed on a DEC-alpha 500 (with machine precision $\varepsilon_M \approx 1.11 \cdot 10^{-16}$).

Table 6. Number of iterations and CPU-time (s) with respect to the mesh size - Example 1.a : Control constraints

N	20	30	40	50	60	70	80	90	100	110	120
#it	3	3	3	3	3	4	4	4	3	4	4
CPU (s)	0.2	0.9	2.5	5.5	10.8	23.8	39.6	62.2	75.0	133.4	186.1

In contrast, the proposed algorithm for state constrained problems is sensitive to the mesh size. We note that the number of iteration grows approximately linearly with N.

Table 7. Number of iterations and CPU-time (s) with respect to the mesh size - Example 2 : State constraints, $\alpha = 10^{-3}$

N	20	30	40	50	60	70	80	90	100	110
#it	13	18	22	27	32	37	41	48	52	56
CPU (s)	1.8	10.5	39	93	222	467	881	1678	2817	4504

4.3   Lack of strict complementarity

In the case of lack of strict complementarity the Lagrange multiplier at the optimal solution and the value of $u - b$ (respectively $y - \varphi$) are equal to zero on a set of nonzero measure. The following examples are constructed such that lack of strict complementarity occurs:

- **Control constraints**

$$z_d \text{ as in Example 1., } b \equiv 0 , u_d = b - \frac{1}{\alpha}[-\Delta^{-1} z_d + \Delta^{-2} b] = \frac{\Delta^{-1} z_d}{\alpha}$$

so that the exact solution to problem $(\mathcal{P}^c)$ is $u^* = b = 0$ and $\lambda^* = 0$.

- **State constraints**

$$\varphi = \beta \left[x_1 x_2 (1 - x_1)(1 - x_2)\right]^5 \, , \ \lambda^\dagger = \begin{cases} 0 & \text{if } 0 < x_1 \leq 1/2 \, , \\ x_1 - \frac{1}{2} & \text{if } 1/2 < x_1 \leq 1 \, , \end{cases}$$

$$u_d \equiv 0 \, , \ z_d = \lambda^\dagger + \varphi + \alpha \, \Delta^2 \varphi \, , \beta > 0 \, .$$

The optimal solution is $y^* = \varphi$ and $\lambda^* = \lambda^\dagger$ so that strict complementarity fails on the subdomain of Ω defined by $0 < x_1 \leq 1/2$.

Due to the fact that the active/inactive sets in the Algorithm are determined by statements involving "$\leq$" and "$=$" difficulties due to finite precision in computer calculus can occur.

In this case one replaces "$> b$" with "$> b - \varepsilon$", where ε depends on the machine precision, to determine the sets $\mathcal{A}_n$ (and $\mathcal{I}_n$). In our numerical tests, the modified algorithm with $\varepsilon = 10^{-10}$ needed only one iteration to find the optimal solution . The Algorithm without modification chattered around the optimal solution. For more details one can refer to [2].

5   Conclusion

The Algorithm that we have presented here is devoted to the treatment of optimal control problems with constraints. The regularity properties of the infinite dimensional solution of the problem have a direct influence on the behavior of the method and on the assumptions for convergence. In the case of control constraints a modified Lagrangian functional serves as a merit function. The algorithm can be described as an exterior method:the iterates are unfeasible so that the cost functional is increasing. For state constraints, on the other hand, the iterates are feasible from a certain index onwards and the cost functional is decreasing. Though the discretized problems are similar from the point of view of finite dimensional optimization, the behavior of the method is quite different for the two cases. A comparison to interior points methods can be found in [2].

References

1. Bergounioux, M., Ito, K., Kunisch, K. (1999) Primal-dual strategy for optimal control problems. SIAM Journal on Control and Optimization **37**, 4, 1176–1194
2. Bergounioux, M., Haddou, M., Hintermüller, M., Kunisch, K. (1998) A comparison of interior point methods and a Moreau-Yosida based active set strategy for constrained optimal control problems. Preprint 98-15, Université d'Orléans
3. Bergounioux, M., Kunisch, K. (1998) Primal-dual Strategy for State-Constrained Optimal Control Problems. Preprint 98-21, Université d'Orléans
4. Ito, K. , Kunisch, K. (1995) Augmented Lagrangian formulation of nonsmooth convex optimization in Hilbert spaces. In: Casas E. (Ed.) Control of Partial Differential Equations and Applications, Lecture Notes in Pure and Applied Mathematics, Marcel Dekker **174** , 107–117

Vector Variational Principles Towards Asymptotically Well Behaved Vector Convex Functions

Serban Bolintinéanu

Université de Perpignan, Mathématiques, 52, Av. de Villeneuve 66860 Perpignan, France, e-mail: `serban@univ-perp.fr`

Abstract. The aim of this paper is to present and analyse several kinds of $\varepsilon-$efficiency related to some metrically consistent vector variational principles as a converse of the asymptotical well behaviour of the vector convex functions.

Key words: vector optimization, multicriteria optimization, ε-efficiency, stationary sequences, minimizing sequences, weakly efficient sequences, Kuhn-Tucker sequences, variational analysis, Ekeland variational principle

1 Introduction

For a convex function a point which is "almost stationary" is not necessarily an "almost minimizing" one. This can be easily seen from the following example given in [2]

$$f(x_1,\, x_2) = \begin{cases} \frac{x_2^2}{2x_1} & \text{if } x_1 > 0 \\ 0 & \text{if } x_1 = x_2 = 0 \\ +\infty & \text{otherwise.} \end{cases}$$

This function (see [13, section 10]) is convex because it is the support function of the parabolic convex set $\{(x_1,\, x_2)|\ x_1 + \frac{x_2^2}{2} \leq 0\}$. On the other hand we have $\nabla f(n^2,\, n) = (-1/2n^2,\, 1/n) \to 0$ (as $n \to +\infty$) but $f(n^2,\, n) = 1/2$ which is not close to $\min f = 0$. This function is not smooth on $\mathbb{R}^2$ but can be transformed, using its Moreau-Yosida regularization, into a smooth (differentiable) function on the whole plane which have the same bad asymptotical behaviour.

Conversely, an "almost minimizing" point is not necessarily an "almost stationary" one, as we can see considering the convex smooth function (see [10]) $f : \mathbb{R}^2 \to \mathbb{R}$, $f(x_1, x_2) = \exp(x_1^2 - x_2)$ and the sequence $n \mapsto x^n = (n,\, n^2 + \ln\sqrt{n})$. We have $f(x^n) = 1/\sqrt{n} \to 0 = \inf f$ but $\nabla f(x^n) = (2\sqrt{n},\, -1/\sqrt{n}) \nrightarrow (0, 0)$.

For the scalar convex optimization problems, there has been characterized a large class of convex functions having the property that every stationary sequence is a minimizing one (see the papers [2,4,3]). These functions are called asymptotically well behaved.

In the paper [6] are given generalizations of these results for extended valued nonsmooth convex vector functions. Other related results for smooth convex vector functions can be found in [5].

Relations between stationary and minimizing sequences have also been considered in [10,19] for constrained scalar mathematical programming problems, where it is also shown the following which is a direct consequence of the Ekeland variational principle.

Theorem 1. *Let $f : \mathbb{R}^n \to \mathbb{R}$ be a continuously differentiable function. Then for any minimizing sequence (x_k), there exists a nearby sequence (y_k) such that*

$(i) \ \lim_{k \to +\infty} (x_k - y_k) = 0, \quad (ii) \ \lim_{k \to +\infty} f(y_k) = \inf_{x \in \mathbb{R}^n} f(x), \quad (iii) \ \lim_{k \to +\infty} \nabla f(y_k) = 0.$

If we assume in addition that the gradient ∇f is uniformly continuous, it is obvious that *any minimizing sequence is a stationary one.*

Some generalizations of this theorem for vector functions have been obtained in [5,7].

For a scalar minimization problem the optimal value is a singleton (eventually $-\infty$). A different situation occurs in a vector minimization problem where the set of "optimal" values (which is the "infimal" set) may be infinite and sometimes unbounded. Also, there are different possibilities to define an approximate solution of a vector optimization problem (which in the scalar case recover the usual definition of an ε-solution), and the related notion of "minimizing" sequence. Thus, in this paper following the definitions given in [7], each kind of ε-solution has the property that the (vector) value associated is located in an ε-neighborhood of the weakly infimal set. The approximate solutions used in the literature may not satisfy this metric quality as it was pointed out in [7].

On the other hand the approximate necessary first order weakly-efficiency condition leads to the notion of *weakly scalarly stationary (or weakly Kuhn-Tucker) sequence.*

In the present paper we will present two vector variational principles which lead to a metrically consistent point. We will also define a new kind of ε-efficiency which allows us to obtain some stronger conclusions for a vector variational principle. We will analyse this concept in connection with the Pareto optimizing sequences and the weakly scalarly stationary ones.

2   Preliminaries

2.1   The weakly infimal set in the policy space

Throughout this paper Y denotes a real Banach space (the "policy space") endowed with a closed convex pointed cone C (i.e., C is a closed set, $\mathbb{R}_+ C +$

$C \subset C$ and $C \cap -C = \{0\}$). This cone defines a (partial) order relation in Y given by : $x \leq_C y \iff y - x \in C$ compatible with vector addition and positive scalar multiplication. Thus Y is a (partially) ordered Banach space. We will assume that the interior $\operatorname{int} C$ is not empty. We will consider the transitive relation

$$x <_C y \iff y - x \in \operatorname{int} C,$$

and we will denote by B the closed unit ball in any normed vector space. The distance between a point $y \in Y$ and a set $T \subset Y$ will be denoted by $d(y; T) = \inf_{z \in T} \|y - z\|$ (with the convention $d(y; \emptyset) = +\infty$).

For a set $T \subset Y$ we define the *weakly infimal set* (see also [5–7]) by

$$\operatorname{INF}_w(T) = \{y \in \bar{T} \mid \not\exists z \in T : z <_C y\}, \tag{1}$$

where $\bar{T}$ denotes the closure of the set T.

In [7] it is shown that $\operatorname{INF}_w(T)$ *is a closed set which is contained in the boundary of* T.

Let us consider the set

$$C^+ = \{\lambda \in Y^* \mid \langle \lambda, y \rangle \geq 0 \ \forall y \in C\},$$

where Y^* is the topological dual of the space Y, and $\langle \cdot, \cdot \rangle$ stands for the duality product.

We recall the following result given in [7].

Theorem 2. *Assume* $\operatorname{int} C^+ \neq \emptyset$ *and let* $T \subset Y$ *be such that there exists* $\lambda \in \operatorname{int} C^+$, *bounded from below on* T. *Then, for every* $y \in T \setminus \operatorname{INF}_w(T|C)$, *there exists* $z \in \operatorname{INF}_w(T|C)$ *such that* $z <_C y$.

2.2 The extended space $\overline{Y}$

We will briefly recall some results presented in [6]. We will extend the space Y to a topological partially ordered space $\overline{Y} = Y \cup \{+\infty_C, -\infty_C\}$ where $+\infty_C$ and $-\infty_C$ are two distinct elements not belonging to Y (some other results about the extended space can be found in [9,20]).

A *neighborhood* of $+\infty_C$ ($-\infty_C$) in $\overline{Y}$ is a set N, ($-N$ resp.) such that $\{+\infty_C\} \cup (y + C) \subset N$ for some $y \in Y$. Thus, the imbedding $Y \subset \overline{Y}$ is continuous, and Y is dense in $\overline{Y}$.

The algebraic operations are extended in the same way as in the convex analysis (see [6]). Note that $+\infty_C - \infty_C = +\infty_C$; $0 \cdot \infty_C = 0_Y$.

We will extend the relations $\leq_C$ and $<_C$ to $\overline{Y}$ by

$$\forall y \in Y, \ -\infty_C \leq_C y \leq +\infty_C, \ -\infty_C <_C y <_C +\infty_C.$$

Note that, despite the analogy with the extended real line $\bar{\mathbb{R}} = \mathbb{R} \cup \{-\infty, +\infty\}$, the space $\overline{Y}$ is in general not compact (even in the case $Y = \mathbb{R}^2$ and $C = \mathbb{R}_+^2$).

For $T \subset \overline{Y}$, $\overline{T}$ stands for the closure in the space $\overline{Y}$, and we define the infimal set $\text{INF}_w(T)$ by the same previous formula (1).

Note that, if $T \neq \{+\infty_C\}$, then

$$\text{INF}_w T = \{y \in \overline{T} \mid -y + T \subset \overline{Y} \setminus (-\text{int}\, C \cup \{-\infty_C\})\}.$$

Also, $\text{INF}_w(T)$ is a closed set in $\overline{Y}$.

Moreover, according to the definition, we obtain easily the following.

Proposition 1. *Let* $T \subset Y \cup \{+\infty_C\}$ *such that* $-\infty_C \notin \overline{T}$ *and* $T \neq \{+\infty_C\}$. *Then*

$$\text{INF}_w(T) = \text{INF}_w(T \setminus \{+\infty_C\}),$$

where $\text{INF}_w(T \setminus \{+\infty_C\})$ *may also be considered as the weakly infimal set associated to* $T \setminus \{+\infty_C\}$ *in the space* Y.

Proposition 2. *[7] Let* (y_n) *be a sequence in* Y. *Then*

$$\lambda \in C^+ \setminus \{0\}, \ y_n \to +\infty_C \implies \langle \lambda, y_n \rangle \to +\infty. \tag{2}$$

Thus, we will extend each element $\lambda \in C^+ \setminus \{0\}$ to a function (denoted λ too) $\lambda : \overline{Y} \to \overline{\mathbb{R}}$ putting

$$\langle \lambda, +\infty_C \rangle = +\infty; \quad \langle \lambda, -\infty_C \rangle = -\infty.$$

We will denote

$$\Lambda = \{\lambda : \overline{Y} \to \overline{\mathbb{R}} \mid \lambda_{|Y} \in C^+; \ \|\lambda_{|Y}\| = 1; \ \lambda(-\infty_C) = -\infty, \ \lambda(+\infty_C) = +\infty\}.$$

2.3   Extended valued vector optimization problem

Unless otherwise stated, X stands for a reflexive Banach space. Let us consider a map $F : X \to \overline{Y}$. We will denote by $\text{dom}\,1(F)$ the effective domain of F, i.e.

$$\text{dom}(F) = \{x \in X \mid F(x) \neq +\infty_C\}.$$

We say that F is *proper* if $\text{dom}(F) \neq \emptyset$ and $F(x) \neq -\infty_C$, $\forall x \in \text{dom}(F)$.

For the vector "minimization" problem:

$$(VMP) \qquad \text{"minimize" } F(x), \quad x \in X$$

we say that $a \in X$ is a *weakly Pareto solution* if there is no $x \in X$ such that $F(x) <_C F(a)$. The set of all the weakly Pareto solutions called *weakly Pareto (or efficient) set* will be denoted by E_w.

Let $F : X \to \overline{Y}$ be a map.

Definition 1. We say that F is *C-convex* if

$$\forall \theta \in [0, 1], \ \forall x, x' \in X, \ F((1 - \theta)x + \theta x') \leq_C (1 - \theta)F(x) + \theta F(x').$$

Definition 2. We say that F is $C-semicontinuous$ if the *level sets*

$$L(F; \alpha) = \{x \in X | \ F(x) \leq_C \alpha\}$$

are closed for all $\alpha \in Y$.

For each $\lambda \in \Lambda$, we will consider the function $F_\lambda : X \to \bar{\mathbb{R}}$ given by

$$F_\lambda = \lambda \circ F.$$

Definition 3. We say that F is *positively lower semicontinuous (continuous)* if for any $\lambda \in \Lambda$, the function F_λ is lower semicontinuous (continuous resp.).

Note that if F is positively lower semicontinuous, then F is $C-$semicontinuous (because $L(F; \alpha) = \cap_{\lambda \in C^+ \backslash \{0\}} \{x \in X | \ F_\lambda(x) \leq \langle \lambda, \alpha \rangle\}$).

Definition 4. Let $F : X \to \overline{Y}$ be a $C-$convex map (Gâteaux differentiable, with $\mathrm{dom}\,(F)$ open resp.). A point $a \in X$ ($a \in \mathrm{dom}\,(F)$ resp.) is called *weakly scalarly stationary*, if there exists $\lambda \in \Lambda$ such that

$$0 \in \partial F_\lambda(a),$$

where

$$\partial F_\lambda(a) = \{x^* \in X^* | \ F_\lambda(x) - F_\lambda(a) \geq \langle x^*, x - a \rangle, \ \forall x \in X\}$$

is the usual subdifferential of the real extended valued convex function F_λ at a (the Gâteaux derivative resp.).

Note that for the Gâteaux derivative at a we also use the notation $F'_\lambda(a)$.
We denote by S_w the set of the weakly scalarly stationary points.

Proposition 3. *(see [6,7]) (i) If $F : X \to \overline{Y}$ is C-convex, then $E_w = S_w$.*
(ii) If $F : X \to \overline{Y}$ is Gâteaux differentiable with $\mathrm{dom}\,(F)$ open, then $E_w \subset S_w$.

Let $F : X \to Y \cup \{+\infty_C\}$ be a proper map.
We will denote

$$\mathrm{INF}_w\,(F) = \mathrm{INF}_w\,(F(X)) = \{y \in \overline{F(X)} | \ F(X) - y \subset \overline{Y} \backslash (-\mathrm{int}\, C \cup \{-\infty_C\})\}$$

Definition 5. (see [7] for more details) The point $a \in X$ is $\varepsilon-$ *weakly efficient* if there exists $e \in B$, such that $F(a) - \varepsilon e \in \mathrm{INF}_w(F)$, where $\varepsilon > 0$.

Remark 1. Note that, for $F(a) \in Y$, we have that a is $\varepsilon-$ weakly efficient iff $\mathrm{d}(F(a); \mathrm{INF}_w(F)) \leq \varepsilon$.

Definition 6. (see [7]) The point $a \in X$ is $\varepsilon-$ *correct weakly efficient* if there exists $e \in C \cap B$, such that $F(a) - \varepsilon e \in \mathrm{INF}_w(F)$, where $\varepsilon > 0$.

Now we will give a new definition, which differs slightly of the preceding one, but permits some stronger conclusions in the vector variational principle which we will obtain in the last section.

Definition 7. Let $\varepsilon > 0$ be a given number. The point $a \in X$ is $\varepsilon-$ *strictly correct weakly efficient* if there exists $e \in (\text{int } C \cup \{0\}) \cap B$, such that $F(a) - \varepsilon e \in \text{INF}_w(F)$.

The following is obvious.

Remark 2. Let $a \in X$ and $\varepsilon > 0$. Then

(i) a is $\varepsilon-$ strictly correct weakly efficient $\implies a$ is $\varepsilon-$ correct weakly efficient $\implies a$ is $\varepsilon-$ weakly efficient.

(ii) a is weakly efficient $\iff a$ is $\varepsilon-$ strictly correct weakly efficient for all $\varepsilon > 0 \iff a$ is $\varepsilon-$ correct weakly efficient for all $\varepsilon > 0 \iff a$ is $\varepsilon-$ weakly efficient for all $\varepsilon > 0$.

Remark 3. If we assume that F_λ is bounded from below for some $\lambda \in \Lambda \cap \text{int } C^+$, then using Theorem 2 (with $T = F(X) \setminus \{+\infty_C\}$) and Proposition 1, we have that for each $a \in \text{dom}(F)$, there exists $y \in \text{int } C \cup \{0\}$ such that $F(a) - y \in \text{INF}_w(F)$. However, when $\text{d}(F(a); \text{INF}_w(F)) < \varepsilon$, we cannot be sure that $\|y\| \leq \varepsilon$. Of course, there exists $z \in \text{INF}_w(F)$ such that $\|F(a) - z\| < \varepsilon$, but not necessarily $z \in C$.

Example 1. Consider $X = Y = \mathbb{R}^2$ with the euclidean structure, $C = \mathbb{R}_+^2$, and $F : X \to \overline{Y}$ such that $F(x) = x$ for all $x \in T = \{(x_1, x_2) \in \mathbb{R}^2 \mid x_2 \geq -10^3; \ x_2 \cdot (x_2 - 10^6 x_1) = 0\}$, and $F(x) = +\infty_C$ elsewhere. Thus $\text{INF}_w(F) = \text{INF}_w(T) =]-\infty, -10^{-3}] \times \{0\} \cup \{(-10^{-3}, -10^3)\}$ and the point $(0; 0)$ is 10^{-3}-correct weakly efficient but it is not 10^{-3}-strictly correct weakly efficient (it is $\varepsilon-$strictly correct weakly efficient iff $\varepsilon \geq 10^3 \sqrt{1 + 10^{-12}}$).

Definition 8. A sequence (x_n) in $\text{dom}(F)$ will be called:

1. *asymptotically weakly Pareto optimizing* (a.w.p.) if

$$\text{d}(F(x_n), \text{INF}_w(F)) \to 0 \quad \text{or} \quad F(x_n) \to -\infty_C$$

2. *correct asymptotically weakly Pareto optimizing (c.a.w.p.)* if there exists a sequence $(\varepsilon_n) \subset \mathbb{R}_+^*$ such that

   $(i) \ \lim_{n \to \infty} \varepsilon_n = 0; \quad (ii) \ \forall n, \ x_n$ is ε_n- correct weakly efficient.

3. *strictly correct asymptotically weakly Pareto optimizing (s.c.a.w.p.)* if there exists a sequence $(\varepsilon_n) \subset \mathbb{R}_+^*$ such that

   $(i) \ \lim_{n \to \infty} \varepsilon_n = 0; \quad (ii) \ \forall n, \ x_n$ is ε_n- strictly correct weakly efficient.

4. *weakly scalarly stationary* (w.s.s.) if F is C−convex or F is Gâteaux differentiable with dom(F) open, and there exists a sequence $(\lambda_n) \subset \Lambda$ verifying the following:

$$\forall n \in \mathbb{N}, \ \exists \xi_n \in \partial F_{\lambda_n}(x_n): \ \|\xi_n\|_{X^*} \to 0, \tag{3}$$

which is equivalent to saying that

$$\forall n \in \mathbb{N}, \ \partial F_{\lambda_n}(x_n) \neq \emptyset, \ d(0; \partial F_{\lambda_n}(x_n)) \to 0.$$

Remark 4. If (x_n) is an a.w.p. sequence, then $\mathrm{INF}_w(F) \neq \emptyset$ and we have two possibilities:

(i) $-\infty_C \notin \mathrm{INF}_w(F)$, and in this case $d(F(x_n), \mathrm{INF}_w(F)) \to 0$.

(ii) $\mathrm{INF}_w(F) = \{-\infty_C\}$, and in this case $F(x_n) \to -\infty_C$.

Note that (i) is equivalent to the following:
(i') *there exists a sequence (ε_n) of positive real numbers tending to 0 such that, for all $n \in \mathbb{N}$, x_n is an ε_n- weakly efficient point.*

3   Vector variational principles

In [6] several characterizations of the class $\mathcal{F}$ of asymptotically well behaved convex vector functions have been given. The class $\mathcal{F}$ consists of all proper, C−semicontinuous convex functions from X to $Y \cup \{+\infty_C\}$ having a nonempty weakly infimal set with the property that *every w.s.s. sequence is an a.w.p. one.* In other words, in [6] was studied some sort of implication *"approximate necessary weakly efficiency condition $\Longrightarrow$ approximate weakly efficiency".*

In this section we will try to give an answer to the converse question : *" what can be said about an approximate weakly efficient point in connection with the necessary weakly efficiency conditions ?".* First we will recall the following vector variational principles.

Theorem 3. *[7] Let X be a finite dimensional euclidean space identified with its dual, and let $F : X \to Y \cup \{+\infty_C\}$ be a proper map, positively lower semicontinuous (l.s.c.) such that $\emptyset \neq \mathrm{INF}_w(F) \subset Y$.*
Then, for any $\varepsilon > 0$, $\gamma > 0$ and $a \in X$ such that $d(F(a); \mathrm{INF}_w(F)) < \varepsilon$ there exist $b \in \mathrm{dom}\, F$ and $e \in B \subset Y$ such that:
(i) $\|a - b\| \leq \gamma$,
(ii) b is a weakly efficient point for the map $G : X \to Y \cup \{+\infty_C\}$, given by

$$x \mapsto G(x) = F(x) - \varepsilon\varphi(1 - \frac{\|x - a\|^2}{\gamma^2})e, \tag{4}$$

where $\varphi : \mathbb{R} \to \mathbb{R}$ is an increasing, convex differentiable function such that

$$\varphi(\mathbb{R}_-) = \{0\}, \quad \varphi'(1) \le 2, \quad \varphi(1) = 1,$$

(for example $\varphi(t) = 0$, if $t < 0$; $\varphi(t) = t^2$ if $t \ge 0$.)
   (iii) for any $x \in \operatorname{dom} F$:

$$\|F(x) - G(x)\| \le \varepsilon. \tag{5}$$

If in addition F is Gâteaux differentiable with $\operatorname{dom} F$ open set, then

$$\exists \lambda \in \Lambda, \quad \|F'_\lambda(b)\| \le 4\frac{\varepsilon}{\gamma}. \tag{6}$$

If the given point a is $\varepsilon-$ correct weakly efficient, a stronger result is obtained.

Theorem 4. *[7] Let X be a reflexive Banach space, and $F : X \to Y \cup \{+\infty\}$. Suppose that F is a proper, positively weakly l.s.c. map. Let $\varepsilon > 0$ and let $a \in X$ be any $\varepsilon-$ correct weakly efficient point. Consider some $e \in C \cap B$ such that $F(a) - \varepsilon e \in \operatorname{INF}_w(F)$.*
   Then, for any $\gamma > 0$, there exists some $b \in X$ such that :
   (i) $\|a - b\| \le \gamma$,

   (ii) b is a weakly efficient point for the perturbed function

$$x \mapsto G(x) = F(x) + \frac{\varepsilon}{\gamma}\|x - b\|e,$$

   (iii) $F(b) \le_C F(a) - \frac{\varepsilon}{\gamma}\|a - b\|e$.
   (iv) If in addition F is $C-$convex or F is Gâteaux differentiable on the set $\operatorname{dom} F$ which is supposed to be open, then there exists some $\lambda \in \Lambda$ such that

$$d(0, \partial F_\lambda(b)) \le \frac{\varepsilon}{\gamma}.$$

Remark 5. An important difference between these principles and the scalar variational principle is that the "new" point b is not necessarily an $\varepsilon'-$weakly efficient one for some small ε'. We will improve the above theorems in order to obtain that the point b is also $\varepsilon'-$weakly efficient.

Lemma 1. *Let X be a reflexive Banach space, and $F : X \to Y \cup \{+\infty_C\}$ be a positively weakly lower semicontinuous function. Let $\varepsilon > 0$, $\gamma > 0$ be given numbers, and let $a \in X$ be an $\varepsilon-$ correct weakly efficient point. Consider some $e \in C \cap B$ such that $F(a) - \varepsilon e \in \operatorname{INF}_w(F)$. Let $S \subset X$ be a weakly closed set such that $a \in S \subset a + \gamma B$. Suppose that there exists a weakly upper semicontinuous function $\psi : X \to \mathbb{R}$ such that* *(i)* $\psi(X) \subset [0, 1]$,

(ii) $\psi(a) = 1$,

(iii) $\psi(X \setminus S) = \{0\}$.

Then, there exists $b \in S$ such that b is a weakly efficient point of the perturbed function

$$X \ni x \mapsto G(x) = F(x) - \varepsilon\psi(x)e,$$

and $G(b) \leq_C G(a)$. In particular

$$\|a - b\| \leq \gamma; \quad F(b) \leq_C F(a); \quad \|F(x) - G(x)\| \leq \varepsilon, \ \forall x \in X.$$

Proof. If $F(a) \in \text{INF}_w(F)$, take $b = a$, $e = 0$, hence $G = F$ and the conclusion follows immediately. Suppose now that $F(a) \notin \text{INF}_w(F)$. Denote $y = F(a) - \varepsilon e \in \text{INF}_w(F)$. For any $\lambda \in \Lambda$, $\langle\lambda, e\rangle \geq 0$, hence the function $G_\lambda = F_\lambda - \langle\lambda, e\rangle\psi$ is weakly lower semicontinuous. Thus G is positively weakly l.s.c., hence $C-$ weakly semicontinuous, hence $L(G; y)$ is weakly closed. Consider some fixed $\lambda \in \Lambda$. Since $S \cap L(G; y)$ is weakly compact and nonempty (it contains the point a because $G(a) = y$), it follows that there exists some $b \in \arg\min_{S \cap L(G;y)} G_\lambda$. To prove that b is weakly efficient for G, let us consider $x \in X$ such that $G(x) <_C G(b)$. Since $b \in L(G; y)$ we must have $x \in L(G; y)$ hence, according to the choice of b, we have $x \notin S$. Thus $G(x) = F(x) <_C G(b) \leq_C y$ which contradicts the fact that $y \in \text{INF}_w(F)$. Hence there is no $x \in X$ such that $G(x) <_C G(b)$, which shows that b is weakly efficient for G. Since $b \in L(G; y)$, we have $G(b) \leq_C y = F(a) - \varepsilon e = G(a)$. Since $\psi(b) \leq 1$ we obtain immediately that $F(b) \leq_C F(a)$. $\qquad\square$

Remark 6. When the function ψ used in the previous lemma is neither convex nor Gâteaux differentiable, to obtain that b is "almost" scalarly weakly stationary we cannot use Proposition 3, and we need to refine the necessary weakly efficiency conditions.

Lemma 2. *Let X be a Banach space, $F : X \to Y \cup \{+\infty_C\}$, $b \in \text{int dom}(F)$ and $e \in C$. Consider a function $\varphi : X \to \mathbb{R}$, Lipschitz near b, such that b is a weakly efficient point for the function $x \mapsto G(x) = F(x) + \varphi(x)e$. If F is Gâteaux differentiable at b, then there exists $\lambda \in \Lambda$ such that*

$$\forall h \in X, \ F'_\lambda(b) \cdot h + \varphi^\circ(b; h)\langle\lambda, e\rangle \geq 0, \tag{7}$$

where

$$\varphi^\circ(b; h) = \limsup_{\substack{x \to b \\ t \downarrow 0}} \frac{1}{t}(\varphi(x + th) - \varphi(x))$$

is the generalized directional derivative of Clarke (see [8, Chapter 2]).

Proof. Consider the upper Dini derivative

$$\varphi'_+(b; h) = \limsup_{t \to 0^+} \frac{1}{t}(\varphi(b + th) - \varphi(b)),$$

which is finite thanks to the Lipschitz assumption. The set $C_1 = Y \setminus (-\operatorname{int} C)$ is a closed cone (nonconvex), and for each $h \in X$, and sufficiently small positive real t we have $G(b + th) - G(b) \in C_1$. Dividing by t and letting $t \to 0^+$ (upon a suitable sequence) we obtain that $F'(b) \cdot h + \varphi'_+(b; h)e \in C_1$. Since $\varphi'_+(b; h) \leq \varphi^\circ(b; h)$ and $e \in C$, it follows easily that

$$F'(b) \cdot h + \varphi^\circ(b; h)e \in C_1.$$

We have that $h \mapsto \varphi^\circ(b; h)$ is positively homogeneous, subadditive (see [8, Chapter 2]). Then the set

$$\{F'(b) \cdot h + \varphi^\circ(b; h)e| \ h \in X\} + C$$

is convex and contained in C_1. Using the separation theorem, and the fact that for each $\lambda \in \Lambda$, $x \in X$ we have

$$\lambda \circ F'(x) = F'_\lambda(x), \tag{8}$$

the conclusion follows easily.   □

Note that from (7), changing h in $-h$, and using the fact that $\varphi^\circ(b; \cdot)$ is homogeneous, we obtain that

$$\|F'_\lambda(b)\| \leq \langle \lambda, e \rangle \sup_{\|h\|=1} |\varphi^\circ(b; h)| \leq \sup_{\|h\|=1} |\varphi^\circ(b; h)| \leq L, \tag{9}$$

where L is the Lipschitz rank of φ near b. Thus, considering $\varphi = -\varepsilon\psi$, we obtain the following.

Proposition 4. *If the hypothesis and notations of Lemma 1 hold, and if we assume in addition that ψ is Lipschitz near b, $b \in \operatorname{int} \operatorname{dom}(F)$, and F is Gâteaux differentiable at b, then there exists a functional $\lambda \in \Lambda$ such that*

$$\|F'_\lambda(b)\| \leq \varepsilon \sup_{\|h\|=1} |\psi^\circ(b; h)|. \tag{10}$$

Remark 7. The quantity $\sup_{\|h\|=1} |\psi^\circ(b; h)|$ may depend on γ or ε.

Next, we will present some variants of vector variational principles where the "new" point b is ε–weakly efficient or ε–correct weakly efficient.

Theorem 5. *Let X, Y be Hilbert spaces identified with their duals, and let $F : X \to Y \cup \{+\infty_C\}$ be a proper map, continuous with respect to the weak topology of X and the strong topology of $\overline{Y}$. Let $\varepsilon > 0$ and $a \in X$ be any ε–*

correct weakly efficient point. Consider some $e \in C \cap B$ such that $F(a) - \varepsilon e \in$
$\text{INF}_w(F)$.

Then, there exists $b \in \text{dom} F$ such that:
(i) $\|a - b\| \le \sqrt{\varepsilon}$,
(ii) b is a weakly efficient point for the map $G : X \to Y \cup \{+\infty_C\}$, given
by

$$X \ni x \mapsto G(x) = F(x) - \varepsilon\varphi(1 - \frac{\|x - a\|^2}{\varepsilon}) \cdot \varphi(1 - \frac{\|F(x) - F(a)\|^2}{\varepsilon})e, \tag{11}$$

where $\varphi : \mathbb{R} \to \mathbb{R}$ is an increasing, convex differentiable function such that

$$\varphi(\mathbb{R}_-) = \{0\}, \quad \varphi'(1) \le 2, \quad \varphi(1) = 1,$$

(for example $\varphi(t) = 0$, if $t < 0$; $\varphi(t) = t^2$ if $t \ge 0$)
(iii) for any $x \in \text{dom} F$:

$$\|F(x) - G(x)\| \le \varepsilon. \tag{12}$$

(iv)

$$\|F(b) - F(a)\| \le \sqrt{\varepsilon}, \tag{13}$$

and b is $\sqrt{\varepsilon}(1 + \sqrt{\varepsilon})$—weakly efficient for F.
If in addition F is Gâteaux differentiable (with respect to the strong topol-
ogy of X) in a neighborhood N of a (note that, from the continuity, $\text{dom} F$
is an open set), ε verifies $a + \sqrt{\varepsilon}B \subset N$, and the map $x \mapsto F'(x)$ from N to
to $\mathcal{L}(X, Y)$ (the Banach space of linear continuous maps from X to Y with
its operator norm topology) is bounded then, there exists a constant $k > 0$
independent on b and ε such that

$$\exists \lambda \in \Lambda, \quad \|F'_\lambda(b)\| \le k\sqrt{\varepsilon}. \tag{14}$$

Proof. Let $S = (a + \sqrt{\varepsilon}B) \cap F^{-1}(F(a) + \sqrt{\varepsilon}B)$, which is obviously a weakly
compact set containing the point a. The function

$$X \ni x \mapsto \psi(x) = \varphi(1 - \frac{\|x - a\|^2}{\varepsilon}) \cdot \varphi(1 - \frac{\|F(x) - F(a)\|^2}{\varepsilon})$$

is weakly upper semicontinuous (as the product between the weakly upper
semicontinuous function $x \mapsto \varphi(1 - \frac{\|x-a\|^2}{\varepsilon})$ and the weakly continuous pos-
itive function $x \mapsto \varphi(1 - \frac{\|F(x)-F(a)\|^2}{\varepsilon})$) and verifies the hypothesis (i), (ii)
and (iii) of Lemma 1 with $\gamma = \sqrt{\varepsilon}$. Hence there exists $b \in S$ which is a weakly
efficient point for G. Thus (i-iv) are obvious.

Suppose now that F is Gâteaux differentiable on N and the map $x \mapsto F'(x)$ is bounded. Hence there exists a constant k_1 independent of b and ε such that, for all $x \in N$, $\|F'(x)\| \leq k_1$. Then G is Gâteaux differentiable on N, and according to Proposition 3, there exists $\lambda \in \Lambda$ such that $G'_\lambda(b) = 0$. Thus computing the derivative of G at b, for each $h \in X$, we have

$$\langle F'_\lambda(b), h \rangle_X + 2\langle \lambda, e \rangle \left(\varphi'(1 - \frac{\|b - a\|^2}{\varepsilon}) \cdot \varphi(1 - \frac{\|F(b) - F(a)\|^2}{\varepsilon}) \cdot \langle b - a, h \rangle_X + \right.$$

$$\left. \varphi(1 - \frac{\|b - a\|^2}{\varepsilon}) \cdot \varphi'(1 - \frac{\|F(b) - F(a)\|^2}{\varepsilon}) \cdot \langle F(b) - F(a), F'(b)h \rangle \right) = 0.$$

Since $\|b - a\| \leq \sqrt{\varepsilon}$, $\|F(b) - F(a)\| \leq \sqrt{\varepsilon}$ and $\|F'(b)\| \leq k_1$, we obtain easily (14). □

Remark 8. It is possible to obtain a similar result considering the nonsmooth function

$$\psi = \left(1 - \frac{\|x - a\|}{\sqrt{\varepsilon}} \right)_+ \cdot \left(1 - \frac{\|F(x) - F(a)\|}{\sqrt{\varepsilon}} \right)_+$$

(where $t_+ = \max(t; 0)$ for any real t) and taking $G(\cdot) = F(\cdot) - \varepsilon\psi(\cdot)e$. In this case, the last part of the proof follows from Proposition 4.

Now we can give a nonsmooth vectorial version of Theorem 1 which shows that given a c.a.w.p. sequence, there exists a nearby w.s.s. sequence which is also an a.w.p. one.

Theorem 6. *Let X, Y be Hilbert spaces identified with their duals, and let $F : X \to Y \cup \{+\infty_C\}$ be a proper map, continuous with respect to the weak topology of X and the strong topology of $\overline{Y}$, Gâteaux differentiable and $x \mapsto F'(x)$ bounded on $\mathrm{dom}(F)$.*

Let (a_n) be a c.a.w.p. sequence. Then there exists a sequence (b_n) such that:

(i) $\|a_n - b_n\| \to 0$,
(ii) (b_n) is a w.s.s. sequence,
(iii) (b_n) is an a.w.p. sequence.

Proof. It is obvious according to Theorem 5. □

If the initial point a is ε−strictly correct weakly efficient we will show in the following that the "new" point b is an ε− correct weakly efficient one.

Theorem 7. *Let X be a reflexive Banach space, and let $F : X \to Y$ be a map continuous with respect to the weak topology of X and the strong topology of Y. Assume that C has the following property (which holds in any Hilbert space Y if $C \subset C^+$, in particular if $Y = \mathbb{R}^n$, and $C = \mathbb{R}^n_+$ is the Pareto cone):*

$$\forall y, y' \in C, \quad y \leq_C y' \implies \|y\| \leq \|y'\|.$$

Let $\varepsilon > 0$ and let $a \in X$ be any ε–strictly correct weakly efficient point. Consider some $e \in (\text{int } C \cup \{0\}) \cap B$ such that $F(a) - \varepsilon e \in \text{INF}_w(F)$.

Then, for any $\gamma > 0$, there exists some $b \in X$ such that :

(i) $\|a - b\| \leq \gamma$

(ii) b is a weakly efficient point for the function $G : X \to Y$ given, for every $x \in X$ by

$$G(x) = F(x) - \varepsilon \left(1 - \frac{\|x - a\|}{\gamma}\right)_+ \cdot \frac{\text{d}(F(x) - F(a) + (\varepsilon/2)e; Y \setminus \text{int } C)}{\|F(x) - F(a)\| + \text{d}((\varepsilon/2)e; Y \setminus \text{int } C)} e, \tag{15}$$

(if $e \neq 0$, otherwise $G = F$), which verifies

$$\forall x \in X, \quad \|F(x) - G(x)\| \leq \varepsilon$$

(iii) $F(a) - (\varepsilon/2)e \leq_C F(b) \leq_C F(a)$, hence $\|F(a) - F(b)\| \leq \varepsilon/2$ and b is an ε–strictly correct weakly efficient point for F.

Proof. If $e = 0$, then a is weakly efficient for F and we can take $b = a$ (and $G = F$). Thus, in the sequel we can suppose that $e \neq 0$. Let $S = (a + \gamma B) \cap \{x \in X \mid y + (\varepsilon/2)e \leq_C F(x)\}$ where $y = F(a) - \varepsilon e$. Note that the set $\{x \in X \mid y + (\varepsilon/2)e \leq_C F(x)\} = \cap_{\lambda \in \Lambda}\{x \in X \mid F_\lambda(x) \geq \langle \lambda, y + (\varepsilon/2)e \rangle\}$ is weakly closed, hence S is a (not empty because $a \in S$) weakly closed and bounded set. Denote $\psi : X \to \mathbb{R}$ the function verifying $\varepsilon\psi(\cdot)e = F(\cdot) - G(\cdot)$. It is obvious that $\psi(a) = 1$, $\psi(X \setminus S) = \{0\}$, and $0 \leq \psi \leq 1$ (because, if $K \subset Y$ is a closed set, we have $\text{d}(a + b; K) \leq \|a\| + \text{d}(b; K)$, for all $a, b \in Y$). ψ is weakly upper semicontinuous since $x \mapsto (1 - \frac{\|x - a\|}{\gamma})_+$ is weakly upper semicontinuous and $x \mapsto \frac{\text{d}(F(x) - F(a) + (\varepsilon/2)e; Y \setminus C)}{\|F(x) - F(a)\| + \text{d}((\varepsilon/2)e; Y \setminus C)}$ is weakly continuous and positive. Then we can apply Lemma 1, hence we obtain (i), (ii) and $G(b) \leq_C G(a)$. The last inequality is equivalent to $F(b) \leq_C F(a) - \varepsilon(1 - \psi(b))e$, and obviously $F(a) - \varepsilon(1 - \psi(b))e \leq_C F(a)$, hence $F(b) \leq_C F(a)$. Moreover, since $b \in S$ we have that $F(a) - (\varepsilon/2)e \leq_C F(b)$. Denote e' the element verifying $\varepsilon e' = F(b) - y$. We have that $0 <_C (\varepsilon/2)e \leq_C \varepsilon e' = F(b) - F(a) + \varepsilon e \leq_C \varepsilon e$, thus $e' \in B \cap \text{int } C$ hence, since $y \in \text{INF}_w(F)$, we obtain that b is an ε–correct weakly efficient point for F. $\square$

Remark 9. The hypothesis: " a is ε–*strictly* correct weakly efficient" is necessary to ensure that the function $x \mapsto \frac{\text{d}(F(x) - F(a) + (\varepsilon/2)e; Y \setminus C)}{\|F(x) - F(a)\| + \text{d}((\varepsilon/2)e; Y \setminus C)}$ is weakly continuous at a.

Remark 10. Unfortunately, despite the fact that ψ is Lipschitz, the Proposition 4 is useless since the supremum $\sup_{\|h\| = 1} |\psi^\circ(b; h)|$ depends "badly" on ε.

References

1. Attouch H., Riahi, R. (1993) Stability results for Ekeland's ε-variational principle and cone extremal solutions. Math. Oper. Res. **18**, 173-201
2. Auslender, A., Crouzeix, J.P. (1989) Well behaved asymptotical convex functions. Analyse non linéaire, Gauthier–Villars, Paris, 101-122
3. Auslender, A. (1997) How to deal with the unbounded in optimization: Theory and algorithms. Mathematical Programming **79**, 3-18
4. Auslender, A., Cominetti R., Crouzeix, P.(1993) Functions with unbounded level sets and applications to duality theory. SIAM J. Optimization **3**, 669-687
5. Bernoussi, B., Bolintinéanu, S., Chou, C.C. (1998) Pareto Optimizing and Scalarly Stationary Sequences. J. Math. Analysis Appl. **220**, 553-561
6. Bolintinéanu, S. Approximate Efficiency and Scalar Stationarity in Unbounded Nonsmooth Convex Vector Optimization Problems (submitted)
7. Bolintinéanu, S. Vector variational principles; ε—efficiency and scalar stationarity (submitted)
8. Clarke, F.H. (1983) Optimization and Nonsmooth Analysis. John Wiley & Sons, New York
9. Borwein, J.M., Penot J.P., Thera, M. (1984) Conjugate Convex Operators. J. Math. Anal. Appl. **102**, 399-414
10. Chou, C.C., Ng, K.F., Pang, J.S. Minimizing and stationary sequences of optimization problems, to appear in SIAM J. Contr. Opt.
11. Dauer, J.P., Stadler, W. (1986) A Survey of Vector Optimization in Infinite-Dimensional Spaces, Part 2. J. Optim. Th. Appl. **51**, 205-241
12. Deville, R., Godefroy, G., Zizler, V. (1993) A smooth variational principle with applications to Hamilton-Jacobi equations in infinite dimensions. J. Funct. Anal. **111**, 197-212
13. Dolecki, S., Malivert, C. (1987) Polarities and stability in vector optimization. Lecture Notes in Economics and Mathematical systems **294**, 96-113
14. Ekeland, I. (1974) On the variational principle. J. Math. Anal. Appl. **47**, 324-353
15. Lemaire, B. (1992) Approximation in multiobjective optimization. J. Global Optim. **2**, 117-132
16. Loridan, P. (1984) ε-solutions in vector minimization problems. J. Optim. Th. Appl. **43**, 265-276
17. Loridan, P. (1996) Recent Developements in Well-Posed Variational Problems. In: Lucchetti R., Revalski J. (Eds.) Kluwer Academic Publishers
18. Luc, D. T. (1989) Theory of vector optimization. Lecture Notes in Economics and Mathematical Systems. **319**, Springer Verlag, Berlin
19. Pang, J.-S. (1997) Error bounds in mathematical programming. Mathematical Programming **79**, 229-332
20. Penot, J. P. (1983) Compact Nets, Filters and Relations. J. Math. Anal. Appl. **93**, 400-417
21. Rockafellar, R.T. (1970) Convex Analysis. Princeton University Press, Princeton, NJ.
22. Staib, T. (1988) On two generalizations of Pareto minimality. J. Optim. Th. Appl. **59**, 289-306
23. Tammer, C. (1992) A generalization of Ekeland's variational principle. Optimization **25**, 129-141

Tangential Regularity of Lipschitz Epigraphic Set-valued Mappings

M. Bounkhel and L. Thibault

Université Montpellier II, Laboratoire d'Analyse Convexe, C.C.: 051, 34095 Montpellier, France

Abstract. A set-valued mapping M from a topological vector space E into a normed vector space F is tangentially regular at a point $(\bar{x}, \bar{y})$ in its graph *gph* M if the Clarke tangent cone to *gph* M at $(\bar{x}, \bar{y})$ is equal to the Bouligand contingent cone to *gph* M at $(\bar{x}, \bar{y})$. Recently, after the work of Burke, Ferris and Qian on directional regularity of distance functions associated with subsets, we characterized in several cases the tangential regularity of a set-valued mapping M as the directional regularity of the scalar function Δ_M defined by $\Delta_M(x, y) := d(y, M(x))$. In this paper we characterize this tangential regularity, for a new class of set-valued mappings, as the directional regularity of Δ_M in a weak sense. We also show that the images of a set-valued mapping are tangentially regular whenever its graph is tangentially regular.

Key words: set-valued mapping, tangent cone, tangential regularity, directional regularity

1991 Mathematics Subject Classification. 49J52, 58C06, 58C20, 90C26

1   Introduction and notations

Let M be a set-valued mapping from a Hausdorff real topological vector space E into a real normed vector space F. The graph *gph* M (resp. the effective domain *dom* M) of $M : E \rightrightarrows F$ is the set *gph* $M := \{(x, y) \in E \times F : y \in M(x)\}$ (resp. *dom* $M := \{x \in E : M(x) \neq \emptyset\}$). The set-valued mapping M is said to be convex if its graph is convex.

Let S be a nonempty closed subset of E and $\bar{x}$ be a point in S. Let us recall the following classical tangent cones $K(S; \bar{x})$ of Bouligand and $T(S; \bar{x})$ of Clarke.

•)The contingent cone $K(S; \bar{x})$ to S at $\bar{x}$ is the set of all $h \in E$ such that for every neigbourhood H of h in E and for every $\epsilon > 0$, there exists $t \in]0, \epsilon[$ such that

$$(\bar{x} + tH) \cap S \neq \emptyset.$$

•)The Clarke tangent cone $T(S; \bar{x})$ to S at $\bar{x}$ is the set of all $h \in E$ such that for every neigbourhood H of h in E there exist a neigbourhood X of $\bar{x}$

in E and a real number $\epsilon > 0$ such that

$$(x + tH) \cap S \neq \emptyset \qquad \text{for all } x \in X \cap S \text{ and } t \in]0, \epsilon[.$$

$T(S; \bar{x})$ is a closed convex cone (see [14]) and $K(S; \bar{x})$ is a closed cone (that may be nonconvex).

Obviously $T(S; \bar{x}) \subset K(S; \bar{x})$. When equality holds in this inclusion i.e. $T(S; \bar{x}) = K(S; \bar{x})$, one says that S is *tangentially regular* at $\bar{x}$.

Let f be a function from E into $\mathbb{R} \cup \{-\infty, +\infty\}$ with $|f(\bar{x})| < +\infty$. The generalized directional derivative $f^\uparrow(\bar{x}; .)$ is defined (see Rockafellar [14]) by

$$
\begin{aligned}
f^\uparrow(\bar{x}; h) &= \limsup_{\substack{(x, \alpha) \downarrow_f \bar{x} \\ t \downarrow 0}} \inf_{h' \to h} t^{-1}\big[f(x + th') - \alpha\big] \\
&:= \sup_{H \in \mathcal{N}(h)} \Big[\limsup_{\substack{(x, \alpha) \downarrow_f \bar{x} \\ t \downarrow 0}} \Big(\inf_{h' \in H} t^{-1}\big[f(x + th') - \alpha\big]\Big)\Big],
\end{aligned}
$$

where $(x, \alpha) \downarrow_f \bar{x}$ means $(x, \alpha) \in epi f := \{(z, \beta) \in E \times \mathbb{R}; f(z) \leq \beta\}$ and $(x, \alpha) \longrightarrow (\bar{x}, f(\bar{x}))$ and $\mathcal{N}(h)$ denotes the filter of neighbourhoods of h.

The lower Hadamard directional derivative of f at $\bar{x}$ is defined (see Penot [13]) by

$$f^H(\bar{x}; h) = \liminf_{\substack{h' \to h \\ t \downarrow 0}} t^{-1}\big[f(\bar{x} + th') - f(\bar{x})\big].$$

When for all $h \in E$ one has

$$f^\uparrow(\bar{x}; h) = f^H(\bar{x}; h),$$

one says (see for example [4,6,10,14]) that f is *directionally regular* at $\bar{x}$.

Now, following Aubin [2], with each notion of tangent cone, a geometric notion of derivative for set-valued mappings is associated in the following natural way.

Definition 1. Let $R(gph\ M; \bar{x}, \bar{y})$ be any tangent cone (for example the contingent cone $K(gph\ M; \bar{x}, \bar{y})$ or the Clarke tangent cone $T(gph\ M; \bar{x}, \bar{y})$) to $gph\ M$ at $(\bar{x}, \bar{y})$. The R-derivative of M at $(\bar{x}, \bar{y}) \in gph\ M$ is the set-valued mapping $D_R M(\bar{x}, \bar{y})$ defined from E into F by

$$gph\ (D_R M(\bar{x}, \bar{y})) := R(gph\ M; \bar{x}, \bar{y}).$$

Now we consider the scalar function Δ_M defined on $E \times F$ by

$$\Delta_M(x, y) = d(y, M(x)),$$

where the right term is $+\infty$ whenever $M(x) = \emptyset$ i.e. $x \notin dom\ M$.

If E is a normed space, it is easily seen that for all $(x, y) \in E \times F$

$$d_{gph\ M}(x, y) \leq \Delta_M(x, y) \leq \psi_M(x, y),$$

where $d_{gph\ M}(x, y) := d((x, y), gph\ M)$ and ψ_M denotes the indicator function of $gph\ M$ i.e. $\psi_M(x, y) = 0$ if $(x, y) \in gph\ M$ and $+\infty$ otherwise.

Several important aspects of the behaviour of a set-valued mapping M have been characterized in terms of the scalar function Δ_M. One knows (see for example, Castaing and Valadier [9]) that, under general assumptions, the measurability of M is equivalent to the measurability of Δ_M. The Rockafellar result, in [15], says that the pseudo-Lipschitzness (see the definition below) of M at $(\bar{x}, \bar{y})$ in $gph\ M$ is equivalent to the Lipschitzness of Δ_M around $(\bar{x}, \bar{y})$. If one assumes that M has closed values, it is not difficult to see that M is convex if and only if the scalar function Δ_M is convex.

In Bounkhel and Thibault [5], we studied the equivalence between the tangential regularity of a set-valued mapping $M : E \rightrightarrows F$, which is defined as the tangential regularity of its graph in the sense recalled above, and the directional regularity of the scalar function Δ_M associated with M. We showed in particular (see Theorem 1 below) that the directional regularity of Δ_M always ensures the tangential regularity of M and the reverse implication holds whenever F is finite dimensional space. This is in the line of earlier results by Burke, Ferris and Qian [8] concerning the connexion between tangential regularity of sets and directional regularity of associated distance functions.

Theorem 1. [5] *Let E be any Hausdorff topological vector space and F be any normed vector space and let $M : E \rightrightarrows F$ be any set-valued mapping defined from E into F with $(\bar{x}, \bar{y}) \in gph\ M$. Assume that Δ_M is directionally regular at $(\bar{x}, \bar{y})$. Then $gph\ M$ is tangentially regular at $(\bar{x}, \bar{y})$. Moreover, if F is finite dimensional space, then the converse is also true.*

When F is an infinite dimensional space, the equivalence in the statement of the previous theorem is not true in the general setting. Indeed, consider the set-valued mapping $M : E \rightrightarrows F$ defined by $M(x) = S$ for any $x \in E$, where S is the closed subset constructed by Borwein and Fabian [3], for which "S is tangentially regular at some point $\bar{y} \in S$ " and " d_S is not directionally regular at the same point $\bar{y}$ ". This set-valued mapping is tangentially regular at $(\bar{x}, \bar{y})$ for each $\bar{x} \in E$ but the scalar function Δ_M associated with this set-valued mapping is easily seen to be not directionally regular at $(\bar{x}, \bar{y})$.

In spite of that, there are several important classes of set-valued mappings $M : E \rightrightarrows F$ (with F infinite dimensional space) for which either both notions of regularity hold or the equivalence between them is true. Let us recall some examples of such classes.

1- For all convex set-valued mappings $M : E \rightrightarrows F$, the tangential regularity of M as well as the directional regularity of Δ_M always hold.

2- The implicit set-valued mapping, defined below, is another example of such a class under some qualification condition. Let E be a topological vector space and F and G be two Banach spaces. Let $f : E \times F \to G$ be a mapping that is strictly Fréchet differentiable at $(\bar{x}, \bar{y})$ with respect to the second variable, that is there exist a neighbourhood X of $\bar{x}$, a continuous linear mapping $\nabla_2 f(\bar{x}, \bar{y})$ from F into G and a mapping $r : X \times F \times F \longrightarrow G$ such that for all $(x, y, y') \in X \times F \times F$

$$f(x, y) - f(x, y') = \nabla_2 f(\bar{x}, \bar{y})(y - y') + \|y - y'\| r(x, y, y')$$

and $\lim_{(x,y,y') \to (\bar{x}, \bar{y}, \bar{y})} r(x, y, y') = 0$. We also assume that f is Gâteaux directionally differentiable at $(\bar{x}, \bar{y})$ with respect to the first variable, that is there exists a (not necessarily linear) continuous mapping $D_1 f(\bar{x}, \bar{y}; .)$ from E into G such that for all $h \in E$

$$\lim_{h' \to h, t \to 0^+} t^{-1} \left[f(\bar{x} + th', \bar{y}) - f(\bar{x}, \bar{y}) \right] = D_1 f(\bar{x}, \bar{y}; h).$$

Consider the implicit set-valued mapping $M : E \rightrightarrows F$ defined by

$$M(x) := \{ y \in F : f(x, y) \in C \}, \tag{1.1}$$

where C is a closed convex subset of G with $\bar{z} := f(\bar{x}, \bar{y}) \in C$. We suppose that the following Robinson qualification condition

$$0 \in core \left[Im \nabla_2 f(\bar{x}, \bar{y}) - (C - \bar{z}) \right] \tag{R.C.}$$

is satisfied. Recall that a point q is in the core of a subset S in G if for each $z \in G$ there exists some real number $\epsilon > 0$ such that $\{ q + t(z - q) : t \in [-\epsilon, \epsilon] \} \subset S$.

Theorem 2. [5] *Under the assumptions above, the tangential regularity of the implicit set-valued mapping M given by (1.1) at $(\bar{x}, \bar{y})$ is equivalent to the directional regularity of the scalar function Δ_M at $(\bar{x}, \bar{y})$.*

In this paper we will provide a general new important class of set-valued mappings for which the equivalence above holds in a weaker sense. For the regularity concept related to normal cones, we refer to [4] and [7].

2   Main results

The class of set-valued mappings that we will explore (in this section) in connexion with the tangential regularity is that of Lipschitz epigraphic set-valued mappings. We owe its consideration here to the part of the proof of Proposition 3.3. in Ioffe [12] establishing that the approximate and the geometric normal cone to an epi-Lipschitzian set (of a Banach space) coincide.

Before defining this class, recall that a closed subset S of F is epi-Lipschitz around a point $\bar{x} \in S$ if it can be represented near $\bar{x}$ as the epigraph of a Lipschitz function. Rockafellar showed in [16] that S is epi-Lipschitz around $\bar{x}$ if and only if there exist some vector $\bar{h} \in F$, neighbourhoods $H \in \mathcal{N}(\bar{h})$, $X \in \mathcal{N}(\bar{x})$ and a real number $\epsilon > 0$ such that

$$S \cap X + tH \subset S,$$

for all $t \in [0, \epsilon]$.

This geometrical characterization allows to adapt the concept for set-valued mappings as follows.

Definition 2. Let $M : E \rightrightarrows F$ be a set-valued mapping with closed values that is lower semicontinuous at $\bar{x}$. We will say that M is *Lipschitz epigraphic* around $(\bar{x}, \bar{y}) \in gph\, M$ if there exist a vector $\bar{h} \in F$, neighbourhoods $H \in \mathcal{N}(\bar{h})$, $X \in \mathcal{N}(\bar{x})$, and $Y \in \mathcal{N}(\bar{y})$ and a real number $\bar{\delta} > 0$ such that

$$M(x) \cap Y + tH \subset M(x),$$

for all $x \in X$ and $t \in [0, \bar{\delta}]$.

An adaptation of the proof of the geometrical characterization of epi-Lipschitzian sets (Theorem 3 in [16]) by Rockafellar allows to obtain a similar characterization for Lipschitz epigraphic set-valued mappings. We give it in the following proposition.

Proposition 1. *Let $M : E \rightrightarrows F$ be a set-valued mapping with closed values that is lower semicontinuous at $\bar{x}$. Then M is Lipschitz epigraphic around $(\bar{x}, \bar{y})$ in gph M if and only if, either $(\bar{x}, \bar{y}) \in int\, gph\, M$, or there exist a topological direct sum $F = G \oplus \mathbb{R}\bar{h}$ with $\bar{y} = \bar{z} + \bar{r}\bar{h}$ ($\bar{z} \in G$ and $\bar{r} \in \mathbb{R}$), a function $f : E \times G \to \mathbb{R}$, neighbourhoods $X \times Z$ of $(\bar{x}, \bar{z})$ in $E \times G$ and I of $\bar{r}$ and a real positive number $l \geq 0$ such that*

i) f is upper semicontinuous at $(\bar{x}, \bar{z})$ and $f(\bar{x}, \bar{z}) = \bar{r}$;

ii) $|f(x, z) - f(x, z')| \leq l\|z - z'\|$ for all $z, z' \in Z$ and $x \in X$ (here $\|.\|$ denotes the norm induced on G by the norm of F);

iii) for $Y := Z + I\bar{h}$ and for all $x \in X$

$$M(x) \cap Y = \{z + r\bar{h} : \quad (z, r) \in (Z \times I) \cap epi\, f(x, .)\}$$

i.e.

$$M(x) \cap Y = \{z + r\bar{h} : \quad (z, r) \in (Z \times I) \text{ and } f(x, z) \leq r\}.$$

Proof. Necessity. This implication can be checked in a straighforward way.

Sufficiency. Assume that $(\bar{x}, \bar{y}) \notin int\ gph\ M$. Let $\bar{h}$, X, Y and $\bar{\delta}$ as given by Definition 2. Choose a topological complement G of the one-dimensional subspace $\mathbb{R}\bar{h}$ of F so that F appears as the topological direct sum $F = G \oplus \mathbb{R}\bar{h}$ and $\bar{y}$ may be written as $\bar{y} = \bar{z} + \bar{r}\bar{h}$ for some $\bar{z} \in G$ and $\bar{r} \in \mathbb{R}$. Choose a symmetric convex neighbourhood V of zero in G endowed with the topology induced by that of F and a positive number $\delta < \min(\bar{\delta}, 1)$ such that $Y' := (\bar{z} + V) \oplus]\bar{r} - \delta, \bar{r} + \delta[\bar{h}$ is included in Y and $H' := V \oplus]1 - \delta, 1 + \delta[\bar{h}$ is included in H. So, Definition 2 says that for all $t \in [0, \delta]$ and $x \in X$

$$M(x) \cap Y' + tH' \subset M(x). \tag{2.1}$$

According to the lower semicontinuity of M at $(\bar{x}, \bar{y})$ there exists a neighbourhood X' of $\bar{x}$ with $X' \subset X$ such that for all $x \in X'$

$$M(x) \cap \left[(\bar{z} + \frac{\delta}{8}V) \oplus (]\bar{r} - \frac{\delta}{4}, \bar{r} + \frac{\delta}{4}[\bar{h}) \right] \neq \emptyset. \tag{2.2}$$

Put $I :=]\bar{r} - \frac{\delta}{2}, \bar{r} + \frac{\delta}{2}[$ and consider the following function f defined on $E \times G$ by

$$f(x, y) := \inf\{r \in I : \quad z + r\bar{h} \in M(x)\}. \tag{2.3}$$

First, we wish to show that for $Z := \bar{z} + \frac{\delta}{8}V$ and for every $(x, z) \in X' \times Z$ the set $\{r \in I : \quad z + r\bar{h} \in M(x)\}$ is nonempty, which will ensure that f takes finite values over $X \times Z$. Fix then any $(x, z) \in X' \times Z$. There exists, by (2.2), some $z' \in \bar{z} + \frac{\delta}{8}V$ and some $r' \in]\bar{r} - \frac{\delta}{4}, \bar{r} + \frac{\delta}{4}[$ such that $z' + r'\bar{h} \in M(x)$. Therefore

$$z + (r' + \frac{\delta}{4})\bar{h} = z' + r'\bar{h} + (z - z') + \frac{\delta}{4}\bar{h} \in M(x) \cap Y' + \frac{\delta}{4}V + \frac{\delta}{4}\bar{h}$$

$$= M(x) \cap Y' + \frac{\delta}{4}[V + \bar{h}]$$

$$\subset M(x) \cap Y' + \frac{\delta}{4}H'$$

$$\subset M(x). \qquad \text{(by (2.1))}$$

This ensures that one has $\{r \in I : \quad z + r\bar{h} \in M(x)\} \neq \emptyset$ and this nonemptiness allows to write

$$f(x, y) = \inf\{r \in \mathbb{R} : \quad r > \bar{r} - \frac{\delta}{2} \quad \text{and} \quad z + r\bar{h} \in M(x)\}. \tag{2.4}$$

We turn now to prove *i)*, *ii)* and *iii)* of the statement of the theorem for the function f defined above. Fix any $(x, z, r) \in X' \times Z \times I$ with $(z, r) \in$

epi $f(x, .)$. Then $f(x, z) \leq r$ and hence for any $s \in I$ with $s > r$ we can find (by (2.3)) $\rho \in I$ with $z + \rho \bar{h} \in M(x)$ and such that $f(x, z) \leq \rho < s$. Thus

$$z + s\bar{h} = z + \rho\bar{h} + (s - \rho)\bar{h} \in M(x) \cap Y' +]0, \delta[H' \subset M(x) \quad \text{(by (2.1))}$$

and as $M(x)$ is closed we obtain $z + r\bar{h} \in M(x)$. So, we have proved that for $Y'' := Z \oplus I\bar{h}$

$$\{z + r\bar{h} : \quad (z, r) \in (Z \times I) \cap epi\ f(x, .)\} \subset M(x) \cap Y''.$$

As the reverse inclusion is obvious, the proof of *iii*) is complete.

Now we show *ii*). It is enough to prove the following

$$f(x, z + tv) \leq f(x, z) + t,$$

for all $x \in X'$, $z \in Z$, $v \in V$ and $t \in [0, \delta]$ with $z + tv \in Z$. Fix $(x, z) \in X' \times Z$ and fix also $r \in I$ with $z + r\bar{h} \in M(x)$. For all $v \in V$ and $t \in [0, \delta]$ with $z + tv \in Z$ one has by (2.1)

$$(z + tv) + (r + t)\bar{h} = (z + r\bar{h}) + t(v + \bar{h}) \in M(x) \cap Y' + tH' \subset M(x).$$

Since $r + t \geq r > \bar{r} - \dfrac{\delta}{2}$, one concludes (by (2.4)) that $f(x, z + tv) \leq r + t$ for all $r \in I$ with $z + r\bar{h} \in M(x)$. Hence it follows from (2.3) that $f(x, z + tv) \leq f(x, z) + t$.

We finish the proof by showing *i*). As $\bar{z} + \bar{r}\bar{h} \in M(\bar{x})$, one has $f(\bar{x}, \bar{z}) \leq \bar{r}$. So, suppose by contradiction that $f(\bar{x}, \bar{z}) < \bar{r}$. Choose (by (2.3)) $r' \in I$ and real numbers $\epsilon > 0$ and $\eta \in]0, \dfrac{\delta}{2}[$ such that $\bar{z} + r'\bar{h} \in M(\bar{x})$, the closed ball Z_ϵ in G centered at $\bar{z}$ with radius ϵ is included in Z and $I_\epsilon :=]r' - \epsilon, r' + \epsilon[\subset I$ and also such that

$$f(\bar{x}, \bar{z}) < r' + \epsilon + 2l\epsilon < \bar{r} - \eta.$$

Then, as M is lower semicontinuous at $\bar{x}$, there exists a neighbourhood $X_\epsilon \subset X'$ of $\bar{x}$ such that $(Z_\epsilon + I_\epsilon\bar{h}) \cap M(x) \neq \emptyset$ for all $x \in X_\epsilon$. For every $(x, z) \in X_\epsilon \times Z_\epsilon$ one can choose $(z'' + r''\bar{h}) \in (Z_\epsilon + I_\epsilon\bar{h}) \cap M(x)$ and by *(ii)* and (2.3) one has

$$f(x, z) \leq f(x, z'') + l\|z - z''\| \leq r'' + 2l\epsilon \leq r' + \epsilon + 2l\epsilon < \bar{r} - \eta$$

and hence, for $I_\eta :=]\bar{r} - \eta, \bar{r} + \eta[$ one gets $X_\epsilon \times Z_\epsilon \times I_\eta \subset epi\ f$ which ensures by *(iii)* that $X_\epsilon \times \left(Z_\epsilon \oplus I_\eta\bar{h}\right) \subset gph\ M$. This contradicts that $(\bar{x}, \bar{z} + \bar{r}\bar{h}) = (\bar{x}, \bar{y}) \notin int\ gph\ M$. So $f(\bar{x}, \bar{z}) = \bar{r}$. Using the same arguments as in what precedes, one can easily see that f is upper semicontinuous at $(\bar{x}, \bar{z})$. So the proof of the theorem is complete. □

In the sequel we will denote by $I\!\!B$ the closed unit ball centered at the origin of a normed vector space.

Theorem 3. *Let $M : E \rightrightarrows F$ be a set-valued mapping that is Lipschitz epigraphic around at a point $(\bar{x}, \bar{y}) \in gph\, M$ and lower semicontinuous at $\bar{x}$. Then M is tangentially regular at $(\bar{x}, \bar{y})$ if and only if there exists some equivalent norm on F such that Δ_M (associated with that norm) is directionally regular at the same point $(\bar{x}, \bar{y})$.*

Proof. We may assume that $(\bar{x}, \bar{y})$ is not in the interior of $gph\, M$ (because otherwise the result is trivial). By Proposition 1 we may also suppose that $F = G \times I\!\!R$ and $\bar{y} = (\bar{z}, \bar{r})$ and that near $(\bar{x}, \bar{y})$ the set-valued mapping M is given for all $x \in X$ by

$$M(x) \cap (Z \times I) = \{(z, r) \in Z \times I : \quad f(x, z) \leq r\},$$

where X, Z, I and f are as in the statement of Proposition 1.

Step 1. We begin by proving the following inequality (which holds for any set-valued mapping defined from a Hausdorff topological vector space E into F endowed with any norm $\|.\|$)

$$\Delta_M^\uparrow(\bar{x}, \bar{z}, \bar{r}; \bar{h}, \bar{k}, \bar{s}) \leq \Delta_{T,M}(\bar{h}, \bar{k}, \bar{s}) \quad \text{for all } (\bar{h}, \bar{k}, \bar{s}) \in E \times G \times I\!\!R.$$

Here we use the notation $\Delta_{T,M}(h, k, s) := d\big((k, s); D_T M(\bar{x}, \bar{z}, \bar{r})(h)\big).$ (2.5) Fix $(\bar{h}, \bar{k}, \bar{s}) \in E \times G \times I\!\!R$. We may assume that $\Delta_{T,M}(\bar{h}, \bar{k}, \bar{s}) < +\infty$. Consider $\epsilon > 0$. There exists $(\bar{v}, \bar{\alpha}) \in D_T M(\bar{x}, \bar{z}, \bar{r})(\bar{h})$ (i.e. $(\bar{h}, \bar{v}, \bar{\alpha}) \in T(gph\, M; \bar{x}, \bar{z}, \bar{r})$) such that

$$\|(\bar{v}, \bar{\alpha}) - (\bar{k}, \bar{s})\| \leq \Delta_{T,M}(\bar{h}, \bar{k}, \bar{s}) + \frac{\epsilon}{2}. \tag{2.6}$$

Let $H \in \mathcal{N}(\bar{h})$, $V \in \mathcal{N}(\bar{v})$, and $J \in \mathcal{N}(\bar{\alpha})$ such that $\|(v, \alpha) - (\bar{v}, \bar{\alpha})\| \leq \frac{\epsilon}{2}$ for all $(v, \alpha) \in V \times J$. Then, by the definition of the Clarke tangent cone, there exist $X \in \mathcal{N}(\bar{x})$, $Z \in \mathcal{N}(\bar{z})$, $\Lambda \in \mathcal{N}(\bar{r})$, and $\delta > 0$ such that

$$\big((x, z, s) + t(H \times V \times J)\big) \cap gph\, M \neq \emptyset$$

for all $(x, z, s) \in (X \times Z \times \Lambda) \cap gph\, M$, and all $t \in]0, \delta[$, and hence there exist $h \in H$, $v \in V$, and $\alpha \in J$ such that

$$(x, z, s) + t(h, v, \alpha) \in gph\, M \quad \text{i.e.} \quad (v, \alpha) \in t^{-1}[M(x + th) - (z, s)]. \tag{2.7}$$

Moreover, we always have

$$t^{-1}\Delta_M(x + th, z + t\bar{k}, s + t\bar{s}) = t^{-1}d\big((z + t\bar{k}, s + t\bar{s}); M(x + th)\big)$$

$$= d\big((\bar{k}, \bar{s}); t^{-1}[M(x+th) - (z, s)]\big).$$

Then for every $(x, z, s) \in (X \times Z \times \Lambda) \cap gph\, M$ and every $t \in]0, \delta[$ there is by (2.7) some $(h, v, \alpha) \in H \times V \times J$ such that

$$t^{-1} \Delta_M (x + th, z + t\bar{k}, s + t\bar{s}) \leq \|(v, \alpha) - (\bar{k}, \bar{s})\|$$

$$\leq \|(v, \alpha) - (\bar{v}, \bar{\alpha})\| + \|(\bar{v}, \bar{\alpha}) - (\bar{k}, \bar{s})\|$$

$$\leq \|(\bar{v}, \bar{\alpha}) - (\bar{k}, \bar{s})\| + \frac{\epsilon}{2},$$

and hence we have by (2.6)

$$t^{-1} \Delta_M (x + th, z + t\bar{k}, s + t\bar{s}) \leq \Delta_{T,M} (\bar{h}, \bar{k}, \bar{s}) + \epsilon.$$

Therefore, we deduce that for $\gamma := \Delta_{T,M} (\bar{h}, \bar{k}, \bar{s})$

$$\sup_{\substack{H \in \mathcal{N}(\bar{h})}} \inf_{\substack{X \times Z \in \mathcal{N}(\bar{x}, \bar{z}) \\ \Lambda \in \mathcal{N}(\bar{r}), \delta > 0}} \sup_{\substack{(x, z, s) \in (X \times Z \times \Lambda) \cap gph\, M \\ t \in]0, \delta[}} \inf_{h \in H} t^{-1} \Delta_M (x + th, z + t\bar{k}, s + t\bar{s}) \leq \gamma + \epsilon,$$

and according to Theorem 3.1 in Bounkhel and Thibault [5] we conclude that

$$\Delta_M^\uparrow (\bar{x}, \bar{z}, \bar{r}; \bar{h}, \bar{k}, \bar{s}) \leq \Delta_{T,M} (\bar{h}, \bar{k}, \bar{s}).$$

So the proof of Step1 is complete.

Step 2. Let l be a Lipschitz constant of f as given by Proposition 1 and let $\|.\|$ be the norm on $G \times \mathbb{R}$ defined by $\|(z, r)\| := l\|z\| + |r|$ for all $(z, r) \in G \times \mathbb{R}$. In this second step, we will prove that, with respect to this norm $\|.\|$, the following equality holds

$$\Delta_M^H (\bar{x}, \bar{z}, \bar{r}; \bar{h}, \bar{k}, \bar{s}) = \Delta_{K,M} (\bar{h}, \bar{k}, \bar{s}) \quad \text{for all } (\bar{h}, \bar{k}, \bar{s}) \in E \times G \times \mathbb{R},$$

where $\Delta_{K,M} (h, k, s)$ is defined in a similar way as in (2.5) with $D_T M (\bar{x}, \bar{z}, \bar{r})(h)$ in place of $D_K M (\bar{x}, \bar{z}, \bar{r})(h)$.

By the same techniques used in the previous step, we can prove (in fact with respect to any norm on $G \times \mathbb{R}$) the following inequality

$$\Delta_M^H (\bar{x}, \bar{z}, \bar{r}; \bar{h}, \bar{k}, \bar{s}) \leq \Delta_{K,M} (\bar{h}, \bar{k}, \bar{s}) \quad \text{for all } (\bar{h}, \bar{k}, \bar{s}) \in E \times G \times \mathbb{R}. \quad (2.8)$$

So, we proceed to showing the reverse inequality

$$\Delta_{K,M} (\bar{h}, \bar{k}, \bar{s}) \leq \Delta_M^H (\bar{x}, \bar{z}, \bar{r}; \bar{h}, \bar{k}, \bar{s}) \quad \text{for all } (\bar{h}, \bar{k}, \bar{s}) \in E \times G \times \mathbb{R}. \quad (2.9)$$

We may assume that the second member is finite. Consider any real number $\rho > 0$ and any $(z, r) \in (\bar{z}, \bar{r}) + \rho \mathbb{B}$. As the set-valued mapping M is lower semicontinuous at $(\bar{x}, \bar{z}, \bar{r}) \in gph\, M$, there exists a neigbourhood $U \in \mathcal{N}(\bar{x})$ such that for all $x \in U$ one has

$$M(x) \cap ((\bar{z}, \bar{r}) + \rho \mathbb{B}) \neq \emptyset.$$

Let any $(z', r') \in M(x)$ with $\|(z', r') - (\bar{z}, \bar{r})\| > 3\rho$. Choose $(z'', r'') \in M(x) \cap ((\bar{z}, \bar{r}) + \rho I\!B)$ and observe that

$$\|(z', r') - (z, r)\| \geq \|(z', r') - (\bar{z}, \bar{r})\| - \|(z, r) - (\bar{z}, \bar{r})\|$$

$$> 3\rho - \rho = 2\rho$$

$$\geq \|(z, r) - (\bar{z}, \bar{r})\| + \|(z'', r'') - (\bar{z}, \bar{r})\|$$

$$\geq \|(z'', r'') - (z, r)\|$$

$$\geq d\Big((z, r), M(x) \cap ((\bar{z}, \bar{r}) + 3\rho I\!B)\Big).$$

One deduces that one has for any $x \in U$

$$d\Big((z, r), M(x)\Big) \geq d\Big((z, r), M(x) \cap ((\bar{z}, \bar{r}) + 3\rho I\!B)\Big),$$

and hence (the reverse inequality being obvious)

$$d\Big((z, r), M(x)\Big) = d\Big((z, r), M(x) \cap ((\bar{z}, \bar{r}) + 3\rho I\!B)\Big). \qquad (2.10)$$

Fix now any neigbourhood $X \times Z \times I \in \mathcal{N}(\bar{x}, \bar{z}, \bar{r})$ as in Proposition 1 and fix also $\rho > 0$ such that $(\bar{z}, \bar{r}) + 3\rho I\!B \subset Y \times I$. Choose $\delta > 0$ and a neigbourhood $H \in \mathcal{N}(\bar{h})$ such that $\bar{x} +]0, \delta[H \subset X \cap U$ (where U is the neigbourhood of $\bar{x}$ given in what precedes) and

$$(\bar{z} +]0, \delta[\bar{k}) \times (\bar{r} +]0, \delta[\bar{s}) \subset (\bar{z}, \bar{r}) + \rho I\!B.$$

Fix $t \in]0, \delta[$ and $h \in H$. Then, for all $z \in Z$ and $r \geq 0$ with

$$(z, f(\bar{x} + th, z) + r) \in (\bar{z}, \bar{r}) + 3\rho I\!B$$

one has

$$\|(\bar{z} + t\bar{k}, f(\bar{x}, \bar{z}) + t\bar{s}) - (z, f(\bar{x} + th, z) + r)\|$$

$$= l\|\bar{z} + t\bar{k} - z\| + |f(\bar{x} + th, z) - f(\bar{x}, \bar{z}) - t\bar{s} + r|$$

$$\geq l\|\bar{z} + t\bar{k} - z\| + f(\bar{x} + th, z) - f(\bar{x}, \bar{z}) - t\bar{s} + r$$

$$\geq (l\|\bar{z} + t\bar{k} - z\| + f(\bar{x} + th, z)) - f(\bar{x}, \bar{z}) - t\bar{s}.$$

From the Lipschitz property of f in $ii)$ of Proposition 1, we deduce that

$$\|(\bar{z} + t\bar{k}, f(\bar{x}, \bar{z}) + t\bar{s}) - (z, f(\bar{x} + th, z) + r)\| \geq f(\bar{x} + th, \bar{z} + t\bar{k}) - f(\bar{x}, \bar{z}) - t\bar{s}.$$

So, (2.10) and this inequality imply that

$$t^{-1}d\Big((\bar{z}+t\bar{k},\bar{r}+t\bar{s}), M(\bar{x}+th)\Big)$$
$$= t^{-1}d\Big((\bar{z}+t\bar{k},\bar{r}+t\bar{s}), (Z\times I)\cap epi\; f(\bar{x}+th,.)\Big)$$
$$\geq t^{-1}[f(\bar{x}+th,\bar{z}+t\bar{k}) - f(\bar{x},\bar{z})] - \bar{s},$$

which gives (taking ii) of Proposition 1 and the definition of $f^H(.;.)$ into account)

$$\Delta_M^H(\bar{x},\bar{z},\bar{r};\bar{h},\bar{k},\bar{s}) \geq f^H(\bar{x},\bar{z};\bar{h},\bar{k}) - \bar{s}. \tag{2.11}$$

To finish this step, it is enough because of (2.11) to show the following inequality

$$f^H(\bar{x},\bar{z};\bar{h},\bar{k}) - \bar{s} \geq \Delta_{K,M}(\bar{h},\bar{k},\bar{s}) \quad \text{for all } (\bar{h},\bar{k},\bar{s}) \notin K(gph\; M;\bar{x},\bar{z},\bar{r}).$$

Indeed, if we take $(\bar{h},\bar{k},\bar{s}) \in K(gph\; M;\bar{x},\bar{z},\bar{r})$, then $(\bar{k},\bar{s}) \in D_K M(\bar{x},\bar{z},\bar{r})(\bar{h})$ and hence $\Delta_{K,M}(\bar{h},\bar{k},\bar{s}) = 0$, which is always not greater than $\Delta_M^H(\bar{x},\bar{z},\bar{r};\bar{h},\bar{k},\bar{s})$. Fix now $(\bar{h},\bar{k},\bar{s}) \notin K(gph\; M;\bar{x},\bar{z},\bar{r})$. As $(X\times Z\times I)\cap gph\; M = (X\times Z\times I)\cap epi\; f$ by Proposition 1, we have

$$K(gph\; M;\bar{x},\bar{z},\bar{r}) = K(epi\; f;\bar{x},\bar{z},f(\bar{x},\bar{z})) = epi\; f^H(\bar{x},\bar{z};.,.). \tag{2.12}$$

So $(\bar{h},\bar{k},\bar{s}) \notin epi\; f^H(\bar{x},\bar{z};.,.)$, that is $f^H(\bar{x},\bar{z};\bar{h},\bar{k}) > \bar{s}$. This implies in particular that $|f^H(\bar{x},\bar{z};\bar{h},\bar{k})| < \infty$ because the first member of (2.11) has been supposed to be finite. Furthermore, one has

$$f^H(\bar{x},\bar{z};\bar{h},\bar{k})-\bar{s} = |f^H(\bar{x},\bar{z};\bar{h},\bar{k})-\bar{s}|+l\|\bar{k}-\bar{k}\| = \|(\bar{k},\bar{s})-(\bar{k},f^H(\bar{x},\bar{z};\bar{h},\bar{k}))\|.$$

By (2.12), we know that $(\bar{k},f^H(\bar{x},\bar{z};\bar{h},\bar{k}))$ lies in $D_K M(\bar{x},\bar{z},f(\bar{x},\bar{z}))(\bar{h})$, which allows to conclude that

$$f^H(\bar{x},\bar{z};\bar{h},\bar{k}) - \bar{s} \geq d\Big((\bar{k},\bar{s}), D_K M(\bar{x},\bar{z},f(\bar{x},\bar{z}))(\bar{h})\Big) = \Delta_{K,M}(\bar{h},\bar{k},\bar{s}).$$

This completes the proof of the second step.

Step 3. We finish the proof of the theorem by proving the following : If M is tangentially regular at $(\bar{x},\bar{z},\bar{r}) \in gph\; M$, then the scalar function Δ_M (associated with the norm $\|.\|$ in Step 2) is directionally regular at the same point. It is sufficient to prove the assertion above because, by Theorem 1, the converse of this assertion holds for any norm on $G\times I\!\!R$.

Assume that M is tangentially regular at $(\bar{x},\bar{z},\bar{r}) \in gph\; M$. Let $(\bar{h},\bar{k},\bar{s}) \in E\times G\times I\!\!R$. Then one has

$$\Delta_{T,M}(\bar{h},\bar{k},\bar{s}) = \Delta_{K,M}(\bar{h},\bar{k},\bar{s}),$$

and by Step1 and Step2 one has

$$\Delta_M^\uparrow(\bar{x},\bar{z},\bar{r};\bar{h},\bar{k},\bar{s}) \leq \Delta_{T,M}(\bar{h},\bar{k},\bar{s}),$$

and
$$\Delta_M^H(\bar{x}, \bar{z}, \bar{r}; \bar{h}, \bar{k}, \bar{s}) = \Delta_{K,M}(\bar{h}, \bar{k}, \bar{s}).$$

Thus , one gets
$$\Delta_M^\uparrow(\bar{x}, \bar{z}, \bar{r}; \bar{h}, \bar{k}, \bar{s}) \leq \Delta_{T,M}(\bar{h}, \bar{k}, \bar{s})$$
$$= \Delta_{K,M}(\bar{h}, \bar{k}, \bar{s})$$
$$= \Delta_M^H(\bar{x}, \bar{z}, \bar{r}; \bar{h}, \bar{k}, \bar{s}).$$

So
$$\Delta_M^\uparrow(\bar{x}, \bar{z}, \bar{r}; \bar{h}, \bar{k}, \bar{s}) = \Delta_M^H(\bar{x}, \bar{z}, \bar{r}; \bar{h}, \bar{k}, \bar{s}),$$

because the reverse inequality is always true. This completes the proof of the theorem. □

Corollary 1. *Let S be a noempty closed subset of a normed vector space F and $\bar{x} \in S$. Assume that S is epi-Lipschitz at $\bar{x}$. Then the two following assertions are equivalent:*

i) S is tangentially regular at $\bar{x}$;

ii) there exists an equivalent norm on F such that the associated distance function d_S is directionally regular at $\bar{x}$.

Proof. The proof of this corollary is a direct application of Theorem 3 with $M(x) = S$ for all $x \in E$. □

3   Tangential regularity of images

Throughout this section E will be a normed vector space. It is well known that any convex set-valued mapping $M : E \rightrightarrows F$ (i.e. $gph\ M$ is convex) has convex image sets $M(x)$, but the converse is not true.

In this section we establish a similar result for the tangential regularity. We show that if the graph of a pseudo-Lipschitz set-valued mapping M is tangentially regular, then it has tangentially regular image sets. The converse is obviously not true in the general setting.

Let us recall the Aubin notion of pseudo-Lipschitz set-valued mappings (see Aubin [1]).

Definition 3. A set-valued mapping $M : E \rightrightarrows F$ is said to be l-pseudo-Lipschitz at $(\bar{x}, \bar{y}) \in gph\ M$ (for a real number $l \geq 0$) if there exist neighbourhoods X of $\bar{x}$ and Y of $\bar{y}$ such that
$$Y \cap M(x') \subset M(x) + l\|x - x'\|\mathbb{B}, \quad \text{for all } x, x' \in X.$$

The following sequential characterizations of the Clarke tangent cone $T(S; \bar{x})$ (see Hiriart-Urruty [11]) and the contingent cone $K(S; \bar{x})$ (see for example Penot [13]) will be needed.

A vector $v \in T(S; \bar{x})$ if and only if for any sequence (x_n) converging to $\bar{x}$ with $x_n \in S$ and any sequence of positive numbers (t_n) converging to 0 there exists a sequence v_n converging to v such that

$$x_n + t_n v_n \in S \quad \text{for each } n \in I\!N.$$

A vector $v \in K(S; \bar{u})$ if and only if there exist a sequence (t_n) of positive real numbers converging to zero and a sequence (v_n) in E converging to v such that

$$\bar{x} + t_n v_n \in S \quad \text{for each } n \in I\!N.$$

Theorem 4. *Let $M : E \rightrightarrows F$ be a set-valued mapping defined between two normed vector spaces E and F and let $(\bar{x}, \bar{y}) \in \mathrm{gph}\, M$. Assume that M is pseudo-Lipschitz at $(\bar{x}, \bar{y})$. If M is tangentially regular at $(\bar{x}, \bar{y})$, then $M(\bar{x})$ is tangentially regular at $\bar{y}$.*

Proof. It is not difficult to see that the following assertions ensure the conclusion of the theorem :

1) $(0, k) \in K(\mathrm{gph}\, M; \bar{x}, \bar{y})$ if and only if $k \in K(M(\bar{x}); \bar{y})$;

2) If $(0, k) \in T(\mathrm{gph}\, M; \bar{x}, \bar{y})$, then $k \in T(M(\bar{x}); \bar{y})$.

So, we begin by showing the first one.

1- Fix $(0, k) \in K(\mathrm{gph}\, M; \bar{x}, \bar{y})$. There exist a sequence $(h_n, k_n) \longrightarrow (0, k)$ in $E \times F$ and a sequence $t_n \downarrow 0$ such that

$$k_n \in t_n^{-1}[M(\bar{x} + t_n h_n) - \bar{y}] \quad \text{for all } n \in I\!N.$$

As M is pseudo-Lipschitz at $(\bar{x}, \bar{y})$, for n sufficiently large we have

$$k_n \in t_n^{-1}[M(\bar{x}) + t_n l \|h_n\| I\!B - \bar{y}],$$

where l is a Lipschitz constant of M as in Definition 3. So, there exists a sequence $b_n \in I\!B$ such that for n sufficiently large

$$\bar{y} + t_n(k_n + l\|h_n\|b_n) \in M(\bar{x}).$$

By taking $w_n = k_n + l\|h_n\|b_n$, which converges to k, we conclude that $k \in K(M(\bar{x}); \bar{y})$.

Assume now that $k \in K(M(\bar{x}); \bar{y})$. Then there exist sequences $k_n \longrightarrow k$ and $t_n \downarrow 0$ such that

$$\bar{y} + t_n k_n \in M(\bar{x}) \quad \text{i.e.} \quad (\bar{x} + t_n h_n, \bar{y} + t_n k_n) \in \mathrm{gph}\, M, \quad \text{with } h_n = 0.$$

Thus $(0, k) \in K(gph\, M; (\bar{x}, \bar{y}))$, which completes the proof of the first assertion.

2- Using the characterization recalled above of the Clarke tangent cone and the same techniques in the proof of 1), it is not difficult to show the second assertion 2).   □

References

1. Aubin, J.P. (1984) Lipschitz behavior of solutions to convex minimization problems. Math. Oper. Res. **9**, 87–111
2. Aubin, J.P. (1981) Contingent derivatives of set-valued maps and existence of solutions to nonlinear inclusions and differential inclusions. Advances in Mathematics, Supplementary studies, Ed. Nachbin L., 160–232
3. Borwein, J.M., Fabian, M. (1994) A note on regularity of sets and of distance functions in Banach space. J. Math. Anal. Appl. **182**, no.2, 566–570
4. Bounkhel, M. (1999) Régularité tangentielle en analyse non lisse. Ph.D. Thesis, University of Montpellier II, France
5. Bounkhel, M., Thibault, L. Scalarization of tangential regularity of set-valued mappings. To appear in Set-Valued Analysis
6. Bounkhel, M., Thibault, L. Subdifferential regularity of directionally Lipschitzian functions. To appear in Canad. Bull. Math.
7. Bounkhel, M., Thibault, L. On various notions of regularity of sets. Preprint. University of Montpellier II, France
8. Burke, J.V., Ferris, M.C., Qian, M. (1992) On the Clarke subdifferential of the distance function of a closed set. J. Math. Anal. Appl. **166**, 199–213
9. Castaing, C., Valadier, M. (1977) Convex Analysis and Measurable Multifunctions. Lecture Notes in Mathematics **580**, Springer, Berlin Heidelberg
10. Clarke, F.H. (1983) Optimization and Nonsmooth Analysis. Wiley-Interscience, New York
11. Hiriart-Urruty, J.B. (1979) Tangent cones, generalized gradients and mathematical programing in Banach spaces. Math. Oper. Res. **4**, 79–97
12. Ioffe, A. (1989) Approximate subdifferentials and applications III: The metric theory. Mathematika **36**, 1–38
13. Penot, J.P. (1978) Calcul sous-différentiel et Optimisation. J. Funct. Anal. **27**, 248–276
14. Rockafellar, R.T. (1980) Generalized directional derivatives and subgradients of nonconvex functions. Canad. J. Math. **39**, 257–280
15. Rockafellar, R.T. Lipschitzian properties of multifunctions. Nonlinear Anal. Th. Meth. Appl. **9**, 867–885
16. Rockafellar, R.T. (1979) Clarke's tangent cones and the boundaries of closed sets in $I\!R^n$. Nonlinear Anal. Th. Meth. Appl. **3**, no. 1, 145–154

Relaxed Assumptions for Iteration Methods

Myrana Brohé and Patricia Tossings

Université de Liège, Institut de Mathématiques, 4000 Liège, Belgium

Abstract. We consider a general inexact iterative method for finding a fixed point of a nonexpansive mapping on a real Hilbert space. Proofs of convergence of sequences generated by such a method generally require at least that the error term goes to zero. The aim of the present paper is to weaken this nonrealistic theoretical assumption. The obtained result is applied to the proximal point algorithm.

Key words: iterative methods, error term, generalized proximal regularization

1 Introduction

Let H be a real Hilbert space with inner product $\langle .,. \rangle$ and associated norm $\|.\|$.

Maximal monotone operators on H have been extensively studied because of their role in convex analysis. In this context, a fundamental problem is the following one :

(P)   "To find $\overline{x} \in H$ such that $0 \in T\overline{x}$,"

where T is a maximal monotone operator defined on H.

A well-known approach for solving problem (P) is to use the *proximal point algorithm* developed by R.T. Rockafellar [10]. This algorithm generates, from any starting point $y_0 \in H$, a sequence (y_n) in H, by the scheme

$$y_n = J_{\lambda_n}^T y_{n-1} + e_n, \quad \forall\, n \in \mathbb{N}^*,$$

where (λ_n) is a sequence of positive real numbers bounded away from zero, (e_n) a sequence in H taking into account a possible inexact computation and $J_\lambda^T = (I + \lambda T)^{-1}$ the resolvent operator associated with T with parameter $\lambda > 0$.

The weak or the strong convergence of a sequence (x_n) generated by this rule, to a solution of problem (P), has been studied by R.T. Rockafellar [10].

In a first time, B. Lemaire [6] studied a perturbed version of this algorithm when $T = \partial f$, the subdifferential of a proper closed convex function defined on H, by replacing, at each iteration, the proper closed convex function f by another one f^n, the sequence (f^n) approaching f in an appropriate manner.

Inspired by H. Attouch and R.J.B. Wets' work [2], P. Tossings [12] introduced the variational metric between two maximal monotone operators and an associated notion of convergence. Thanks to this notion of convergence, P. Tossings [13] studied a perturbed version of the proximal point algorithm. She replaced so, at each iteration, the maximal monotone operator T by another one T^n, the sequence (T^n) approaching T in a certain sense, tied to the variational metric.

After, P. Alexandre, V.H. Nguyen and P. Tossings [1] replaced, at each iteration, the square of the norm associated with the inner product by a variable metric and introduced some relaxation too.

All of these authors obtained convergence results which present similar assumptions to R.T. Rockafellar's one. They added an assumption which specifies how the sequence (f^n) or (T^n) approaches the function f or the operator T.

However, in most contexts, the authors have to assume that $\sum \|e_n\|$ is convergent. In [6], B. Lemaire gave a less weaker assumption : he assumed that $\|e_n\|$ goes to zero.

But, even this assumption is not very realistic. Indeed, in practice, we can at best suppose that (e_n) remains bounded. This fact opens an important question.

Recently, several papers dealt with this problem. In [8], B. lemaire gives a partial answer. He has notably to impose conditions on the perturbed recursive operators. Another reply [11] is proposed by associating proximal and projection methods, projection on a hyperplane separating strictly the current iterate from the solution set of the problem.

We formulate, in the third section, one more partial answer to this question. Reversing by B. Lemaire, we impose conditions on the nonperturbed recursive operator. After that, we apply this result to the perturbed proximal point algorithm with nonquadratic kernel. So, we recover, in part, the other works above-mentioned. Before that, let us recall, in the next section, some definitions given in B. Lemaire, [7].

2   Preliminaries

Consider some problem with data d and solution set S, a subset of H. We call

- the *basic method* any iteration scheme

$$\begin{cases} \xi_0 = x \in H, \\ \xi_n = P_n \xi_{n-1}, \quad \forall\, n \in \mathbb{N}^*, \end{cases}$$

where $P_n = P(d, n)$ is some mapping defined from H into H, depending on the data d and the iteration index n. Moreover we suppose that S is the set of fixed points of all iterative mappings P_n.

With such a basic method, we associate
- the *perturbed method, i.e.*

$$\begin{cases} \xi_0 = x \in H, \\ \xi_n = Q_n \xi_{n-1}, \quad \forall\, n \in \mathbb{N}^*, \end{cases}$$

where the iteration mapping $Q_n = P(d_n, n)$ from H to H is dependent on the (perturbed) data d_n (rather than on data d).
- the *approximative basic method* defined by the iteration scheme

$$\begin{cases} x_0 = x \in H, \\ x_n = P_n x_{n-1} + e_n, \quad \forall\, n \in \mathbb{N}^*. \end{cases}$$

where e_n is the error term which interests us.
- the *perturbed approximative basic method* defined by the following rule

$$\begin{cases} x_0 = x \in H, \\ x_n = Q_n x_{n-1} + e_n, \quad \forall\, n \in \mathbb{N}^*. \end{cases}$$

Definition 1. An iterative method is said to be *convergent* if, from any starting point, the sequence generated by the basic method has a strong limit in S.

An iterative method is said to be *stable with respect to the perturbed data* $\{d_n\}$ if the associated perturbed method is convergent.

Definition 2. Let $\rho \geq 0$. The *approximation error between d and d_n*, is defined by

$$\Delta_{n,\rho} = \sup_{\|x\| \leq \rho} \|Q_n x - P_n x\|, \quad \forall\, n \in \mathbb{N}^*.$$

3   Main results

3.1   General result

The following lemma is inspired from B. Lemaire [7], lemma 2.1, and [8], lemma 2.1.

Lemma 1. *Suppose that problem* (P) *has at least one solution. Consider some basic method defined by the mappings P_n and the associated perturbed approximative basic method with error terms e_n. Assume that*

(i)   *there is $\varepsilon > 0$ such that $\|e_n\| < \varepsilon$, $\forall\, n \in \mathbb{N}^*$,*

(ii) *the sequence (x_n) generated by the perturbed approximative basic method is bounded,*

(iii) *there is $0 \leq \sigma < 1$ such that, for all $n \in \mathbb{N}^*$,*

$$\|P_n x - P_n y\| \leq \sigma \|x - y\|, \quad \forall\, x, y \in H,$$

(iv) $\displaystyle\sum_{n=1}^{+\infty} \Delta_{n,\rho} < +\infty,\ \forall\, \rho \geq 0.$

Then the solution set S is reduced to a singleton $\{\overline{x}\}$. Furthermore, $\overline{x}$ satisfies the following property : for all $\varepsilon' > 0$, there is a range $N \in I\!N^*$ from which

$$x_n \in B\left(\overline{x}, \varepsilon' + \frac{\varepsilon}{1-\sigma}\right),$$

where $B(\overline{x}, R)$ denotes the ball centred on $\overline{x}$ of radius $R > 0$; in other words,

$$d\left(x_n, B\left(\overline{x}, \frac{\varepsilon}{1-\sigma}\right)\right) \to 0 \quad \text{when} \quad n \to +\infty.$$

Proof. Assumption (iii) implies obviously the reduction of S to a singleton $\{\overline{x}\}$.

Set, using hypothesis (ii),

$$\rho = \sup_{n \in I\!N} \|x_n\|.$$

We get, by construction,

$$\|x_n - \overline{x}\| \leq \sigma^n \|x_0 - \overline{x}\| + \sum_{k=0}^{n-1} \sigma^k \left[\Delta_{n-k,\rho} + \|e_{n-k}\|\right].$$

Moreover, assumption (iii) and Hardy-Cesaro's theorem ensure that, for all $\varepsilon' > 0$, there exists of a range $N \in I\!N^*$ such that, for all $n \geq N$,

$$\sigma^n \leq \frac{\varepsilon'}{2\|x_0 - \overline{x}\|} \quad \text{and} \quad \sum_{k=0}^{n-1} \sigma^k \Delta_{n-k,\rho} \leq \frac{\varepsilon'}{2}.$$

The conclusion arises immediately, by using assumption (i). $\qquad\square$

3.2 Perturbed proximal point algorithm with nonquadratic kernel

Recently, more and more authors decided to change the square of the norm appearing in the proximal regularization. One of the first is G. Cohen, who introduced the Auxiliary Problem Principle [4] which generates a sequence (x_n) by the iterative scheme

$$x_n = (\nabla h + \lambda_n T)^{-1} \nabla h(x_{n-1}), \quad \forall\, n \in I\!N^*, \tag{1}$$

where h denotes a strongly convex real-valued function, assumed be Gateaux differentiable. On his side, J. Eckstein [5] studied the convergence of a sequence (x_n) generated by this rule too. But he asked to the function h to be a Bregman's one.

Using the nonlinear change of coordinates $\nabla h(x_n) = u_n$ $(n \in \mathbb{N})$, the scheme (1) may be rewritten as

$$u_n = \left(I + \lambda_n T(\nabla h)^{-1} \right)^{-1} u_{n-1}, \quad \forall\, n \in \mathbb{N}^*. \tag{2}$$

At this point, scheme (2) is nothing else but the proximal point algorithm, applied to the operator $T(\nabla h)^{-1}$. However, the composition of monotone operators fails to be monotone, generally. In consequence, the results obtained in the context of the proximal point algorithm are not directly applicable.

In a previous paper [3], we introduced, for a real-valued, strongly convex function h assumed to be Gateaux differentiable, which gradient ∇h Lipschitz but also weakly continuous, an error term and a perturbation in the Cohen's rule. So, we studied the convergence of a sequence (x_n) generated by the rule

$$x_n = (\nabla h + \lambda T^n)^{-1} \nabla h(x_{n-1}) + e_n, \quad \forall\, n \in \mathbb{N}^*, \tag{3}$$

called the *perturbed proximal point algorithm with nonquadratic kernel.*

The aim of this section is to apply lemma 1 to the sequence (x_n) generated by the relationship (3) for which we have $P_n = (\nabla h + \lambda T)^{-1} \nabla h$ and $Q_n = (\nabla h + \lambda T^n)^{-1} \nabla h$. In order to interpret assumption (iv) in this particular case, we have to introduce a notion of variational metric.

Definition 3. Let T^1 and T^2 be two maximal monotone operators on H, $\lambda > 0$ and $\rho \geq 0$. The *variational metric between T^1 and T^2, associated with h, with parameters λ, ρ* [1] is the semi-distance

$$\delta^h_{\lambda,\rho}(T^1, T^2) = \sup_{\|x\| \leq \rho} \left\| J^{h,T^1}_\lambda x - J^{h,T^2}_\lambda x \right\|,$$

where $J^{h,T}_\lambda = (\nabla h + \lambda T)^{-1} \nabla h$ is called *the resolvent operator with nonquadratic kernel, associated with the maximal monotone operator T, with parameter h, λ.* [2]

If the function h is the half-square of the norm associated with the inner product, then all of the above notions coincide with the classical ones (see for example P. Tossings [12,7]).

Futhermore, this new variational metric can be compared to the classical one, as shown in the following proposition which is proved in [3].

Proposition 1. *Let T^1 and T^2 be two maximal monotone operators on H, ∇h be Lipschitz continuous with constant $M > 0$ and $\rho \geq 0$. If $\lambda, \mu > 0$ satisfy $\dfrac{\mu}{\lambda} < \dfrac{2}{M}$, then, for all $\bar{b} \in \left[\dfrac{\mu}{\lambda}, \dfrac{2}{M} \right[$, there is a constant $C > 0$, depending only on α, M et $\bar{b}$, such that*

$$\delta^h_{\lambda,\rho}(T^1, T^2) \leq C \frac{\lambda}{\mu} \delta_{\mu,\rho_0}(T^1, T^2),$$

---

[1] We simply say generalized variational metric.

[2] In the same way, we simply say generalized resolvent operator.

for all

$$\rho_0 \geq \left[\frac{\mu}{\lambda}M + \left(\frac{\mu}{\lambda}M + 1\right)\frac{M}{\alpha}\right]\rho + \left(\frac{\mu}{\lambda}M + 1\right)\left\|J_\lambda^{h,T^1}0\right\|.$$

Moreover, if T^1 has at least one zero x^*, then the minimal value imposed on ρ_0 can be replaced by

$$\left(\frac{\mu}{\lambda}M + \frac{M}{\alpha} + \frac{\mu}{\lambda}\frac{M^2}{\alpha}\right)\rho + \left(1 + \frac{\mu}{\lambda}M\right)\left(\frac{M}{\alpha} + 1\right)\|x^*\|.$$

Now, we can apply lemma 1 to the proximal point algorithm with non-quadratic kernel.

Theorem 1. *Assume that*

(i) $0 < \underline{\lambda} \leq \lambda_n, \forall n \in I\!N^*$, where $\underline{\lambda} > (M - \alpha)a$, a being defined as below,

(ii)   the sequence (x_n) generated by the recursive rule (3) is bounded,

(iii)   there is $\varepsilon > 0$ such that $\|e_n\| < \varepsilon, \forall n \in I\!N^*$,

(iv) $\displaystyle\sum_{n=1}^{+\infty} \delta_{\lambda_n,\rho}^h(T^n, T) < +\infty, \forall \rho \geq 0,$

(v)   the operator T is strongly monotone with constant $\dfrac{1}{a}$.

Then problem (P) *has a unique solution* $\overline{x}$ *and the sequence* (x_n) *generated by the recursive rule satisfies the following property : for all* $\varepsilon' > 0$, *there is a range* $N \in I\!N^*$ *from which*

$$x_n \in B\left(\overline{x}, \varepsilon' + \varepsilon\frac{\alpha a + \underline{\lambda}}{(\alpha - M)a + \underline{\lambda}}\right);$$

in other words,

$$d\left(x_n, B\left(\overline{x}, \varepsilon\frac{\alpha a + \underline{\lambda}}{(\alpha - M)a + \underline{\lambda}}\right)\right) \to 0 \quad when \quad n \to +\infty.$$

Proof. Since

$$\frac{\nabla h(x) - \nabla h(J_\lambda^{h,T}x)}{\lambda} \in T\left(J_\lambda^{h,T}x\right), \quad \forall x \in H,$$

the strong monotonicity of operator T involves

$$\left\langle (\nabla h(x) - \nabla h(y)) - (\nabla h(J_{\lambda_n}^{h,T}x) - \nabla h(J_{\lambda_n}^{h,T}y)), J_{\lambda_n}^{h,T}x - J_{\lambda_n}^{h,T}y \right\rangle$$

$$\geq \frac{\lambda_n}{a}\left\|J_{\lambda_n}^{h,T}x - J_{\lambda_n}^{h,T}y\right\|^2, \quad \forall n \in I\!N^*, \forall x, y \in H.$$

Consequently, by the strong monotonicity of ∇h, the Cauchy-Schwarz inequality and the Lipschitz assumption on ∇h, we get

$$\left\| J_{\lambda_n}^{h,T} x - J_{\lambda_n}^{h,T} y \right\| \leq \frac{aM}{\alpha a + \underline{\lambda}} \| x - y \|, \quad \forall\, n \in I\!N^*,\ \forall\, x, y \in H.$$

$\square$

Remark 1. Considering the proposition 1, the assumption (iv) is notably satisfies if

$$\sum_{n=1}^{+\infty} \lambda_n \delta_{\underline{\lambda},\rho}(T^n, T) < \infty.$$

This formulation presents the advantage to consider the well-known classical variational metric between two maximal monotone operators with fixed parameter λ.

Remark 2. The lemma 1 can obviously be applied to the classical proximal point algorithm by taking as function h the half-square of the norm associated with the inner product.

References

1. Alexandre, P., Nguyen, V.H., Tossings, P. (1998) The Perturbed Generalized Proximal Point Algorithm. Mathematical Modelling and Numerical Analysis. **32**, 2, 223–253
2. Attouch, H., Wets, R.J.B. (19986) Isometries for Legendre-Fenchel Transform. Trans. A.M.S. **296**, 1, 33–60
3. Brohé, M., Tossings, P. Perturbed Proximal Point Algorithm with Nonquadratic Kernel. Submitted to Set-Valued Analysis
4. Cohen, G. (1980) Auxiliary Problem Principle and Decomposition of Optimization Problems. Journal of Optimization Theory and Applications **32**, 277–305
5. Eckstein, J. (1993) Nonlinear proximal point algorithms using Bregman functions, with applications to convex programming. Mathematics of Operations Research **18**, 1, 202–226
6. Lemaire, B. (1988) Coupling Optimization Methods and Variational Convergence. In: Hoffmann K.H., Hiriart-Urruty J.B., Lemarechal C., Zowe J. (Eds.) Trends in Mathematical Optimization. International Series of Num. Math., Birkhäuser Verlag, Basel **84**, 163–179
7. Lemaire, B. (1996) Stability of the Iteration Method for non expansive Mappings. Serdica Mathematical Journal **22**, 1001–1010
8. Lemaire, B. (1998) Itération et approximation. Equations aux dérivées partielles et application. Articles dédiés à J.L. Lions, Gauthier-Villars
9. Rockafellar, R.T. (1970) Convex Analysis. Princeteon University Press, Princeton, New Jersey
10. Rockafellar, R.T. (1976) Monotone Operators and the Proximal Point Algorithm. SIAM J. Control and Optimization **14**, N5, 877–898

11. Solodov, M.V., Svaiter, B.F. (1999) Projection-Proximal Point Algorithm. Journal of Convex Analysis **6**, 1

12. Tossings, P. (1991) Convergence Variationnelle et Opérateurs Maximaux Monotones d'un Espace de Hilbert réel. Bulletin de la Société Royale des Sciences de Liège, 60e année **2-3**, 103–132

13. Tossings P. (1994) The perturbed Proximal Point Algorithm and Some its Applications. Applied Mathematics and Optimization **29**, pp.125–159

Approximation Methods in Equilibrium and Fixed Point Theory
Proximally nondegenerate sets

Bernard Cornet[1] and Marc-Olivier Czarnecki[2]

[1] CERMSEM, Université de Paris 1, 106-112 Boulevard de l'Hopital,
75013 Paris, France. E-mail: cornet@univ-paris1.fr
[2] Laboratoire d'Analyse Convexe, Université de Montpellier 2,
place Eugène Bataillon, 34095 Montpellier cedex 5, France.
E-mail: marco@math.univ-montp2.fr

Abstract. The aim of this paper is to show how approximation methods allow to extend the Equilibrium and Fixed Point Theory to the nonsmooth and nonconvex setting. Our approach relies heavily on the use of the distance function d_M to a set M. We show that, for our purpose, Clarke's subdifferential ∂d_M is too big, and that the appropriate notion is the upper limit $\partial_+ d_M(\overline{x}) \overset{def}{=} \limsup_{x \to \overline{x}, x \notin M} \partial d_M(x)$.

Key words: approximation, generalized equilibria, proximally nondegenerate sets, epi-Lipschitzian sets, Clarke's normal cone

1 Introduction[3]

We first recall some definitions. A correspondence F from M to $I\!R^n$ is a map from M to the set of all the subsets of $I\!R^n$. If $X \subset I\!R^n$ and $\overline{x} \in I\!R^n$ we define:

---

[3] We let $I\!R_+ = \{x \in I\!R | x \geq 0\}$ and sgn $x = x/|x|$ if $x \in I\!R \setminus \{0\}$. If $x = (x_1, ..., x_n)$ and $y = (y_1, ..., y_n)$ belong to $I\!R^n$, we denote $(x|y) = \sum_{i=1}^{n} x_i y_i$, the scalar product of $I\!R^n$, $\|x\| = \sqrt{(x|x)}$, the Euclidean norm; we denote $B(x, r) = \{y \in I\!R^n | \|x - y\| < r\}$, $\overline{B}(x, r) = \{y \in I\!R^n | \|x - y\| \leq r\}$, $S(x, r) = \{y \in I\!R^n | \|x - y\| = r\}$, and $S = S(0, 1)$. If $X \subset I\!R^n$, $Y \subset I\!R^n$, and $x \in I\!R^n$, we let $d_X(x) = \inf_{y \in X} \|x - y\|$, $X \setminus Y = \{x \in X | x \notin Y\}$ the set-difference of the sets X and Y, $X + Y = \{x + y | x \in X, y \in Y\}$, the sum of the sets X and Y, $B(X, r) = X + B(0, r)$, $\overline{B}(X, r) = X + \overline{B}(0, r)$, cl$X$, the closure of X, int$X$, the interior of X, bd$X$ =cl$X \setminus$ int$X$, the boundary of X, $X^\circ = \{y \in I\!R^n | \forall x \in X, (y|x) \leq 0\}$, the negative polar cone of X, co$X$, the convex hull of X. A map $f : X \to I\!R$ is locally Lipschitzian if, for every $x \in X$, there is $\varepsilon > 0$ and $L > 0$ such that $\|f(x_1) - f(x_2)\| \leq L\|x_1 - x_2\|$ for every x_1 and x_2 in $B(x, \varepsilon)$. A correspondence F from X to $I\!R^n$ is a map from X to the set of all the subsets of $I\!R^n$; the correspondence F is said to be upper semicontinuous (u.s.c.), resp. lower semicontinuous (l.s.c.), if the set $\{x \in X | F(x) \subset V\}$, resp. the set $\{x \in X | F(x) \cap V \neq \emptyset\}$, is open in X for every open set $V \subset I\!R^n$. If F is a correspondence from X to $I\!R^n$, its graph, denoted G(F), is defined by G$(F) = \{(x, y) \in X \times I\!R^n | y \in F(x)\}$.

$\limsup_{x \to \overline{x}, x \in X} F(x)$ as the set

$$\{v \in I\!\!R^n | \exists (x_k) \subset X, \exists (v_k) \subset I\!\!R^n, x_k \to \overline{x}, v_k \in F(x_k), v_k \to v\}.$$

If $f : I\!\!R^n \to I\!\!R$ is differentiable at $x \in I\!\!R^n$, we denote by $\nabla f(x)$ the gradient of f at x. If $f : I\!\!R^n \to I\!\!R$ is locally Lipschitzian and $\overline{x} \in I\!\!R^n$, its Clarke's subdifferential at $\overline{x}$, denoted $\partial f(\overline{x})$, is defined by:

$$\partial f(\overline{x}) = \mathrm{co} \limsup_{x \to \overline{x}, x \in \mathrm{dom}\,(\nabla f)} \{\nabla f(x)\},$$

where $\mathrm{dom}\,(\nabla f)$ is the set on which f is differentiable.[4]

Let $M \subset I\!\!R^n$ be closed. Noticing that d_M, the distance function to M, is Lipschitzian, from above, one can define, at $x \in M$, Clarke's normal cone $N_M(x)$ by:

$$N_M(x) = \mathrm{cl}\,(\cup_{\lambda \geq 0} \lambda \partial d_M(x)).$$

The aim of this paper is to show how approximation methods allow to extend the Equilibrium and Fixed Point Theory to the nonsmooth and non-convex setting. Precisely, we look for conditions on a nonempty, compact set $M \subset I\!\!R^n$ so that the following assertions hold for $N(x) = N_M(x)$ (Clarke's normal cone to M at x), in which case $N(x)^\circ = T_M(x)$ (Clarke's tangent cone to M at x). In the definitions below of Assertions $(E; N)$ and $(GE; N)$, we only assume that N is a correspondence defined on M, with values in $I\!\!R^n$, such that $N(x)$ is a closed cone (of vertex 0), for every $x \in M$. This will allow us later to consider other normal cones smaller than Clarke's one.

Assertion $(E; N)$ [Equilibria] *For every u.s.c. correspondence F from M to $I\!\!R^n$, with nonempty, convex, compact values, such that $F(x) \cap N(x)^\circ \neq \emptyset$ for every $x \in M$, there exists a N-equilibrium $x^* \in M$ of F in the sense that:*

$$x^* \in M \text{ and } 0 \in F(x^*).$$

Assertion $(GE; N)$ [Generalized Equilibria] *For every u.s.c. correspondence F from M to $I\!\!R^n$, with nonempty, convex, compact values, there exists a generalized N-equilibrium $x^* \in M$ of F in the sense that:*

$$x^* \in M \text{ and } 0 \in F(x^*) - N(x^*).$$

The relations between the assertions $(E; N)$ and $(GE; N)$ are extensively studied in Section 5.

---

[4] We recall that $I\!\!R^n \setminus \mathrm{dom}\,(\nabla f)$ is of Lebesgue measure zero, from Rademacher's theorem.

The convex and smooth case

The assertions $(E; N)$ and $(GE; N)$ are respectively deduced from Brouwer-Kakutani and Poincaré-Hopf theorems if M is convex or smooth, as recalled below.

Let $M \subset I\!\!R^n$ be nonempty, compact, and convex, let F be a u.s.c. correspondence from M to $I\!\!R^n$, with nonempty, convex, compact values. Then:

$$[F(M) \subset M] \Rightarrow [\exists x^* \in M, \, x^* \in F(x^*)];$$
$$[\forall x \in M, F(x) \cap T_M(x) \neq \emptyset] \Rightarrow [\exists x^* \in M, \, 0 \in F(x^*)];$$
$$[\text{no condition on } F] \qquad\qquad [\exists x^* \in M, \, 0 \in F(x^*) - N_M(x^*)].$$

Let $M \subset I\!\!R^n$ be nonempty, compact, and smooth[5]. Following Milnor [28], the Euler characteristic $\chi(M)$ of M is an integer equal to:

$$\chi(M) = \deg(g_M, \operatorname{int} M, 0),$$

where $g_M = N_M \cap S$ is the Gauss map of M. See the examples in $I\!\!R^2$ illustrated in Fig. 1.

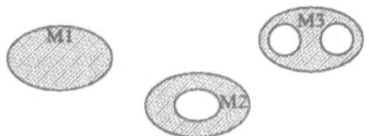

Fig. 1. $\chi(M_1) = 1$, $\chi(M_2) = 0$, $\chi(M_3) = -1$

Then $\chi(M) = \deg(F, \operatorname{int} M, 0)$, if F points outward, i.e., $F(x) \cap -T_M(x) \neq \emptyset$ and $0 \notin F(x)$, for every $x \in \operatorname{bd} M$. Hence one deduces that:

(GE) $[\chi(M) \neq 0$ and no condition on $F] \Rightarrow [\exists x^* \in M, 0 \in F(x^*) - N_M(x^*)]$.

2 Existence of (generalized) equilibria

A natural idea to obtain further results in the nonsmooth and nonconvex setting is to approximate a set M by a sequence of smooth sets or by a sequence of convex sets. Convex sets are not good candidates for approximating a set M, since the set $M = \liminf M_k$ is convex if the sets M_k are convex. So we will use hereafter the smooth approximation.

The approximation approach of this paper can be summarized by the following result. We recall that, if (M_k) is a sequence of subsets of $I\!\!R^n$, then:

$$\limsup M_k = \{x \in I\!\!R^n | \exists (x_k) \subset I\!\!R^n, \exists \varphi \in \mathcal{I}, x_k \to x, x_k \in M_{\varphi(k)} \text{ for all } k\},$$

where $\mathcal{I}$ is the set of all increasing maps $\varphi : I\!\!N \to I\!\!N$.

---

[5] i.e., a C^2 submanifold of $I\!\!R^n$, with a boundary and of full dimension.

Theorem 1 ([15]). *Let $M \subset \mathbb{R}^n$ be compact, admitting a smooth normal approximation in the sense that there is a sequence (M_k) of closed subsets of $\mathbb{R}^n$ such that:*

(sc) $M = \limsup_{k \to \infty} M_k$;

(s) M_k *is smooth and* $M_k \subset B(M, 1)$, *for every* $k \in \mathbb{N}$;

(nc) $\limsup_{k \to \infty} G(N_{M_k}) \subset G(N_M)$.

Assume in addition that:

(χ_k) $\chi(M_k) \neq 0$ *for every* $k \in \mathbb{N}$;

Then the assertions $(E; N_M)$ and $(GE; N_M)$ hold.

Proof. **Assertion** $(GE; N_M)$. Let F be an u.s.c. correspondence from M to $\mathbb{R}^n$, with nonempty, convex, compact values. For every $k \in \mathbb{N}$, since M_k is smooth and $\chi(M_k) \neq 0$, there is a generalized equilibrium x_k of F on M_k:

$$x_k \in M_k \text{ and } 0 \in F(x_k) - N_{M_k}(x_k),$$

i.e., there is $y_k \in F(x_k)$ such that $(x_k, y_k) \in G(N_{M_k})$. Without any loss of generality, by a compactness argument, there is (x^*, y^*) such that:

$$(x^*, y^*) = \lim(x_k, y_k) \in \limsup G(N_{M_k}) \subset G(N_M),$$
$$(x^*, y^*) = \lim(x_k, y_k) \in \mathrm{cl}\,[G(F)] = G(F),$$

Hence $x^* \in M$ and $0 \in F(x^*) - N_M(x^*)$.

Assertion $(E; N_M)$. First define the correspondence $\limsup N_{M_k}$ on the set M by:

$$(\limsup N_{M_k})(x) = \limsup_{k \to \infty, x' \to x} N_{M_k}(x').$$

Note that we have shown above that the assertion $(GE; \limsup_{k \to \infty} N_{M_k})$ holds. The implication:

$$(GE; \limsup_{k \to \infty} N_{M_k}) \Rightarrow (E; \limsup_{k \to \infty} N_{M_k})$$

is a consequence of Lemma 1 proved in Section 5, noticing that the correspondence $x \mapsto \limsup_{k \to \infty} N_{M_k}(x)$ has a closed graph. From the inclusion $\limsup_{k \to \infty} G(N_{M_k}) \subset G(N_M)$, we deduce that:

$$\forall x \in M, (\limsup N_{M_k})(x) \subset N_M(x),$$

hence the implication $[(E; \limsup_{k \to \infty} N_{M_k}) \Rightarrow (E; N_M)]$ holds. $\qquad\square$

Remark 1. Theorem 1 may not hold without the assumption (nc). Indeed, consider, in $\mathbb{R}^2$:

$$M = \overline{B}(0, 1) \setminus B(0, 1/2);$$
$$M_k = M \cup \overline{B}((0, 1 + \tfrac{1}{2^k}), \tfrac{1}{2^{k+1}}).$$

The sequence (M_k) satisfies the assumptions (s) (for $k \geq 1$), (χ_k) and (sc) of Theorem 1, but the assumption (nc) is not satisfied. In this case both assertions $(E; N_M)$ and $(GE; N_M)$ do not hold.

Theorem 1 is very general since many classes of sets do satisfy its assumptions, but the main drawback of these assumptions is that they are not expressed in terms of the set M and do depend upon the chosen approximation (M_k). Thus the following questions arise:

Question 1. Can we replace the previous topological assumption:

$$(\chi_k) \quad \chi(M_k) \neq 0 \text{ for every } k \in I\!N,$$

by an intrinsic topological assumption on M? For example, can we replace it by the assumption $\chi(M) \neq 0$?

Recall that, in general, the Euler characteristic of M is defined by (see [21]):

$$\chi(M) = \sum_{k \in I\!N} (-1)^k \mathrm{rg} H_k M,$$

where $(H_k M)_{k \in I\!N}$ are the (singular) homology groups of M, whenever the sum has a meaning.

A first answer is given by the two following examples in which $\chi(M)$, the Euler characteristic of M, is not defined:

Example 1. In $I\!R^2$, take $M = \{0\} \cup \{(\frac{1}{2^n}, 0) | n \in I\!N\}$ (Fig. 2) . Then M clearly satisfies the assumptions of Theorem 1, with an approximation (M_k) such that $\chi(M_k) = k$, but the Euler characteristic of M is not defined.

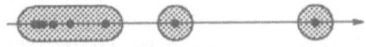

Fig. 2. $M = \{0\} \cup \{(\frac{1}{2^n}, 0) | n \in I\!N\}$

Example 2. In $I\!R^2$, take $M = \cup_{n \in I\!N} S\left((\frac{1}{2^n}, 0), \frac{1}{2^n}\right)$, which is arc-wise connected (Fig. 3). Then M clearly satisfies the assumptions of Theorem 1, with an approximation (M_k) such that $\chi(M_k) = -k$, but the Euler characteristic of M is not defined.

Question 2. Which classes of sets satisfy the assumptions of Theorem 1, with possibly only intrinsic assumptions on the set M?

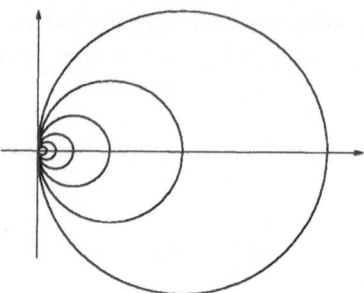

Fig. 3. $M = \cup_{n \in \mathbb{N}} S\left(\left(\frac{1}{2^n}, 0\right), \frac{1}{2^n}\right)$

Since we want that the same topological properties hold for M and M_k (at least to get the equality $\chi(M) = \chi(M_k)$, which is the case in particular if M and M_k are homeomorphic), a natural idea is to write:

$$M = \{x \in \mathbb{R}^n | f_M(x) \leq 0\} \text{ and } M_k = \{x \in \mathbb{R}^n | f_M(x) \leq 1/k\},$$

with a "good" function f_M (say locally Lipschitzian) which also represents the normal cone N_M in the sense that $N_M(x) = \mathrm{cl}\left(\cup_{\lambda \geq 0} \lambda \partial f_M(x)\right)$ (as for $f_M = d_M$).

First try: the distance function d_M. But the following non-degeneracy condition, which seems essential to relate the topological properties of the sets M and M_k:

$$(nd) \quad 0 \notin \partial f(x) \text{ if } f(x) = 0,$$

is never satisfied with $f = d_M$.

Second try: the function:

$$\Delta_M = d_M - d_{\mathbb{R}^n \setminus M}.$$

(introduced and studied in optimization by Hiriart-Urruty [23] and [24]) If the corresponding nondegeneracy condition:

$$0 \notin \partial \Delta_M(x) \text{ if } \Delta_M(x) = 0$$

is satisfied, then the set M is epi-Lipschitzian, in the sense that its Clarke's normal cone $N_M(x)$ is pointed (i.e., if $N_M(x) \cap -N_M(x) = \{0\}$) at every $x \in M$.[6]

---

[6] The above definition is not the original one introduced by Rockafellar [32], but is equivalent to it. Originally, Rockafellar defined epi-Lipschitzian sets as sets that can be locally written as the epigraph of a Lipschitzian function.

In fact, a closed subset M of $I\!\!R^n$ is epi-Lipschitzian if and only if there is a Lipschitzian function $f : I\!\!R^n \to I\!\!R$ such that $M = \{x \in I\!\!R^n | f(x) \leq 0\}$ and such that (nd) $0 \notin \partial f(x)$ if $f(x) = 0$. The sufficient part is classical and the necessary part follows from Proposition 1 below. This justifies that we devote the next section to the epi-Lipschitzian case.

3   The epi-Lipschitzian case

The class of epi-Lipschitzian sets, introduced in optimization by Rockafellar, is of particular importance since it includes both (i) closed convex sets with a nonempty interior, and (ii) sets defined by finite smooth inequality constraints satisfying a nondegeneracy assumption (independence of the gradients of the binding constraints).

3.1   Representation by Δ_M

The function Δ_M provides a first way to "represent" the set M, and N_M, by a Lipschitzian function.

Proposition 1 (Lipschitzian representation, [13]). *Let $M \subset I\!\!R^n$ be closed such that $M \neq \emptyset$ and $M \neq I\!\!R^n$, then the three following conditions are equivalent:*
(i) M is epi-Lipschitzian;
(ii) for every $x \in \operatorname{bd} M$, $0 \notin \partial \Delta_M(x)$;
(iii) Δ_M is a Lipschitzian representation of M, in the following sense:

   (l) Δ_M is Lipschitzian on $I\!\!R^n$;
   (r) $M = \{x \in I\!\!R^n | \Delta_M(x) \leq 0\}$;
   (nr) $N_M(x) = \cup_{\lambda \geq 0} \lambda \partial \Delta_M(x)$ for every $x \in \operatorname{bd} M$;
   (nd) $0 \notin \partial \Delta_M(x)$ if $\Delta_M(x) = 0$.
Moreover, if (i), (ii), (iii) holds, then: $\forall x \in \operatorname{bd} M, \partial \Delta_M(x) \subset \operatorname{co}(N_M(x) \cap S)$.

For the proof of Proposition 1, we refer to [13]. In general, the function Δ_M is **not differentiable**. Thus the sets:

$$M_k = \{x \in I\!\!R^n | \Delta_M(x) \leq 1/k\}$$

may not be smooth, even though they are epi-Lipschitzian if M is epi-Lipschitzian. To recover smoothness, we shall need to "smooth" Δ_M with a convolution type argument.

3.2   Quasi-smooth representation

Theorem 2 (quasi-smooth representation, [13]). *Let $M \subset I\!\!R^n$ be closed, then the two following conditions are equivalent:*
(a) M is epi-Lipschitzian;

(b) there is a Lipschitzian map $f_M : I\!\!R^n \to I\!\!R$ such that:
- (s) f_M is C^∞ on $I\!\!R^n \setminus \mathrm{bd}\,M$;
- (r) $M = \{x \in I\!\!R^n | f_M(x) \le 0\}$ and $f_M^{-1}((-\infty, 1/2]) \subset B(M, 1)$;
- (nd) $0 \notin \partial f_M(x)$ if $f_M(x) = 0$;
- $(\partial - nr)$ $\partial f_M(x) = \partial \Delta_M(x)$ if $f_M(x) = 0$;
- (nr) $N_M(x) = \cup_{\lambda \ge 0} \lambda \partial f_M(x)$ for every $x \in \mathrm{bd}\,M$.

The implication $[(b) \Rightarrow (a)]$ is classical (see Clarke [2]). We now sketch the proof of the implication $[(a) \Rightarrow (b)]$. For the details, we refer to [13].
Idea of the proof of the implication $[(a) \Rightarrow (b)]$: "smoothing" Δ_M.
From Proposition 1, the function Δ_M satisfies all the assertions of (b) but the smoothness one. We now "smooth" Δ_M as follows. Let $\rho : I\!\!R^n \to I\!\!R_+$ be a given C^1 function. We define the function $f_\rho : I\!\!R^n \to I\!\!R$ by:

$$f_\rho(x) = \int_{\overline{B}(0,1)} \theta(t) \Delta_M(x - \rho(x)t) dt, \text{ for every } x \in I\!\!R^n,$$

where θ is a mollifier. A first idea is to take $\rho = |\Delta_M|$, but the corresponding function $f_{|\Delta_M|}$ may not be smooth in general. Then the proof of $[(a) \Rightarrow (b)]$ goes in two steps.

Step 1. We smooth $|\Delta_M|$. Precisely, we show that there is a C^1 function $\rho : I\!\!R^n \to I\!\!R$ such that $0 \le \rho \le |\Delta_M|$, $\|\nabla \rho\| \le |\Delta_M|$, and $\rho(x) = 0$ if and only if $\Delta_M(x) = 0$.

Step 2. We show that $f_M = f_\rho$ satisfies the assertion (b) if in the above step 1, we consider (instead of $|\Delta_M|$) the map $\Delta = \min\{1, \Delta_M^2, \frac{1}{2}|\Delta_M|\}$.

3.3   Smooth normal approximation

Let $M \subset I\!\!R^n$ be compact and epi-Lipschitzian, and let $f_M : I\!\!R^n \to I\!\!R$ be given by Theorem 2. Then the sets:

$$M_k = \{x \in I\!\!R^n | f_M(x) \le 1/k\}$$

define a smooth normal external approximation[7] of M in the following sense (stronger that the one in Theorem 1):
- (sc) $M = \cap_{k \ge 1} M_k$ and $M_{k+1} \subset \mathrm{int}\, M_k$ for every $k \ge 1$;
- (s) M_k is smooth and $M_k \subset B(M, 1)$ for every $k \ge 1$;
- (nc) $\limsup_{k \to \infty} \mathrm{G}(N_{M_k}) \subset \mathrm{G}(N_M)$;
and furthermore they satisfy the following topological property:
- (homeo) M and M_k are homeomorphic for every $k \ge 1$.

---

[7] In the same way, the sets $M^k = \{x \in R^n | f_M(x) \le -1/k\}$ define a smooth normal internal approximation of M. We refer to [14] for a general study of normal approximations of epi-Lipschitzian sets

The function f_M allows us to construct a retraction of M_k on M, by following the vector field $dx/dt = -\nabla f_M(x(t))$, as illustrated in Fig. 4.

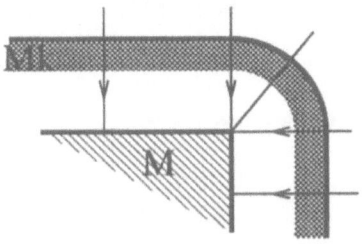

Fig. 4.

By slightly perturbing the vector field ∇f_M, one moreover shows (Bonnisseau and Cornet [4]) that the sets M and M_k are homeomorphic. This result precises previous results by relating the topological properties of M and its approximation.

Other results in the literature

There are stated without the condition (nc), and without topological properties of the approximation (Nečas [29], Massari and Pepe [25], Doktor [20], Benoist [2], Clarke, Ledyaev & Stern [8]).

3.4   A necessary and sufficient condition for the existence of equilibria

In the epi-Lipschitzian case, Theorem 1 can be precised, and we provide a necessary and sufficient condition.

Theorem 3. *Let $M \subset \mathbb{R}^n$ be nonempty, compact and epi-Lipschitzian. Then the assertions $(E; N_M)$ and $(GE; N_M)$ (and to their single-valued versions) are equivalent to the following condition:*

Assertion (χ): *There is a connected component M_i of M such that $\chi(M_i) \neq 0$.*

Proof. **Sufficient part.** Assume first that M is connected; then $\chi(M) \neq 0$. Let (M_k) be a smooth normal approximation of M. Since M_k is homeomorphic to M, $\chi(M_k) \neq 0$ and the assumptions of Theorem 1 are satisfied, hence the assertions $(E; N_M)$ and $(GE; N_M)$ are satisfied. To get the result in the case where M is not connected, one only needs to check that M has a finite number of epi-Lipschitzian connected components and that $\chi(M) = \sum \chi(M_i)$ (see [15]).

Necessary part. It relies on the existence of a smooth internal approximation of M and on the corresponding classical result for smooth sets (see, for example, Milnor [27]). We refer to [13]. $\square$

Remark 2, on Assertion (χ) : Assertion (χ) is satisfied if $\chi(M) \neq 0$ (since $\chi(M) = \sum_{i \in I} \chi(M_i)$), but it cannot be replaced by the assertion $\chi(M) \neq 0$. Take $M = M_1 \cup M_2$, with M_1 and M_2 smooth and connected, $M_1 \cap M_2 = \emptyset$, $\chi(M_1) = 1$, and $\chi(M_2) = -1$. Then $\chi(M) = 0$, and the assertions $(E; N_M)$ and $(GE; N_M)$ hold true.

Other results in the literature

In these results, we assume that $M \subset I\!R^n$ is nonempty, compact and epi-Lipschitzian.

Theorem [Cornet [12]] $(\chi(M) \neq 0) \Rightarrow (E; N_M) \Rightarrow (GE; N_M)$.

Since $(\chi(M) \neq 0) \Rightarrow (\chi)$, it is a consequence of Theorem 3.

Theorem [Clarke, Ledyaev, Stern [7]] *Assume that M is homeomorphic to a compact convex subset of $I\!R^n$. Then Assertion $(E; N_M)$ holds true.*

This result is a consequence of Theorem 3 and of the fact that $\chi(M) = 1$ if M is homeomorphic to a compact convex subset of $I\!R^n$. [7] shows another existence result in the more general setting of infinite dimension.

4   The proximally nondegenerate case

In this section, we introduce a class of sets, wider than the class of epi-Lipschitzian sets, which also satisfies the assumptions of Theorem 1, with intrinsic conditions on the set M. For the proofs of the results contained in this section, we refer to [16] and [18].

4.1   Definition and examples

Noticing that, at points $x \in \operatorname{bd} M$, Clarke's subdifferential of the distance function d_M is always "too big", since:

$$0 \in \partial d_M(x),$$

we define for $x \in \operatorname{bd} M$ (this notion can be defined on $I\!R^n$, see [18]) the set:

$$\partial_+ d_M(x) = \limsup_{x' \to x, \ x' \notin M} \partial d_M(x'),$$

which is clearly contained in $\partial d_M(x)$ (from the u.s.c. of the correspondence ∂d_M). The following proposition gives a characterization of epi-Lipschitzian sets in terms of this new notion.

Proposition 2 ([18]). *A closed subset M of $I\!R^n$ is epi-Lipschitzian if and only if:*

$$0 \notin \operatorname{co} \partial_+ d_M(x), \text{ for every } x \in \operatorname{bd} M.$$

It is then natural to extend the class of epi-Lipschitzian sets by replacing the set $\operatorname{co} \partial_+ d_M(x)$ with the set $\partial_+ d_M(x)$ as it is done in the following definition.

Definition 1 ([18]). A closed subset M of $I\!R^n$ is proximally nondegenerate if, at every $x \in M$, one of the following equivalent assertions is satisfied:

$$0 \notin \partial_+ d_M(x); \tag{1a}$$

$$\exists \alpha > 0, \forall x' \in B(x, \alpha) \setminus M, \forall p \in \partial d_M(x'), \|p\| \geq \alpha. \tag{1b}$$

The class of proximally nondegenerate sets is very broad, as shown by the following proposition.

Proposition 3 ([18]). *A closed set $M \subset I\!R^n$ is proximally nondegenerate if it satisfies one of the following conditions:*

(i) M is convex;
(ii) $I\!R^n \setminus M$ is convex;
(iii) M is a C^1 submanifold of $I\!R^n$, with or without a boundary, with or without corners;
(iv) M is epi-Lipschitzian;
(v) M is proximally smooth (Clarke, Stern, and Wolenski [9]);
(vi) M is proximally regular (Poliquin, Rockafellar, Thibault, [30] and [31])

4.2   Quasi-smooth representation and smooth approximation

We now smooth the distance function to a set M (in the same way as we did for Δ_M, in Theorem 2), to get a quasi-smooth representation of the set M, as precisely stated in the following result.

Theorem 4 ([16]). *Let M be a closed subset of $I\!R^n$, then there is a Lipschitzian function $f_M : I\!R^n \to I\!R_+$ such that:*

(a) f_M is C^∞ on $I\!R^n \setminus M$;
(b) $M = \{x \in I\!R^n | f_M(x) = 0\}$, and $f_M^{-1}((0, 1/2]) \subset B(\operatorname{bd} M, 1)$;
(c) $\forall x \notin M, \|\nabla f_M(x)\| \geq (1/2) \min\{\|v\| \, | v \in \partial d_M(\overline{B}(x, d_M(x)/2))\}$;
(d) $\forall x \in M, \partial_+ f_M(x) \subset \partial_+ d_M(x)$.

Let $M \subset I\!R^n$ be compact and proximally nondegenerate, and let $f_M : I\!R^n \to I\!R$ be given by Theorem 4. Then the sets:

$$M_k = \{x \in I\!R^n | f_M(x) \leq 1/k\}$$

define a smooth normal external approximation of M in the same sense as before, and they satisfy the weaker topological property:

(ret) for every $k \geq 1$, M is a deformation retract of M_k, i.e., there is a continuous map $H : [0,1] \times M_k \rightarrow M_k$ such that, for every $x \in M_k$, $H(0,x) = x$, and $H(1,x) \in M$, and, for every $x \in M$, $H(1,x) = x$. We note that in general, the sets M and M_k are not homeomorphic.

4.3   Existence of equilibria

In the proximally nondegenerate case, Theorem 1 can be precised and we get the following result, stated intrinsicly in terms of the set M.

Theorem 5 ([18]). *Let $M \subset \mathbb{R}^n$ be nonempty, compact, proximally nondegenerate, and assume* **Assertion** *(χ). Then the assertions $(E; N_M)$, $(GE; N_M)$, and $(GE; \widetilde{N}_M)$ hold, where $\widetilde{N}_M(x) = \cup_{\lambda \geq 0} \lambda \partial_+ d_M(x)$ for every $x \in M$.*

We point out that the cone $\widetilde{N}_M(x)$ is smaller than Clarke's normal cone $N_M(x)$, and may not be convex (Fig.5).

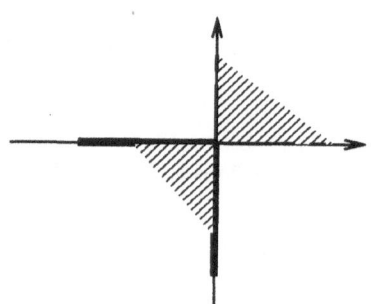

Fig. 5. $M = [-1,0] \times \{0\} \cup \{0\} \times [-1,0]$, $\widetilde{N}_M(0) = \mathbb{R}^2_+ \cup \mathbb{R}^2_-$, $N_M(0) = \mathbb{R}^2$.

Proof. Assume first that M is connected; then $\chi(M) \neq 0$. Let M_k be a smooth normal approximation of M. Since M is a deformation retract of M_k, $\chi(M_k) = \chi(M) \neq 0$ and the assumptions of Theorem 1 are satisfied, hence the assertions $(E; N_M)$ and $(GE; N_M)$ are satisfied. Assertion $(GE; \widetilde{N}_M)$ is a consequence of the inclusion:

$$\limsup_{k \to \infty} \mathrm{G}(N_{M_k}) \subset \mathrm{G}(\widetilde{N}_M).$$

To get the result in the case where M is not connected, one only need to check that M has a finite number of connected component and that $\chi(M) = \sum \chi(M_i)$ (see [18]).   □

Remark 3. Theorem 5 may no longer be true if the set M is not proximally nondegenerate. Consider the following connected continuous submanifold with a boundary, illustrated in Figure 6:

$$M = \left\{(x,y) \in I\!\!R^2 \middle| \begin{array}{l} |x| \geq 1/4, \ (x - (1/4)\mathrm{sgn}\,x)^2 + (y - \mathrm{sgn}\,y)^2 \in [1/4, 1], \\ \text{or } |x| < 1/4, \ y \in [-2, -3/2] \cup [-1/2, 1/2] \cup [3/2, 2] \end{array}\right\}.$$

Assertion (χ) holds true, since $\chi(M) = -1$. But, since there is a continuous map $f : M \to I\!\!R^2$, such that $f(x) \in T_M(x) \setminus \{0\}$ for every $x \in M$, the assertions $(GE; N_M)$ and $(E; N_M)$ do not hold.

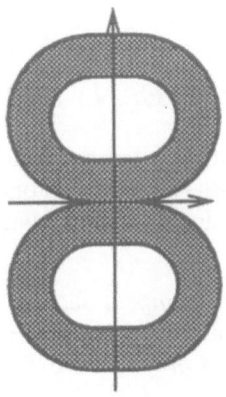

Fig. 6.

Remark 4. In Theorem 5, Assertion (χ) is sufficient, but not necessary, for the assertions $(GE; N_M)$ and $(E; N_M)$ to hold. In $I\!\!R^2$, consider the square:

$$M = \{(x,y) \in I\!\!R^2 \,|\, \sup\{|x|, |y|\} = 1\}.$$

Then M is proximally nondegenerate. Assertion (χ) does not hold, since M is connected and $\chi(M) = 0$. the assertions $(GE; N_M)$ and $(E; N_M)$ hold, since $N_M(1, 1) = I\!\!R^2$ and $T_M(1, 1) = \{0\}$.

Other results in the literature

Ben-El-Mechaiekh and Kryszewski [1] show the implication $[(\chi(M) \neq 0) \Rightarrow (E; N_M)]$ in infinite dimension, for the wider class of $\mathcal{L}$-retracts. Ćwiszewski and Kryszewski [19] show the implication $[(\chi(M) \neq 0) \Rightarrow (GE; N_M)]$ for the class of $\mathcal{L}$-retracts. Both papers use the Čech cohomology and their key-tool is Lefschetz fixed-point theorem.

Remark 5. One could think of enlarging the class of proximally nondegenerate sets by considering the class $\mathcal{M}$ of closed subsets M of $I\!\!R^n$ such that:

$$\forall x \in M, \exists \alpha > 0, \forall x' \in B(x, \alpha) \setminus M, 0 \notin \partial d_M(x') \ (1a').$$

But Theorem 5 may not be true for sets in $\mathcal{M}$. In $I\!\!R^2$, consider the spiral defined by:

$$M = S(0,1) \cup \{((1+1/\theta)\cos\theta, (1+1/\theta)\sin\theta)|\theta \in [3,\infty)\},$$

and the tangent field $f : M \to I\!\!R^2$ defined by $f((1+1/\theta)\cos\theta, (1+1/\theta)\sin\theta) = (-\sin\theta - \cos\theta/\theta^2, \cos\theta - \sin\theta/\theta^2)$ and $f(\cos\theta, \sin\theta) = (-\sin\theta, \cos\theta)$. Then $M \in \mathcal{M}$, $\chi(M) = 1$, and $f(x) \in T_M(x) \setminus \{0\}$ for every $x \in M$, hence none of the assertions $(GE; \widetilde{N}_M)$, $(GE; N_M)$, $(E; N_M)$ is satisfied. Notice also that M is not a neighborhood retract.

5 Relations between $(GE; N)$ and $(E; N)$

In this section, we state and prove a lemma which is a keytool in the proof of Theorem 1. This lemma precises the relations between the assertions $(GE; N)$ and $(E; N)$, and their single-valued versions, denoted respectively $(GE_{sv}; N)$, and $(E_{sv}; N)$.

Lemma 1. Let $M \subset I\!\!R^n$ be a compact set, let N be a correspondence from M to $I\!\!R^n$, such that $N(x)$ is a cone for every $x \in M$:
(a) assume that N has a closed graph, then:

$$(GE_{sv}; N) \Leftrightarrow (GE; N) \Rightarrow (E; N) \Rightarrow (E_{sv}; N);$$

(b) additionally assume that N has convex values, then:

$$(GE; N) \Leftrightarrow (GE_{sv}; N) \Leftrightarrow (E; N) \Leftrightarrow (E_{sv}; N).$$

Remark 6. If the correspondence N does not have convex values, then the implication $[(E; N) \Rightarrow (GE; N)]$. See Remarks 10 and 11 below, with $N = \widehat{N}_M$, the limiting normal cone.

Remark 7. If $\operatorname{co} N$ does not have a closed graph (even if N has a closed graph), the implication $[(E_{sv}; N) \Rightarrow (E; N)](iv)$ may not be true. In $I\!\!R^3$, consider $M = M_1 \cup M_2 \cup \{0\}$, with $M_1 = (0,1] \times \{0\} \times \{0\}$, $M_2 = \{0\} \times (0,1] \times \{0\}$, define the correspondence N by $N(0) = I\!\!R(0,0,1)$, $N((x,0,0)) = I\!\!R(x,0,1) \cup I\!\!R(-x,0,1)$ for $(x,0,0) \in M_1$ and $N((0,y,0)) = I\!\!R(0,y,1) \cup I\!\!R(0,-y,1)$ for $(0,y,0) \in M_2$. Then $(E_{sv}; N)$ holds, but $(E; N)$ does not hold (consider the correspondence F defined by $F(0) = \operatorname{co}\{(0,1,0),(1,0,0)\}$, $F(x) = (0,1,0)$ for $x \in M_1$ and $F(x) = (1,0,0)$ for $x \in M_2$.

Before proving Lemma 1, we shall need a claim, the proof of which is (more or less) classical , and is left to the reader.

Claim (1). Let F be a correspondence from M to $\mathbb{R}^n$, and let T be a l.s.c. correspondence from M to $\mathbb{R}^n$, with convex values, such that $F(x) \cap T(x) \neq \emptyset$ for every $x \in M$. We let, for $x \in M$ and $k \in \mathbb{N} \setminus \{0\}$:

$$F_k(x) = \operatorname{co} B\Big(F(B(x, 1/k) \cap M), 1/k\Big).$$

(a) The correspondence $F_k \cap T$ is l.s.c., with nonempty convex values, hence admits a continuous selection (from Michael's selection theorem).

(b) Let (x_k, y_k) be a sequence in $M \times \mathbb{R}^n$ converging to $(x, y) \in M \times \mathbb{R}^n$, such that $y_k \in F_k(x_k)$ for every k. Assume that F is bounded on a neighborhood of x, then $(x, y) \in \operatorname{co}(\limsup(F)(x))$

We now come back to the proof of Lemma 1.

Proof. Assertion (c) is a clear consequence of the assertions (a) and (b).

Part (a). The implications $[(GE; N) \Rightarrow (GE_{sv}; N)]$ and $[(E; N) \Rightarrow (E_{sv}; N)]$ are immediate.

$[(GE_{sv}; N) \Rightarrow (GE; N)]$. Let F be a u.s.c. correspondence from M to $\mathbb{R}^n$, with convex compact values. For $k \in \mathbb{N} \setminus \{0\}$, the correspondence F_k is defined as above. From Claim 1 (taking $T(x) = \mathbb{R}^n$), the correspondence F_k admits a continuous selection $f_k : M \to \mathbb{R}^n$, i.e., such that:

$$\forall x \in M, f_k(x) \in F_k(x).$$

From $(GE_{sv}; N)$ there is $x_k \in M$ such that $f_k(x_k) \in N(x_k)$. Without any loss of generality, we may assume that the sequence $(x_k, f_k(x_k))$ converges to some (x^*, y^*) in the compact set $M \times \operatorname{cl} \operatorname{co} B(F(M), 1)$. Since the correspondence N has a closed graph, we get that $y^* \in N(x^*)$. Since F is u.s.c., with compact convex values, applying Claim 1 to the sequence $(x_k, f_k(x_k))$, we get that $y^* \in \operatorname{co}(\limsup F)(x^*) \subset F(x^*)$, hence that $0 \in F(x^*) - N(x^*)$, i.e., x^* is a generalized equilibrium of F, hence $(GE; N)$ holds..

$[(GE_{sv}; N) \Rightarrow (E; N)]$. Let F be an u.s.c. correspondence from M to $\mathbb{R}^n$, with convex compact values, such that $F(x) \cap N(x)^\circ \neq \emptyset$ for every $x \in M$. For $k \in \mathbb{N} \setminus \{0\}$, the correspondence F_k is defined as above, and we let, for every $x \in M$:

$$T_k(x) = \{y \in \mathbb{R}^n | \forall p \in N(\dot{x}) \cap S, (y|p) < 1/k\}.$$

Noticing that, for all x, $\emptyset \neq F(x) \cap N(x)^\circ \subset F(x) \cap T_k(x)$, and that the correspondence T_k has convex values, and is lower semicontinuous, since the set $\{x \in M | y \in T_k(x)\}$ is open in M (for its relative topology), for every $y \in \mathbb{R}^n$. From Claim 1, the correspondence $F_k \cap T_k$ admits a continuous selection $f_k : M \to \mathbb{R}^n$, i.e., such that:

$$\forall x \in M, f_k(x) \in F_k(x) \cap T_k(x).$$

From $(GE_{sv}; N)$ there is $x_k \in M$ such that $0 \in f_k(x_k) - N(x_k)$. Without any loss of generality, we may assume that the sequence (x_k) converges to some element $x^* \in M$, and we prove that $(f_k(x_k))$ converges to 0. Indeed, $f_k(x_k) \in N(x_k) \cap T_k(x_k)$, hence $\|f_k(x_k)\| < 1/k$. Applying Claim 1 to the sequence $(x_k, f_k(x_k))$, we get that $0 \in \operatorname{co}(\limsup F)(x^*)) \subset F(x^*)$, i.e., x^* is an equilibrium of F, hence $(E; N)$ holds.

Part (b). $[(E_{sv}; N) \Rightarrow (E; N)]$. Let F be an u.s.c. correspondence from M to $I\!\!R^n$, with convex compact values, such that $F(x) \cap N(x)^\circ \neq \emptyset$ for every $x \in M$. For $k \in I\!\!N \setminus \{0\}$, the correspondence F_k is defined as above. Since the correspondence N has a closed graph with convex values, one easily shows that the correspondence $x \mapsto N(x)^\circ$ is l.s.c.. Hence, from Claim 1, the correspondence $F_k \cap N^\circ$ admits a continuous selection $f_k : M \to I\!\!R^n$, i.e., such that:

$$\forall x \in M, f_k(x) \in F_k(x) \cap N(x)^\circ.$$

From $(E_{sv}; N)$ there is $x_k \in M$ such that $f_k(x_k) = 0$. Without any loss of generality, we may assume that the sequence (x_k) converges to some element $x^* \in M$. Applying Claim 1 to the sequence $(x_k, f_k(x_k)) = (x_k, 0)$, we get that $0 \in \operatorname{co}(\limsup F)(x^*) \subset F(x^*)$, i.e., x^* is an equilibrium of F, hence $(E; N)$ holds.

$[(E; N) \Rightarrow (GE; N)]$. Let F be an u.s.c. correspondence from M to $I\!\!R^n$, with nonempty convex compact values, then F is bounded on the compact set M, i.e., there exists $k > 0$ such that $F(x) \subset \overline{B}(0, k)$ for every $x \in M$. We define the correspondence F from M to $I\!\!R^n$ by:

$$\Phi(x) = F(x) - N(x) \cap \overline{B}(0, k).$$

The correspondence Φ is clearly u.s.c. with nonempty convex compact values and we now show that:

$$\forall x \in M, \quad \Phi(x) \cap N(x)^\circ \neq \emptyset.$$

Indeed, let $x \in M$ and let $y \in F(x)$. Since $N(x)$ is a closed convex cone, we recall that there exist a unique element $y_N \in N(x)$ and a unique element $y_T \in N(x)^\circ$ such that $y = y_N + y_T$ and $(y_N | y_T) = 0$. Consequently, the element $y_T = y - y_N$ belongs to $\Phi(x)$ since $\|y_N\|^2 \leq \|y\|^2 \leq k$. Consequently, from Assertion $(E; N)$, there exists $x^* \in M$ such that $0 \in \Phi(x^*) \subset (x^*) - N(x^*)$, i.e., x^* is a generalized equilibrium of F, hence $(GE; N)$ holds. $\square$

Remark 8, on Theorem 3. In the epi-Lipschitzian case, since the correspondence N_M has a closed graph and convex values, Lemma 1 directly (without using Theorem 1) gives the equivalence:

$$(E; N_M) \Leftrightarrow (GE; N_M).$$

Remark 9, on Theorem 5. In the proximally nondegenerate case, the assertions $(GE; N_M)$ and $(E; N_M)$ are both direct consequences of the assertion $(GE; \widetilde{N}_M)$, using Lemma 1. Indeed, the correspondence $\widetilde{N}_M$ has a closed graph (but, in general, not the correspondence N_M!), hence Lemma 1 gives the implication:

$$(GE; \widetilde{N}_M) \Rightarrow (E; \widetilde{N}_M),$$

and we notice the equivalence $[(E; \widetilde{N}_M) \Leftrightarrow (E; N_M)]$, since $\mathrm{cl\,co}\,\widetilde{N}_M(x) = N_M(x)$. The last implication $[(GE; \widetilde{N}_M) \Rightarrow (GE; N_M)]$ is a direct consequence of the inclusion $\widetilde{N}_M(x) \subset N_M(x)$.

6   Other notions of normal cones

Up to now, we have considered the two cases $N = N_M$ (Clarke's normal cone) and $N = \widetilde{N}_M$. The following remarks discuss the existence problem of equilibria and generalized equilibria, when one considers other notions of normal cones such as $N = \widehat{N}_M$, the limiting normal cone [8] and $N = N_M^B$, Bouligand normal cone. [9]

The case $N = \widehat{N}_M$, the limiting normal cone

Remark 10 (generalized equilibria). The implication $[(\chi) \Rightarrow (GE; \widehat{N}_M)]$ may not be true, even if the set M is compact and epi-Lipschitzian. In $I\!\!R^3$, consider the counterexample in Clarke, Ledyaev and Stern [7], $\chi(M) = 1$ (for example, M is homeomorphic to a convex set).

Remark 11 (equilibria). However, the implication $[(\chi) \Rightarrow (E; \widehat{N}_M)]$ holds under the assumption of Theorem 5, since $\mathrm{cl\,co}\,\widehat{N}_M(x) = N_M(x)$, which implies the equivalence $[(E; \widehat{N}_M) \Leftrightarrow (E; N_M)]$.

The case $N = N_M^B$, Bouligand normal cone

Remark 12 (generalized equilibria). The implication $[(\chi) \Rightarrow (GE; N_M^B)]$ may not be true, even if the set M is compact and epi-Lipschitzian. Consider the counterexample in [7] and note that $N_M^B(0) = \{0\}$.

---

[8] We recall that the limiting normal cone to M at x is defined by $\widehat{N}_M(x) = \cup_{\lambda \geq 0} \lambda \nabla_+ d_M(x) \cup \{0\}$.

[9] We recall that Bouligand normal cone to M at x is defined by $N_M^B(x) = T_M^B(x)^\circ$, where $T_M^B(x) = \{v \in I\!\!R^n | \exists (\lambda_k)_{k \in N}, \lambda_k > 0, \exists (y_k)_{k \in I\!\!N}, y_k \in M, y_k \to x, v = \lim_{k \to \infty} \lambda_k(y_k - x)\}$.

Remark 13 (equilibria in the multi-valued case). The implication $[(\chi) \Rightarrow (E; N^B_M)]$ may not be true, even if the set M is compact and epi-Lipschitzian. Consider the set M in the counterexample in [7].

Remark 14 (equilibria in the single-valued case). The implication $[(\chi) \Rightarrow (E_{sv}; N^B_M)]$ holds under the assumptions of Theorem 5. Indeed, let M be a closed subset of $\mathbb{R}^n$ and let $f : M \to \mathbb{R}^n$ be a continuous map; then from [11], the two following assertions are equivalent:
(i) $f(x) \in T^B_M(x)$ for every $x \in M$;
(ii) $f(x) \in T_M(x)$ for every $x \in M$.
Hence, the equivalence $[(E_{sv}; N_M) \Leftrightarrow (E_{sv}; N^B_M)]$ holds.

The case of the upper-limit cone $\limsup N_{M_k}$

Let M be a closed subset of $\mathbb{R}^n$ and let (M_k) be a sequence of closed subsets of $\mathbb{R}^n$ such that $M = \limsup_{k\to\infty} M_k$.

Remark 15. Assume that, for every k, the assertion $(GE; N_{M_k})$ holds. Then the assertions $(GE; \limsup N_{M_k})$ and $(E; \limsup N_{M_k})$ hold. If additionally $G(\limsup N_{M_k}) \subset G(N_M)$ (resp. $G(\limsup N_{M_k}) \subset G(\widehat{N}_M)$), then $(GE; N)$ and $(E; N)$ hold (resp. $(GE; \widehat{N})$).

Remark 16. Conversely, notice that (see [14] and [2]):

$$G(\widehat{N}) \subset G(\limsup N_{M_k});$$
$$G(N) \subset G(\mathrm{clco} \limsup N_{M_k}),\,^{10}$$

Then we get the implications $[(GE; \widehat{N}_M) \Rightarrow (GE; \limsup N_{M_k})]$ and $[(E; N_M) \Rightarrow (E; \limsup N_{M_k})]$. The converse of these implications may not be true. Consider, in $\mathbb{R}^2$, $M = \overline{B}(0,1) \setminus B(0,1/2)$ and $M_k = M \cup \overline{B}((0, 1 + 1 + \frac{1}{2^k}), \frac{1}{2^{k+1}})$.

References

1. Ben-El-Mechaiekh, H., Kryszewski, W. (1997) Equilibria of set-valued maps on nonconvex domains. Trans. Amer. Math. Soc. **349**, 10, 4159–4179 (French abridged version in: C. R. Acad. Sci. Sér. 1 **320**, 573–576, (1995))
2. Benoist, J. (1994) Approximation and regularization of arbitrary sets in finite dimension. Set Valued Anal. **2**, 1-2, 95–115
3. Bergman, G., Halpern, B. (1968) A fixed point theorem for inward and outward maps. Trans. Amer. Math. Soc. **130**, 353–358

---

[10] The correspondence $\mathrm{clco} \limsup N_{M_k}$ is defined by: $(\mathrm{clco} \limsup N_{M_k})(x) = \mathrm{cl}(\mathrm{co}(\limsup N_{M_k}(x)))$.

4. Bonnisseau, J.-M., Cornet, B. (1990) Fixed-point theorem and Morse's lemma for Lipschitzian functions. J. Math. Anal. Appl. **146**, 2, 318–322
5. Cellina, A. (1969) A theorem on the approximation of compact multi-valued mappings. Atti Accad. Naz. Lincei Cl. Sci. Fis., Mat. Nat. Rend. (8) **47**, 429–433
6. Clarke, F.H. (1983) Optimization and nonsmooth analysis. John Wiley, New York. (Reprinted as Vol. 5 of the series Classics in Applied Mathematics, SIAM, Philadelphia (1989))
7. Clarke, F.H., Ledyaev, Y.S., Stern, R.J. (1995) Fixed points and equilibria in nonconvex sets. Nonlinear Analysis **25**, 145–161
8. Clarke, F.H., Ledyaev, Y.S., Stern, R.J. (1997) Complements, approximations, smoothing and invariance properties. J. Convex Anal. **2**, 189–219
9. Clarke, F.H., Stern, R.J., Wolenski, P.R. (1995) Proximal smoothness and the lower-C^2 property. J. Convex Anal. **2**, 117–144
10. Cornet, B. (1995) Paris avec handicap et théorèmes de surjectivité de correspondances. C.R. Acad. Sci. Paris Sér. 1 **281**, 479–482
11. Cornet, B. (1981) Contributions à la théorie mathématique des mécanismes dynamiques d'allocation des ressources, Thèse d'Etat, Université de Paris 9-Dauphine
12. Cornet, B. (1992) Euler characteristics and fixed-point theorems, manuscript
13. Cornet, B., Czarnecki, M.-0. (1999) Smooth representations of epi-Lipschitzian subsets of $I\!R^n$, Nonlinear Anal. **37**, 139–160 (French abridged version in: C. R. Acad. Sci. Sér. 1 **325**, 475–480, (1997))
14. Cornet, B., Czarnecki, M.-O. (1998) Approximation of epi-Lipschitzian subsets of $I\!R^n$, SIAM J. Control Optim. **37**, 710–730 (French abridged version in: C. R. Acad. Sci. Sér. 1 **325**, 583–588, (1997))
15. Cornet, B., Czarnecki, M.-O. (1999) Existence of (generalized) equilibria: necessary and sufficient conditions, Commun. Appl. Nonlinear Anal. **6**, (previous version in Cahier Eco-Maths no 95-55, Université de Paris 1)
16. Cornet, B., Czarnecki, M.-O. (1998) Smoothing the distance function to a closed subset of $I\!R^n$ and applications, manuscript
17. Cornet, B., Czarnecki, M.-O. (1998) Proximally nondegenerate subsets of $I\!R^n$, manuscript
18. Cornet, B., Czarnecki, M.-O. (1998) Existence of generalized equilibria, to appear in Nonlinear Anal. (French abridged version in: C. R. Acad. Sci. **327**, Série 1, 917–922)
19. Ćwiszewski, A., Kryszewski, W. (1998), Equilibria of set-valued maps: a variational approach, manuscript
20. Doktor, P. (1976) Approximation of domains with Lipschitzian boundary. Casopsis Pest. Mat. **101**, 3, 237–255
21. Dold, A. (1972) Lectures on algebraic topology. Springer-Verlag, Berlin, Heidelberg
22. Dugundji, J. (1966) Topology. Allyn and Bacon, Boston
23. Hiriart-Urruty, J.-B. (1979) New concepts in nondifferentiable programming. Bull. Soc. Math. France **60**, 57–85
24. Hiriart-Urruty, J.-B. (1979) Tangent cones, generalized gradients and mathematical programming in Banach spaces. Math. Oper. Res. **4**, 1, 79–97

25. Massari, U., Pepe, L. (1974) Sull'approssimazione degli aperti lipschitziani di $I\!R^n$ con varietà differenziabili. Boll. U.M.I. , Ser. 4, **10**, 532–544
26. Michael, E. (1956) Continuous selections I. Ann. of Math. (2) **63**, 361–382
27. Milnor, J. (1963) Morse theory. Annal of Mathematics Studies. Princeton Univ. Press, Princeton
28. Milnor, J. (1965) Topology from the differentiable viewpoint. University Press of Virginia, Charlottesville
29. Nečas, J. (1962) On type $\mathcal{M}$ domains (in Russian). Czechoslovak Math. J. **12**, 87, 274–287.
30. Poliquin, R. A., Rockafellar, R. T. (1996) Prox-regular functions in variational analysis. Trans. Amer. Math. Soc. **348**, 1805–1838
31. Poliquin, R. A., Rockafellar, R. T., Thibault, L. (1998) Local differentiability of distance functions, to appear in Trans. Amer. Math. Soc.
32. Rockafellar, R.T. (1979) Clarke's tangent cones and the boundaries of closed sets in $I\!R^n$. Nonlinear Anal. **3**, 145–154

Eigenvalue Localization for Multivalued Operators[*]

Rafael Correa[1] and Alberto Seeger[2]

[1] Universidad de Chile, Facultad de Ciencias Físicas y Matemáticas,
Casilla 2777/3-Correo 3, Santiago, Chile
[2] University of Avignon, Department of Mathematics, 33, rue Louis Pasteur,
84000 Avignon, France

Abstract. Let $(H, \langle \cdot, \cdot \rangle)$ be a Hilbert space, and $F : H \to H$ be an operator with closed convex values. Denote by $\sigma(F)$ the set of all (real) eigenvalues of F. As shown in this note, if F satisfies a so-called "upper-normality" assumption, then it is possible to derive a simple variational formula for the supremum of $\sigma(F)$. Lower-normality of F yields, of course, an analogous formula for the infimum of $\sigma(F)$.

Key words: eigenvalue localization, multivalued operator, variational formula

Mathematics Subject Classification (1991): 47H04, 47H12, 58C40

1 Introduction and Preliminary Results

The spectral theory of linear operators can be extended to a much broader setting. For a multivalued operator $F : H \to H$ on a Hilbert space $(H, \langle \cdot, \cdot \rangle)$, eigenvalues and eigenvectors are defined in the following natural way:

Definition 1. The number $\lambda \in \mathbb{R}$ is an *eigenvalue* of $F : H \to H$ if the set

$$P_F(\lambda) := \{x \in H : \lambda x \in F(x)\}$$

contains a nonzero vector. The term *eigenvector* of $F : H \to H$ refers to a nonzero vector $x \in H$ such that

$$\Lambda_F(x) := \{\lambda \in \mathbb{R} : \lambda x \in F(x)\}$$

is nonempty. The *point spectrum* of F is the set

$$\sigma(F) := \sqcup \{\Lambda_F(x) : x \in H \setminus \{0\}\}$$

of all eigenvalues of F.

It is clear that the mappings $P_F : \mathbb{R} \to H$ and $\Lambda_F : H \to \mathbb{R}$ are inverse to each other, in the sense that

$$x \in P_F(\lambda) \quad \Leftrightarrow \quad \lambda \in \Lambda_F(x).$$

[*] Supported by Fondecyt Project 1961107

The topological and convexity properties of these mappings depend, of course, on the assumptions made on F. To warm up and introduce some notation, we record below some basic facts. Recall that a set Q in a linear space is said to be a cone if $tQ \subset Q$ for all $t > 0$.

Proposition 1.
(a) If F is closed-valued (respectively, convex-valued), then Λ_F is closed-valued (respectively, convex-valued);
(b) if $GrF := \{(x,y) \in H \times H : y \in F(x)\}$ is a closed set, then $GrP_F := \{(\lambda,x) \in \mathbb{R} \times H : x \in P_F(\lambda)\}$ and $Gr\Lambda_F := \{(x,\lambda) \in H \times \mathbb{R} : \lambda \in \Lambda_F(x)\}$ are also closed sets;
(c) if GrF is a convex set, then P_F is convex-valued;
(d) if GrF a cone, then P_F is conic-valued (i.e. $\forall \lambda \in \mathbb{R}, \quad P_F(\lambda)$ is a cone).

Proof. All these facts can be proven in a straightforward manner. The conditions (c) and (d) appear already in Seeger [6].   □

Spectral results for special classes of multivalued operators can be found already in the literature. For instance, Aubin, Frankowska and Olech [2], Aubin and Frankowska [1], and Leizarowitz [4] discuss the case of closed convex processes. Seeger [7] studies the spectral properties of equilibrium processes defined by linear complementarity conditions.

The purpose of this paper is to estimate the least upper bound

$$\sup \sigma(F) := \sup\{\lambda \in \mathbb{R} : \lambda \in \sigma(F)\}$$

of the point spectrum $\sigma(F)$. The greatest lower bound

$$\inf \sigma(F) := \inf\{\lambda \in \mathbb{R} : \lambda \in \sigma(F)\}$$

can be worked-out in a similar way, and therefore the details of this case are omitted. Our approach relies on the "nested maximization principle"

$$\sup \sigma(F) = \sup_{x \neq 0} \sup \Lambda_F(x) = \sup_{x \neq 0} \sup_{\lambda \in \Lambda_F(x)} \lambda, \tag{1}$$

in which the chief role is played by the inner problem

$$\text{maximize } \{\lambda : \lambda \in \Lambda_F(x)\}.$$

Most of the particular examples found in practice show that, for each fixed $x \neq 0$, the feasible set $\Lambda_F(x)$ has a fairly simple structure. For instance, if F has closed convex values, then each $\Lambda_F(x)$ is a closed interval (cf. Proposition 1). This very favorable case will be explored in detail in the next section.

2   Eigenvalue Localization

Throughout this section it is assumed that $F : H \rightarrow H$ has closed convex values. Under this assumption, the set $\Lambda_F(x)$ is entirely determined by its support function

$$\alpha \in I\!\!R \mapsto s[\alpha, \Lambda_F(x)] := \sup\{\lambda\alpha : \lambda \in \Lambda_F(x)\}.$$

In turns, $s[\cdot, \Lambda_F(x)]$ can be expressed in terms of the support function

$$q \in H \mapsto s[q, F(x)] := \sup\{\langle y, q \rangle : y \in F(x)\}$$

of the closed convex set $F(x)$.

Proposition 2. *Let* $F : H \rightarrow H$ *be a multivalued operator with closed convex values. If* $x \in H \setminus \{0\}$ *is an eigenvector of* F, *then the equality*

$$s[\alpha, \Lambda_F(x)] = \inf_{\langle x,q \rangle = \alpha} s[q, F(x)] \tag{2}$$

holds for any $\alpha \in I\!\!R$.

Proof. For notational convenience, it is helpful to introduce the function

$$\alpha \in I\!\!R \mapsto v(\alpha) := \inf_{\langle x,q \rangle = \alpha} s[q, F(x)]. \tag{3}$$

A standard rule of convex analysis (cf. [5], Theorem 16.3) allows us to obtain the explicit formula

$$v^*(\lambda) = \Psi[\lambda x, F(x)] := \begin{cases} 0 & \text{if } \lambda x \in F(x), \\ +\infty & \text{if } \lambda x \notin F(x) \end{cases}$$

for the Legendre-Fenchel conjugate

$$\lambda \in I\!\!R \mapsto v^*(\lambda) := \sup_{\alpha \in I\!\!R}\{\lambda\alpha - v(\alpha)\}$$

of v. Thus the conjugate v^{**} of v^* takes the form

$$v^{**}(\alpha) = \sup_{\lambda \in I\!\!R}\{\lambda\alpha - v^*(\lambda)\} = \sup\{\lambda\alpha : \lambda x \in F(x)\} \quad \forall \alpha \in I\!\!R.$$

In short, $v^{**} = s[\cdot, \Lambda_F(x)]$. To complete the proof of the proposition, it remains to show that $v^{**} = v$. Such an equality holds if the function v happens to be convex, proper, and lower-semicontinuous (cf. [3], p. 97). The convexity of v can be checked in a direct way, so we just focuss on the two other properties. Pick up any $\lambda \in I\!\!R$ in $\Lambda_F(x)$. Since

$$\alpha\lambda \leq s[\alpha, \Lambda_F(x)] = v^{**}(\alpha) \leq v(\alpha) \quad \forall \alpha \in I\!\!R$$

and

$$v(0) = \inf_{\langle x,q \rangle =0} s[q, F(x)] \leq s[0, F(x)] \leq 0,$$

it follows that v is a proper convex function. As a matter of fact, one has

$$v(\alpha) = \begin{cases} \alpha v(1) & \text{if } \alpha > 0, \\ 0 & \text{if } \alpha = 0, \\ -\alpha v(-1) & \text{if } \alpha < 0, \end{cases} \tag{4}$$

with $v(1) \neq -\infty$ and $v(-1) \neq -\infty$. The positive homogeneity condition (3) implies not only the properness, but also the lower-semicontinuity of the function v. □

The formula obtained in Proposition 2 can be exploited in many different ways. For instance, one can estimate the extrema of the interval $\Lambda_F(x)$. In what follows, we use also the notation

$$r[q, F(x)] := \inf\{\langle y, q \rangle : y \in F(x)\}.$$

Theorem 1. *Let $F : H \to H$ be a multivalued operator with closed convex values. Let $x \in H \setminus \{0\}$ be an eigenvector of F. Then, one has the estimates*

$$\sup \Lambda_F(x) = \inf_{\langle x,q \rangle =1} s[q, F(x)] \quad and \quad \inf \Lambda_F(x) = \sup_{\langle x,q \rangle =1} r[q, F(x)] . \tag{5}$$

As a consequence, the set of all eigenvalues associated to x is the (possibly unbounded) closed interval

$$\Lambda_F(x) = \{\lambda \in \mathbb{R} : \sup_{\langle x,q \rangle =1} r[q, F(x)] \leq \lambda \leq \inf_{\langle x,q \rangle =1} s[q, F(x)]\} . \tag{6}$$

Proof. To obtain (4) it suffices to write $\sup \Lambda_F(x) = s[1, \Lambda_F(x)] = v(1)$, and $\inf \Lambda_F(x) = -s[-1, \Lambda_F(x)] = -v(-1)$. According to Proposition 2, the closed interval $\Lambda_F(x)$ is the support set of the function v, i.e.

$$\Lambda_F(x) = \{\lambda \in \mathbb{R} : \lambda \alpha \leq v(\alpha) \quad \forall \alpha \in \mathbb{R}\}.$$

This yields the equality (5). □

We discuss now the possibility of extending Theorem 1 to the case in which $x \in H \setminus \{0\}$ is not necessarily an eigenvector of F.

Example 1. Let $F : H \to H$ be a single-valued operator. If $x \in H \setminus \{0\}$ is an eigenvector of F, then

$$\sup \Lambda_F(x) = \inf_{\langle x,q \rangle =1} \langle q, F(x) \rangle = \frac{\langle x, F(x) \rangle}{\langle x, x \rangle},$$

$$\inf \Lambda_F(x) = \sup_{\langle x,q \rangle =1} \langle q, F(x) \rangle = \frac{\langle x, F(x) \rangle}{\langle x, x \rangle}.$$

If $x \in H \setminus \{0\}$ is not an eigenvector of F, then

$$\sup \Lambda_F(x) = \inf_{\langle x,q \rangle = 1} \langle q, F(x) \rangle = -\infty,$$

$$\inf \Lambda_F(x) = \sup_{\langle x,q \rangle = 1} \langle q, F(x) \rangle = +\infty.$$

So, in the single-valued case, (4) and (5) remain true even if $\Lambda_F(x) = \phi$. However, the following example prevents us from being overoptimistic.

Example 2. Consider the multivalued operator $F : \mathbb{R}^2 \to \mathbb{R}^2$ defined by $F(x) = \{y \in \mathbb{R}^2 : y_2 \le -x_1\}$. Observe that $x = (1, 0)$ is not an eigenvector of F, and thus $s[\cdot, \Lambda_F(x)]$ takes always the value $-\infty$. On the other hand,

$$s[q, F(x)] = \begin{cases} -q_2 & \text{if } q_1 = 0 \text{ and } q_2 \ge 0 \\ +\infty & \text{otherwise,} \end{cases}$$

$$v(\alpha) = \begin{cases} +\infty & \text{if } \alpha \ne 0 \\ -\infty & \text{if } \alpha = 0. \end{cases}$$

Thus, the formula (2) fails for $\alpha \ne 0$. In particular, both estimates in (4) are wrong.

The effective domain $\text{dom} v := \{\alpha \in \mathbb{R} : v(\alpha) < +\infty\}$ of the function (3) is given by $\text{dom} v = \{\langle x, q \rangle : q \in \text{bar} F(x)\}$, where $\text{bar} F(x) := \{q \in H : s[q, F(x)] < +\infty\}$ stands for the barrier cone of the set $F(x)$. It is easy to see that $\alpha \in \text{dom} v$ if and only if, $\text{bar} F(x)$ intersects the closed hyperplane

$$\{\langle x, \cdot \rangle = \alpha\} := \{q \in H : \langle x, q \rangle = \alpha\}.$$

Without further ado we state:

Proposition 3. *Assume that $F : H \to H$ has closed convex values, and that $x \in H \setminus \{0\}$ is not an eigenvector of F. Let $\alpha \ne 0$ be a real number for which the consistency condition*

$$\{\langle x, \cdot \rangle = \alpha\} \cap \text{bar} F(x) \ne \phi \tag{7}$$

holds. Then, both terms in (2) are equal to $-\infty$.

Proof. Obviously $s[\alpha, \Lambda_F(x)] = -\infty$. Suppose to the contrary that $v(\alpha) \ne -\infty$. Since the assumption (6) rules out the case $v(\alpha) = +\infty$, the value $v(\alpha)$ must be finite. Hence,

$$v(t\alpha) = tv(\alpha) \in \mathbb{R} \quad \forall t > 0.$$

By convexity, one has necessarily $v(0) = 0$, and

$$v(t\alpha) \ne -\infty \quad \forall t < 0.$$

As a consequence, the convex function v is proper and lower-semicontinuous, and therefore $v(\alpha) = v^{**}(\alpha) = s[\alpha, \Lambda_F(x)]$. This contradicts the assumption $v(\alpha) \neq -\infty$. □

The following definition has a very technical flavour, but it emerges in a natural way from the assumptions of Proposition 3.

Definition 2. The multivalued operator $F : H \rightarrow H$ is declared *upper-normal* if the consistency condition $\{\langle x, \cdot \rangle = 1\} \cap barF(x) \neq \phi$ holds, whenever $x \in H \setminus \{0\}$ is not an eigenvector of F. The definition of *lower-normality* is analogous: the consistency assumption is stated with $\alpha = -1$.

Remark 1. If $F : H \rightarrow H$ is an operator with bounded closed convex values, then F is upper-normal and lower-normal. To see this, recall that the barrier cone of a bounded closed convex set is the whole space.

Now we are ready to state:

Theorem 2. *Let $F : H \rightarrow H$ be an upper-normal multivalued operator with closed convex values. Then, the following conditions on $x \in H \setminus \{0\}$ are equivalent:*
(a) x is an eigenvector of F.
(b) $s[\cdot, F(x)]$ is bounded from below on $\{\langle x, \cdot \rangle = 1\}$.
Moreover, one can write the variational formula

$$\sup \sigma(F) = \lambda^F := \sup_{x \neq 0} \inf_{\langle x, q \rangle = 1} s[q, F(x)]. \tag{8}$$

Proof. According to Proposition 2, the implication $(a) \Rightarrow (b)$ is true even if F fails to be upper-normal. The reverse implication follows from Proposition 3. To obtain the formula (7), one can apply the nested maximization principle (1). Due to the upper-normality assumption, the estimate for $\sup \Lambda_F(x)$ obtained in Theorem 1 remains true even if x is not an eigenvector of F. □

For the sake of completeness, we write below an analogous result for the greatest lower bound of $\sigma(F)$.

Theorem 3. *Let $F : H \rightarrow H$ be a lower-normal multivalued operator with closed convex values. Then, the following conditions on $x \in H \setminus \{0\}$ are equivalent:*
(a) x is an eigenvector of F;
(b) $r[\cdot, F(x)]$ is bounded from above on $\{\langle x, \cdot \rangle = 1\}$.
Moreover,

$$\inf \sigma(F) = \lambda_F := \inf_{x \neq 0} \sup_{\langle x, q \rangle = 1} r[q, F(x)]. \tag{9}$$

3   Conclusions

For an operator $F : H \to H$ with closed convex values, the following conclusions can be drawn:

(a) $\lambda_F \leq \inf \sigma(F)$ and $\sup \sigma(F) \leq \lambda^F$. As a consequence, all the eigenvalues of F lie in the (possible unbounded) closed interval

$$I(F) := \{\lambda \in \mathbb{R} : \lambda_F \leq \lambda \leq \lambda^F\}.$$

(b) If F is upper-normal and lower-normal, then $I(F)$ is the smallest closed interval containing all the eigenvalues of F. More precisely,

$$I(F) = \text{ closed convex hull of } \sigma(F).$$

The above equality applies even if F has no eigenvalues.

(c) If one of the normality assumptions fail, then the inclusion $\sigma(F) \subset I(F)$ can be strict. This fact can be checked-out with the operator $F : \mathbb{R}^2 \to \mathbb{R}^2$ defined by

$$F(x) = \begin{cases} \{y \in \mathbb{R}^2_+ : y_1 y_2 \geq (x_1)^2\} & \text{if } x_2 \leq 0, \\ \phi & \text{if } x_2 > 0. \end{cases}$$

This operator is lower-normal, but it is not upper-normal. In this case $\sigma(F) = \mathbb{R}_-$ and $I(F) = \mathbb{R}$.

(d) If GrF is a cone, then it is possible to write

$$\lambda^F = \sup_{\|x\|=1} \inf_{\langle x,q \rangle=1} s[q, F(x)],$$
$$\lambda_F = \inf_{\|x\|=1} \sup_{\langle x,q \rangle=1} r[q, F(x)],$$

where $\| \cdot \|$ is the norm associated to the inner-product in H.

(e) If $F : H \to H$ is a closed convex process (i.e. GrF is a closed convex cone), then one can write also

$$\lambda^F = \sup_{\|x\|=1} \inf_{\langle x,q \rangle=1} r[x, F^\triangle(q)],$$
$$\lambda_F = \inf_{\|x\|=1} \sup_{\langle x,q \rangle=1} s[x, F^\triangle(q)].$$

Here $F^\triangle$ denotes the negative adjoint of F, that is to say, $F^\triangle : H \to H$ is the closed convex process defined by

$$(q,p) \in GrF^\triangle \Leftrightarrow \langle q, y \rangle \leq \langle p, x \rangle \quad \forall (x,y) \in GrF.$$

References

1. Aubin, J. P., Frankowska, H. (1990) Set–Valued Analysis. Birkhauser, Boston
2. Aubin, J. P., Frankowska, H., Olech, C. (1986) Controllability of convex processes. SIAM J. Control and Optimization **24**, 1192–1211
3. Barbu, V., Precupanu, T. (1986) Convexity and optimization in Banach spaces. Mathematics and Its Applications Series. D. Reidel Publ. Co., Dordrecht
4. Leizarowitz, A. (1994) Eigenvalues of convex processes and convergence properties of differential inclusions. Set–Valued Analysis **2**, 505–527
5. Rockafellar, R.T. (1970) Convex analysis, Princeton Univ. Press, Princeton, New Jersey
6. Seeger, A. (1998) Spectral analysis of set-valued mappings. Acta Mathematica Vietnamica **23**, 49–63
7. Seeger, A. (1999) Eigenvalue analysis of equilibrium processes defined by linear complementarity conditions. Linear Algebra and its Applications, in press

Arrow-Barankin-Blackwell Theorems and Related Results in Cone Duality: A Survey

Aris Daniilidis

Laboratoire des Mathématiques Appliquées, Université de Pau et des Pays de l' Adour, Avenue de l' Université, 64000 PAU, France.
E-mail: Aris.Daniilidis@univ-pau.fr

Abstract. We attempt a brief survey on the cone duality and on the density theorems of Arrow-Barankin & Blackwell's type. Concerning the latter aspect we show the equivalence of two recent and ostensibly different results. We follow a unified approach which provides in particular a simple way of surveying these results and their proofs.

Key words: vector optimization, efficient points, positive functionals, denting points

1 Introduction

In 1953, Arrow, Barankin and Blackwell stated an interesting density result in multicriteria optimization (see [1]) concerning the approximation of the Pareto efficient points of a compact convex subset of R^n by points that are maximizers of some strictly positive functional on this set. This theorem was extended to cover more general notions of efficiency that are defined via an abstract cone, see [2], [19] and was subsequently generalized to an infinite dimensional setting, involving either weakly or norm compact sets.

In this article we endeavour a survey on these density results of Arrow, Barankin and Blackwell's type. Our aim is to survey the state of the art and to set in detail the relations among ostensibly different results. To this end, we shall adopt a unified approach available nowadays and, in doing so, we shall slightly improve some norm approximation results concerning weakly compact subsets of a Banach space. Finally we shall show the equivalence of a recent result of Gong [16] with a well-known earlier one of Petschke [32].

2 Notation

Throughout this paper, X will always be a Banach space and X^* its (topological) dual. However for most of what follows this is not essential and one can also consider a more general setting (for instance that of a locally convex space). In the sequel, we shall focus our interest in the norm and the weak topology of X, which will be denoted respectively by $\|\cdot\|$-topology and w-topology.

If $\varepsilon > 0$ and $x \in X$, we denote by $B_\varepsilon(x)$ the closed ball centered at x with radious ε. For any $x, y \in X$, we define by $[x, y]$ the closed segment $\{tx + (1 - t)y : 0 \leq t \leq 1\}$, while the segments (x, y), $(x, y]$ and $[x, y)$ are defined analogously. For any subset $A \subseteq X$, we denote by $int(A)$ the norm interior of the set A, by $cl(A)$ (resp. $w - cl(A)$) the norm (resp. the weak) closure of A and by $co(A)$ its convex hull. It is well known that for convex subsets of X the norm and the weak topological closures coincide (see [7] e.g.).

Let now K be a nonempty subset of X. A point $x_0 \in K$ is said to belong to the *algebraic interior* $algint(K)$ of the set K, if for every $y \in X$, the intersection of the set K with the line joining x_0 and y, contains an open interval around the point x_0. It is easily seen that $int(K) \subseteq algint(K)$. Moreover if $x_0 \in algint(K)$, then one has $\bigcup_{\lambda>0} \lambda(K - \{x_0\}) = X$. If K is closed and convex, then using Baire's theorem one can deduce from the latter relation that $intK \neq \emptyset$ and $int(K) = algint(K)$.

We further recall the definition of a *quasi-relative interior* point, see [5, Def. 2.3], or *inner* point, according to the terminology used in [18].

Definition 1. Let K be a nonempty closed convex subset of X and let $x_0 \in K$. The point x_0 is called a quasi-relative interior (or inner) point of the set K, if the set $cl(\bigcup_{\lambda>0} \lambda(K - \{x_0\}))$ is a subspace of X.

We shall keep the simple term 'inner point' in order to refer to this notion. We further denote by $innK$ the set of all inner points of K. The following proposition (see [5, Prop. 2.16]) reveals an interesting and characteristic property of these points. This property was actually used as the definition of inner points in [18].

Proposition 1. Let K be a nonempty closed convex subset of X. Then $x_0 \in innK$ if and only if x_0 is a nonsupport point of K, in the sense that the following implication is true for every $x^* \in X^*$:

$$(x^*, x - x_0) \leq 0, \forall x \in K \implies (x^*, x - x_0) = 0, \forall x \in K \tag{1}$$

It is easy to see that $intK \subseteq algintK \subseteq innK$. If K is closed and convex, each of the previous inequalities becomes equality whenever the smaller set is nonempty. We further recall from [18, Prop. 2.1] the following proposition:

Proposition 2. If K is a (nonempty, closed, convex and) separable subset of X, then $innK \neq \emptyset$.

Recently, inner points met important applications in variational inequality problems, see [18], [9] and [23]. In the following paragraph we shall see that this concept fits naturally also in the cone duality.

3   Order relations in Banach spaces

A nonempty subset C of a Banach space X is called a *cone*, if for every $x \in C$ the whole semiline $\{\lambda x : \lambda > 0\}$ is contained in C. A cone C is called *pointed*, if it does not contain whole lines, or equivalently if 0 is an extreme point of C. We recall here that a point x_0 is said to be an *extreme* point for the set A, if $x_0 \in A$ and x_0 is not contained in any open segment (x, y) lying in A. In the sequel we shall assume that the cone C is always closed, convex and pointed.

It is well known (see for instance [31] or [22]) that the cone C induces a partial order relation $\preceq$ on X by means of the following formula:

$$x \preceq y \Leftrightarrow y - x \in C \tag{2}$$

Setting $x = 0$ in the above formula (4) we see that the cone C itself corresponds to the set of nonnegative elements.

Let further A be a nonempty subset of X. The set A inherits naturally from X the aforementioned order relation $\preceq$. Consequently one can consider the set $Max(A, C)$ of *maximal* (or *efficient*) points of A (with respect to the cone C) as follows:

$$Max(A, C) = \{x_0 \in A : \{x_0\} = A \cap (x_0 + C)\} \tag{3}$$

The *dual cone* C^* of C is defined by

$$C^* = \{f \in X^* : f(x) \geq 0, \forall x \in C\} \tag{4}$$

The dual cone C^* corresponds to the set of all *positive functionals*. It is easily seen that C^* is always a nonempty closed convex cone of X^*.
We further denote by

$$C^\sharp = \{f \in Y^* : f(x) > 0, \forall x \in C, x \neq 0\} \tag{5}$$

the set of all *strictly positive functionals*. This set is also a cone; however in some cases it may be empty (see the example that follows Proposition 19). In fact one can show (see [5] e.g.) that $C^\sharp$ actually coincides with the set of inner points $innC^*$ of the closed convex set C^*, so its nonemptiness is assured if the space X^* is separable (see Proposition 11 above). The importance of the strictly positive functionals stems from the fact that they are closely related to the notion of a *cone base*. The definition of the latter is recalled below:

Definition 2. A closed convex subset V of C is said to be a (cone) base, if for every $x \in C$, $x \neq 0$, there exist unique $\lambda > 0$, $b \in V$ such that $y = \lambda b$.

The existence of a cone base for a given cone C is in fact equivalent to the nonemptiness of the set $C^\sharp = innC^*$, see also [22]. Indeed, if $C^\sharp \neq \emptyset$, then for any $f \in C^\sharp$ the set $\{x \in C : f(x) = 1\}$ defines a cone base on C. Conversely, if the cone C has a base V, then separating V from 0 (by the Hahn-Banach theorem), one immediately obtains a functional $f \in C^\sharp$.

It follows directly from Proposition 11 that if X is a separable Asplund space (i.e. X^* is separable), then every cone has a base. This result can be refined even further, as shows the following proposition in [5, Th. 2.19].

Proposition 3. *Assume that X is a separable Banach space. Then every (closed, convex pointed) cone C on X has a base.*

The separability assumption is indispensable in the statement of Proposition 19. Indeed, without this assumption nice cones may not have a base, as shows the following example taken from [18].

Example:

Let I be any uncountable set and $Y = \ell^2(I)$ be the Hilbert space of all square integrable (with respect to the counting measure) functions $f : I \to R$. Consider the cone C of all non-negative real valued functions of Y. One easily sees that $C^* = C$. However this cone has no inner points, hence C has no base.

We further consider the interior $intC^*$ of the cone C^*, which is a (possibly empty) convex cone. One obviously has $intC^* \subseteq innC^* = C^\sharp$, the equality holding whenever $intC^* \neq \emptyset$. In particular, the latter is equivalent with the existence of a bounded base for the cone C, as states the following proposition, see [22].

Proposition 4. *Let C be a closed, convex, pointed cone of Y. The following are equivalent:*
(i) The dual cone C^* has a non-empty interior $intC^*$.
(ii) The cone C has a bounded base V.

However it is possible to have $intC^* = \emptyset$ and $innC^* \neq \emptyset$. In fact this is very often the case. To enlighten further the above situation we present below some standard examples of Banach spaces possessing a natural ordering structure.

Examples:

1. Let $X = R^n = X^*$, and $C = C^* = R^n_+$. In this case the cone has a bounded base, defined for instance by the strictly positive linear form $y = (1, 1, ..., 1) \in R^n$.

2. Let $X = \ell^1(N)$ be the space of all absolutely summable sequences and $C = \ell^1_+(N)$ be the corresponding cone of all nonnegative sequences of

$\ell^1(N)$. One can easily see that the dual cone C^* (which is the set $\ell^\infty(N)_+$ of all nonnegative bounded sequences of the dual space $X = \ell^\infty(N)$) has a nonempty interior, which coincides with the set of all positive bounded sequences. We conclude from Proposition 4 that C has a bounded base.

3. Let $X = \ell^p(N)$, $X^* = \ell^q(N)$ where $\frac{1}{p} + \frac{1}{q} = 1$ and $1 < p, q < +\infty$. Consider the cone $C = \ell^p(N)_+$. It follows from Proposition 11 (or Proposition 19) that the dual cone $C^* = \ell^q(N)_+$ has inner points, hence C has a base. However since C^* has an empty interior, every base of C is unbounded.

4. Let X be the space $c_0(N)$ of all null sequences and let $C = c_0(N)_+$ be the cone of all nonnegative null sequences. Then $X^* = \ell^1(N)$ and $C^* = \ell^1(N)_+$. As in the previous case we conclude that the cone $c_0(N)_+$ has a base, but not a bounded base.

5. Let X be the space $C([0,1])$ of the real continuous functions equipped with the topology of the uniform convergence. Since X is separable, applying Proposition 19 we conclude that the cone $C([0,1])_+$ of the nonnegative valued functions has a base. However in this case X^* coincides with the set $BV([0,1])$ of all regural Borel (signed) measures on $[0,1]$ and C^* with the set $BV([0,1])_+$ of all regural Borel positive measures. Since the latter set has an empty interior, we conclude that the cone $C([0,1])_+$ has no bounded base.

From the previous examples it becomes clear that the existence of a cone base is a natural assumption in vector optimization, which is always fulfilled if X is separable. On the other hand this is not the case for the assumption of the boundedness of the base: Among the classical Banach spaces, this condition is fulfilled only in $\ell^1(N)$ (or in general in $L^1(\mu)$) and in the finite dimensional spaces. We summarize below our main conclusions from the above discussion:

Proposition 5. *Let C be a closed, convex, pointed cone of X. Then*
  (i) C has a base iff $innC^* \neq \emptyset$
  (ii) C has a bounded base iff $intC^* \neq \emptyset$
  (iii) If X is separable, then $innC^* \neq \emptyset$

4 Positive (or proper efficient) points. Arrow-Barankin-Blackwell Theorem

In the sequel we shall deal with a closed, convex pointed cone C with a base V in a Banach space X. In this case one has $C^\sharp = innC^* \neq \emptyset$, hence for any subset $A \subseteq X$ one can define the set of positive points $Pos(A,C)$ of A as follows:

$$Pos(A,C) = \{x_0 \in A : \exists f \in innC^*, f(x_0) = \sup f(A)\} \qquad (6)$$

where $\sup f(A)$ denotes the supremum of the functional f on the set A. We mention here that also other (more restrictive) notions of efficiency have been defined in the literature, as for instance the notion of "superefficiency" introduced in [6], see also [17] for a survey.

It is straightforward from relations (3), (6) and the definition of $C^\sharp$ (relation (5)) that $Pos(A, C) \subseteq Max(A, C)$. However simple examples even in two-dimensionsal spaces certify that in general this inclusion is strict.

In the special case $X = R^n$ and $C = R^n_+$ relations (3) and (6) have a certain interpretation in Economics in terms of the Pareto efficient commodity bundles and the supporting system of prices. This has motivated Arrow, Barankin and Blackwell in 1953 to show the following density result [1] (see also [30] for an alternative approach).

Theorem 1. *Let A be a compact convex subset of R^n and $C = R^n_+$. Then $Pos(A, C)$ is dense in $Max(A, C)$.*

In [19] and independently in [2] the preceding theorem has been extended to cover the case of more general cones C in R^n. Theorem 1 was also generalized to an infinite dimensional setting. The particular case of $\ell^\infty(N)$ has a certain significance in Economics involving models with an infinite horizon production, and has been studied in [33], [27], [29] and [12]. However the statement of Theorem 1 itself as a density result had an independent interest and generated pure mathematical extensions to arbitrary Banach spaces. Many authors have worked in this direction, see for instance [34], [4], [21], [10] etc.

In infinite dimensions there are two topologies that enter naturally into consideration, the weak and the norm topology. The result that follows was originally proved in [15]. Nowadays an easy and direct proof of it is available, that uses the technique of 'dilating cones' (see [20]). This technique is now classical and has already been repeated several times in density results of this kind in [35], [16], [28] and [13]; see also [14] for a more general approach in a locally convex setting. However we give here a sketch of this proof, since it will help the presentation of the forthcoming density results.

In the following statement one can consider $\mathfrak{S}$ to be either the norm or the weak topology of X.

Theorem 2. *Let X be a Banach space, X^* its dual and $\mathfrak{S}$ any topology of the dual system (X, X^*). Let A be a $\mathfrak{S}$-compact, convex subset of X and C a closed, convex, pointed cone with a base V. Then*

$$Max(A, C) \subseteq \overline{Pos(A, C)}^{\mathfrak{S}} \tag{7}$$

Proof. (Sketch) Let $x_0 \in Max(A, C)$, i.e. $\{x_0\} = A \cap (x_0 + C)$.
We first observe that $C = \overline{cone}(V)$, where $\overline{cone}(V)$ denotes the closed cone generating by V. Moreover, it is no loss of generality to assume that the distance $d(0, V)$ of the cone base V from 0, is greater than $1/2$.

Step 1: For every $n \geq 2$, consider the (closed, convex, pointed, based) cone $C_n = \overline{cone}(V + B_{\frac{1}{n}}(0))$. Then we obviously have $C = \bigcap_{n \geq 2} C_n$. Note that in general x_0 does not remain a maximal point of A for the larger cone C_n.

Step 2: For each $n \geq 2$, choose a maximal (with respect to C_n) point $x_n \in Max(A, C_n)$, such that $x_n \in A_n := (x_0 + C_n) \cap A$. This is always possible (see for instance [26, Cor. 3.6]), since the set A_n is $\Im$-compact. Since the relation $C = \bigcap_{n \geq 2} C_n$ implies that $\{x_0\} = \bigcap_{n \geq 2} A_n$, we easily conclude that $x_n \to x_0$ in the $\Im$-topology.

Step 3: Since $\{x_n\} = A \cap (x_n + C_n)$ and the cone C_n has a nonempty interior, there exists a functional $x^* \in C_n^*$ that supports the set A at the point x_n. The proof now finishes by the observation that x^* is actually a strictly positive functional for the original cone C.   □

A careful investigation of the previous proof leads easily to the forthcoming corollary. We will first need the following definition.

Definition 3. We say that $x_0 \in A$ is a *point of continuity* of the set A, if the identity mapping $id : (A, w) \to (A, \|\|)$ is continuous at x_0.

The proof of the following corollary is straighforward. However this result will be useful in the sequel. Let us recall from the proof of Theorem 2 that for $n \geq 2$, $C_n := \overline{cone}(V + B_{\frac{1}{n}}(0))$ and $A_n := (x_0 + C_n) \cap A$.

Corollary 1. *Let A be a w-compact, convex subset of X. Assume that $x_0 \in Max(A, C)$ and that for some $n_0 \geq 2$, x_0 is a point of continuity of the set A_{n_0}. Then $x_0 \in \overline{Pos(A, C)}^{\|\cdot\|}$.*

Proof. Repeating the proof of Theorem 2 we produce a sequence $(x_n)_n \subset Pos(A, C)$ that is weakly converging to x_0. We note that this sequence is eventually contained in A_{n_0}, hence in view of Definition 3, it is actually norm converging to x_0.   □

Theorem 2 expresses simultaneously two different density results, one for the norm and one for the weak topology. However in the first case, the norm compactness assumption imposed on the convex set A is very restrictive in infinite dimensions. On the other hand the approximation result that we obtain in the second case is rather weak. It is desirable to obtain a strong approximation result involving weakly compact subsets of X, as for example does (in a local way) Corollary 1. To this end, Jahn [21] was the first to derive a norm approximation result for weakly compact subsets, by assuming that the cone C was of a 'Bishop-Phelps type'. Subsequently Petschke [32] (see also [15] for a different approach) refined Jahn's proof to conclude the same result, using - more general - any cone having a bounded base. We state below Petsche's result [32].

Theorem 3. *Let A be a w-compact convex subset of X and assume that C has a bounded base. Then*

$$Max(A, C) \subseteq \overline{Pos(A, C)}^{\|\cdot\|} \tag{8}$$

However, as we have already discussed in the previous section, the assumption of a bounded based cone is unpleasant. Recently Gong [16] tried to deal with this inconvenience by relaxing this assumption to an apparently weaker one. Before we proceed to this result, we shall need the following definition.

Definition 4. Let A be a closed convex subset of X and $x_0 \in A$.
(i) x_0 is called a *denting* point of A, if for every $\varepsilon > 0$, we have $x_0 \notin \overline{co}(A \setminus B_\varepsilon(x_0))$, where $\overline{co}(A \setminus B_\varepsilon(x_0))$ denotes the closed convex hull of the set $(A \setminus B_\varepsilon(x_0))$.
(ii) x_0 is called a *strongly exposed* point of A by the functional $x^* \in X^*$, if for every sequence $(x_n)_n \subset A$, the relation $x^*(x_n) \to x^*(x_0)$ implies the norm convergence of the sequence $(x_n)_n$ to x_0.

It follows easily from Definitions 3 and 4 that every denting point of A is a point of continuity for this set. Moreover every strongly exposed point of A is denting. It is worthmentioning that these last two notions coincide if $A = C$ and $x_0 = 0$, since in that case they are both equivalent to the boundedness of the cone base. This is the content of the following proposition in [22] (see also [16] for the equivalence of (ii) and (iii)).

Proposition 6. *The following statements are equivalent:*
(i) 0 is a strongly exposed point of the cone C.
(ii) 0 is a denting point of C.
(iii) C has a bounded base

We are now ready to state Gong's density result, see [16].

Theorem 4. *Let A be a w-compact convex subset of X. Assume that one of the following two conditions is fulfilled.*
(i) Every maximal point of A is denting.
(ii) 0 is a point of continuity of the cone C, i.e.

$$\forall \varepsilon > 0, \quad 0 \notin \overline{K \setminus B(0, \varepsilon)}^w \tag{9}$$

Then the following approximation result holds:

$$Max(A, C) \subseteq \overline{Pos(A, C)}^{\|\cdot\|} \tag{10}$$

Condition (i) of Theorem 4 is satisfied if for example A is taken to be the unit ball of ℓ^p, for $1 < p < +\infty$, see [16]. In the next section we shall see that this condition can be remplaced by a weaker one that would only require that every maximal point of A is a point of continuity. However even this latter condition remains undesirable, since it imposes an a priori assumption on the set of maximal points of A.

On the other hand, in view of Proposition 6 and of Definition 4(i) and its subsequent comments, it follows that condition (9) holds trivially whenever the cone C has a bounded base. In that sense the result of Theorem 4(ii) appears to be more general than the one in Theorem 3. In [16], the author queries (and states it as an open question) whether Theorem 4(ii) is indeed a real extension of Theorem 3. In next section we shall answer this question to the negative, by means of a characterization of the denting points of the closed convex subsets of a Banach space.

5   Equivalence of Petscke's and Gong's theorems

In this section we show that if 0 is a point of continuity of a pointed cone C, then it is also a denting point of C. Consequently, it will follow that Theorems 3 and 4(ii) are equivalent.

Let K be a closed convex subset of X and $x_0 \in K$. As already partially seen in the previous section, every denting point is both an extreme and a point of continuity of K. In [24] (see also [25]) it has been proved that these two properties actually characterize denting points, in case of a closed convex and bounded subset K. The following proposition extends this result to the class of all closed convex subsets of X.

Proposition 7. *Let x_0 be a point of a closed convex subset K of a Banach space. Then x_0 is denting if and only if x_0 is an extreme point and a point of continuity.*

Proof. Let us assume that x_0 is both an extreme and a point of continuity of the set K. Take any $R > 0$ and consider the set $K_R = \{x \in K : \|x - x_0\| \leq R\}$. Since $K_R \subseteq K$ and $x_0 \in K_R$, it follows easily that x_0 remains an extreme point and a point of continuity for the set K_R. Since the latter set is bounded, it follows from [25] that x_0 is a denting point of it. The following claim finishes the proof.

Claim: x_0 remains a denting point for the set K.
[Indeed, take any $\varepsilon > 0$. With no loss of generality we can assume that $R > \varepsilon$. Since x_0 is a denting point of the set K_R, we have $x_0 \notin \overline{co}(K_R \setminus B_\varepsilon(x_0))$, hence there exist $x^* \in X^*$ and $\alpha \in R$ such that $x^*(x_0) < \alpha < x^*(x')$, $\forall x' \in \overline{co}(K_R \setminus B_\varepsilon(x_0))$. Set $W = \{x \in X : x^*(x) < \alpha\}$ and observe that since W is a half-space and K is convex, we have $W \cap K \subset B_\varepsilon(x_0) \cap K$. Note now that $W \cap K$ is a neighborhood of x_0 for the (relative) weak topology of K. It now follows that $\overline{co}(K \setminus B_\varepsilon(x_0)) \subseteq K \setminus W$, hence in particular $x_0 \notin \overline{co}(K \setminus B_\varepsilon(x_0))$. The claim is proved.]

$\square$

Remark: It is interesting to observe that the previous result has the following interesting restatement:

$$\forall \varepsilon > 0, x \notin \overline{co}\,(K \backslash B(x,\varepsilon)) \Leftrightarrow \forall \varepsilon > 0, x \notin co(K \backslash B(x,\varepsilon)) \text{ and } x \notin \overline{K \backslash B(x,\varepsilon)}^w$$

i.e. the convex and the weak topological hull of the set $(K \backslash B(x,\varepsilon))$ can be considered separately.

In the special case of a closed, convex pointed cone C, since the point $x_0 = 0$ is extreme, we infer the following corollary.

Corollary 2. *Let C be a closed convex pointed cone of X. The following statements are equivalent:*
(i) 0 is a denting point of C
(ii) 0 is a point of continuity of C

The above corollary together with Proposition 6 shows in particular that Petschke's result (Theorem 3) and Gong's result (Theorem 4(ii)) are equivalent. Consequently, it remains widely open whether we can efficiently relax (or omit) the assumption of a bounded cone in Theorem 3, without giving up the norm approximation result theorem.

In the following theorem we survey the statements of Proposition 4, of Proposition 6 and of the previous corollary in the following theorem, see also [8]. The equivalence of (ii) and (iv) has also been observed in [16].

Theorem 5. *Let C be a closed convex pointed cone of Y. The following statements are equivalent:*
(i) 0 is a strongly exposed point of C
(ii) 0 is a denting point of C
(iii) 0 is a point of continuity of C
(iv) $\exists \varepsilon > 0, \quad x \notin \overline{co}(K \backslash B(x,\varepsilon))$
(v) C has a bounded base
(vi) $intC^* \neq \emptyset$

The following proposition is a local density result which extends in particular Theorem 4(i). The essence of this result comes actually from Corollary 1. We recall that a norm is said to have the *Kadec-Klee property* ([11] eg.), if the relative norm and the relative weak topologies on the unit ball B_X coincide at any point of the unit sphere $S_X := \{x \in X : \|x\| = 1\}$. We also recall that every reflexive Banach space admits a Kadec-Klee renorming.

Proposition 8. *Let A be a w-compact convex subset of X and x_0 belongs to $Max(A, C)$. Consider the following conditions:*
(i) x_0 is a point of continuity of the set A.
(ii) 0 is a point of continuity of the cone C

(iii) There exists $y \in X$, such that for some $n_0 \geq 2$, x_0 is the farthest point of y for the set $A_{n_0} := (x_0 + C_{n_0}) \cap A$, (i.e. $\|y - x_0\| \geq \|y - x\|$, for all $x \in A_{n_0}$), with respect to an equivalent norm $\|\cdot\|$ of X having the Kadec-Klee property.
(iv) For some $n_0 \geq 2$, x_0 is a point of continuity of the set A_{n_0}.
If any of the conditions (i)-(iv) holds, we have

$$x_0 \in \overline{Pos(A, C)}^{\|\cdot\|}$$

Proof. In virtue of the Corollary 1, it suffices to show that each of the conditions (i)-(iii) implies condition (iv).
Since $A_{n_0} \subseteq A$, it follows directly that condition (i) implies (iv).
Let us now assume that (ii) holds. Then from Theorem 5 it follows that C has a bounded base V. Following the construction of the proof of Theorem 2, we observe that the cones C_n also have a bounded base, hence applying again Theorem 5 we conclude that (iv) holds.
Let us finally assume that (iii) holds. Then x_0 is a boundary point of the closed ball $B_r(y)$ centered at y with radius $r = \|y - x_0\|$. Since the norm $\|\cdot\|$ has the Kadec-Klee property, it follows that x_0 is a point of continuity of the set $B_r(y)$. Since $A_{n_0} \subseteq B_r(y)$ it follows that (iv) holds.

□

Remark: Since condition (ii) is equivalent to the existence of a bounded base (see Theorem 5), the above proposition gives in particular an alternative (and simpler) way to prove Theorem 3 of Petschke.

Acknowledgement: The author acknowledges N. Hadjisavvas, I. Polyrakis and M. Zissis for useful discussions. He also acknowledges L. Barbet, N. Hadjisavvas and J.P. Penot for their careful reading on a preliminary version of this manuscript. The author wishes to express his gratitude to S. Argyros for illuminating discussions related to this work and for communicating him the references [24] and [25].

References

1. Arrow, K., Barankin, E. et al. (1953) Admissible points of convex sets. In: Kuhn & Tucker (Eds.) Contributions to the Theory of Games, Princeton University Press, Princeton, NJ
2. Bitran, G., Magnanti, T. (1979) The structure of Admissible Points with Respect to Cone Dominance. J. Optimization Theory and Appl. **29**, 573–614
3. Borwein, J. (1977) Proper Efficient points for Maximization with respect to cones. SIAM J. Control and Optim. **15**, 57–63
4. Borwein, J. (1980) The geometry of Pareto efficiency over cones. Math. Operationsforch. Statist. Ser. Optim. **11**, 235–248
5. Borwein, J., Lewis, A. (1992) Partially finite convex programming, Part I: Quasi relative interiors and duality theory. Math. Program. **57**, 15–48

6. Borwein, J., Zhuang, D. (1993) Super Efficiency in Vector Optimization. Trans. Am. Math. Soc. **338**, 105–122
7. Brezis, H. (1983) Analyse fonctionelle, Théorie et Applications. Masson, Paris
8. Daniilidis, A. (1997) Applications of Generalized Monotonicity and Generalized Convexity in Variational Inequalities and Vector Optimization. PhD Thesis, University of the Aegean, Greece
9. Daniilidis, A., Hadjisavvas, N. (1996) Existence Theorems for Vector Variational Inequalities. Bull. Austral. Math. Soc. **54**, 473–481
10. Dauer, J., Gallagher, R. (1990) Positive Proper Efficient Points and related cone results in vector optimization theory. SIAM J. Control and Optim. **28**, 158–172
11. Deville, R., Godefroy, G. et al. (1993) Smoothness and renormings in Banach spaces. Pitman Monographs and Surveys in Pure and Applied Mathematics 64, Longman Scientific & Technical, John Wiley & Sons, New York
12. Ferro, F. (1993) A general form of the Arrow-Barankin-Blackwell Theorem in Normed Spaces and the l^∞ case, J. Optimization Theory and Appl. **79**, 127–138
13. Ferro, F. (1998) A new ABB Theorem in Banach Spaces. Preprint, 11 p, University of Genova, Italy.
14. Fu, W. (1996) On the density of Proper Efficient Points. Proc. Am. Math. Soc. **124**, 1213–1217
15. Gallagher, R., Saleh, O. (1993) Two Generalizations of a Theorem of Arrow, Barankin and Blackwell. SIAM J. Control Optim. **31**, 217–256
16. Gong, X. (1995) Density of the Set of Positive Proper Minimal points in the Set of Minimal Points, J. Optimization Theory and Appl. **86**, 609–630
17. Guerraggio, A., Molho, E. et al. (1994) On the Notion of Proper Efficiency in Vector Optimization, J. Optimization Theory and Appl. **82**, 1–21
18. Hadjisavvas, N. & Schaible, S. (1996) Quasimonotone Variational Inequalities in Banach spaces. J. Optimization Theory and Appl. **90**, 95–111
19. Hartley, R. (1978) On Cone Efficiency, Cone Convexity and Cone Compactness. SIAM J. Appl. Math. **34**, 211–222
20. Henig, M. (1982) Proper efficiency with respect to cones. J. Optimization Theory and Appl. **36**, 387–407
21. Jahn, J. (1988) A Generalization of a Theorem of Arrow, Barankin and Blackwell. SIAM J. Control Optim. **26**, 999–1005
22. Jameson, G. (1970) Ordered Linear Spaces. Springer-Verlag, Berlin.
23. Konnov, I. (1998) On Quasimonotone Variational Inequalities. J. Optimization Theory and Appl. **99**, 165–181
24. Lin, B-L, Lin, P-K et al. (1985-1986) A characterization of denting points of a closed, bounded, convex set. Longhorn Notes. Y.T. Functional Analysis Seminar. The University of Texas, Austin, 99–101
25. Lin, B-L, Lin, P-K et al. (1986) Some geometric and topological properties of the unit sphere in Banach spaces. Math. Annalen **274**, 613–616
26. Luc, D-T. (1988) Theory of Vector Optimization. Lecture Notes in Economics and Mathematical Systems 319. Springer-Verlag, Berlin
27. Majumdar, M. (1970) Some Approximation Theorems on Efficiency Prices for Infinite Programs. J. Econom. Theory **2**, 399–410
28. Makarov, E., Rachkovski, N. (1996) Density Theorems for Generalized Henig Proper Efficiency. J. Optimization Theory and Appl. **91**, 419–437
29. Peleg, B. (1972) Efficiency Prices for Optimal Consumption Plans. J. Math. Anal. Appl. **35**, 531–536

30. Peleg, B. (1972) Topological properties of the efficient point set. Proc. Am. Math. Soc. **35**, 531–536
31. Peressini, A. (1967) Ordered Topological Vector Spaces. Harper & Row. New York
32. Petschke, M. (1990) On A Theorem of Arrow, Barankin and Blackwell. SIAM J. Control and Optim. **28**, 395–401
33. Radner, R. (1967) Efficiency Prices for Infinite Horizon Production Programmes. Rev. Econom. Stud. **34**, 51–66
34. Salz, W. (1976) Eine topologische eigenschaft der effizienten Punkte konvexer Mengen. Operat. Res. Verfahren **XXIII**, 197–202
35. Zhuang, D. (1994) Density Results for Proper Efficiencies. SIAM J. Control and Optim. **32**, 51–58

Formulating and Solving Nonlinear Programs as Mixed Complementarity Problems*

Michael C. Ferris and Krung Sinapiromsaran

University of Wisconsin, Computer Sciences Department,
1210 West Dayton Street, Madison, Wisconsin 53706

Abstract. We consider a primal-dual approach to solve nonlinear programming problems within the AMPL modeling language, via a mixed complementarity formulation. The modeling language supplies the first order and second order derivative information of the Lagrangian function of the nonlinear problem using automatic differentiation. The PATH solver finds the solution of the first order conditions which are generated automatically from this derivative information. In addition, the link incorporates the objective function into a new merit function for the PATH solver to improve the capability of the complementarity algorithm for finding optimal solutions of the nonlinear program. We test the new solver on various test suites from the literature and compare with other available nonlinear programming solvers.

Key words: complementarity problems, nonlinear programs, automatic differentiation, modeling languages

1   Introduction

While the use of the simplex algorithm for linear programs in the 1940's heralded the inception of operations research as a practical discipline, the extension of the field to nonlinear programs (NLP) has been much more recent. The theory of NLP was extensively developed in the 1950's and 60's, culminating perhaps with the landmark books [12,25]. Leaving aside unconstrained optimization, practical algorithms for constrained nonlinear optimization rivaling the simplex algorithm were much slower to develop. In fact, the MINOS code [26] released in 1976 was the first code that could deal reliably with problems of relatively large size.

The advent of modeling languages [3,14] allowed these solvers to be used by modelers that were not operation research or numerical analysis specialists. Modeling languages allow optimization problems to be communicated to solvers in an efficient form, carrying out data manipulations, generation of multiple sets of indexed equations, exploiting simple constraint types and

---

* This research was partially supported by National Science Foundation Grant CCR-9619765 and Air Force Office of Scientific Research Grant F49620-98-1-0417.

converting problems to the format required by a solver without modeler intervention. Furthermore, computational advances such as the use of automatic differentiation techniques [19,18,28] to generate the first order derivatives of the nonlinear functions can be used directly in a solver implementation. Currently, GAMS [3] and AMPL [14] are used in a large variety of applications. Most of the commercially available solvers for linear and nonlinear programs can be used directly from one or both of these systems.

The 1980's and 1990's have generated two significant algorithmic changes to the field. The first major change was the introduction of interior point methods for linear programming by Karmarkar [21] in 1984, as a practical alternative to the theoretically important polynomial time ellipsoid algorithm of Khachian [23]. The idea has been considerably developed; currently it appears that primal-dual methods are the most effective in large scale linear programming settings [34].

In nonlinear programming, a significant improvement has been observed for non-convex problems by using second order information. While Quasi-Newton methods can be used for problems whose feasible region lies in a relatively small dimension subspace, and limited memory methods are effective for unconstrained and bound constrained problems, it is becoming increasingly clear that methods that exploit second order information (either using negative curvature within a trust region or line search framework) are more efficient and robust. Unfortunately, it is only recently [15] that second order information has become available from a modeling language, namely AMPL.

This paper is an attempt to combine some of the features of these last two improvements. The idea is to use a primal-dual framework for NLP in conjunction with second order information. We first start with the first order conditions of the original NLP model in Section 2.1, which we cast as a mixed complementarity problem (MCP) in Section 2.2. In Section 3, we explain the PATH solver implementation for MCP and its requirements and describe the use of a merit function to solve the MCP problem. Then we introduce a new merit function associated with solving NLP's. Section 4 gives details of our NLP solver, PATHNLP, with the MCP function evaluation and its Jacobian being evaluated by AMPL. In particular, we show how second order information of the NLP is utilized via solver link libraries in Section 4.2.

Section 5 gives numerical results for our approach on a set of nonlinear test problems extracted from the AMPL web site. Specifically, we test all models in the Hock/Schittkowski test suite [20] and compare the results of the PATH solver with LANCELOT [4], MINOS [27], NPSOL [17] and SNOPT [16]. Other large scale examples, including problems from portfolio and structural optimization are also tested. We believe these results indicate this is already a promising approach and warrants further investigation in the future.

2   Mathematical formulation

In this paper, we concentrate on the following constrained nonlinear program

$$\begin{aligned} \text{minimize } & f(x) \\ \text{subject to } & g(x) \leq 0, \ h(x) = 0, \ x \in B, \end{aligned} \tag{1}$$

where $f : \mathbf{R}^n \mapsto \mathbf{R}, g : \mathbf{R}^n \mapsto \mathbf{R}^m$ and $h : \mathbf{R}^n \mapsto \mathbf{R}^p$ are twice continuously differentiable, and $B := \{x \in \mathbf{R}^n | r \leq x \leq s\}$ with $r_i \in [-\infty, \infty]$ and $s_i \in [r_i, \infty]$. Let $S := \{x \in B | g(x) \leq 0, h(x) = 0\}$ denote the feasible region. We will focus on finding a point that satisfies the first order conditions of the NLP (1).

2.1   The first order conditions of NLP

The concept of the Lagrangian function and the Lagrange multipliers play a crucial role in defining a first order point for the NLP (1). The Lagrangian function is a weighted summation of the objective function and the constraint functions, defined as follows

$$L(x, \lambda, \nu) := f(x) - \lambda^T g(x) - \nu^T h(x),$$

where λ and ν denote the Lagrange multipliers (dual variables) corresponding to the inequality and equality constraints, respectively.

The first order necessary conditions for the NLP (1) are

$$\begin{aligned} 0 &\in \nabla_x L(x, \lambda, \nu) + N_B(x) \\ 0 &\geq \lambda \perp g(x) \leq 0 \\ h(x) &= 0, \end{aligned} \tag{2}$$

where $N_B(x) = \{z \in \mathbf{R}^n | (y - x)^T z \leq 0, \forall y \in B\}$ is the normal cone [32] to B at x.

In the case that r_i or s_i is finite, the definition of the normal cone allows the first equation of (2), to be rewritten in the following manner. If $x_i = r_i$, then

$$(\nabla_x L(x, \lambda, \nu))_i \geq 0,$$

while if $x_i = s_i$, then

$$(\nabla_x L(x, \lambda, \nu))_i \leq 0$$

and for any values of r_i and s_i, if $r_i < x_i < s_i$, then

$$(\nabla_x L(x, \lambda, \nu))_i = 0.$$

These conditions coupled with the regularity condition on the point x establish the necessary conditions for NLP which are normally called the Karush-Kuhn-Tucker (KKT) conditions [22,24]. Whenever the Hessian matrix of the Lagrangian function is positive definite at (x^*, λ^*, ν^*), the first order conditions are also sufficient for x^* to be a strict local minimizer of NLP.

2.2   Primal-dual formulation of NLP

The standard mixed complementarity problem (MCP) is defined as the problem of finding a point $z \in \mathbf{R}^n$ inside the box $B = \{z| -\infty \leq l < z < u \leq \infty\}$ that is complementary to a nonlinear function $F : \mathbf{R}^n \to \mathbf{R}^n$. We assume without loss of generality that $l_i < u_i$ for all $i = 1, 2, \ldots, n$.

The point z is complementarity to $F(z)$ when

$$
\begin{array}{ll}
\text{either } z_i = l_i & \text{and } F_i(z) \geq 0 \\
\text{or } z_i = u_i & \text{and } F_i(z) \leq 0 \text{ for } i = 1, \ldots, n \\
\text{or } l_i < z_i < u_i & \text{and } F_i(z) = 0.
\end{array}
$$

If $l \equiv -\infty$ and $u \equiv \infty$, MCP becomes the problem of finding a zero of a system of nonlinear equations, that is to find $z \in \mathbf{R}^n$ such that $F(z) = 0$, while if $l = 0$ and $u \equiv \infty$, the problem is the Nonlinear Complementarity Problem (NCP) of finding $z \in \mathbf{R}^n$ such that $z_i \geq 0, F_i(z) \geq 0$ and $z_i F_i(z) = 0$, for all $i = 1, \ldots, n$. The latter property $z_i F_i(z) = 0$ is often called complementarity between z_i and $F_i(z)$.

Let z be composed of the primal variable x and the dual variables λ and ν of the NLP (1). The nonlinear MCP function can be written as a vector function of the first order derivative evaluation of the Lagrangian function with respect to the corresponding primal and dual variables that is

$$
F(z) := \begin{bmatrix} \nabla_x L(z) \\ -\nabla_\lambda L(z) \\ -\nabla_\nu L(z) \end{bmatrix}.
$$

The nonlinear MCP model is to find $z = (x, \lambda, \nu) \in \mathbf{R}^q$ where $q = n+m+p$ that is complementary to the nonlinear vector function F from $\mathbf{R}^q \mapsto \mathbf{R}^q$ given above along with lower bounds l and upper bounds u

$$
F(z) = \begin{bmatrix} \nabla_x L(z) \\ g(x) \\ h(x) \end{bmatrix}, l := \begin{bmatrix} r \\ -\infty \\ -\infty \end{bmatrix}, u := \begin{bmatrix} s \\ 0 \\ \infty \end{bmatrix}. \tag{3}
$$

Here $\nabla_x L(z) = \nabla_x f(x) - \lambda^T \nabla_x g(x) - \nu^T \nabla_x h(x)$
$= \nabla_x f(x) - \sum_{i=1}^m \lambda_i \nabla_x g_i(x) - \sum_{j=1}^p \nu_j \nabla_x h_j(x)$.

By comparing the MCP (3) to the KKT conditions (2), it is clear that this formulation is equivalent to the first order conditions of the NLP (1). This simple observation allows us to solve the NLP problem using an MCP solver, which is the subject of Section 4.

3 The PATH solver and merit functions

The PATH solver [6] is a nonsmooth Newton type algorithm [31] which finds a zero of the normal map [30]

$$F_+(x) := F(\pi(x)) + x - \pi(x),$$

where $\pi(x)$ is the closest point in B to the variable x in the Euclidean norm. It is well known [30] that finding a zero of this normal map is equivalent to solving MCP. In particular if x is a zero of the normal map, then $\pi(x)$ solves MCP, while if z solves MCP then $z - F(z)$ is a zero of the normal map.

3.1 Overview of the algorithm

The essential idea of the code is to linearize the normal map $F_+(x)$ about the current iterate to obtain a piecewise linear map whose zero is sought using a homotopy approach [7]. To monitor progress in the nonlinear model, a nonmonotone path-search is used [29]. Recent extensions [9] have introduced a function Ψ to be used in conjunction with the code, both as a residual and a merit function.

The following pseudo code shows the main algorithm steps of the PATH solver to find a KKT point

Loop until $\Psi(x)$ is less than a convergence tolerance {

Solve the linearization of the MCP problem to obtain the Newton point;

Search the path between the current point and the Newton point.

If the new point gives rise to a better value for the merit function then accept it.

Otherwise use the merit function to find a descent direction and search along this direction for a new point.

}

Details on the solution of linearization and the path-search mechanism can be found in [6,10]. In this paper, we just indicate the changes specific to solving NLP's. The Newton-type PATH solver uses the Jacobian matrix of the MCP function (3) to find its path-searching direction. In the above context, the Jacobian matrix is computed by finding the derivative of the MCP function. It uses the first and second order derivatives of the original NLP objective function and constraints as

$$\nabla_z F(z) := \begin{bmatrix} \nabla_{xx}^2 L(x, \lambda, \nu) & -\nabla_x^T g(x) & -\nabla_x^T h(x) \\ \nabla_x g(x) & 0 & 0 \\ \nabla_x h(x) & 0 & 0 \end{bmatrix},$$

where $\nabla_{xx}^2 L(x, \lambda, \nu) = \nabla_{xx}^2 f(x) - \sum_{i=1}^m \lambda_i \nabla_{xx}^2 g_i(x) - \sum_{j=1}^p \nu_j \nabla_{xx}^2 h_j(x)$.

3.2   The merit function for the PATH solver

The most recent version of the PATH solver [9] does not use the residual of the normal map for a merit function. Instead, it utilizes the Fischer-Burmeister function [13] defined as the mapping $\phi : \mathbf{R}^2 \to \mathbf{R}$,

$$\phi(p, q) := \sqrt{p^2 + q^2} - p - q,$$

where p and q are scalar variables. This function exhibits the complementarity property when the function value is zero, that is

$$\phi(p, q) = 0 \text{ if and only if } p \geq 0, q \geq 0 \text{ and } pq = 0.$$

For the MCP problem, the residual and merit function used is $\Psi : \mathbf{R}^n \to \mathbf{R}$,

$$\Psi(x) := \frac{1}{2}\psi(x)^T\psi(x),$$

where $\psi(x)$ is the Fischer operator [1] defined in (4) from $\mathbf{R}^n$ to $\mathbf{R}^n$ that maps x_i and $F_i(x)$ as parameters to the Fischer-Burmeister function component-wise as follows:

$$\psi_i(x) := \begin{cases} \phi(x_i - l_i, F_i(x)) & \text{if } -\infty < l_i \leq x_i < \infty, \\ -\phi(u_i - x_i, -F_i(x)) & \text{if } -\infty < x_i \leq u_i < \infty, \\ \phi(x_i - l_i, \phi(u_i - x_i, -F_i(x))) & \text{if } -\infty < l_i \leq x_i \leq u_i < \infty, \\ -F_i(x) & \text{if } -\infty < x_i < \infty. \end{cases} \quad (4)$$

This function is nonnegative and is zero at the solution point. A key feature for its use as a merit function is its continuously differentiability. It allows gradient steps to be used when the path-searching direction does not lead to a descent direction.

The nonlinear MCP function (3) from Section 2.2 contains only the first order derivatives of the objective function and constraints. The formulation exhibits the deficiency of finding KKT points for NLP. In an effort to avoid this deficiency, we introduce a new merit function for the PATH solver that explicitly incorporates the objective function. We now describe the implementation of the new merit function and give some computational results in Section 5.

The PATH solver uses a merit function to find a gradient descent direction when its Newton direction fails to find a descent direction. It uses the residual function $\Psi(x)$ to identify the stopping criteria. We define a new merit function for the PATH solver applied to NLP's which is a weighted average of the residual function Ψ and the objective function f as

$$\varphi(x) = (1 - \gamma)\Psi(x) + \gamma f(x),$$

where $\gamma \in [0, 1]$.

When γ is equal to zero, $\varphi(x) = \Psi(x)$ entreating the original PATH solver to satisfy the first order conditions of the NLP problem. For $\gamma > 0$, the objective function affects the search direction. However, if the weighted value of the objective function reaches 1, then a solution is not guaranteed to satisfy the first order conditions. With appropriate choice of γ, our new merit function guides the path-searching algorithm to escape KKT points that are not local minimizers of the original NLP. After our experimentation with the value of γ, we decided to take a fixed value of $\gamma = 0.3$ for the purposes of the results given in Section 5.

In the next section, we show how the NLP model in AMPL is automatically modified and transformed into the MCP formulation. The MCP function (3) and its Jacobian evaluation are specified in more detail.

4   The PATHNLP solver for AMPL nonlinear programs

To solve the NLP problem in AMPL, a user could specify the complementarity formulation directly using the AMPL language [8]. This would require a modeler to write down explicitly the first order conditions as detailed in Section 2.2. This process is very cumbersome and prone to error. In this paper, we propose to use the AMPL solver library to take an NLP specified directly in AMPL and form the required F and its Jacobian matrix for the PATH solver automatically within the solver link. This means that a modeler simply has to change the solver name in order to use the approach outlined in this paper.

4.1   MCP formulation from AMPL

The NLP problem passed to a solver from the AMPL environment is defined as

$$\begin{aligned} \text{minimize} \quad & f(x) \\ \text{subject to } a \leq c(x) \leq b, \; r \leq x \leq s, \end{aligned}$$

where $f : \mathbf{R}^n \mapsto \mathbf{R}, c : \mathbf{R}^n \mapsto \mathbf{R}^m$ with $a, b \in \mathbf{R}^m$ and $x, r, s \in \mathbf{R}^n$.

We now show how to recover the NLP format (1) as described in Section 2 from the data given above. We define five mutually exclusive index subsets of an index set $I = \{1, 2, \ldots, m\}$ of the constraint function c as

$$\mathcal{L} := \{i \in I| -\infty < a_i \text{ and } b_i \equiv \infty\}$$
$$\mathcal{U} := \{i \in I| a_i \equiv -\infty \text{ and } b_i < \infty\}$$
$$\mathcal{E} := \{i \in I| -\infty < a_i = b_i < \infty\}$$
$$\mathcal{R} := \{i \in I| -\infty < a_i < b_i < \infty\}$$
$$\mathcal{F} := \{i \in I| a_i \equiv -\infty \text{ and } b_i \equiv \infty\},$$

where $\mathcal{L}$ is the index set of lower bound constraints, $\mathcal{U}$ is the index set of upper bound constraints, $\mathcal{E}$ is the index set of equality constraints, $\mathcal{R}$ is the index set of range constraints, and $\mathcal{F}$ is the index set of free constraints.

The NLP model from AMPL is therefore rewritten as

$$
\begin{aligned}
\text{minimize} \quad & f(x) \\
\text{subject to } a_i \leq & c_i(x) & i \in \mathcal{L} \\
& c_i(x) \leq b_i & i \in \mathcal{U} \\
& c_i(x) = a_i & i \in \mathcal{E} \\
a_i \leq & c_i(x) \leq b_i & i \in \mathcal{R} \\
& c_i(x) \text{ is free } i \in \mathcal{F} \\
r \leq & x \leq s.
\end{aligned}
$$

Define $y \in \mathbf{R}^{|\mathcal{R}|}$ as artificial variables for each range constraint, where $|\mathcal{R}|$ is the number of range constraints. Then by dropping the free constraints, the model is equivalent to

$$
\begin{aligned}
\text{minimize} \quad & f(x) \\
\text{subject to} \quad & a_i - c_i(x) & \leq 0 \, i \in \mathcal{L} \\
& c_i(x) - b_i & \leq 0 \, i \in \mathcal{U} \\
& c_i(x) - a_i & = 0 \, i \in \mathcal{E} \\
& c_i(x) - y_{j_i} & = 0 \, i \in \mathcal{R} \\
& a_i \leq y_{j_i} \leq b_i & i \in \mathcal{R} \\
& r \leq x \leq s,
\end{aligned}
$$

where j_i is the index from 1 to $|\mathcal{R}|$, corresponding to the order of index $i \in \mathcal{R}$. We write the constraint function g and h of the NLP (1) as

$$g(x) = \begin{cases} a_i - c_i(x) & \text{if } i \in \mathcal{L} \\ c_i(x) - b_i & \text{if } i \in \mathcal{U} \end{cases}$$

and

$$h(x) = \begin{cases} c_i(x) - a_i & \text{if } i \in \mathcal{E} \\ c_i(x) - y_{j_i} & \text{if } i \in \mathcal{R}. \end{cases}$$

The new Lagrangian function for this model is

$$rlL(x, \lambda, \nu, y) = f(x) - \lambda_{\mathcal{L}}^T(a_{\mathcal{L}} - c_{\mathcal{L}}(x)) - \lambda_{\mathcal{U}}^T(c_{\mathcal{U}}(x) - b_{\mathcal{U}})$$
$$- \nu_{\mathcal{E}}^T(c_{\mathcal{E}}(x) - a_{\mathcal{E}}) - \nu_{\mathcal{R}}^T(c_{\mathcal{R}}(x) - y).$$

Defining $\lambda = (\lambda_{\mathcal{L}}, \lambda_{\mathcal{U}})$ and $\nu = (\nu_{\mathcal{E}}, \nu_{\mathcal{R}})$, the corresponding MCP model is to find $z = (x, \lambda, \nu, y) \in \mathbf{R}^q$ (where $q = n + m + |\mathcal{R}|$) that is complementary to a nonlinear vector function F from $\mathbf{R}^q \to \mathbf{R}^q$ defined as

$$F(z) := \begin{bmatrix} \nabla_x L(z) \\ a_{\mathcal{L}} - c_{\mathcal{L}}(x) \\ c_{\mathcal{U}}(x) - b_{\mathcal{U}} \\ c_{\mathcal{E}}(x) - a_{\mathcal{E}} \\ c_{\mathcal{R}}(x) - y \\ \nu_{\mathcal{R}} \end{bmatrix},$$

where $\nabla_x L(z) = \nabla_x f(x) - \lambda^T \nabla_x g(x) - \nu^T \nabla_x h(x)$, and

$$\begin{bmatrix} r \\ -\infty \\ -\infty \\ a_{\mathcal{R}} \end{bmatrix} \le z = \begin{bmatrix} x \\ \lambda \\ \nu \\ y \end{bmatrix} \le \begin{bmatrix} s \\ 0 \\ \infty \\ b_{\mathcal{R}} \end{bmatrix}.$$

4.2   Solver links in AMPL

AMPL executes the NLP solver as a separate program and communicates with it using the file system. Files with extension .nl contain a description of the model whereas files with extension .sol contain a termination message and the final solution written by the solver. The AMPL system uses information from these files to allocate space, generate the ASL structure and set global variable values. These values are used to identify the problem dimension, the value of objective function at the current point, the gradient evaluation, the constraint evaluation and its derivatives in sparse format.

Useful global variables are

n_var the total number of variables,
n_obj the total number of objective functions,
n_con the total number of constraints,
nzc   the number of nonzeros in the Jacobian matrix and
nzo   the number of nonzeros of the objective gradient.

The ASL structure is made up of two main components, Edagpars and Edaginfo. The Edagpars contains information to evaluate the objective function, constraint functions and their first and second order derivatives. The Edaginfo contains the upper and lower bounds, the initial point, the compressed column structure of the Jacobian matrix of the constraint functions, the pointer structure of the first order derivatives of the objective function and constraints, and information about the NLP problem. For a complete listing of all global variables and the ASL structure, the reader should consult the AMPL manual [15].

A detailed description of our implementation, called `pathnlp`, now follows. After the `solve` command is invoked in AMPL, the AMPL system generates associated NLP problem files and communicates to the `pathnlp` solver. This solver, written in the C language, automatically constructs the primal-dual formulation of the original NLP problem. It calls the PATH solver with additional options if necessary. The PATH solver runs and returns the status of the solution point via the `Path_Solved` variable and the final solution z using the `Path_FinalZ(p)` routine. The link returns these results to the AMPL system by calling `write_sol`. AMPL reports the solution back to the user who further analyzes and manipulates the model.

We now give details of how F and $\nabla_z F$ are evaluated in the link.

- Our program allocates the ASL structure by calling `ASL_alloc` with parameter `ASL_read_pfgh` which requests the AMPL to generate all first order and second order derivatives of the objective function and constraints. In addition, the flag, `want_xpi0 = 1` is set to 1 to request the initial point. The flag, `want_deriv = 1` is set to 1 to request Jacobian evaluations and Hessian evaluations.

- Our program initializes all NLP variables by calling `getstub`. It calls `jacdim` to obtain information about the Jacobian and Hessian of the objective function and constraints.

- Our program defines the MCP variable z as (x, λ, ν, y) and sets up the lower bound as $(r, -\infty, -\infty, a_{\mathcal{R}})$ and the upper bound as $(s, 0, \infty, b_{\mathcal{R}})$.

- The function evaluation of the MCP model is defined as

$$
F(z) := \begin{bmatrix} \nabla_x L(z) \\ a_{\mathcal{L}} - c_{\mathcal{L}}(x) \\ c_{\mathcal{U}}(x) - b_{\mathcal{U}} \\ c_{\mathcal{E}}(x) - a_{\mathcal{E}} \\ c_{\mathcal{R}}(x) - y \\ \nu_{\mathcal{R}} \end{bmatrix}.
$$

The value of this function at the current point is kept in the vector F. To compute $\nabla_x L(z) = \nabla_x f(x) - \lambda^T \nabla_x g(x) - \nu^T \nabla_x h(x)$, the program first evaluates $\nabla_x f(x)$ at the current point by calling `objgrd`. It retrieves the sparse Jacobian matrix of c by calling `jacval` and uses `Cgrad` as the sparse matrix structures. This produces values of $c(x)$. Then it multiplies the sparse Jacobian matrix with the corresponding Lagrange multipliers and subtracts these from $\nabla_x f(x)$. The rest of the vector is computed by calling `conval` and using the appropriate multipliers of 1, -1 or 0 to generate the vector F. Then it copies the values of $\nu_{\mathcal{R}}$ for the last $|\mathcal{R}|$ elements.

- The Jacobian evaluation of the MCP (3) is given as

$$\begin{bmatrix} \nabla_{xx}^2 L(z) & +\nabla_x c_{\mathcal{L}}(x) & -\nabla_x c_{\mathcal{U}}(x) & -\nabla_x c_{\mathcal{E}}(x) & -\nabla_x c_{\mathcal{R}}(x) & 0 \\ -\nabla_x c_{\mathcal{L}}(x) & 0 & 0 & 0 & 0 & 0 \\ +\nabla_x c_{\mathcal{U}}(x) & 0 & 0 & 0 & 0 & 0 \\ +\nabla_x c_{\mathcal{E}}(x) & 0 & 0 & 0 & 0 & 0 \\ +\nabla_x c_{\mathcal{R}}(x) & 0 & 0 & 0 & 0 & -I \\ 0 & 0 & 0 & I & 0 & 0 \end{bmatrix}.$$

This computation uses the Hessian of the Lagrangian evaluation implemented in AMPL using the following form

$$\nabla_{xx}^2 L(x) = \nabla_{xx}^2 \left[\sum_{i=0}^{n\_obj-1} OW[i] f_i(x) + \sigma \sum_{i=0}^{n\_con-1} Y[i] c_i(x) \right],$$

where f_i is the objective function, c_i is the constraint function, σ is a scaling factor commonly set to +1 or -1, $OW[i]$ is a scaling factor for objective function f_i, and $Y[i]$ is Lagrange multiplier for each c_i and equals to zero when c_i is a free constraint.

To call this routine, our program sets up the scale multiplier to be 1, $OW[0] = 1$, and the scale multiplier for the sum of constraints to be negative one, $\sigma = -1$. It copies the appropriate Lagrange multipliers to Y and calls the function sphes. The result returns in the structure variable named sputinfo which is already in the compressed column vector format used by PATH. The matrix is stored as the top left corner of the MCP Jacobian matrix. The rest of the matrix is constructed using jacval and put it in an appropriate place in the MCP Jacobian matrix. Note that our program uses FORTRAN indices, which is a requirement for the PATH solver.

5   Results using the PATHNLP solver

We assume that a user has created a nonlinear problem using the AMPL syntax and solves it by issuing the following commands:

```
option solver pathnlp;
solve;
```

A user can guide the PATH solver using an option file, path.opt identified by

```
options pathnlp_options "optfile=path.opt";
```

Alternatively, the user can specify the options directly using the following syntax

```
options pathnlp_options "option_name=option_value";
```

Note that option_name must be a valid option of the PATH solver (see [10]). For example, to see the warning messages and current option settings of the PATH solver, a user can specify the following:

```
options pathnlp_options "output_warn=yes output_options=yes";
```

To increase the number of iterations, a user can specify

```
options pathnlp_options
"major_iteration_limit=1000 minor_iteration_limit=10000";
```

To decrease the convergence tolerance from 1×10^{-6} to 1×10^{-8}, a user can specify

```
options pathnlp_options "convergence_tolerance=1E-8";
```

Consult [10,11] for details on these and other options.

5.1   The Hock/Schittkowski test suite

We tested pathnlp with and without the new merit function using the Hock/Schittkowski [20] test suite. This test used 113 NLP problems, since two of the suite are incompletely specified. All problems are retrieved from the AMPL web site, http://www.ampl.com/ampl. The Hock/Schittkowski test suite was implemented in AMPL by Professor Robert Vanderbei.

¿From the 113 NLP problems, 59 problems are unconstrained nonlinear program, 48 problems have only equality constraints, while 3 problems contain range constraints and 3 problems have both equality and range constraints. We compare our results with four different NLP solvers available in AMPL, LANCELOT [4], MINOS [27], NPSOL [17] and SNOPT [16]. All solvers run using their default options. The PATH solver with the new merit function uses the weight $\gamma = 0.30$.

Table 1 shows details of these test runs on the Hock/Schittkowski test suite.

Here Fail identifies the number of errors that occur because of an unexpected break from the solver, Infea identifies the number of solutions that are termed by the solver to be infeasible, No prog identifies the number of solutions that cannot be improved upon the current point by the solver, Iter identifies the number of solutions that the solver reached its default iteration limits, Local indicates the number of solutions that the solver found solutions that are different from reported global solutions, Optimal identifies the number of optimal solutions that are the same as reported optimal solutions, and KKT identifies the sum of Local and Optimal, which are KKT solutions.

The PATHNLP solver with the new merit function is very effective for solving this problem suite, solving 108 out of 113 problems. It is certainly

Table 1. Number of final solutions reported from each solver

Solver	Fail	Infea	No prog	Iter	Local	Optimal	KKT
LANCELOT	1	2	9	8	2	91	93
MINOS	0	1	0	7	11	94	105
NPSOL	7	0	2	0	8	96	104
PATH	0	0	10	0	21	82	103
PATH (merit)	0	0	5	0	20	88	108
SNOPT	0	0	2	12	4	95	99
Total	8	3	28	27	66	546	

comparable to the other NLP solvers listed here. Furthermore, the new merit function improves the robustness of the PATH code over the default version.

The test suite provides an indication of the global solution for each of the problems. Comparing these values to those found by our algorithms, the columns labeled Local and Optimal can be generated. As one can see from the local solution column, the PATHNLP solver is more likely to find first order points that are not globally optimal for this given test problems. A more complete breakdown of the failures is given in Table 2.

Table 2. Number of nonoptimal solutions reported from each solver.

Solver	Unconstrained	Equalities	Ranges	Both	Total
LANCELOT	19	3	0	0	22
MINOS	9	10	0	0	19
NPSOL	11	5	0	0	16
PATH	23	8	0	0	31
PATH (merit)	16	6	3	0	25
SNOPT	15	3	0	0	18
Total	59	48	3	3	

It is clear that for finding globally optimal solutions, the NPSOL solver is the most effective solver, failing only 16 times.

Table 3 reports the total timing of nonoptimal and optimal solutions from each solver in seconds. Results were tested on the Sparc machine with 64 MB RAM running SunOS version 5.6.

Table 3 shows that SNOPT uses less time to solve this problem suite. It spends only 20.95% of the total times to detect nonoptimal solutions or failures. MINOS consumes the largest times to find nonoptimal solutions but comparable to SNOPT for finding globally optimal solutions. Our PATHNLP solver with the merit function reduces the total time by 31.36% from the default version of PATH. Clearly, these problems are too small to derive many definitive conclusions on speed.

Table 3. Total timing of nonoptimal and optimal solutions from each solver in seconds.

Solver	Nonoptimal	Optimal	Total
LANCELOT	127.52	123.24	250.76
MINOS	352.47	39.66	392.13
NPSOL	130.91	60.10	191.01
PATH	107.15	100.30	207.45
PATH (merit)	63.43	78.95	142.38
SNOPT	9.23	34.83	44.06
Total	790.71	437.08	1227.79

5.2   Large nonlinear programs

We selected 4 other problems as representative large scale examples from portfolio optimization, minimal surface design, nonnegative least squares and structural optimization. All problems were retrieved from the AMPL web site, http://www.ampl.com/ampl. Some information regarding size and numbers of (equality) constraints is given in Table 4.

Table 4. Problem dimension statistics.

Problem	Variables	Constraints	Optimal Value
Markowitz	1200	201	-0.526165
Minimal	1681	0	7.611023
NonnegLS	543	393	32.644706
Structural	13448	13488	1039.825620

Table 5 summarizes the result of our test runs on large problem sets. Results were tested on Sparc machine with 245 MB RAM running SunOS version 5.5.1.

Table 5. Total timing from each solver in seconds.

Solver	Markowitz	Minimal	Nonnegative	Structural
LANCELOT	503	106	3	*mem*
MINOS	*sup*	*sup*	*sup*	*inf*
NPSOL	538	657	191	*mem*
PATH	84	333	2	*res*
PATH (merit)	123	221	4	18,375
SNOPT	*itr*	*sup*	*sup*	*ini*

Here a keyword in the table identifies that the solver has difficulty solving this problem, where *mem* identifies that the solver could not allocate enough spaces, *sup* identifies that the solver reported the superbasics limit is too small, *itr* identifies that the solver reached its iteration limits, *inf* identifies that the solver reported problem is unbounded, *res* identifies that the solver exceeded the resource limits and *ini* identifies that the solver found the problem is infeasible due to a bad starting point. Optimal solution values from all successfully solved problems are the same for all solvers, and are reported in Table 4. Note that MINOS and SNOPT failed to solve each of these large problems, while PATHNLP with merit function solved all of them. This shows the ability of our code for handle large problem sets which is essential for solving the real world models.

6   Conclusion

It is clear from the results presented here that forming the KKT system and solving this as a complementarity problem is a viable approach for nonlinear programming. Further experimentation is required to ascertain whether a primal-dual formulation or the use of the second order information is the critical aspect. Moreover, by adapting the link code we have described in this paper, we can solve the NLP problem using other MCP solvers such as semismooth [5], or an interior point approach [33]. This will be subject of further research.

Currently the PATH solver uses a proximal point perturbation [2] to overcome singularity problems in the Jacobian matrix. This has the tendency to remove any negative curvature and may hinder progress on non-convex problems. The improvement in performance by using the composite merit function leads us to believe that further progress on this front can be achieved by (i) modification and tuning of the merit function and (ii) exploitation of negative curvature instead of using proximal point perturbation.

References

1. Billups, S. C. (1995) Algorithms for Complementarity Problems and Generalized Equations. PhD thesis, University of Wisconsin–Madison, Madison, Wisconsin
2. Billups, S. C., Ferris, M. C. (1997) QPCOMP: A quadratic program based solver for mixed complementarity problems. Mathematical Programming **76**, 533–562
3. Brooke, A., Kendrick, D., Meeraus, A. (1988) GAMS: A User's Guide. The Scientific Press, South San Francisco, CA
4. Conn, A. R., Gould, N. I. M., Toint, Ph. L. (1992) LANCELOT: A Fortran package for Large–Scale Nonlinear Optimization (Release A). Number 17 in Springer Series in Computational Mathematics. Springer Verlag, Heidelberg, Berlin

5. De Luca, T., Facchinei, F., Kanzow, C. (1996) A semismooth equation approach to the solution of nonlinear complementarity problems. Mathematical Programming **75**, 407–439
6. Dirkse, S.P., Ferris, M.C. (1995) The PATH solver: A non-monotone stabilization scheme for mixed complementarity problems. Optimization Methods and Software **5**, 123–156
7. Eaves, B. C. (1976) A short course in solving equations with PL homotopies. In: Cottle R. W., Lemke C. E. (Eds.) Nonlinear Programming. American Mathematical Society, SIAM–AMS Proceedings. Providence, RI, 73–143
8. Ferris, M. C., Fourer, R., Gay, D.M. (1999) Expressing complementarity problems and communicating them to solvers. SIAM Journal on Optimization, forthcoming
9. Ferris, M.C., Kanzow, C., Munson, T.S. (1999) Feasible descent algorithms for mixed complementarity problems. Mathematical Programming, forthcoming
10. Ferris, M.C., Munson, T.S. (1999) Complementarity problems in GAMS and the PATH solver. Journal of Economic Dynamics and Control, forthcoming
11. Ferris, M.C., Munson, T.S. (1999) Interfaces to PATH 3.0: Design, implementation and usage. Computational Optimization and Applications, **12**, 207–227
12. Fiacco, A.V., McCormick, G.P. (1968) Nonlinear Programming: Sequential Unconstrained Minimization Techniques. John Wiley & Sons, New York. (SIAM Classics in Applied Mathematics 4, SIAM, Philadelphia, 1990)
13. Fischer, A. (1992) A special Newton–type optimization method. Optimization **24**, 269–284
14. Fourer, R., Gay, D.M., Kernighan, B.W. (1993) AMPL: A Modeling Language for Mathematical Programming. Duxbury Press
15. Gay, D.M. (1997) Hooking your solver to AMPL. Technical report, Bell Laboratories, Murray Hill, New Jersey
16. Gill, P.E., Murray, W., Saunders, M.A. (1997) SNOPT: An SQP algorithm for large-scale constrained optimization. Report NA 97-2, Department of Mathematics, University of California, San Diego, San Diego, California
17. Gill, P.E., Murray, W., Saunders, M.A., Wright, Margaret H. (1986) User's Guide for NPSOL (Version 4.0): A Fortran Package for Nonlinear Programming. Technical Report SOL 86-2, Department of Operations Research, Stanford University, Stanford, California
18. Griewank, A., Corliss, G.F. (1991) Automatic differentiation of algorithms: Theory, implementation, and application. SIAM, Philadelphia, Pennsylvania
19. Griewank, A., Juedes, D., Utke, J. (1996) ADOL-C: A package for the automatic differentiation of algorithms written in C/C++. ACM Transactions on Mathematical Software **20**, 131–167
20. Hock, W., Schittkowski, K. (1981) Test Examples for Nonlinear Programming Codes. Lecture Notes in Economics and Mathematical Systems **187**, Springer Verlag, Berlin
21. Karmarkar, N. (1984) A new polynomial time algorithm for linear programming. Combinatorica **4**, 373–395
22. Karush, W. (1939) Minima of functions of several variables with inequalities as side conditions. Master's thesis, Department of Mathematics, University of Chicago
23. Khachian, L.G. (1979) A polynomial algorithm for linear programming. Soviet Mathematics Doklady **20**, 191–194

24. H. W. Kuhn, H.W., Tucker, A.W. (1951) Nonlinear programming. In: Neyman J. (Ed.) Proceedings of the Second Berkeley Symposium on Mathematical Statistics and Probability, University of California Press, Berkeley and Los Angeles, 481–492
25. Mangasarian, O.L. (1969) Nonlinear Programming. McGraw–Hill, New York, 1969. (SIAM Classics in Applied Mathematics 10, SIAM, Philadelphia, 1994)
26. Murtagh, B.A., Saunders, M.A. (1978) Large-scale linearly constrained optimization. Mathematical Programming **14**, 41–72
27. Murtagh, B.A., Saunders, M.A. (1983) MINOS 5.0 user's guide. Technical Report SOL 83.20, Stanford University, Stanford, California
28. Rall, L.B. (1981) Automatic Differentiation: Techniques and Applications **120**, Springer Verlag, Berlin
29. Ralph, D. (1994) Global convergence of damped Newton's method for nonsmooth equations, via the path search. Mathematics of Operations Research **19**, 352–389
30. Robinson, S.M. (1992) Normal maps induced by linear transformations. Mathematics of Operations Research **17**, 691–714
31. Robinson, S.M. (1994) Newton's method for a class of nonsmooth functions. Set Valued Analysis **2**, 291–305
32. Rockafellar, R.T. (1970) Convex Analysis. Princeton University Press, Princeton, New Jersey
33. Shanno, D., Simantiraki, E. (1997) Interior point methods for linear and nonlinear programming. In: Duff I. S., Watson G.A. (Eds.) State of the Art in Numerical Analysis, Oxford University Press, Oxford
34. Wright, S.J. (1997) Primal–Dual Interior–Point Methods. SIAM, Philadelphia, Pennsylvania

Convergence Analysis of a Perturbed Version of the $\theta-$Scheme of Glowinski-Le Tallec

Sylvianne Haubruge, Van Hien Nguyen, and Jean-Jacques Strodiot

Facultés Universitaires Notre-Dame de la Paix, Département de Mathématique, Unité d'Optimisation, Namur, Belgium

Abstract. Many problems of convex programming can be reduced to that of finding a zero of the sum of two maximal monotone operators. Different splitting methods have been proposed to solve this problem, for example the Forward-Backward scheme, and the θ-scheme.

In this paper, we consider the θ-scheme. Our purpose is to give a convergence result for a perturbed version of it. Then, we consider a convex programming problem and we present some classes of penalty functions associated to this problem which could be used to apply the perturbed θ-scheme to it.

Key words: maximal monotone operator, splitting method, $\theta-$scheme, perturbation, semi-variational convergence, penalty function

1 Introduction

Many problems of convex programming can be reduced to that of finding a zero z of a maximal monotone operator T on $I\!\!R^n$, i.e., $0 \in Tz$. This problem can be solved via the proximal point algorithm, which generates a sequence $\{z^k\}$ as follows:

$$\begin{cases} z^{k+1} = J_{\lambda T} z^k, \ k = 0, 1, \dots \\ z^0 \in I\!\!R^n \end{cases}$$

where $J_{\lambda T} = (I + \lambda T)^{-1}$ is called the resolvent of T. The major drawback of this algorithm is that it requires the evaluation of the resolvent, which can be difficult.

However, in some cases, the operator T can be splitted in the sum of two maximal monotone operators A and B whose resolvents $J_{\lambda A}$ and $J_{\lambda B}$ are easier to evaluate than $J_{\lambda T}$. In this case, a strategy consists in finding a zero of $T = A + B$ by using only the resolvents $J_{\lambda A}$ and $J_{\lambda B}$ of A and B, rather than $J_{\lambda T}$. Such a method is called an operator splitting method.

In this paper, we will consider splitting methods for solving the problem

(P) $\qquad\qquad$ Find $z \in I\!\!R^n$ such that $0 \in (A + B)z$,

where both A and B are maximal monotone operators defined on the Euclidean space $I\!\!R^n$.

A large variety of operator splitting methods can be found in the literature. The first one is the forward-backward scheme where a forward step for B is alternated with a backward step for A as follows:

$$z^{k+1} = J_{\lambda_k A}(I - \lambda_k B)z^k$$

with $\{\lambda_k\}$ some sequence of positive stepsizes. This algorithm has been extensively studied by Passty [21], Gabay [10], Tseng [24], [25], Chen and Rockafellar [3], [4], [5], [6] and Zhu [26]. It converges to a zero of $A + B$ if B^{-1} is strongly monotone. Another one is the $\theta-$scheme, which has been introduced by Glowinski and Le Tallec [11], [12], [18], and whose the iteration can be written as follows:

$$z^{k+1} = J_{\lambda_1 A}(I - \lambda_1 B)J_{\lambda_2 B}(I - \lambda_2 A)J_{\lambda_1 A}(I - \lambda_1 B)z^k \qquad (1)$$

where $0 < \lambda_2 \leq \lambda_1$. It has been proven in [14] that it converges also to a zero of $A + B$ if B^{-1} is strongly monotone.

In this paper, we consider a perturbed version of the $\theta-$scheme. At each iteration k, we replace the operators A and B by some approximations A^k and B^k. The iteration becomes then:

$$z^{k+1} = J_{\lambda_1 A^k}(I - \lambda_1 B^k)J_{\lambda_2 B^k}(I - \lambda_2 A^k)J_{\lambda_1 A^k}(I - \lambda_1 B^k)z^k. \qquad (2)$$

This kind of approximation has been done in the literature for other methods. For example, Lemaire in [16] considered a perturbation of the proximal point algorithm in the case where T is the subdifferential of a closed proper convex function f, i.e. $T = \partial f$. The idea was to replace, at each iteration k, the function f by another closed proper convex function f^k, which approaches f in a suitable sense. The iteration was then given by:

$$z^{k+1} = J_{\lambda \partial f^k}z^k.$$

In 1990, Tossings [23] extended the study of the perturbed proximal point algorithm to the case where T is a general maximal monotone operator. The operator T was then replaced, at each iteration k, by another maximal monotone operator T^k approaching T in some sense, giving then the following iteration:

$$z^{k+1} = J_{\lambda T^k}z^k.$$

More recently, Lemaire [17] presented a general result saying that, under some convergence assumptions on the perturbed operators, if a basic method is convergent, then so is also the perturbed method. He applied this result to the Forward-Backward scheme to give a convergence result for a perturbed version of it in which both A and B are perturbed, i.e.

$$z^{k+1} = J_{\lambda_k A^k}(I - \lambda_k B^k)z^k.$$

In this paper, we will apply the result of Lemaire for proving the convergence of our perturbed version of the θ–scheme (2). We prove that if the sequences $\{A^k\}$ and $\{B^k\}$ converge to A and B respectively, in the sense that

$$\sum_{k \geq 1} \delta_{\lambda_1, \rho}(A^k, A) < \infty, \quad \forall \rho \geq 0$$

and

$$\sum_{k \geq 1} \delta_{\lambda_2, \rho}(B^k, B) < \infty, \quad \forall \rho \geq 0,$$

where $\delta_{\lambda, \rho}(T^1, T^2) = \sup_{\|x\| \leq \rho} \|J_{\lambda T^1} x - J_{\lambda T^2} x\|$ is called the variational metric between the two maximal monotone operators T^1 and T^2, with parameters λ and ρ (see [23]), then the sequence $\{z^k\}$ generated by (2) converges to a solution of (P).

Then, we consider the following convex programming problem:

$$\begin{cases} \min f(x) \\ x \in C \end{cases}$$

where $C = \{x \in I\!\!R^n : g_i(x) \leq 0, 1 \leq i \leq m\}$. We give, for this problem, some examples of functions, called "penalty functions", satisfying the proximity assumption.

The paper is organised as follows. Section 2 is devoted to preliminary results. We recall the result obtained by Lemaire concerning the convergence of a general perturbed method (see [17]), and the convergence result for the nonperturbed θ–scheme (see [14]). In Section 3, we prove the convergence of the perturbed θ–scheme. Finally, in Section 4, we consider a convex programming problem. We present some penalty functions associated to it and we see that they satisfy the proximity assumption. Throughout this paper, $\| \cdot \|$ denotes the l_2 norm for vectors or matrices, and $\langle \cdot, \cdot \rangle$ denotes the Euclidean scalar product of $I\!\!R^n$. Let $T : I\!\!R^n \to I\!\!R^n$ be a multivalued operator. We will make no difference between T and its graph, that is, the set

$$Gr(T) = \{(x, y) \in I\!\!R^n \times I\!\!R^n \,|\, y \in Tx\}.$$

The effective domain of T, which will be denoted by $D(T)$, is the projection of T onto the first coordinate

$$D(T) = \{x \in I\!\!R^n \,|\, \exists y \in I\!\!R^n \text{ such that } (x, y) \in T\} = \{x \in I\!\!R^n \,|\, Tx \neq \emptyset\}.$$

The inverse T^{-1} of T is defined as follows

$$T^{-1} = \{(x, y) \in I\!\!R^n \times I\!\!R^n \,|\, (y, x) \in T\}.$$

An operator T is said to be monotone if

$$\langle x - x', y - y' \rangle \geq 0 \quad \forall (x, y), (x', y') \in T.$$

We say that T is maximal monotone if, in addition, its graph is not strictly contained in the graph of any other monotone operator. It is well known that T is maximal monotone if and only if T^{-1} is maximal monotone. The most familiar example of a maximal monotone operator is the subdifferential ∂f of a closed proper convex function $f : \mathbb{R}^n \to \mathbb{R} \cup \{+\infty\}$, that is,

$$\partial f(x) = \{y \in \mathbb{R}^n \mid f(z) \geq f(x) + \langle y, z - x \rangle \quad \forall z \in \mathbb{R}^n\}.$$

Let us recall that the resolvent J_{cT} of a maximal monotone operator T is single-valued, nonexpansive and everywhere defined (see, for example, [2], [7]).

T is strongly monotone with modulus $\sigma > 0$ if

$$\langle x - x', y - y' \rangle \geq \sigma \|x - x'\|^2 \quad \forall (x, y), (x', y') \in T.$$

2   Preliminary results

In this section, we present the results which will be used in the sequel for proving the convergence of the perturbed θ-scheme. The first one concerns the convergence of a general perturbed scheme and the second one concerns the convergence of the nonperturbed θ-scheme.

Firstly, let us introduce the following iteration, which will be called basic method, devoted to solve our problem:

$$\begin{cases} z^k = P_k z^{k-1} , k = 1, 2, \dots \\ z^0 \in \mathbb{R}^n \end{cases}$$

where the iteration mappings $P_k : \mathbb{R}^n \to \mathbb{R}^n$ are built from the operators A and B, and may also depend on the iteration index k through a sequence of parameters λ_k.

To this basic method, we associate the following perturbed one:

$$\begin{cases} z^k = Q_k z^{k-1} , k = 1, 2, \dots \\ z^0 \in \mathbb{R}^n \end{cases}$$

where the sequence of operators Q_k is defined on the same manner than in the basic method, except in the fact that, at each step k, the operators A and B are replaced by perturbed ones, denoted by A^k and B^k respectively.

The following theorem (see [17]) establishes the convergence of the perturbed method:

Theorem 1. *Let us consider the basic method with iteration mappings P_k, and the associated perturbed method with iteration mappings Q_k. Let us assume that, $\forall k \in \mathbb{N}, \forall x, y \in \mathbb{R}^n, \forall \rho \geq 0$,*

- *P_k is nonexpansive*
- *$\bar{x}$ is a fixed point of P_k*

- $\|Q_k(x) - Q_k(y)\| \leq (1 + \epsilon_k)\|x - y\|$, where $\epsilon_k \geq 0 \ \forall k$ and $\sum_{k \geq 1} \epsilon_k < \infty$
- $\sum_{k \geq 1} \Delta_{k,\rho} < \infty$, where $\Delta_{k,\rho} = \sup_{\|x\| \leq \rho} \|Q_k(x) - P_k(x)\|$

We suppose also that the iteration mappings P_k depend on the iteration index only through a sequence of parameters and that the property needed on this sequence for the convergence of the basic method is invariant by translation. Then, if the basic method converges to a solution of (P), the sequence generated by the perturbed method converges also to a solution of (P).

In the sequel of this paper, we will take, $\forall k \geq 0$,

$$P_k = J_{\lambda_1 A}(I - \lambda_1 B)J_{\lambda_2 B}(I - \lambda_2 A)J_{\lambda_1 A}(I - \lambda_1 B) \qquad (3)$$

and

$$Q_k = J_{\lambda_1 A^k}(I - \lambda_1 B^k)J_{\lambda_2 B^k}(I - \lambda_2 A^k)J_{\lambda_1 A^k}(I - \lambda_1 B^k) \qquad (4)$$

and Theorem 1 will then allow us to prove the convergence of the perturbed θ—scheme. To this effect, we will also need the following theorem, which establishes the convergence of the θ—iteration in the nonperturbed case.

Theorem 2. *([14]) Let z^* be any solution of problem (P). Suppose that B^{-1} is strongly monotone with modulus $\sigma > 0$ and $0 < \lambda_2 \leq \lambda_1 < 2\sigma$. Then the sequence $\{z^k\}$ generated by the θ—iteration*

$$z^{k+1} = J_{\lambda_1 A}(I - \lambda_1 B)J_{\lambda_2 B}(I - \lambda_2 A)J_{\lambda_1 A}(I - \lambda_1 B)z^k$$

is such that, for all k,

$$\|z^{k+1} - z^*\| \leq \|z^k - z^*\|.$$

Moreover the sequence $\{z^k\}$ converges to a solution of (P).

Finally, let us also recall the following lemma, which concerns the Lipschitz character of the operator $\gamma J_{\mu T} - I$.

Lemma 1. *([14]) Let T be a maximal monotone operator defined on $\mathbb{R}^n$ and let μ and ν be two positive real numbers. Set $\gamma = 1 + (\mu/\nu)$. Then the operator $\gamma J_{\mu T} - I$ is Lipschitz continuous with constant $L = \max\{1, (\mu/\nu)\}$. In particular, when $\mu = \nu$, the operator $2J_{\mu T} - I$ is nonexpansive.*

3   Convergence of the perturbed θ—scheme

In this section, we establish the convergence of a perturbed θ—iteration, in which both A and B are perturbed.

Theorem 3. *Let A and B be two maximal monotone operators. Assume that B^{-1} is strongly monotone with modulus $\sigma > 0$ and that $0 < \lambda_2 \leq \lambda_1 < 2\sigma$.*
Let $\{A^k\}$ and $\{B^k\}$ be two sequences of maximal monotone operators such that:

$$\sum_{k \geq 1} \delta_{\lambda_1, \rho}(A^k, A) < \infty, \quad \forall \rho \geq 0 \tag{5}$$

and

$$\sum_{k \geq 1} \delta_{\lambda_2, \rho}(B^k, B) < \infty, \quad \forall \rho \geq 0. \tag{6}$$

Suppose, in addition, that, for all k, $B^k = B + G^k$, where G^k is a single-valued operator on $\mathbb{R}^n$, satisfying the following conditions:

$$\|G^k(x) - G^k(y)\| \leq \eta_k \|x - y\|, \text{ where } \eta_k \geq 0 \text{ and } \sum_{k \geq 1} \eta_k < \infty, \tag{7a}$$

$$\exists x^0 \in \mathbb{R}^n : G^k(x^0) = 0 \ \forall k. \tag{7b}$$

Then the sequence generated by the following iteration

$$z^{k+1} = J_{\lambda_1 A^k}(I - \lambda_1 B^k) J_{\lambda_2 B^k}(I - \lambda_2 A^k) J_{\lambda_1 A^k}(I - \lambda_1 B^k) z^k \tag{8}$$

converges to a solution of (P).

Proof. Let P_k and Q_k be defined by (4) and (5) respectively. We will check that the assumptions of Theorem 1 are satisfied. To this effect, let us firstly remark that, by using the definitions of P_k and Q_k both with the following identity (see [14]):

$$I - \nu T = \frac{\nu}{\mu}(\gamma J_{\mu T} - I)(I + \mu T), \tag{9}$$

where T is a maximal monotone operator and $\gamma = 1 + \mu/\nu$, the operators P_k and Q_k can be rewritten equivalently as:

$$P_k = J_{\lambda_1 A}(I - \lambda_1 B) J_{\lambda_2 B} \frac{\lambda_2}{\lambda_1}(\alpha J_{\lambda_1 A} - I)(I - \lambda_1 B) \ \forall k \geq 0 \tag{10}$$

$$Q_k = J_{\lambda_1 A^k}(I - \lambda_1 B^k) J_{\lambda_2 B^k} \frac{\lambda_2}{\lambda_1}(\alpha J_{\lambda_1 A^k} - I)(I - \lambda_1 B^k) \ \forall k \geq 0, \tag{11}$$

where $\alpha = 1 + (\lambda_1/\lambda_2)$.

We will begin by proving that P_k is nonexpansive. B^{-1} being strongly monotone, the operator $(I - \lambda_1 B)$ is nonexpansive. On the other hand, it follows from Lemma 1 that $(\lambda_2/\lambda_1)(\alpha J_{\lambda_1 A} - I)$ is nonexpansive too. Moreover, $J_{\lambda_1 A}$ and $J_{\lambda_2 B}$ are contractions. The announced result then follows from (10).

We have now to prove that $\|Q_k(x) - Q_k(y)\| \leq (1 + \epsilon_k)\|x - y\|$, where $\epsilon_k \geq 0 \; \forall k$ and $\sum_{k \geq 1} \epsilon_k < \infty$. From (37) and the nonexpansiveness of $J_{\lambda_1 A^k}$, we have:

$$\|Q_k(x) - Q_k(y)\| \leq \|(I - \lambda_1 B^k)J_{\lambda_2 B^k}\frac{\lambda_2}{\lambda_1}(\alpha J_{\lambda_1 A^k} - I)(I - \lambda_1 B^k)x$$
$$-(I - \lambda_1 B^k)J_{\lambda_2 B^k}\frac{\lambda_2}{\lambda_1}(\alpha J_{\lambda_1 A^k} - I)(I - \lambda_1 B^k)y\|.$$

Since $B^k = B + G^k$, where G^k is Lipschitz with constant η_k, this implies:

$$\|Q_k(x) - Q_k(y)\| \leq \|(I - \lambda_1 B)J_{\lambda_2 B^k}\frac{\lambda_2}{\lambda_1}(\alpha J_{\lambda_1 A^k} - I)(I - \lambda_1 B^k)x$$
$$-(I - \lambda_1 B)J_{\lambda_2 B^k}\frac{\lambda_2}{\lambda_1}(\alpha J_{\lambda_1 A^k} - I)(I - \lambda_1 B^k)y\|$$
$$+\lambda_1 \eta_k \|J_{\lambda_2 B^k}\frac{\lambda_2}{\lambda_1}(\alpha J_{\lambda_1 A^k} - I)(I - \lambda_1 B^k)x$$
$$-J_{\lambda_2 B^k}\frac{\lambda_2}{\lambda_1}(\alpha J_{\lambda_1 A^k} - I)(I - \lambda_1 B^k)y\|.$$

The operators $(I - \lambda_1 B)$, $J_{\lambda_2 B^k}$, $\frac{\lambda_2}{\lambda_1}(\alpha J_{\lambda_1 A^k} - I)$ being nonexpansive, it follows that:

$$\|Q_k(x) - Q_k(y)\| \leq (1 + \lambda_1 \eta_k)\|(I - \lambda_1 B^k)x - (I - \lambda_1 B^k)y\|.$$

Using once again the properties of B^k and G^k, both with the nonexpansiveness of $(I - \lambda_1 B)$, we get:

$$\|Q_k(x) - Q_k(y)\| \leq (1 + \lambda_1 \eta_k)^2 \|x - y\|$$
$$= (1 + 2\lambda_1 \eta_k + \lambda_1^2 \eta_k^2)\|x - y\|.$$

Let $\epsilon_k = 2\lambda_1 \eta_k + \lambda_1^2 \eta_k^2 \geq 0$. Since $\sum_{k \geq 1} \eta_k < \infty$, we have immediately that $\sum_{k \geq 1} \epsilon_k < \infty$ as required.

Finally, it remains to prove that $\sum_{k \geq 1} \Delta_{k,\rho} < \infty$, where

$$\Delta_{k,\rho} = \sup_{\|x\| \leq \rho} \|Q_k(x) - P_k(x)\|.$$

Let x be such that $\|x\| \leq \rho$. Using (10), (37), the identity (9) and setting $\beta = 1 + (\lambda_2/\lambda_1)$, we get:

$$\|Q_k(x) - P_k(x)\| = \|J_{\lambda_1 A^k}\frac{\lambda_1}{\lambda_2}(\beta J_{\lambda_2 B^k} - I)\frac{\lambda_2}{\lambda_1}(\alpha J_{\lambda_1 A^k} - I)(I - \lambda_1 B^k)x$$
$$-J_{\lambda_1 A}\frac{\lambda_1}{\lambda_2}(\beta J_{\lambda_2 B} - I)\frac{\lambda_2}{\lambda_1}(\alpha J_{\lambda_1 A} - I)(I - \lambda_1 B)x\|$$
$$\leq \|J_{\lambda_1 A^k}\frac{\lambda_1}{\lambda_2}(\beta J_{\lambda_2 B^k} - I)\frac{\lambda_2}{\lambda_1}(\alpha J_{\lambda_1 A^k} - I)(I - \lambda_1 B^k)x$$
$$-J_{\lambda_1 A^k}\frac{\lambda_1}{\lambda_2}(\beta J_{\lambda_2 B} - I)\frac{\lambda_2}{\lambda_1}(\alpha J_{\lambda_1 A} - I)(I - \lambda_1 B)x\|$$
$$+\|J_{\lambda_1 A^k}\frac{\lambda_1}{\lambda_2}(\beta J_{\lambda_2 B} - I)\frac{\lambda_2}{\lambda_1}(\alpha J_{\lambda_1 A} - I)(I - \lambda_1 B)x$$
$$-J_{\lambda_1 A}\frac{\lambda_1}{\lambda_2}(\beta J_{\lambda_2 B} - I)\frac{\lambda_2}{\lambda_1}(\alpha J_{\lambda_1 A} - I)(I - \lambda_1 B)x\|.$$

Let $\rho_1 = \sup_{\|x\| \leq \rho} \|\frac{\lambda_1}{\lambda_2}(\beta J_{\lambda_2 B} - I)\frac{\lambda_2}{\lambda_1}(\alpha J_{\lambda_1 A} - I)(I - \lambda_1 B)x\|$.
By using the definition of the variational metric, we can then write:

$$\|Q_k(x) - P_k(x)\| \leq \|\frac{\lambda_1}{\lambda_2}(\beta J_{\lambda_2 B^k} - I)\frac{\lambda_2}{\lambda_1}(\alpha J_{\lambda_1 A^k} - I)(I - \lambda_1 B^k)x$$
$$-\frac{\lambda_1}{\lambda_2}(\beta J_{\lambda_2 B} - I)\frac{\lambda_2}{\lambda_1}(\alpha J_{\lambda_1 A} - I)(I - \lambda_1 B)x\|$$
$$+\delta_{\lambda_1,\rho_1}(A^k, A)$$
$$\leq \|\frac{\lambda_1}{\lambda_2}(\beta J_{\lambda_2 B^k} - I)\frac{\lambda_2}{\lambda_1}(\alpha J_{\lambda_1 A^k} - I)(I - \lambda_1 B^k)x$$
$$-\frac{\lambda_1}{\lambda_2}(\beta J_{\lambda_2 B^k} - I)\frac{\lambda_2}{\lambda_1}(\alpha J_{\lambda_1 A} - I)(I - \lambda_1 B)x\|$$
$$+\|\frac{\lambda_1}{\lambda_2}(\beta J_{\lambda_2 B^k} - I)\frac{\lambda_2}{\lambda_1}(\alpha J_{\lambda_1 A} - I)(I - \lambda_1 B)x$$
$$-\frac{\lambda_1}{\lambda_2}(\beta J_{\lambda_2 B} - I)\frac{\lambda_2}{\lambda_1}(\alpha J_{\lambda_1 A} - I)(I - \lambda_1 B)x\|$$
$$+\delta_{\lambda_1,\rho_1}(A^k, A).$$

We let now $\rho_2 = \sup_{\|x\| \leq \rho} \|(\lambda_2/\lambda_1)(\alpha J_{\lambda_1 A} - I)(I - \lambda_1 B)x\|$.

As previously, we obtain then, by using also the nonexpansiveness of

$\beta J_{\lambda_2 B^k} - I$ (see Lemma 1):

$\|Q_k(x) - P_k(x)\|$

$$\leq \tfrac{\lambda_1}{\lambda_2}\|\tfrac{\lambda_2}{\lambda_1}(\alpha J_{\lambda_1 A^k} - I)(I - \lambda_1 B^k)x - \tfrac{\lambda_2}{\lambda_1}(\alpha J_{\lambda_1 A} - I)(I - \lambda_1 B)x\|$$

$$+\tfrac{\lambda_1}{\lambda_2}\beta\delta_{\lambda_2,\rho_2}(B^k, B) + \delta_{\lambda_1,\rho_1}(A^k, A)$$

$$\leq \tfrac{\lambda_1}{\lambda_2}\|\tfrac{\lambda_2}{\lambda_1}(\alpha J_{\lambda_1 A^k} - I)(I - \lambda_1 B^k)x - \tfrac{\lambda_2}{\lambda_1}(\alpha J_{\lambda_1 A^k} - I)(I - \lambda_1 B)x\|$$

$$+\tfrac{\lambda_1}{\lambda_2}\|\tfrac{\lambda_2}{\lambda_1}(\alpha J_{\lambda_1 A^k} - I)(I - \lambda_1 B)x - \tfrac{\lambda_2}{\lambda_1}(\alpha J_{\lambda_1 A} - I)(I - \lambda_1 B)x\|$$

$$+\tfrac{\lambda_1}{\lambda_2}\beta\delta_{\lambda_2,\rho_2}(B^k, B) + \delta_{\lambda_1,\rho_1}(A^k, A).$$

By repeating once more the same reasonment, we get:

$$\|Q_k(x) - P_k(x)\| \leq \frac{\lambda_1^2}{\lambda_2}\|B^k x - Bx\| + \alpha\delta_{\lambda_1,\rho_3}(A^k, A) + \frac{\lambda_1}{\lambda_2}\beta\delta_{\lambda_2,\rho_2}(B^k, B)$$
$$+\delta_{\lambda_1,\rho_1}(A^k, A),$$

where $\rho_3 = \sup_{\|x\|\leq\rho}\|(I - \lambda_1 B)x\|$.
As $B^k = B + G^k$, we have, by using also the properties of G^k:

$$\|Q_k(x) - P_k(x)\| \leq \frac{\lambda_1^2}{\lambda_2}\eta_k(\|x\| + \|x^0\|) + \alpha\delta_{\lambda_1,\rho_3}(A^k, A) + \frac{\lambda_1}{\lambda_2}\beta\delta_{\lambda_2,\rho_2}(B^k, B)$$
$$+\delta_{\lambda_1,\rho_1}(A^k, A).$$

Consequently, since $\|x\| \leq \rho$, this yields to:

$$\Delta_{k,\rho} \leq \frac{\lambda_1^2}{\lambda_2}\eta_k(\rho + \|x^0\|) + \alpha\delta_{\lambda_1,\rho_3}(A^k, A) + \frac{\lambda_1}{\lambda_2}\beta\delta_{\lambda_2,\rho_2}(B^k, B)$$
$$+\delta_{\lambda_1,\rho_1}(A^k, A).$$

Since, by assumption, the series $\sum_{k\geq 1}\eta_k, \sum_{k\geq 1}\delta_{\lambda_1,\rho}(A^k, A)$ and $\sum_{k\geq 1}\delta_{\lambda_2,\rho}(B^k, B)$ are finite for all $\rho \geq 0$, so is also $\sum_{k\geq 1}\Delta_{k,\rho}$, as required. The convergence of the sequence $\{z^k\}$ to a solution of (P) follows then from Theorems 1 and 2. $\square$

4   Application to convex programming

In this section, we consider the following problem:

$$(P')\begin{cases} \min f(x) \\ \text{s.t. } x \in C. \end{cases}$$

where $f, g_i : I\!R^n \to I\!R \cup \{+\infty\}(1 \leq i \leq m)$ are closed proper convex functions and $C = \{x \in I\!R^n : g_i(x) \leq 0,\ 1 \leq i \leq m\}$.

We assume that C satisfies the Slater condition and that (P') admits at least one solution. (P') is clearly equivalent to the following problem, where ψ_C denotes the indicator function of the subset C of $I\!R^n$:

$$(P'') \begin{cases} \min\ f(x) + \Psi_C(x) \\ \text{s.t. } x \in I\!R^n. \end{cases}$$

or, in terms of operators,

$$0 \in \partial f(x) + \partial \Psi_C(x).$$

This problem is then of the form of (P) and we will give for it some examples of operators satisfying the assumption:

$$\sum_{k \geq 1} \delta_{\lambda,\rho}(T^k, T) < \infty,\ \ \forall \lambda > 0,\ \forall \rho \geq 0.$$

In fact, we will replace, at each iteration k, the function ψ_C by a penalty function $\varphi_k : I\!R^n \to I\!R$, which will be built from the constraint set C. A lot of such functions has been studied in the literature (see e.g. [8], [23]). So, for example, let us present the class of outer penalties (see [9], [1], [16]). A sequence of functions $\{\varphi_k\}$ belongs to this class if it satisfies the following three conditions:

1. $0 \leq \varphi_k(x) \leq \varphi_{k+1}(x),\ \forall x \in I\!R^n,\ \forall k \in I\!N_0$
2. $\varphi_k(x) = 0,\ \forall k \in I\!N_0$ if $x \in C$
3. $\lim_{k \to \infty} \varphi_k(x) = +\infty$ else.

We can mention the classical outer penalties, defined by:

$$\varphi_k(x) = \frac{s_k}{2} \sum_{i=1}^{m} (g_i^+(x))^2,\ \ \forall x \in I\!R^n,\ \forall k \in I\!N_0 \tag{12}$$

and the exact outer penalties:

$$\varphi_k(x) = s_k \sum_{i=1}^{m} g_i^+(x),\ \ \ \forall x \in I\!R^n,\ \forall k \in I\!N_0, \tag{13}$$

where, in both cases, $\{s_k\}$ denotes an increasing sequence of positive numbers satisfying $\lim_{k \to \infty} s_k = +\infty$ The following results say that the classical outer penalties and the exact outer penalties satisfy the proximity assumption required in Theorem 3.

Theorem 4. (El Bachari [8]) *Let φ_k be defined by (12).*
If $\sum_{k \geq 1} \frac{1}{\sqrt{s_k}} < \infty$, then $\sum_{k \geq 1} \delta_{\lambda,\rho}(\partial \varphi_k, \partial \Psi_C) < \infty,\ \forall \lambda > 0, \forall \rho \geq 0$.

Theorem 5. (El Bachari [8]) *Let φ_k be defined by (8). Then*
$\sum_{k \geq 1} \delta_{\lambda,\rho}(\partial \varphi_k, \partial \Psi_C) < \infty, \; \forall \lambda > 0, \forall \rho \geq 0.$

Let us now present an other class of penalty functions, introduced by Kort and Bertsekas [15], and which contains as a particular case the class of exponential penalties (see [20], [13], [22], [19]):

$$\varphi_k(x) = \frac{1}{s_k} \sum_{i=1}^{m} \phi(r_k g_i(x)) \tag{14}$$

where $\phi : \mathbb{R} \to \mathbb{R}$ is a twice continuously differentiable function satisfying the following assumptions:

$$\phi(0) = 1 \tag{15a}$$
$$\lim_{t \to -\infty} \phi(t) = 0 \tag{15b}$$
$$\phi'(0) = 1 \tag{15c}$$
$$\lim_{t \to +\infty} \phi'(t) = +\infty \tag{15d}$$
$$\phi''(t) > 0 \; \forall t \in \mathbb{R} \tag{15e}$$

Remark 1. 1. Note that assuptions (15) imply that ϕ' is everywhere positive and that $\lim_{t \to -\infty} \phi'(t) = 0$.
  2. Assumptions (15) are satisfied by the following functions:
  (a)
$$\phi(t) = exp(t) \; \forall t \in \mathbb{R}$$

  (b)
$$\phi(t) = \begin{cases} t + t^2 + 1 & \text{if } t \geq 0 \\ \frac{t}{1-t} + 1 & \text{if } t < 0 \end{cases}$$

  (c)
$$\phi(t) = \begin{cases} t + 3\frac{t^2}{4} + 1 & \text{if } t \geq 0 \\ \frac{1}{(1-\frac{t}{2})^2} & \text{if } t < 0 \end{cases}$$

Note that these functions are all of exponential type.

We will now prove that the sequences of penalties $\{\varphi_k\}$ defined by (9), (15) approach ψ_C in the sense defined above. To this effect, we will need the following results:

Proposition 1. (Tossings [23]) *Let $f^1 : \mathbb{R}^n \to \mathbb{R} \cup \{+\infty\}$ and $f^2 : \mathbb{R}^n \to \mathbb{R} \cup \{+\infty\}$ be two closed proper convex functions, $\lambda > 0$ and $\rho \geq 0$.
Then*
$$\delta_{\lambda,\rho}(\partial f^1, \partial f^2) \leq (1+\lambda)\sqrt{2d_{\lambda,\rho_0}(f^1, f^2)}$$
for all $\rho_0 \geq (1 + \frac{1}{\lambda})\rho + \frac{1}{\lambda}(\|J_{\lambda \partial f^1} 0\| + \|J_{\lambda \partial f^2} 0\|).$

Lemma 2. *Let* $r : \mathbb{R} \to \mathbb{R}^+$ *be an application,* $p, q \in \mathbb{R}_0^+$, *and* $\phi : \mathbb{R} \to \mathbb{R}$ *be a twice continuously differentiable function satisfying (15). Suppose that*

$$r(x) \leq pq, \quad \forall x \in \mathbb{R}. \tag{16}$$

We have then

$$r(x)y \leq p\phi(qy), \quad \forall x, y \in \mathbb{R}. \tag{17}$$

Proof. The result is trivial in the case where $r(x) = 0$.
Let us assume that $r(x) \neq 0$. Let us consider the function $h_x(y) : \mathbb{R} \to \mathbb{R}$, defined by:

$$h_x(y) = p\phi(qy) - r(x)y, \quad \forall y \in \mathbb{R}.$$

It follows from the convexity of ϕ that $h_x(y)$ is also a convex function of the variable y. Its derivative $pq\phi'(qy) - r(x)$ is equal to zero at a unique point denoted by $\bar{y}$, which is well-defined since, as mentioned above, $\phi'' > 0$, $\lim_{t \to -\infty} \phi'(t) = 0$ and $\phi'(0) = 1$. $\bar{y}$ is then the minimum of h_x and

$$h_x(\bar{y}) = p\phi(q\bar{y}) - r(x)\bar{y}. \tag{18}$$

As by assumption $(r(x)/pq) \leq 1$, it follows from (15c-e) that:

$$\bar{y} \leq 0. \tag{19}$$

By using (15b) both with the fact that ϕ is an increasing function, we have also:

$$\phi(q\bar{y}) > 0. \tag{20}$$

(19), (20), (21) together with the positiveness of $p, r(x)$ imply that $h_x(\bar{y}) > 0$ and consequently,

$$h_x(y) = p\phi(qy) - r(x)y > 0, \quad \forall y \in \mathbb{R}$$

as required.   □

We will also need the following proposition, which is an extention of ([23], Proposition VI.5.5.):

Proposition 2. *Let* φ_k *be defined by (9,15). Then*

$$\forall \lambda_0 > 0 \ \forall \rho \geq 0 \ \exists K \in \mathbb{N}_0 : d_{\lambda,\rho}(\varphi^k, \Psi_C) \leq \frac{m}{s_k}, \quad \forall k \geq K, \ \forall \lambda \geq \lambda_0$$

where

$$d_{\lambda,\rho}(\varphi^k, \Psi_C) = \sup_{\|x\| \leq \rho} \|\varphi_\lambda^k(x) - \Psi_{C,\lambda}(x)\|,$$

$\Psi_{C,\lambda}$ *(resp.* φ_λ^k*) denoting the Moreau-Yosida approximation of* Ψ_C *(resp.* φ^k*), i.e.*

$$\Psi_{C,\lambda}(x) = \inf_{y \in \mathbb{R}^n} \{\Psi_C(y) + \frac{1}{2\lambda}\|y - x\|^2\}, \quad \forall x \in \mathbb{R}^n.$$

Proposition 3. *Let φ_k be defined by (9,15).*
If $\lim_{k\to+\infty} s_k = +\infty$, then

$$\lim_{k\to+\infty} d_{\lambda,\rho}(\varphi^k, \Psi_C) = 0, \quad \forall\lambda > 0, \quad \forall\rho \geq 0$$

and

$$J_{\lambda\partial\varphi^k}x \to J_{\lambda\partial\Psi_C}x \ as \ k \to +\infty, \quad \forall x \in I\!\!R^n, \forall\lambda > 0.$$

Proof. This result follows directly from Proposition 2 and ([23], Propositions V.2.21 and V.2.19). $\qquad\square$

We can now prove the following result:

Theorem 6. *Let φ_k be defined by (9,15).*
If $\sum_{k\geq 1} \frac{1}{\sqrt{s_k}} < +\infty$, then $\sum_{k\geq 1} \delta_{\lambda,\rho}(\partial\varphi^k, \partial\Psi_C) < +\infty$, $\forall\lambda > 0$, $\forall\rho \geq 0$.

Proof. In view of Proposition 1, we have, for all $\lambda > 0$ and $\rho \geq 0$:

$$\delta_{\lambda,\rho}(\partial\varphi^k, \partial\Psi_C) \leq (1 + \lambda)\sqrt{2d_{\lambda,\rho'}(\varphi^k, \Psi_C)}$$

for all $\rho' \geq \rho_k = (1 + \frac{1}{\lambda})\rho + \frac{1}{\lambda}(\|J_{\lambda\partial\varphi^k}0\| + \|J_{\lambda\partial\Psi_C}0\|)$. By Proposition 3, the sequence $\{\rho_k\}$ is bounded. So let $\bar\rho = \sup_k \|\rho_k\|$. We have then:

$$\delta_{\lambda,\rho}(\partial\varphi^k, \partial\Psi_C) \leq (1 + \lambda)\sqrt{2d_{\lambda,\bar\rho}(\varphi^k, \Psi_C)}. \tag{21}$$

On the other hand, it follows from Proposition 2 that

$$\exists K \in I\!\!N_0 : d_{\lambda,\bar\rho}(\varphi^k, \Psi_C) \leq \frac{m}{s_k}, \quad \forall k \geq K. \tag{22}$$

Combining (22) and (23) , we get:

$$\delta_{\lambda,\rho}(\partial\varphi^k, \partial\Psi_C) \leq \sqrt{2}(1 + \lambda)\sqrt{\frac{m}{s_k}}, \quad \forall k \geq K.$$

Consequently, since by assumption

$$\sum_{k\geq 1} \frac{1}{\sqrt{s_k}} < +\infty,$$

we have

$$\sum_{k\geq 1} \delta_{\lambda,\rho}(\partial\varphi^k, \partial\Psi_C) < +\infty, \quad \forall\lambda > 0, \quad \forall\rho \geq 0.$$

$\qquad\square$

References

1. Auslender, A., Crouzeix, J.P., Fedit, P. (1987) Penalty proximal methods in convex programming. J. Optim. Theory Appl. **55**, 1–21
2. Brézis, H. (1973) Opérateurs maximaux monotones et semi-groupes de contraction dans les espaces de Hilbert. North-Holland, Amsterdam, Holland
3. Chen, G., Rockafellar, R. T. (1990) Application of a splitting algorithm to optimal control and extended linear-quadratic programming. Technical Report. Department of Applied Mathematics, University of Washington
4. Chen, G., Rockafellar, R. T. (1990) Convergence and structure of Forward–Backward splitting methods. Technical Report. Department of Applied Mathematics, University of Washington
5. Chen, G., Rockafellar, R. T. (1990) Extended Forward-Backward splitting methods and convergence. Technical Report. Department of Applied Mathematics, University of Washington
6. Chen, G., Rockafellar, R. T. (1992) Forward-Backward splitting methods in lagrangian optimization. Technical Report. Department of Applied Mathematics, University of Washington
7. Eckstein, J., Bertsekas, D.P. (1992) On the Douglas-Rachford splitting method and the proximal point algorithm for maximal monotone operators. Math. Program. **55**, 293–318
8. El Bachari, R. (1996) Contribution à l'étude des algorithmes proximaux : décomposition et perturbation variationnelle. Thèse d'état, Université de Rouen
9. Fedit, P. (1985) Contribution aux méthodes numériques en programmation mathématique non différentiable. Thèse de troisième cycle, Université de Clermont II
10. Gabay, D. (1983) Application of the method of multipliers to variational inequalities. In: Fortin M., Glowinski R. (Eds.) Application to the solution of boundary-valued problems. North-Holland, Amsterdam, Holland, 299–331
11. Glowinski, R. (1986) Splitting methods for the numerical solution of the incompressible Navier-Stokes equations. In: Balakrishnan A.V., Dorodnitsyn A.A., Lions J.L. (Eds.) Vistas in applied mathematics: Atmospheric sciences, Immunology. Optimization software, New York, New York, 57–95
12. Glowinski, R., Le Tallec, P. (1989) Augmented Lagrangian and operator splitting methods in nonlinear mechanics. SIAM, Philadelphia, Pennsylvania
13. Hartung, J. (1980) On exponential penalty function methods. Math. Operationstorsch. Statist., Ser. Optimization **11**, 71–84
14. Haubruge, S., Nguyen, V.H., Strodiot, J.J. (1998) Convergence analysis and applications of the Glowinski-Le Tallec splitting method for finding a zero of the sum of two maximal monotone operators. J. Optim. Theory Appl. **97**, 645–673
15. Kort, B.W., Bertsekas, D.P. (1972): A new penalty function method for constrained minimization. Proc. 1972 I.E.E.E. Conf. on Decision and Control, New Orleans, LA, 162–166
16. Lemaire, B. (1988) Coupling optimization methods and variational convergence. In: Hoffmann K.H., Hiriart-Urruty J.B., Lemaréchal C., Zowe J. (Eds.) Trends in Mathematical Optimization International Series of Num. Math. **84**, Birkhäuser Verlag, Basel, 163–179

17. Lemaire, B. (1996) Stability of the iteration method for nonexpansive mappings. Serdica Math. J. **22**, 331–340
18. Le Tallec, P. (1990) Numerical analysis of viscoelastic problems. Masson, Paris
19. Mouallif, K., Tossings, P. (1987) Une méthode de pénalisation exponentielle associée à une régularisation proximale. Bull. Soc. Roy. Sc. de Liège **56**, 181–192
20. Murphy, F. (1974) A class of exponential penalty functions. SIAM J. Control. **12**, 679–687
21. Passty, G.B. (1979) Ergodic convergence to a zero of the sum of monotone operators in Hilbert space. J. Math. Anal. Appl. **72**, 383–390
22. Strodiot, J.J., Nguyen, V.H. (1979) An exponential penalty method for non-differentiable minimax problems with general constraints. J. Optim. Theory Appl. **27**, 205–219
23. Tossings, P. (1990) Sur les zéros des opérateurs maximaux monotones et Applications. Thèse d'Etat, Université de Liège
24. Tseng, P. (1990) Applications of a splitting algorithm to decomposition in variational inequalities and convex programming. Math. Program. **48**, 249–263
25. Tseng, P. (1991) Applications of a splitting algorithm to decomposition in convex programming and variational inequalities. SIAM J. Control and Optimization. **29**, 119–138
26. Zhu, C. (1995) Asymptotic convergence analysis of the Forward-Backward splitting algorithm. Math. Oper. Res. **20**, 449–464

Qualitative Stability of Convex Programs with Probabilistic Constraints

René Henrion

Weierstrass Institute Berlin, 10117 Berlin, Germany

Abstract. We consider convex stochastic optimization problems with probabilistic constraints which are defined by so-called r-concave probability measures. Since the true measure is unknown in general, the problem is usually solved on the basis of estimated approximations, hence the issue of perturbation analysis arises in a natural way. For the solution set mapping and for the optimal value function, stability results are derived. In order to include the important class of empirical estimators, the perturbations are allowed to be arbitrary in the space of probability measures (in contrast to the convexity property of the original measure). All assumptions relate to the original problem.

Key words: stochastic programming, probabilistic constraints, qualitative stability, r-concave measures

1 Introduction

Most constraint sets in optimization problems can be described by an inclusion $0 \in H(x)$, where H is some multifunction. In a large class of applied problems, the constraints are subject to uncertainty such that their description changes to $\xi \in H(x)$, where ξ is some random variable. Usually, the optimization of x- variables has to be carried out without or with partial knowledge only about the realizations of the random variable. Then, of course, the above formulation has to be replaced by some reasonable deterministic equivalent. One possible way is to define an admissible x as to satisfiy the inclusion $\xi \in H(x)$ with high probability: $\mu(\xi \mid \xi \in H(x)) \geq p$ or briefly $\mu(H(x)) \geq p$, where μ is the probability distribution of ξ and $p \in (0,1)$ is some specified probability level. We shall refer to such constraints as to probabilistic constraints. In the following, we shall consider optimization problems of the type

$$(P) \qquad \min\{g(x) \mid x \in X, \ \mu(H(x)) \geq p\}.$$

Here, g is a cost function on $\mathbb{R}^m$, $X \subset \mathbb{R}^m$ is a non-specified set of deterministic constraints, $H : \mathbb{R}^m \longrightarrow \mathbb{R}^s$ is a multifunction and μ is the probability distribution of an s-dimensional random variable ξ, i.e. $\mu \in \mathcal{P}(\mathbb{R}^s)$, where $\mathcal{P}(\mathbb{R}^s)$ denotes the space of probability measures on $\mathbb{R}^s$. Throughout this paper, we shall make the following basic assumptions for problem (P):

$g : \mathbb{R}^m \to \mathbb{R}$ is convex. $\qquad(1)$

$X \subseteq \mathbb{R}^m$ is closed and convex. $\qquad(2)$

$H : \mathbb{R}^m \longrightarrow \mathbb{R}^s$ has closed and convex graph. $\qquad(3)$

$\mu \in \mathcal{P}(\mathbb{R}^s)$ is r- concave for some $r < 0$. $\qquad(4)$

We note, that (3) is equivalent to a description

$$H(x) = \{z \in \mathbb{R}^s \mid h(x, z) \leq 0\},$$

where $h : \mathbb{R}^m \times \mathbb{R}^s \to \mathbb{R}^k$ is convex and lower semicontinuous (in both variables). Concerning assumption (4), we refer to Section 2.1.

A peculiarity of stochastic optimization problems of type (P) is that usually there is no or only partial information on the measure μ available. In solution procedures, μ is therefore replaced by sample-based estimators which for increasing sampling size are supposed to approximate μ. In this context, the question of stability arises in quite a natural way: When the sampling size tends to infinity, do the optimal solutions and their cost function values of the approximate problems converge towards an optimal solution and its cost function value, respectively, of the original problem? This issue is intimately related with the qualitative stability of the solution set mapping and of the optimal value function, both depending on perturbed probability measures in a neighbourhood of the original one. For a list of papers dealing with stability aspects in stochastic programming problems with probabilistic constraints, we refer to e.g. [1], [4],[6],[7],[10],[14],[16] and references therein.

At this point, it is emphasized that we allow for arbitrary perturbations of μ in the space $\mathcal{P}(\mathbb{R}^s)$ of probability measures on $\mathbb{R}^s$. In particular, the important class of empirical measures is included as approximation. Hence, although the original measure μ is supposed to have some nice convexity property (assumption 4), the perturbations are allowed even to be discontinuous. The purpose of this paper is to provide a fairly complete characterization of qualitative stability in the settings introduced above by means of verifiable conditions for the unperturbed problem.

2   Preliminaries

In this section, we collect some basic definitions and facts which are necessary for the following analysis.

2.1 r-concave probability measures

Here we recall the notion of an r-concave probability measure for some $r \in [-\infty, \infty]$ which was imposed as a basic assumption to the problem we are going to analyze (see (4)). We start with the definition of the generalized mean function m_r on $I\!R_+ \times I\!R_+ \times [0,1]$:

$$
m_r(a,b;\lambda) = \begin{cases}
(\lambda a^r + (1-\lambda)b^r)^{1/r} & \text{if } r \in (0,\infty) \text{ or } r \in (-\infty,0), ab > 0 \\
0 & \text{if } ab = 0, r \in (-\infty,0) \\
a^\lambda b^{1-\lambda} & \text{if } r = 0 \\
\max\{a,b\} & \text{if } r = \infty \\
\min\{a,b\} & \text{if } r = -\infty
\end{cases}
\tag{5}
$$

The measure $\mu \in \mathcal{P}(I\!R^s)$ is called r-concave ([3]) for some $r \in [-\infty, \infty]$, if the inequality

$$
\mu(\lambda B_1 + (1-\lambda)B_2) \geq m_r(\mu(B_1), \mu(B_2); \lambda)
\tag{6}
$$

holds for all Borel measurable, convex subsets B_1, B_2 of $I\!R^s$ and all $\lambda \in [0,1]$ for which the convex combination $\lambda B_1 + (1-\lambda)B_2$ is Borel measurable as well (note that convex sets need not be Borel measurable, see [5]). For $r = 0$ and $r = -\infty$, μ is also called log-concave and quasi-concave, respectively. Since $m_r(a,b;\lambda)$ is increasing in r if all the other variables are fixed, the sets $\mathcal{M}_r(I\!R^s)$ of all r-concave probability measures are increasing if r is decreasing, i.e., we have for all $-\infty < r_1 \leq r_2 < \infty$ that

$$
\mathcal{M}_{-\infty}(I\!R^s) \supseteq \mathcal{M}_{r_1}(I\!R^s) \supseteq \mathcal{M}_{r_2}(I\!R^s) \supseteq \mathcal{M}_{\infty}(I\!R^s).
\tag{7}
$$

Recall, that the distribution function F_μ corresponding to some $\mu \in \mathcal{P}(I\!R^s)$ is defined by

$$
F_\mu(z) = \mu(\xi \mid \xi_i \leq z_i \ (i = 1, \dots, s)) = \mu(z + I\!R^s_-).
$$

For this particular case of cells $B = z + I\!R^s_-$, $z \in I\!R^s$, and for $r \in (-\infty, 0)$, the inequality (6) implies the distribution function F_μ to have the property that the extended-real-valued function F_μ^r is convex on $I\!R^s$. Moreover, (6) and (7) entail that F_μ is quasi-concave on $I\!R^s$.

As a consequence of a Theorem by Prékopa ([12], Th. 4.2.1.), the probability measure μ induced by a log-concave density f (i.e. a density the logarithm of which is concave) is log-concave as well, in particular it is r-concave for all $r < 0$ in view of (7). Examples of distributions having log-concave densities are the uniform distribution (on any bounded convex subset of $I\!R^s$ with non-zero Lebesgue measure), the (nondegenerate) multivariate normal distribution, the Dirichlet distribution, the multivariate Student and Pareto distributions. These examples qualify our basic assumption (4) as being not very restrictive. For more information on this issue, proofs and details we refer to Chapter 4 of [12].

2.2 The parametric problem and $\mathcal{B}$-discrepancy between probability measures

In order to study qualitative stability, we imbed problem (P) into the parametric problem

$$(P_\nu) \qquad \min\{g(x) \mid x \in \Phi(\nu)\} \quad (\nu \in \mathcal{P}(I\!\!R^s)),$$

where the constraint set mapping $\Phi : \mathcal{P}(I\!\!R^s) \longrightarrow I\!\!R^m$ is defined as $\Phi(\nu) = \{x \in X \mid \nu(H(x)) \geq p\}$. Clearly, $(P) = (P_\mu)$. We are interested in the behaviour of the solution set and value function corresponding to this parametric problem. For technical reasons, we introduce the slightly more general localized concepts for some open $V \subseteq I\!\!R^m$ and $\nu \in \mathcal{P}(I\!\!R^s)$:

$$\varphi_V(\nu) = \inf\{g(x) \mid x \in X \cap \mathrm{cl}V, \nu(H(x)) \geq p\}$$
$$\Psi_V(\nu) = \mathrm{argmin}\{g(x) \mid x \in X \cap \mathrm{cl}V, \nu(H(x)) \geq p\}$$

We recall the following elementary fact:

$$\emptyset \neq \Psi(\nu) \subseteq V \Longrightarrow \Psi(\nu) = \Psi_V(\nu), \varphi(\nu) = \varphi_V(\nu) \tag{8}$$

By φ, Ψ without index, we refer to the usual optimal value function and set of optimal solutions respectively (i.e. $V = I\!\!R^m$).

Before stating any stability result for optimal solutions and values as functions of perturbed probability measures $\nu \in \mathcal{P}(I\!\!R^s)$ in a neighbourhood of the original measure μ, we want to specify a distance on $\mathcal{P}(I\!\!R^s)$ which is suitable for our purposes (cf. discussion in [15]):

$$\alpha_\mathcal{B}(\nu_1, \nu_2) = \sup_{B \in \mathcal{B}} |\nu_1(B) - \nu_2(B)|, \quad \nu_1, \nu_2 \in \mathcal{P}(I\!\!R^s) \tag{9}$$

Here, $\mathcal{B}$ is a system of closed subsets of $I\!\!R^s$ such that it contains all sets $\{H(x) \mid x \in X\}$ (with H and X as introduced in (2) and (3)) and that it forms a determining class, i.e. whenever $\nu_1|_\mathcal{B} = \nu_2|_\mathcal{B}$, then $\nu_1 = \nu_2$. This last condition implies $\alpha_\mathcal{B}$ to be a distance, which is also referred to as the $\mathcal{B}$-discrepancy. A useful choice in the setting of our problem (P) under the stated assumptions is $\mathcal{B} = \{H(x) \mid x \in X\} \cup \{z + I\!\!R^s_- \mid z \in I\!\!R^s\}$, where the second part of the union serves to turn $\mathcal{B}$ into a determining class, while the first part is essential to obtain the important observations of the following Proposition:

Proposition 1. *In problem (P_ν), it holds that*

1. *The multifunction $\Phi : (\mathcal{P}(I\!\!R^s), \alpha_\mathcal{B}) \longrightarrow I\!\!R^m$ has closed graph.*
2. *For $\nu \in \mathcal{P}(I\!\!R^s)$, define $w_\nu(x) := \nu(H(x))$. Assume that there exists some subset $Q \subseteq X$, such that $w_\nu(x) \geq \rho > 0$ for all $x \in Q$. Then, for all $r < 0$, there exist constants $c, \delta > 0$ such that*

$$|w_\nu^r(x) - w_{\nu'}^r(x)| \leq c\alpha_\mathcal{B}(\nu, \nu') \quad \forall x \in Q \forall \nu' \in \mathcal{P}(I\!\!R^s), \alpha_\mathcal{B}(\nu, \nu') < \delta.$$

Proof. 1. is shown in [14] (Prop. 3.1). For the second assertion, note that

$$|u^r - v^r| \leq |r| \max\{u^{r-1}, v^{r-1}\} |u - v| \quad \forall u, v > 0.$$

Then, choosing $\delta := \rho/2$, one has

$$w_{\nu'}(x) \geq \rho/2 > 0 \quad \forall x \in Q \, \forall \nu' \in \mathcal{P}(\mathbb{R}^s), \, \alpha_{\mathcal{B}}(\nu, \nu') < \delta.$$

Fix c as $|r| (\rho/2)^{r-1}$. □

As a consequence of assertion 1. in the last proposition, all constraint sets are closed and, hence, so are all solution sets $\Psi(\nu)$.

3   Qualitative stability

In this section we study qualitative stability in terms of upper and lower semicontinuity of the solution set mapping and (upper Lipschitz) continuity of the optimal value function in the parametric version (P_ν) of the problem (P). The following theorem gives the main result in this direction.

Theorem 1. *Consider the parametric problem (P_ν) under the basic assumptions (1-4). Let additionally the following assumptions be satisfied at μ:*

1. *$\Psi(\mu)$ is nonempty and bounded.*
2. *There exists some $\hat{x} \in X$ such that $\mu(H(\hat{x})) > p$ (Slater condition).*

Then, the multifunction $\Psi : (\mathcal{P}(\mathbb{R}^s), \alpha_{\mathcal{B}}) \longrightarrow \mathbb{R}^m$ is upper semicontinuous at μ, and there exist constants $L, \delta > 0$, such that

$$\Psi(\nu) \neq \emptyset \text{ and } |\varphi(\nu) - \varphi(\mu)| \leq L\alpha_{\mathcal{B}}(\nu, \mu) \text{ for all } \nu \in \mathcal{P}(\mathbb{R}^s), \, \alpha_{\mathcal{B}}(\nu, \mu) < \delta.$$

Proof. We define $f(x) := \mu^r(H(x)) - p^r$ with the improper value ∞ allowed in case of $\mu(H(x)) = 0$. Then, in view of $r < 0$, the unperturbed constraint set may be written as $\Phi(\mu) = \{x \in X \mid f(x) \leq 0\}$, where f is convex due to (3) and (4). Furthermore, $f(\hat{x}) = \mu^r(H(\hat{x})) - p^r < 0$ by assumption 2., i.e. $\hat{x}$ is a Slater point of f w.r.t. X. Using a result by Klatte [11], it was shown in [14] (Cor. 3.7.) that under the assumptions made here, the desired continuity properties at μ hold in the <u>localized</u> case. More precisely, with V being some bounded, open neighbourhood of $\Psi(\mu)$ (see assumption 1.), one has that Ψ_V is upper semicontinuous at μ and there exist constants $L_1, \delta_1 > 0$, such that $|\varphi_V(\nu) - \varphi(\mu)| \leq L_1 \alpha_{\mathcal{B}}(\nu, \mu)$ for all $\nu \in \mathcal{P}(\mathbb{R}^s)$ with $\alpha_{\mathcal{B}}(\nu, \mu) < \delta_1$. Note that, by definition, one has $\varphi_V(\mu) = \varphi(\mu)$ and $\Psi_V(\mu) = \Psi(\mu)$ since $\emptyset \neq \Psi(\mu) \subseteq V$ (see (8)).

Suppose now that Ψ was not upper semicontinuous at μ. Then, by the compactness of $\Psi(\mu)$ (see assumption 1. and recall the closedness of $\Psi(\mu)$), there exists some $\varepsilon > 0$ as well as sequences ν_n, x_n such that $\alpha_{\mathcal{B}}(\nu_n, \mu) \to 0$,

$x_n \in \Psi(\nu_n)$ and $d(x_n, \Psi(\mu)) \geq \varepsilon$. On the other hand, in case that local nonemptiness of Ψ is violated, $\Psi(\nu_n) = \emptyset$ would hold for a sequence ν_n with $\alpha_B(\nu_n, \mu) \to 0$. Since $\Psi(\mu) \neq \emptyset$ by assumption 1., there is some $x^* \in \Psi(\mu)$, hence $x^* \in X \cap V$ and $f(x^*) \leq 0$. With the Slater point $\hat{x}$ from assumption 2., select $\lambda \in (0,1]$ such that $\tilde{x} := \lambda\hat{x} + (1-\lambda)x^* \in X \cap V$ (by convexity of X). Since f is convex, it follows that $f(\tilde{x}) \leq \lambda f(\hat{x}) + (1-\lambda)f(x^*) < 0$. Hence, $\mu(H(\tilde{x})) > p$ and $\nu(H(\tilde{x})) \geq p$ for $\nu \in \mathcal{P}(\mathbb{R}^s)$ with $\alpha_B(\nu, \mu) < \delta := 1/2(\mu(H(\tilde{x})) - p)$. In particular, the localized perturbed constraint sets $\Phi(\nu_n) \cap \text{cl}V$ are non-empty (they contain $\tilde{x}$) and compact for n large enough. Consequently, $\Psi_V(\nu_n) \neq \emptyset$ (since g is continuous as a convex function which is finite-valued on $\mathbb{R}^m$, see (1)). But then, $\Psi(\nu_n) = \emptyset$ means the existence of a sequence $x_n \in \Phi(\nu_n) \setminus \text{cl}V$ with $g(x_n) \leq \varphi_V(\nu_n)$.

Summarizing, if the upper semicontinuity or the local non-emptiness of Ψ is violated at μ, then there are sequences ν_n, x_n such that (with some $\varepsilon > 0$)

$$\alpha_B(\nu_n, \mu) \to 0, \; x_n \in \Phi(\nu_n), \; g(x_n) \leq \varphi_V(\nu_n) \text{ and } d(x_n, \Psi(\mu)) \geq \varepsilon. \tag{10}$$

In the following, we lead these relations to a contradiction. We define the set

$$A := \begin{cases} \mathbb{R}^m & \text{if } \Psi(\mu) = \Phi(\mu) \\ g^{-1}(-\infty, g(x')] & \text{if there is some } x' \in \Phi(\mu) \setminus \Psi(\mu), \; f(x') < 0. \end{cases} \tag{11}$$

Note first, that the case distinction above is complete. Indeed, assume that $f(x) = 0$ for all $x \in \Phi(\mu) \setminus \Psi(\mu)$ and choose an arbitrary such x. Then, $f(\lambda\hat{x} + (1-\lambda)x) \leq \lambda f(\hat{x}) + (1-\lambda)f(x) < 0$ for $\lambda \in (0,1]$, hence, due to $\hat{x}, x \in \Phi(\mu)$ and to convexity of $\Phi(\mu)$, one gets $\lambda\hat{x} + (1-\lambda)x \in \Psi(\mu)$. This, however, entails $\Psi(\mu) = \Phi(\mu)$.

We note that $A \cap \Phi(\mu)$ is convex and compact. The convexity being evident from the convexity of g, the compactness follows in the first case above from the compactness of $\Psi(\mu) = A \cap \Phi(\mu)$ due to assumption 1. In the second case, one may use the fact that the level set $g^{-1}(-\infty, \varphi(\mu)]$ of the convex function g intersected with the convex set $\Phi(\mu)$ equals $\Psi(\mu)$ which is compact by assumption 1. Then, according to [13] (Cor. 8.7.1), the intersection $\Phi(\mu) \cap g^{-1}(-\infty, \alpha)]$ has to be compact for all levels α, hence the compactness of $A \cap \Phi(\mu)$ follows with $\alpha := g(x')$.

Next, we verify that the sequence x_n in (10) satisfies $x_n \in A$ for n large enough. While this is trivial in the first case of (11), assuming the contrary in the second case would yield the existence of subsequences ν_{n_k}, x_{n_k} such that $\alpha_B(\nu_{n_k}, \mu) \to_k 0$ and

$$\varphi_V(\mu) = \varphi(\mu) < g(x') < g(x_{n_k}) \leq \varphi_V(\nu_{n_k}) \xrightarrow[k\to\infty]{} \varphi_V(\mu)$$

which contradicts the already stated continuity of φ_V at μ.

Now, we claim that $f(x_n) > 0$ holds for the sequence x_n in (10) with n large enough. Indeed, otherwise $f(x_{n_k}) \leq 0$ holds for some subsequence.

Since then $x_{n_k} \in A \cap \Phi(\mu)$ by definition of f and by the statement proven just before (also recall that $x_{n_k} \in \Phi(\nu_{n_k}) \subseteq X$) and, since $A \cap \Phi(\mu)$ was shown above to be compact, one has $x_{n_{k_l}} \to x^* \in A \cap \Phi(\mu)$ for another subsequence. Now, because of $g(x_{n_{k_l}}) \leq \varphi_V(\nu_{n_{k_l}})$ (see (10)), the continuity of φ_V at μ and that of g as a convex function provide $g(x^*) \leq \varphi_V(\mu) = \varphi(\mu)$ which entails the contradiction $x^* \in \Psi(\mu)$ to $d(x_{n_{k_l}}, \Psi(\mu)) \geq \varepsilon$ in (10).

Since $x_n \in \Phi(\nu_n)$, one has $\nu_n(H(x_n)) \geq p$, hence $\mu(H(x_n)) \geq p - \alpha_B(\mu, \nu_n) \geq p/2 > 0$ for n large enough. Therefore, statement 2. in Proposition 1 yields the existence of some $c > 0$, such that $\mu^r(H(x_n)) - \nu_n^r(H(x_n)) \leq c\alpha_B(\mu, \nu_n)$ for n large enough. Hence, $f(x_n) \leq c\alpha_B(\mu, \nu_n)$. Next, define

$$\bar{x} := \begin{cases} \hat{x} & \text{in the first case of } (11) \\ x' & \text{in the second case of } (11). \end{cases}$$

Set $y_n := \tau_n \bar{x} + (1 - \tau_n)x_n$, where $\tau_n \in [0, 1]$ is chosen such that $f(y_n) = 0$ (recall that $f(\bar{x}) < 0$ and $0 < f(x_n) \leq c\alpha_B(\mu, \nu_n) < \infty$). Then, $0 \leq \tau_n f(\bar{x}) + (1 - \tau_n)c\alpha_B(\mu, \nu_n)$ by convexity of f. Since $(1 - \tau_n)c\alpha_B(\mu, \nu_n) \to 0$ and $f(\bar{x}) < 0$, one derives that $\tau_n \to 0$. Furthermore, one has $\|y_n - \bar{x}\| = (1 - \tau_n)\|x_n - \bar{x}\|$. Now, $y_n \in \Phi(\mu)$ (since $f(y_n) = 0$ and $y_n \in X$ due to $\bar{x}, x_n \in X$ and to convexity of X). Finally, one has $y_n \in A$ which is trivial in the first case of (11) and which follows in the second case of (11) from $x', x_n \in A$ since A is convex. Knowing that $A \cap \Phi(\mu)$ is compact, the sequence y_n must be bounded, which, by the relations above, entails the sequence x_n to be bounded as well. Observing that $\|y_n - x_n\| = \tau_n\|\bar{x} - x_n\|$, we conclude $\|y_n - x_n\| \to 0$. It follows that $d(x_n, \Phi(\mu) \cap A) \to 0$ and, hence, $x_{n_k} \to x^* \in \Phi(\mu) \cap A$ for some subsequence. Now, similarly to an argumentation above, the relation $g(x_{n_k}) \leq \varphi_V(\nu_{n_k})$ along with continuity of φ_V at μ yields $g(x^*) \leq \varphi(\mu)$ and, hence, the contradiction $x^* \in \Psi(\mu)$ to $d(x_{n_k}, \Psi(\mu)) \geq \varepsilon$ in (10).

It remains to verify the statement on φ. Up to now, we have shown that Ψ is upper semicontinuous at μ and nonempty-valued close to μ. Accordingly, there is some $\delta > 0$ such that $\emptyset \neq \Psi(\nu) \subseteq V$ for all $\nu \in \mathcal{P}(\mathbb{R}^s)$ with $\alpha_B(\nu, \mu) < \delta$. But then $\varphi(\nu) = \varphi_V(\nu)$ for these ν (see (8)) and the formulated continuity property of φ results from the same property of φ_V already stated in the beginning of this proof. $\qquad \square$

As it was mentioned in the proof of the preceding Theorem, a corresponding version with localized mappings Ψ_V and φ_V was shown in an earlier work. A restriction to localizations seemed to be necessary due to allowing for non-convex perturbations of the probabilistic constraint. However, the uniformity property of the B-discrepancy introduced in Section 2.2 permits to obtain results even for the non-localized mappings.

The necessity of each single of the assumptions made in Theorem 1 in order to arrive at the assertions stated there is shown by a series of counter-examples in a preprint version of this paper [9]. The same results as in Theorem 1 were shown in [7] (Th. 1) to hold for the localized mappings Ψ_V and

φ_V in a more general setting of problem (P), namely for locally Lipschitzian g, closed X, H with closed graph and $\mu \in \mathcal{P}(\mathbb{R}^s)$ arbitrary. Then, however, assumption 2. of Theorem 1 has to be replaced by the so-called metric regularity w.r.t. X of the probabilistic constraint, which in the setting of Theorem 1 is equivalent with the Slater condition, but which is a stronger requirement in the non-convex context.

Another example in [9] demonstrates that the assumptions of Theorem 1 do not suffice to derive the lower semicontinuity of Ψ. It turns out that a lack of curvature in the level set of the probabilistic constraint may cause the solution set to collapse after arbitrary small perturbations of the measure. Therefore it seems natural to require some strict convexity property of the probability measure. Before stating a corresponding result, some auxiliary facts are needed.

In the following, the multifunction H is specified as $H(x) = \{z \in \mathbb{R}^s \mid \xi \leq h(x)\}$, where $h : \mathbb{R}^m \to \mathbb{R}^s$ is supposed to have concave components h_i in order to satisfy the basic assumption (3). This specific system of inequalities, where the realizations z of the random vector ξ occur explicitly on the left-hand side, typically reflects some supply/demand relationship, where the random demand z of some good has to be met by the supply $h(x)$ depending on the decision variable x. The assumption of this specific structure is crucial for the following, since it allows to write the constraint function as a composition of two single-valued mappings in contrast to the set-valued formulation of (P). Indeed, by definition of the distribution function, (P) now writes as

$$(P') \qquad \min\{g(x) \mid x \in X, F_\mu(h(x)) \geq p\}.$$

Furthermore, the system $\mathcal{B}$ of closed sets figuring in the definition of the discrepancy distance $\alpha_\mathcal{B}$ reduce to the system of cells $z + \mathbb{R}^s_-$, $z \in \mathbb{R}^s$, since the sets $H(x)$ now are cells themselves (cf. section 2.2). Accordingly, the discrepancy $\alpha_\mathcal{B}$ turns into the Kolmogorov distance

$$d_K(\nu_1, \nu_2) = \sup_{z \in \mathbb{R}^s} |F_{\nu_1}(z) - F_{\nu_2}(z)| \quad (\nu_1, \nu_2 \in \mathcal{P}(\mathbb{R}^s)),$$

which equals the uniform distance between the distribution functions induced by the corresponding measures. Therefore, from now on, the stability results are formulated by using d_K.

The following lemma opens a way by means of decomposition separately to study properties of F_μ and h in the context of lower semicontinuity of Ψ.

Lemma 1. *Under the assumptions of Theorem 1 let V be an open (in the maximum norm) ball containing $\Psi(\mu)$. Set*

$$Y_V = [h(X \cap clV) + \mathbb{R}^s_-] \cap F_\mu^{-1}([p/2, 1])$$
$$Y(\nu) = argmin\{\pi(y) \mid y \in Y_V, F_\nu(y) \geq p\} \quad (\nu \in \mathcal{P}(\mathbb{R}^s))$$
$$\pi(y) = \inf\{g(x) \mid x \in X \cap clV, h(x) \geq y\},$$
$$\sigma(y) = argmin\{g(x) \mid x \in X \cap clV, h(x) \geq y\} \quad (y \in Y_V).$$

Then it holds that

1. Y_V *is convex and compact.*
2. π *is convex, finite and lower semicontinuous on* Y_V.
3. *There is some* $\delta > 0$ *such that for all* $\nu \in \mathcal{P}(\mathbb{R}^s)$ *with* $d_K(\mu, \nu) < \delta$

$$\varphi(\nu) = \inf\{\pi(y) \mid y \in Y_V, F_\nu(y) \geq p\} \tag{12}$$
$$\Psi(\nu) = \sigma(Y(\nu)) \tag{13}$$

4. $Y : \mathcal{P}(\mathbb{R}^s) \longrightarrow \mathbb{R}^s$ *is upper semicontinuous at* μ.

Proof. ad 1.

The convexity of Y_V follows from the assumed convexity of X and V along with h having concave components and μ being r-concave (note that $F_\mu \leq 1$ since $\mu \in \mathcal{P}(\mathbb{R}^s)$). The compactness of $X \cap \mathrm{cl}\,V$ implies closedness of $h(X \cap \mathrm{cl}\,V) + \mathbb{R}^s_-$ and, hence, closedness of Y_V due to F_μ being upper semicontinuous as a distribution function. If Y_V was not bounded, there would be a sequence $y_n \in Y_V$ with $\|y_n\| \to \infty$. By definition, $y_n \leq h(x_n)$ for $x_n \in X \cap \mathrm{cl}\,V$. Since h is continuous (having concave components which are finite-valued on $\mathbb{R}^m$), each component of y_n is bounded from above. On the other hand, the condition $F_\mu(y_n) \geq p/2 > 0$ (due to $y_n \in Y_V$) implies all components of y_n to be bounded from below, since F_μ is a distribution function of some probability measure. This contradicts $\|y_n\| \to \infty$.

ad 2.

The convexity and finiteness of π on Y_V are immediate from the properties of g, h, X and Y_V. Now, consider any sequence $y_n \to \bar{y}$ with $y_n, \bar{y} \in Y_V$ and $\pi(y_n) \to \alpha \in \mathbb{R} \cup \{\infty\}$. Then, there are $x_n \in X \cap \mathrm{cl}\,V$ such that $y_n \leq h(x_n)$ and $g(x_n) = \pi(y_n)$ (note that the constraint set in the definition of $\pi(y)$ is nonempty and compact for $y \in Y_V$). By compactness of $X \cap \mathrm{cl}\,V$, one has $x_{n_k} \to_k \bar{x}$ for some subsequence, where $\bar{x} \in X \cap \mathrm{cl}\,V$ and $h(\bar{x}) \geq \bar{y}$ due to continuity of h. Consequently,

$$\pi(\bar{y}) \leq g(\bar{x}) = \lim_{k \to \infty} g(x_{n_k}) = \lim_{k \to \infty} \pi(y_{n_k}) = \alpha$$

by continuity of g. In particular, $\alpha > -\infty$. It results that π is lower semicontinuous on Y_V.

ad 3.

¿From the local nonemptiness and upper semicontinuity at μ of Ψ, which was stated in Theorem 1, one derives the existence of some $\delta > 0$ such that

$$\emptyset \neq \Psi(\nu) \subseteq V \quad \forall \nu \in \mathcal{P}(\mathbb{R}^s), \, d_K(\mu, \nu) < \delta. \tag{14}$$

Fix an arbitrary such ν. Then (14) and (8) yield that $\Psi_V(\nu) = \Psi(\nu) \neq \emptyset$ and $\varphi_V(\nu) = \varphi(\nu)$. Select some $\bar{x} \in \Psi(\nu)$. Then, since $\bar{x} \in \Psi_V(\nu) \subseteq \{x \in X \cap \mathrm{cl}\,V \mid F_\nu(h(x)) \geq p\}$, it follows that

$$\varphi(\nu) = g(\bar{x}) \geq \pi(h(\bar{x})) \geq \inf\{\pi(y) \mid y \in Y_V, F_\nu(y) \geq p\},$$

where the last inequality relies on $F_\mu(h(\bar{x})) \geq p/2$, which is true if δ in (14) is chosen smaller than $p/2$. For the reverse direction of (12), consider an arbitrary $\bar{y} \in Y_V$ with $F_\nu(\bar{y}) \geq p$. By definition, there is an $\tilde{x} \in X \cap \mathrm{cl}\, V$ with $h(\tilde{x}) \geq \bar{y}$. Choose some $x^* \in \sigma(\bar{y})$ (note that the constraint set in the definition of $\sigma(\bar{y})$ contains $\tilde{x}$ and, hence, is nonempty and compact such that $\sigma(\bar{y})$ is nonempty). We continue by $\pi(\bar{y}) = g(x^*) \geq \varphi_V(\nu) = \varphi(\nu)$, where the inequality follows from $x^* \in X \cap \mathrm{cl}\, V$, $h(x^*) \geq \bar{y}$ as well as $F_\nu(h(x^*)) \geq F_\nu(\bar{y}) \geq p$ (recall that distribution functions are nondecreasing). Since $\bar{y}$ was arbitrary, this establishes (12).

Concerning (13), note that $\pi(h(x)) \leq g(x) = \varphi(\nu)$ holds for all $x \in \Psi_V(\nu) = \Psi(\nu)$, hence (12) implies

$$\pi(h(x)) = \inf\{\pi(y) \mid y \in Y_V, F_\nu(y) \geq p\} = g(x) \quad \forall x \in \Psi(\nu).$$

Therefore, $x \in \sigma(h(x))$ and $h(x) \in Y(\nu)$ for all these x. Consequently, $\Psi(\nu) \subseteq \sigma(Y(\nu))$. For the reverse inclusion, let $x \in \sigma(Y(\nu))$ be arbitrary, i.e., $x \in \sigma(y)$ for some $y \in Y(\nu)$. Then,

$$g(x) = \pi(y) \leq \pi(h(x')) \leq g(x') \quad \forall x' \in X \cap \mathrm{cl}\, V, F_\nu(h(x')) \geq p,$$

which amounts to $x \in \Psi_V(\nu) = \Psi(\nu)$.

ad 4.

Although it seems tempting to proove 4. via Theorem 1 by setting $g := \pi$, $h := id$, $X := Y_V$, this is not justified since the domain $h(X \cap \mathrm{cl}\, V) + \mathbb{R}^s_-$ of π is not the whole space in general and, hence, π - although convex on this domain - cannot be assumed to be locally Lipschitzian. Instead, we write

$$Y(\nu) = \{y \in \mathbb{R}^s \mid F_\nu(y) \geq p\} \cap \{y \in Y_V \mid \pi(y) \leq \varphi(\nu)\},$$

where the first part is a closed multifunction (compare 1. in Prop. 1 with $h := id$ and $X := \mathbb{R}^s$) and the second part too is closed at μ: In fact, if $\nu_n \in \mathcal{P}(\mathbb{R}^s)$, $y_n \in Y_V$ are sequences with $y_n \to \bar{y}$, $d_K(\mu, \nu_n) \to 0$ and $\pi(y_n) \leq \varphi(\nu_n)$, then by continuity of φ at μ, closedness of Y_V (see 1.) and lower semicontinuity of π on Y_V (see 2.) one gets

$$\bar{y} \in Y_V \text{ and } \pi(\bar{y}) \leq \liminf_{n \to \infty} \pi(y_n) \leq \varphi(\mu),$$

which is the desired closedness property. As a consequence, Y itself is closed at μ (as an intersection of closed multifunctions). On the other hand, for all ν, $Y(\nu)$ is contained in the compact set Y_V (see 1.) by definition, hence Y must be upper semicontinuous at μ. $\square$

Relation (13) suggests that lower semicontinuity of Ψ may be formulated in terms of the same property for the two constituents σ and Y:

Proposition 2. *The solution set mapping Ψ of problem (P') is lower semicontinuous at $\mu \in \mathcal{P}(\mathbb{R}^s)$ provided that the following two assumptions hold:*

1. $\sigma : Y_V \longrightarrow I\!\!R^m$ is lower semicontinuous at each $y \in Y_V$.
2. $Y : \mathcal{P}(I\!\!R^s) \longrightarrow I\!\!R^s$ is lower semicontinuous at μ.

Proof. Let $U \subseteq I\!\!R^m$ be an arbitrary open set with $U \cap \Psi(\mu) \neq \emptyset$. By (13), there exists some $y \in Y(\mu) \subseteq Y_V$ such that $U \cap \sigma(y) \neq \emptyset$. According to assumption 1., there exists an open neighbourhood V of y such that

$$U \cap \sigma(y') \neq \emptyset \quad \text{for all } y' \in V \cap Y_V. \tag{15}$$

On the other hand, since $y \in Y(\mu) \cap V$, assumption 2. provides the existence of some $\delta > 0$ such that $Y(\nu) \cap V \neq \emptyset$ for all $\nu \in \mathcal{P}(I\!\!R^s)$ with $d_K(\mu, \nu) < \delta$. Combining this with (15) yields that (due to $Y(\nu) \subseteq Y_V$)

$$U \cap \Psi(\nu) = U \cap \sigma(Y(\nu)) \supseteq U \cap \sigma(Y(\nu) \cap V) \neq \emptyset$$

for $\nu \in \mathcal{P}(I\!\!R^s)$ with $d_K(\mu, \nu) < \delta$. This, however, is the asserted lower semi-continuity of Ψ at μ. □

We continue by deriving verifiable conditions for the two assumptions in Proposition 2. First we make use of the following concept introduced in [2]: A function $\alpha : I\!\!R^n \to I\!\!R$ is called *weakly analytic* if for any $a, b \in I\!\!R^n$ with $a \neq b$, one has

$$\alpha \text{ is constant on a line segment conv } \{a, b\} \Longrightarrow$$
$$\alpha \text{ is constant on the entire line lin } \{a, b\}.$$

Accordingly, a subset $Q \subseteq I\!\!R^n$ has a convex, weakly analytic description, if $Q = \{x \in I\!\!R^n \mid \alpha(x) \leq 0\}$ for some mapping $\alpha : I\!\!R^n \to I\!\!R^{n'}$ with convex, weakly analytic components α_i. In particular, all analytic or strictly convex functions are weakly analytic. Furthermore, each polyhedral set Q has a convex, weakly analytic description by means of the affine linear mapping $\alpha(x) := Ax + b$ for some matrix A and vector b.

Proposition 3. *Assumption 1. of Proposition 2 is satisfied if in problem (P') the functions g and h_i are weakly analytic and the set X has a convex, weakly analytic description.*

Proof. Consider the multifunction $M : \Lambda \subseteq I\!\!R^{n_1} \longrightarrow I\!\!R^{n_2}$ defined by $M(\lambda) = \{x \in I\!\!R^{n_2} \mid \alpha(x) \leq \beta(\lambda)\}$, with functions $\alpha : I\!\!R^{n_2} \to I\!\!R^{n_3}$ and $\beta : \Lambda \to I\!\!R^{n_3}$. The results in [2] (Th. 3.2.1., Th. 3.2.2. and Cor. 3.2.2.1) imply M to be lower semicontinuous at all $\lambda \in \Lambda$ under the following assumptions:

- $M(\lambda) \neq \emptyset \, \forall \lambda \in \Lambda$.
- The α_i are convex and weakly analytic.
- β is continuous.

First, put in the context of the definitions of Lemma 1:

$$\Lambda := Y_V, \quad \left.\begin{array}{l} \alpha_i(x) := -h_i(x) \\ \beta_i(y) := -y_i \end{array}\right\} i = 1, \ldots, s \quad \left.\begin{array}{l} \alpha_{s+i}(x) := \gamma_i(x) \\ \beta_i(y) \quad := 0 \end{array}\right\} i = 1, \ldots, n'.$$

Here, γ refers to a convex, weakly analytic description of the set $X \cap \operatorname{cl} V$ (recall that $\operatorname{cl} V$ is a closed ball in the maximum norm and hence a polyhedral set). Obviously, M represents the multifunction $y \mapsto \{x \in X \cap \operatorname{cl} V \mid h(x) \geq y\}$ here. Now, the assumptions above are satisfied due to the definition of Y_V, and to the h_i being concave and weakly analytic. Consequently, M is lower semicontinuous at all $y \in Y_V$. Therefore, again by an argument of parametric optimization ([2], Th. 4.2.2), π defined in Lemma 1 is upper semicontinuous on Y_V due to the continuity of g. This yields the continuity of π on Y_V along with statement 2. of Lemma 1. Now, we apply the result cited in the beginning of this proof, a second time: In addition to the settings above, put $\alpha_{s+n'+1} := g$ and $\beta_{s+n'+1} := \pi$. Then, M is exactly the multifunction σ and again, the three assumptions above are met by $M = \sigma$: since, for any $y \in Y_V$ the set $\{x \in X \cap \operatorname{cl} V \mid h(x) \geq y\}$ is nonempty by the definition of Y_V and compact by $\operatorname{cl} V$ being a closed ball, the set $\sigma(y)$ of global minima of g on this set must be nonempty as well due to continuity of g. The remaining two assumptions are valid due to $\alpha_{s+n'+1} = g$ being convex (see (1)) and to the continuity of $\beta_{s+n'+1} = \pi$ just shown before. As a consequence, σ is lower semicontinuous at all $y \in Y_V$ as was to be shown. $\qquad \square$

A counter-example in [2] (Ex. 3.3.1.) shows that the weak analyticity assumptions in the last Proposition cannot be dispensed with.

Next, we turn to the second assumption in Proposition 2. At this point, the strict convexity property of the probability measure, mentioned before as a necessary requirement for lower semicontinuity of the solution set, comes into play.

Lemma 2. *Assumption 2. of Lemma 2 is satisfied in problem (P') if, in addition to the assumptions of Theorem 1, there exists some open convex neighbourhood U of $Y(\mu)$ such that F_μ^r is strictly convex on U (with $r < 0$ being the modulus of r- concavity of μ).*

Proof. Setting $b_\nu(y) := F_\nu^r(y) - p^r$ for $\nu \in \mathcal{P}(\mathbb{R}^s)$, the original problem $(\widetilde{P}_\mu)$ may be written as

$$(\widetilde{P}_\mu) \quad \min \{\pi(y) \mid y \in Y_V, b_\mu(y) \leq 0\}.$$

Clearly, $(\widetilde{P}_\mu)$ is a convex program (since the r- concavity of μ implies F_μ^r to be convex (see Section 2.1) and due to 1. and 2. in Lemma 1). Also, $(\widetilde{P}_\mu)$ satisfies the Slater condition $b_\mu(\hat{y}) < 0$ for some $\hat{y} \in Y_V$. Indeed, in the proof of Theorem 1, the existence of some $\tilde{x} \in X \cap V$ with $F_\mu^r(h(\tilde{x})) = \mu^r(H(\tilde{x})) < p^r$ was shown, hence one may take $\hat{y} := h(\tilde{x})$. We proceed by case distinction with respect to the relation between $Y(\mu)$ and the solution set $Q := \arg\min\{\pi(y) \mid y \in Y_V\}$ of the relaxed problem:

<u>case 1:</u> $Y(\mu) \cap Q = \emptyset$.

Choose some $y^* \in Y(\mu)$ (recall that $Y(\mu) \neq \emptyset$ due to $\Psi(\mu) \neq \emptyset$ and to (13)). Since π and b_μ are finite-valued on Y_V and $\varphi(\mu) = \pi(y^*) > -\infty$, the Slater condition ensures the existence of a Lagrange multiplier $\lambda^* \geq 0$ such that (cf. [13], Cor. 28.2.1)

$$\pi(y^*) = \min\ \{\pi(y) + \lambda^* b_\mu(y) \mid y \in Y_V\} \text{ and } \lambda^* b_\mu(y^*) = 0.$$

By the case 1- assumption, one has $\lambda^* \neq 0$, hence $\lambda^* > 0$ and $\pi + \lambda^* b_\mu$ is strictly convex on $Y_V \cap U$ due to the additional assumption in this lemma. Accordingly,

$$\pi(y) + \lambda^* b_\mu(y) > \pi(y^*) + \lambda^* b_\mu(y^*) = \pi(y^*)\quad \forall y \in Y_V \cap U,\, y \neq y^*,$$

which implies that y^* is the unique minimizer of $(\widetilde{P}_\mu)$, i.e., $Y(\mu) = \{y^*\}$. Because of this uniqueness, the upper semicontinuity of Y at μ (statement 4. of Lemma 1) entails the desired lower semicontinuity of Y at μ.

<u>case 2:</u> $Y(\mu) \cap Q \neq \emptyset$.

In this case, $Y(\mu)$ has the simple representation

$$Y(\mu) = \{y \in Q \mid b_\mu(y) \leq 0\}. \tag{16}$$

Note also, that Q is closed and convex by the properties of π and Y_V stated in Lemma 1.

<u>case 2.1</u> $\exists\, \bar{y} \in Y(\mu),\, b_\mu(\bar{y}) < 0$.

Since $\bar{y} \in Y_V$, one has $F_\mu(\bar{y}) \geq p/2 > 0$ such that statement 2. of Proposition 1 yields the existence of $c, \delta > 0$ with

$$|b_\mu(\bar{y}) - b_\nu(\bar{y})| < c d_K(\mu, \nu)\quad \forall \nu \in \mathcal{P}(I\!R^s),\, d_K(\mu, \nu) < \delta.$$

In particular, choosing $\delta_0 := \min\{|b_\mu(\bar{y})|c^{-1}, \delta\}$, one gets $b_\nu^r(\bar{y}) < 0$ for all $\nu \in \mathcal{P}(I\!R^s)$ with $d_K(\mu, \nu) < \delta_0$. Since on the other hand, $\bar{y} \in Q$ by (16), one derives that $\bar{y} \in Y(\nu)$, hence $Y(\nu) \cap Q \neq \emptyset$ for all these ν. Consequently, the representation (16) holds as

$$Y(\nu) = \{y \in Q \mid b_\nu(y) \leq 0\}$$

for all ν close to μ.

Now, $\bar{y}$ is a Slater point of the constraint $b_\mu(y) \leq 0$ with respect to Q. According to the results in [14] (Cor. 3.7 and Th. 3.2), the multifunction Y satisfies a so-called upper Pseudo-Lipschitzian property at all $(\mu, y) \in \mathrm{Gph}\, Y$. This means in particular, that each $y \in Y(\mu)$ is supplied with neighbourhoods V_y of y and U_y of μ and with a constant $L_y > 0$ such that

$$d(y', Y(\nu)) \leq L_y d(\mu, \nu)\quad \forall \nu \in U_y\ \forall y' \in Y(\mu) \cap V_y.$$

The compactness of $Y(\mu) \subseteq Y_V$ (see Lemma 1) then allows to extract a neighbourhood $\tilde{U}$ of μ, an open set $\tilde{V}$ containing $Y(\mu)$ and a constant $L > 0$ such that

$$d(y, Y(\nu)) \leq Ld(\mu, \nu) \quad \forall \nu \in \tilde{U} \; \forall y \in Y(\mu).$$

This, however, implies the lower semicontinuity of Y at μ.

case 2.2 $b_\mu(y) = 0 \; \forall y \in Y(\mu)$.

The convexity of $Y(\mu)$ along with the strict convexity of b_μ on $U \supseteq Y_V$ imply that $Y(\mu)$ reduces to a singleton. Then, as in case 1., the upper semicontinuity of Y at μ yields the lower semicontinuity at μ. $\Box$

Combining the results of Theorem 1, Proposition 2, Proposition 3 and Lemma 2 one gets the following statement on <u>continuity</u> (i.e. upper- and lower semicontinuity at the same time) of the solution set mapping to problem (P'):

Theorem 2. *Consider problem (P') under the following assumptions:*

1. *g is convex and weakly analytic.*
2. *The h_i are concave and weakly analytic.*
3. *X has a convex, weakly analytic description.*
4. *μ is r-concave for some $r < 0$.*
5. *$\Psi(\mu)$ is nonempty and bounded.*
6. *There exists some $\hat{x} \in X$ such that $F_\mu(h(\hat{x})) > p$.*
7. *F_μ ist strictly convex on some open convex neighbourhood U of the compact set Σ with*
   $$Y(\mu) \subseteq \Sigma := [h(\Psi(\mu)) + I\!R^s_-] \cap \{y \in I\!R^s \mid F_\mu(y) \geq p\} \subseteq Y_V.$$

Then, the multifunction $\Psi : (\mathcal{P}(I\!R^s), d_K) \longrightarrow I\!R^m$ is continuous at μ.

Note that all assumptions of the Theorem relate to the original data of the problem (no assumptions with perturbed measures ν are involved). We recall, that assumptions 1. and 2. are satisfied, for instance, if g and the h_i are convex (concave, respectively) and analytic (e.g. linear) or stricly convex (concave, respectively). Assumption 3. is met among others by polyhedral sets X or balls (in any of the p-norms). Assumption 4. has already been qualified in section 2.1 to hold for most of the common multivariate probability measures. Assumptions 5. and 6. were shown above to be indispensable in the context of upper semicontinuity of solutions, and they are common in general parametric programming problems. Finally, the strict convexity property of the measure μ assumed in 7. was found to be necessary when passing from upper to lower semicontinuity of solutions. In case that h is a linear mapping, it is sufficient to require U in assumption 7. to be a convex, open neighbourhood of the simpler set $\Sigma := h(\Psi(\mu))$.

Having the qualitative results obtained so far, one might ask about quantitative stability of the solution set in program (P'). This question was investigated in [7] in terms of relating the Hausdorff distance between the solution sets of the original and the perturbed problems to the Kolmogorov distance between the original and the perturbed measure. Example 1 below demonstrates, that the assumptions of Theorem 1 allone are not sufficient in order to derive any Hölder rate of upper semicontinuity for solutions. The basic additional argument is to strengthen assumption 7. of Theorem 2 towards a strong convexity property of F_μ^r. Of course, this raises the question whether such strong convexity properties of probability measures are still as common as simple convexity of F_μ^r (coming from the r-concavity of μ). So far, some partial results have been obtained in this direction, for example the multivariate normal distribution with independent components or the uniform distribution on rectangles in $\mathbb{R}^s$ satisfy a strong convexity property.

Example 1. In the program (P'), set $m = s = 2$, $h = \mathrm{id}$, $g(x_1, x_2) = x_2$, $X = [-1, 1] \times [1/2, 1]$, $p = 1/2$. We are going to construct a probability measure μ on $\mathbb{R}^2$ such that all three assumptions of Theorem 1 are satisfied but Ψ fails to be upper Hölder continuous at μ with any rate. To this aim, put

$$f_1(x) := \begin{cases} \alpha \cdot e^{1/x - x^2} & x < 0 \\ 0 & x \geq 0 \end{cases}, \quad (\alpha \text{ such that } \int_{-\infty}^{\infty} f_1(x)dx = 1).$$

$$f_2(x) := \begin{cases} 1 \ x \in [0, 1] \\ 0 \text{ else} \end{cases}$$

Clearly, both f_1 and f_2 are log-concave probability densities on $\mathbb{R}$. Hence, $f(x_1, x_2) := f_1(x_1)f_2(x_2)$ is a log-concave probability density on $\mathbb{R}^2$. As a consequence of a Theorem by Prékopa ([12], Th. 4.2.1.), the probability measure μ induced by f is log-concave as well, in particular it is r-concave for all $r < 0$ (see section 2.1), hence the third assumption of Theorem 1 is met. Now, denote by F_1, F_2 the one-dimensional distribution functions induced by f_1, f_2, respectively. Obviously, the distribution function F_μ belonging to μ is then given by $F_\mu(x_1, x_2) = F_1(x_1)F_2(x_2)$. Setting $c := F_1^{-1}(1/2)$ and $\varphi(x) := 1/2/F_1(x)$, it is elementary to check that the constraint set is defined by

$$\{(x_1, x_2) \in [-1, 1] \times [1/2, 1] \mid x_1 \geq c \ \text{ if } \ x_2 \geq 1$$
$$x_2 \geq \varphi(x_1) \ \text{ if } \ x_1 \geq c\}.$$

Obviously, the solution set is $\Psi(\mu) = \{(x, 1/2) \mid x \in [0, 1]\}$. Finally, $(0, 3/4)$ is a possible candidate for a Slater point. Summarizing, all three assumptions of Theorem 1 are met.

Now, suppose that Ψ was upper semicontinuous at μ with some Hölder rate $1/k$ ($k \in N$). Then, there exist constants $L, \delta > 0$ such that

$$\sup_{y \in \Psi(\nu)} d(y, \Psi(\mu)) \leq L[d(\mu, \nu)]^{1/k} \quad \forall \nu \in \mathcal{P}(\mathbb{R}^s), \ d(\mu, \nu) < \delta. \qquad (17)$$

In order to lead this assumption to a contradiction, we may assume without loss of generality that k is an odd number. We define perturbed probability measures $\nu_\varepsilon \in \mathcal{P}(I\!\!R^s)$ via a perturbed density by $f^\varepsilon(x_1, x_2) = f_1(x_1) f_2^\varepsilon(x_2)$, where

$$f_2^\varepsilon(x) := \begin{cases} 1 & x \in [-\varepsilon, 1-\varepsilon] \\ 0 & \text{else} \end{cases} \quad (\varepsilon > 0)$$

The induced perturbed distribution function F_{ν_ε} satisfies $\|F_{\nu_\varepsilon} - F_\mu\|_\infty < \varepsilon$ and, consequently, $d_K(\mu, \nu_\varepsilon) < \varepsilon$. The perturbed constraint set now becomes

$$\{(x_1, x_2) \in [-1, 1] \times [1/2, 1] \mid x_1 \geq c \quad \text{if} \quad x_2 \geq 1 - \varepsilon$$
$$x_2 \geq \varphi(x_1) - \varepsilon \quad \text{if} \quad x_1 \geq c\}.$$

Accordingly, the solution set of the perturbed problem becomes $\Psi(\nu_\varepsilon) = \{(x, 1/2) \mid x \in [q, 1]\}$, where $q = \varphi^{-1}(\varepsilon + 1/2)$, hence $\sup_{y \in \Psi(\nu_\varepsilon)} d(y, \Psi(\mu)) = |q|$.

Since, by definition, one has $f_1 \in \mathcal{C}^\infty(I\!\!R)$ with $f_1^{(k)}(0) = 0$ for $k = 0, 1, 2, \ldots$, it follows that $F_1 \in \mathcal{C}^\infty(I\!\!R)$ with $F_1^{(k)}(0) = 0$ for $k = 1, 2, \ldots$ and, hence, $\varphi \in \mathcal{C}^\infty(I\!\!R)$ with $\varphi^{(k)}(0) = 0$ for $k = 1, 2, \ldots$. Consequently,

$$\left((\cdot)^{k+1} + 1/2 - \varphi\right)^{(j)}(0) = \begin{cases} 0 & j = 0, \ldots, k \\ > 0 & j = k+1 \end{cases}$$

and one gets the relation $x^{k+1} + 1/2 \geq \varphi(x)$ for x close to 0. In particular, we may insert the point $x := -\varepsilon^{1/(k+1)}$ for ε sufficiently close to 0, and it follows that $\varphi(-\varepsilon^{1/(k+1)}) \leq \varepsilon + 1/2$ (recall that k was odd). More generally, one has

$$\varphi(x) \leq x^{k+1} + 1/2 \leq \varepsilon + 1/2 \quad \forall x \in [-\varepsilon^{1/(k+1)}, 0],$$

which implies that $q = \varphi^{-1}(\varepsilon + 1/2) < -\varepsilon^{1/(k+1)}$, whence for all small enough $\varepsilon > 0$ the contradiction (see (17))

$$\varepsilon^{1/(k+1)} < |q| = \sup_{y \in \Psi(\nu_\varepsilon)} d(y, \Psi(\mu)) \leq L[d(\mu, \nu)]^{1/k} < L\varepsilon^{1/k}.$$

Acknowledgment. The author wishes to thank W. Römisch (Humboldt University Berlin) for instructive discussion which essentially influenced this paper.

References

1. Artstein, Z. (1994) Sensitivity with respect to the underlying information in stochastic programs. Journal of Computational and Applied Mathematics **56**, 127–136
2. Bank, B., Guddat, J., Klatte, D., Kummer, B., Tammer, K. (1982) Nonlinear Parametric Optimization. Akademie-Verlag, Berlin
3. Borell, C. (1975) Convex set functions in d-space. Periodica Mathematica Hungarica **6**, 111–136
4. Dupačová, J. (1990) Stability and Sensitivity Analysis for Stochastic Programming. Annals of Operations Research **27**, 115–142

5. Elstrodt, J. (1996) Maß- und Integrationstheorie. Springer, Berlin
6. Gröwe, N. (1997) Estimated stochastic programs with chance constraints. European Journal of Operational Research **101**, 285–305
7. Henrion, R., Römisch, W. (1999) Metric regularity and quantitative stability in stochastic programs with probabilistic constraints. Mathematical Programming **84**, 55–88
8. Henrion, R., Römisch, W. (1998) Stability of solutions to chance constrained stochastic programs. Preprint No. 397, Weierstrass Institute Berlin
9. Henrion, R. (1998) Qualitative Stability of Convex Programs with Probabilistic Constraints. Preprint No. 460, Weierstrass Institute Berlin
10. Kaňková, V. (1997) On the Stability in Stochastic Programming: The Case of Individual Probability Constraints. Kybernetika **33**, 525–546
11. Klatte, D. (1987) A note on quantitative stability results in nonlinear optimization. In: Lommatzsch K. (Ed.) Proceedings of the 19. Jahrestagung Mathematische Optimierung, Seminarbericht No. 90, Humboldt University Berlin, Department of Mathematics, 77–86
12. Prékopa, A. (1995) Stochastic Programming. Kluwer, Dordrecht
13. Rockafellar, R.T. (1970) Convex Analysis. Princeton University Press, Princeton
14. Römisch, W., Schultz, R. (1991) Stability analysis for stochastic programs. Annals of Operations Research **30**, 241–266
15. Römisch, W., Schultz, R. (1991) Distribution sensitivity in stochastic programming. Mathematical Programming **50**, 197–226
16. Wets, R.J.-B. (1989) Stochastic programming. In: Nemhauser G.L., Rinnoy Kan A.H.G., Todd M.J. (Eds.) Handbooks in operations research and management science, Volume 1: Optimization, North-Holland, 573–629

An Approximation Framework
for Infinite Horizon Optimization Problems
in a Mathematical Programming Setting*

Lisa A. Korf

IBM T.J. Watson Research Center, Yorktown Heights, NY

Abstract. Dynamic optimization problems, including optimal control problems, have typically relied on the solution techniques of dynamic programming, involving the sequential solution of certain optimality equations. However, many problems cannot be handled this way, due to complex constraints, a continuous state space, and other complicating factors. When recast as mathematical programs relying on the powerful tools of optimization, especially duality, and decomposability to deal with very large problems, the boundaries imposed by dynamic programming are lifted. This paper develops approximation techniques for stationary infinite horizon problems with discounted costs, in the framework of mathematical programming. A reference is given for parallel results for stochastic dynamic optimization problems.

Key words: mathematical programming, dynamic programming, infinite horizon optimization, optimal control, epi-convergence

1   Introduction

Since the 40's, *mathematical programming* techniques based on approximation, convexity, duality, and decomposition principles, have developed into powerful tools for solving large scale optimization problems. However, large scale *dynamic* optimization problems that involve a sequence of decisions, have generally been dealt with via *dynamic programming*. Dynamic programming emerged in the late 40's and early 50's out of a need to handle very large dynamic decision problems in fields including statistics, c.f. Wald [15], water resource management, c.f. Massé [10], inventory theory, c.f. Arrow, Blackwell, and Girshick [1], and later optimal control in the 60's, c.f. Bellman [2]. Given the technology of that time, the problems facing researchers were computationally unmanageable using available methods, including linear and nonlinear programming. The dynamic programming approach is computationally attractive because it is based on making a sequence of smaller decisions and then combining them to arrive at a solution. However, while this sequential approach leads to numerically manageable subproblems, it works only for a very special class of problems and can therefore be very restrictive.

---

* Research supported in part by a grant of the National Science Foundation

For example, except for very simple cases in which explicit solutions may be obtained, the *system state* must be discrete since at each stage it is treated as a parameter in a smaller optimization problem involving one or only a few variables. And consequently, insurmountable troubles can arise when more than one variable is involved at each stage, particularly in the presence of constraints, and especially so when those constraints involve the system states or are imposed jointly on both states and decision variables.

With the advent of parallel processors, and faster and more powerful computers, today we have the ability to solve much larger mathematical programs with much greater speed than in the past. Thus the need for a "sequential" approach for dynamic optimization problems is somewhat diminished, and might now be replaced at least in part by mathematical programming approaches that allow much greater flexibility. For example in the mathematical programming setting, constraints are naturally dealt with whether they be on decision variables or states (tracking variables) or both. The state space need no longer be discrete. Though one still doesn't have the computational ability to solve every problem, one does have the powerful tools of convexity, duality, and decomposability at one's disposal, and the ability to adapt these tools to today's technology. This, along with good approximation techniques, leads to computationally manageable models that could not even have been considered in the past.

This paper develops an approximation framework for a general class of *stationary infinite horizon optimization problems with discounted costs.* Such infinite horizon optimization problems arise naturally in economic and environmental planning, industrial, and other important application areas. Previously, dynamic programming was the primary tool used to deal with such problems. They were generally therefore subject to the same restrictions on the model structure as those just mentioned. To model and solve these problems in a more flexible mathematical programming setting, a first step is to approximate them by computationally more tractable ones, and that is the focus here.

Section 2 introduces a recursively defined *value function* associated with such a problem. This differs from the value function of the dynamic programming and optimal control literature [3] in that here, infinite values are permissible (and identifiable with constraints). In addition, the eventual goal is not the pointwise evaluation of this value function as it would be in the dynamic programming setting. Instead, the focus will be to approximate the value function so that it may serve as an "end term" for a finite horizon approximation of the original problem that may eventually be solved using mathematical programming techniques. Existence and optimality results are obtained that relate the value function to the original problem.

The major contributions of Sect. 3 are the approximation theorems. Here an iterative procedure is set up, and it is shown that one may approximate the value function via these iterations to obtain lower bounds that converge almost monotonically (see Sect. 3.1) to the value function. The almost monotonic convergence provides a means of gauging the fitness of approximate solutions and values based on prior ones in an iterative solution scheme. The convergence is shown to hold in the sense of epi-convergence, which in turn ensures the convergence of solutions to a solution of the original problem. Lower semicontinuity properties of the value function under certain assumptions on the cost function are also derived.

Section 4 is devoted to finite horizon approximations to the original infinite horizon problem. The first technique introduces rough lower bounds that do not take the extended future into consideration, but then proposes using the approximation theorems in Sect. 3 to obtain better and better bounds (in the epigraphical sense) that progressively take the future into account. The second technique addresses the case of convex costs, and extends the approximation methods of Grinold [6], and Flåm and Wets [5] which take the future into account via taking convex combinations. The new approximation framework set up in Sect. 3 is used to prove the convergence of optimal policies for these approximations.

Section 5 is devoted to a particular class of infinite horizon optimization problems in which the cost function is *piecewise linear-quadratic*. Such problems are quite flexible, yet have a highly exploitable, decomposable structure, as is brought out in [8] and [12]. The main result here shows that approximation of an infinite horizon problem with piecewise linear-quadratic costs via the approximation theorems preserves the piecewise linear-quadraticity of the problem. The theoretical implication is that one can keep the number of stages of a problem low, and still obtain *explicit* bounds as close as one would like to the original problem, with little increased computational difficulty.

The paper closes in Sect. 6 with an illustration of an inventory management problem, in which a store manager wants to minimize his cost of inventory replenishment while assuring that his demand is met at each review period, over an infinite time horizon. This type of problem is often dealt with in the dynamic programming setting, but here new cost structures, constraints, and penalties are incorporated which would not have fit the dynamic programming "template." The results from the preceding sections are applied to this problem in the piecewise linear-quadratic setting. This serves as a motivating model, and sets the stage for decomposition and solution schemes like those developed in [4], [8], [7], [12], and [13].

The results of this paper may be generalized to take into account *stochastic* dynamic optimization models as well. This has been carried out in detail in [9] and [8], for a class of stationary infinite horizon stochastic optimization problems.

2   The Value Function

We work in the following setting. Let $c : \mathbb{R}^s \times \mathbb{R}^n \to \overline{\mathbb{R}}$ be a proper ($\neq -\infty$, $\neq +\infty$), lower semicontinuous (lsc) function, bounded on its domain (i.e. where it is finite-valued), and $\delta \in (0,1)$ a discount factor. The *stationary infinite horizon optimization problem with discounted cost* is given by $P(x)$:

$$\text{minimize} \quad \sum_{t=1}^{\infty} \delta^{t-1} c(x_{t-1}, u_t)$$

$$\text{s.t. } x_t = A x_{t-1} + B u_t + b \text{ for } t = 1, 2, \ldots$$

$$x_0 = x$$

By $\sum_{t=1}^{\infty} \delta^{t-1} c(x_{t-1}, u_t)$ we mean $\lim_{T \to \infty} \sum_{t=1}^{T} \delta^{t-1} c(x_{t-1}, u_t)$, which will always exist (possibly $= +\infty$) by the assumption that c is bounded on its domain. A, B, and b are all matrices of appropriate dimensions. We can think of the u_t's as the primary decisions, or controls, at each time period, while the x_t's keep track of the evolution of the state of the system. The x_t's may be thought of in tandem both as problem variables and as a tracking mechanism for the dynamics of the system. We let min $P(x)$ denote the optimal value of the problem $P(x)$. Similarly, we let feas P denote the set of feasible states, or the set of $x \in \mathbb{R}^s$ such that min $P(x) < \infty$.

A solution $(u_1, u_2, \ldots)$ to $P(x)$ is *stationary* with respect to shifts in time, if for any corresponding trajectory $(x_0, x_1, \ldots)$, $u_t = u_s$ whenever $x_{t-1} = x_{s-1}$ for any $s, t \in \mathbb{N}$. That stationary solutions exist when P is feasible will follow straightforwardly from the fact that A, B, b and c do not depend on time. An *optimal policy* for $P(\cdot)$ is then a function $u : \text{feas } P \to \mathbb{R}^n$ such that the corresponding solution $(u_1, u_2, \ldots)$ and trajectory $(x_0, x_1, \ldots)$ defined by

$$x_0 = x, \qquad\qquad u_1 = u(x_0)$$
$$x_t = A x_{t-1} + B u_t + b, \qquad u_{t+1} = u(x_t) \tag{1}$$

is optimal for $P(x)$ for every $x \in \text{feas } P$. Note that such a solution $(u_1, u_2, \ldots)$ is stationary. Now let us consider the following recursively defined function,

$$Q(x) = \inf_u \left\{ c(x, u) + \delta Q(Ax + Bu + b) \right\}.$$

This looks similar to the "value function" of the optimal control literature, c.f. [3], the primary distinction being that c and therefore Q have possibly infinite values. Note also that no smoothness assumptions have been imposed on c. Our first goal is to set up the correspondence between P and Q, and in the process verify the existence of Q.

Theorem 1. (existence of recursive value function). *For each* $x \in \mathbb{R}^s$, *let* $Q(x) = \min P(x)$, *the value of the problem* $P(x)$ *at optimality (note* $Q(x)$ *could be* $+\infty$). *Then* $Q(x) = \inf_u \left\{ c(x, u) + \delta Q(Ax + Bu + b) \right\}.$

Proof. We can first express $\inf_u \{c(x, u) + \delta Q(Ax + Bu + b)\}$ as the optimal value of the problem

$$\text{minimize } c(x_0, u) + \delta Q(x_1)$$
$$\text{s.t. } x_1 = Ax_0 + Bu + b$$
$$x_0 = x.$$

Then, under the definition, $Q(x) = \min P(x)$, it suffices to show that this problem is equivalent to $P(x)$. It may again be rewritten as

$$\text{minimize } c(x_0, u_1) + \delta \min_{u_t, t \geq 2} \lim_{T \to \infty} \sum_{t=2}^{T} \delta^{t-2} c(x_{t-1}, u_t)$$
$$\text{s.t. } x_t = Ax_{t-1} + Bu_t + b \text{ for } t = 1, 2, \ldots$$
$$x_0 = x$$

which is the same as $P(x)$. $\qquad\square$

We have established the existence of a particular recursively defined function Q which we will from now on refer to as the *value function* for P. The next theorem establishes the equivalence between optimal policies of P and functions u that "solve" Q.

Theorem 2. (equivalence between P and Q). *u is an optimal policy for P if and only if for all $x \in$ feas P, $u(x) \in \text{argmin}_u \{c(x, u) + \delta Q(Ax + Bu + b)\}$.*

Proof. Suppose first that u is an optimal policy for P. For fixed x, consider a sequence $(u_1, u_2, \ldots)$ defined by u via the identifications in (1). Then, using the fact that Q is the value function for P, i.e. $Q(x) = \min P(x)$, we obtain,

$$\inf_u \{c(x, u) + \delta Q(Ax + Bu + b)\} = Q(x) = \sum_{t=1}^{\infty} \delta^{t-1} c(x_{t-1}, u_t)$$

$$= c(x_0, u_1) + \delta Q(x_1) = c(x, u(x)) + \delta Q(Ax + Bu(x) + b),$$

whereby $u(x) \in \text{argmin}_u \{c(x, u) + \delta Q(Ax + Bu + b)\}$. To proceed in the other direction, assuming now that $u(x) \in \text{argmin}_u \{c(x, u) + \delta Q(Ax + Bu + b)\}$, and letting $(u_1, u_2, \ldots)$ be a sequence obtained by the same identifications as in (1), we have that

$$\min P(x) = Q(x) = c(x_0, u_1) + \delta Q(Ax_0 + Bu_1 + b) = \sum_{t=1}^{\infty} \delta^{t-1} c(x_{t-1}, u_t),$$

whereby u is an optimal policy for P. $\qquad\square$

The next theorem establishes the existence of optimal policies (and therefore stationary solutions) when P is feasible.

Theorem 3. (optimal policies from solutions). *Suppose that for each $x \in$ feas P, $P(x)$ has an optimal solution $(u_1^x, u_2^x, \ldots)$, with an associated trajectory $(x_0^x, x_1^x, \ldots)$. Let $u :$ feas $P \to \mathbb{R}^n$ be defined by $u(x) = u_1^x$. Then u is an optimal policy for P.*

Proof. The proof relies on the previous development by observing that the function u minimizes $c(x, u(x)) + \delta Q(Ax + Bu(x) + b)$, for all $x \in$ feas P, through the fact that $u(x) = u_1^x$. Therefore u is an optimal policy for P by Theorem 2. $\qquad\Box$

This theorem brings up the important question of when optimal policies (or equivalently solutions) for P exist. We next address an important criterion that will guarantee that P will have a solution.

Definition 1. (uniform level-boundedness). A function $f : \mathbb{R}^s \times \mathbb{R}^n \to \overline{\mathbb{R}}$ with values $f(x, u)$ is *level-bounded in u locally uniformly in x* if for each $\bar{x} \in \mathbb{R}^s$ and $\alpha \in \mathbb{R}$ there is a neighborhood V of $\bar{x}$ along with a bounded set $B \subset \mathbb{R}^s$ such that $\{u \mid f(x, u) \leq \alpha\} \subset B$ for all $x \in V$; or equivalently, there is a neighborhood V of $\bar{x}$ such that the set $\{(x, u) \mid x \in V, \ f(x, u) \leq \alpha\}$ is bounded in $\mathbb{R}^s \times \mathbb{R}^n$.

We make use of the following Theorem from [13, Theorem 1.17].

Theorem 4. (parametric minimization). *Consider*

$$p(x) := \inf_u f(x, u), \qquad U(x) := \operatorname{argmin}_u f(x, u),$$

in the case of a proper, lsc function $f : \mathbb{R}^s \times \mathbb{R}^n \to \overline{\mathbb{R}}$ such that $f(x, u)$ is level-bounded in u locally uniformly in x. Then the function p is proper and lsc on $\mathbb{R}^s$, and for each $x \in \operatorname{dom} p$ the set $U(x)$ is nonempty and compact, whereas $U(x) = \emptyset$ when $x \notin \operatorname{dom} p$.

Lemma 1. (boundedness of value function). *If the cost function, $c : \mathbb{R}^s \times \mathbb{R}^n \to \overline{\mathbb{R}}$ is bounded on its domain so that $\sup_{(x,u) \in \operatorname{dom} c} |c(x, u)| \leq K$, then the value function $Q : \mathbb{R}^s \to \overline{\mathbb{R}}$ is also bounded on its domain; in particular*

$$\sup_{x \in \operatorname{dom} Q} |Q(x)| \leq \frac{K}{1 - \delta}.$$

Proof. Using the fact that Q is the value function for P,

$$|Q(x)| = |\min P(x)| \leq \sum_{t=1}^{\infty} \delta^{t-1} \sup_{(x,u) \in \operatorname{dom} c} |c(x, u)| \leq \frac{K}{1 - \delta},$$

which provides the desired bound. $\qquad\Box$

Theorem 5. (attainment of minimum). *Suppose c is level bounded in u locally uniformly in x, feas $P \neq \emptyset$, and Q is lsc. Then there exists an optimal policy $u :$ feas $P \to I\!\!R^n$.*

Proof. This applies Theorem 4 to the function $g : I\!\!R^s \times I\!\!R^n \to \overline{I\!\!R}$ defined by $g(x, u) = c(x, u) + \delta Q(Ax + Bu + b)$, once we can show that g is uniformly level-bounded, lsc and proper. The lower semicontinuity comes out of that of c and Q. The properness comes out of the observation that for any u such that $c(x, u) < \infty$, dom $c(\cdot, u) \supset$ dom $Q =$ feas P. For the uniform level-boundedness, fix $\bar{x} \in I\!\!R^s$, $\alpha \in I\!\!R$, and let V be a neighborhood of $\bar{x}$, B a bounded set, such that $\{u \mid c(x, u) \leq \alpha + \frac{K}{1-\delta}\} \subset B$ for all $x \in V$, which is possible by the uniform level-boundedness of c. Next, observe through Lemma 2 that

$$
\{u \mid g(x, u) \leq \alpha\} = \{u \mid c(x, u) + \delta Q(Ax + Bu + b) \leq \alpha\}
$$
$$
\subset \left\{u \mid c(x, u) \leq \alpha + \frac{K}{1-\delta}\right\} \subset B.
$$

We have shown that g is lsc, proper, and uniformly level-bounded. Hence it satisfies the assumptions of Theorem 4, which implies that an optimal policy for P exists by the fact that $\operatorname{argmin}_u g(x, u)$ is nonempty (and compact) for each x in dom Q. $\qquad\square$

3   Approximation Framework

Now that we have established the existence of the value function Q and its relation to P, we may proceed with approximation theorems for Q. In particular, our interest is in obtaining lower bounds for Q to aid in the development of finite horizon approximations of P.

The results of Sect. 2 have shown that the finite horizon problem, $P_h^T(x)$:

$$
\text{minimize } \sum_{t=1}^{T} \delta^{t-1} c(x_{t-1}, u_t) + \delta^T h(x_T)
$$
$$
\text{s.t. } x_t = Ax_{t-1} + Bu_t + b \text{ for } t = 1, \dots, T
$$
$$
x_0 = x
$$

is equivalent to P when $h = Q$ for any T (in particular $T = 1$) in the sense that min $P(x)$ is equal to min $P_Q^T(x)$, and an optimal policy for P also solves P_Q^T, i.e. if u is an optimal policy for P, and for fixed $x \in$ feas P, $(u_1, u_2, \dots)$ and $(x_0, x_1, \dots)$ are the corresponding solution and trajectory, then $(u_1, \dots, u_T)$ with trajectory $(x_0, \dots, x_T)$ solves $P_Q^T(x)$. So, we have an exact finite horizon representation of P that theoretically could be amenable to computational schemes, if we had an explicit representation for Q. The goal is to obtain an explicit function Q^ν that approximates Q in the right sense, to obtain the problem $P^T_{Q^\nu}$. Of particular interest are *bounds* Q^ν to Q. In summary, the approximations will have the greatest utility if they satisfy the conditions

i) Optimal policies of $P^T_{Q^\nu}$ converge to optimal policies of P as $\nu \to \infty$ (mandatory!),

ii) Q^ν is a lower (upper) bound for Q for each $\nu \in I\!N$,

iii) Q^ν monotonically increases (decreases) to Q as $\nu \to \infty$,

iv) Q^ν is explicitly available or easily computed.

We will see that item iii) may be relaxed somewhat but in a way that still provides the same utility of assuring that solutions and values are increasingly better estimates of the true solutions and values as $\nu \to \infty$.

3.1   Epi-convergence

When referring to "approximation" for a minimization problem, the appropriate notion of convergence is *epi-convergence*, which ensures the convergence of infima and solutions to those of the original problem. A sequence of functions, $f^\nu : I\!R^n \to \overline{I\!R}$ is said to *epi-converge* to $f : I\!R^n \to \overline{I\!R}$, written $f^\nu \xrightarrow{e} f$, if

i) $\forall\, x^\nu \to x$, $\liminf_\nu f^\nu(x^\nu) \geq f(x)$,

ii) $\exists\, x^\nu \to x$, $\limsup_\nu f^\nu(x^\nu) \leq f(x)$.

Epi-convergence is so-named because it corresponds to set-convergence of the *epigraphs* of sequences of functions. A basic theorem relating epi-convergence to the convergence of infima and solutions is given below.

Theorem 6. (convergence in minimization). *If $\left\{ f, f^\nu : I\!R^n \to \overline{I\!R} \,\middle|\, \nu \in I\!N \right\}$ be such that $f^\nu \xrightarrow{e} f$, then $\limsup_\nu (\inf f^\nu) \leq \inf f$. Moreover, if there exists $x^k \to x$ and a subsequence $\left\{ x^{\nu_k} \right\}_{k \in N}$ such that $x^{\nu_k} \in \operatorname{argmin} f^{\nu_k}$, $k \in I\!N$, then $x \in \operatorname{argmin} f$ and $\inf f^{\nu_k} \to \inf f$.*

These results are well-known. For a proof one could consult [13, Theorem 7.30]. We begin with some useful properties of epi-convergence, the proofs of which can also be found in [13, Chap. 7].

Theorem 7. (properties of epi-limits). *The following properties hold for any sequence $\{f^\nu\}_{\nu \in N}$ of functions on $I\!R^n$.*

(a) *The functions $\text{e-}\liminf_\nu f^\nu$, $\text{e-}\limsup_\nu f^\nu$ are lower semicontinuous, and so too is $\text{e-}\lim_\nu f^\nu$ when it exists.*

(b) *If the sequence $\{f^\nu\}_{\nu \in N}$ is nonincreasing, then $\text{e-}\lim_\nu f^\nu$ exists and equals $\operatorname{cl}[\inf_\nu f^\nu]$;*

(c) *If the sequence $\{f^\nu\}_{\nu \in N}$ is nondecreasing, then $\text{e-}\lim_\nu f^\nu$ exists and equals $\sup_\nu [\operatorname{cl} f^\nu]$.*

Theorem 8. (epi-limits of sums of functions). *For sequences of functions f_1^ν and f_2^ν on $\mathbb{R}^n$, if $f_i^\nu \to f_i$ and $f_i^\nu \xrightarrow{e} f_i$ for $i = 1, 2$, then $f_1^\nu + f_2^\nu \xrightarrow{e} f_1 + f_2$.*

The result presented next is new, and provides a test for epi-convergence when a sequence of functions is *almost monotonic*. A sequence of functions $f^\nu : \mathbb{R}^n \to \overline{\mathbb{R}}$ is said to be *almost nonincreasing* if there exists a nonnegative sequence $\{\alpha^\nu\}_{\nu \in N}$ such that $\sum_{k=1}^\infty \alpha^k < \infty$, and for all $\nu \in N$, $f^\nu \geq f^{\nu+1} - \alpha^\nu$. A sequence of functions $f^\nu : \mathbb{R}^n \to \overline{\mathbb{R}}$ is said to be *almost nondecreasing* if there exists a nonnegative sequence $\{\alpha^\nu\}_{\nu \in N}$ such that $\sum_{k=1}^\infty \alpha^k < \infty$, and for all $\nu \in N$, $f^\nu \leq f^{\nu+1} + \alpha^\nu$.

Theorem 9. (epi-limits of almost monotonic functions). *Let $f^\nu : \mathbb{R}^n \to \overline{\mathbb{R}}$ be a sequence of lsc functions that converges pointwise to $f : \mathbb{R}^n \to \overline{\mathbb{R}}$.*

  (a) *If $\{f^\nu\}_{\nu \in N}$ is almost nonincreasing, and f is lsc, then $f^\nu \xrightarrow{e} f$.*
  (b) *If $\{f^\nu\}_{\nu \in N}$ is almost nondecreasing, then f is lsc and $f^\nu \xrightarrow{e} f$.*

Proof. For part (a), let $g^\nu = f^\nu - \sum_{k=1}^{\nu-1} \alpha^k$, and $g = f - \sum_{k=1}^\infty \alpha^k$. Then g^ν is nonincreasing since

$$g^\nu = f^\nu - \sum_{k=1}^{\nu-1} \alpha^k \geq f^{\nu+1} - \sum_{k=1}^\nu \alpha^k = g^{\nu+1}.$$

By Theorem 7 (b) and the lower semicontinuity of f, e-$\lim_\nu g^\nu$ exists and

$$\text{e-}\lim_\nu g^\nu = \text{cl}\left[\inf_\nu g^\nu\right] = f - \sum_{k=1}^\infty \alpha^k = g.$$

Now observe that $f^\nu = g^\nu + \sum_{k=1}^{\nu-1} \alpha^k$ and $f = g + \sum_{k=1}^\infty \alpha^k$, where now g^ν converges to g both epigraphically and pointwise, and $\sum_{k=1}^{\nu-1} \alpha^k \to \sum_{k=1}^\infty \alpha^k$ (both epigraphically and pointwise when considered as constant functions). Applying Theorem 8 for epi-limits of sums of functions gives us that $f^\nu \xrightarrow{e} f$.

In part (b), the approach is similar. Let $g^\nu = f^\nu + \sum_{k=1}^{\nu-1} \alpha^k$, and $g = f + \sum_{k=1}^\infty \alpha^k$. Then g^ν is nondecreasing since

$$g^\nu = f^\nu + \sum_{k=1}^{\nu-1} \alpha^k \leq f^{\nu+1} + \sum_{k=1}^\nu \alpha^k = g^{\nu+1}.$$

Theorem 7 (c) says that e-$\lim_\nu g^\nu$ exists and is equal to $\sup_\nu g^\nu = f + \sum_{k=1}^\infty \alpha^k = g$. We have that $f^\nu = g^\nu - \sum_{k=1}^{\nu-1} \alpha^k$ and $f = g - \sum_{k=1}^\infty \alpha^k$, and g^ν converges to g both pointwise and epigraphically. Also, because $-\sum_{k=1}^{\nu-1} \alpha^k \to -\sum_{k=1}^\infty \alpha^k$, applying Theorem 8 gives us that $f^\nu \xrightarrow{e} f$, and Theorem 7 (a) implies that f is lsc. $\qquad\square$

3.2 Approximation Theorems

We proceed now to develop approximation theorems for the infinite horizon problem P and its associated value function Q, which do not depend on knowing the set of feasible initial points (equivalently $\operatorname{dom} Q$). Tighter bounds are obtainable in the case that the domain of Q is known ahead of time, and these results can be found in [8]. Many problems are not so simple however, and the subject of determining the domain a priori is an important area to investigate in its own right. We restrict our attention to approximations from below, with an emphasis on drawing out the almost monotonic convergence (see Sect. 3.1 for the definition) that is inherent in the approximations.

Our setting is the space C of functions $h : \mathbb{R}^s \to \overline{\mathbb{R}}$ that are bounded by $\frac{K}{1-\delta}$ on their domains, and also bounded above by Q (i.e. $\sup_{x \in \operatorname{dom} h} |h(x)| \le \frac{K}{1-\delta}$ and $h \le Q$). Define the operator W for $h \in C$ by

$$Wh(x) = \inf_u \left\{ c(x, u) + \delta h(Ax + Bu + b) \right\}.$$

If we begin with a given function $h \in C$, every iteration $W^\nu h$ will be a lower bound of Q. In addition, we can obtain the almost monotonicity of these iterates.

Theorem 10. (almost nondecreasing iterates). *For any $h \in C$, we have $Wh \in C$ and $\{W^\nu h\}_{\nu \in N}$ is almost nondecreasing; specifically, for $\alpha^\nu = \delta^\nu \frac{2K}{1-\delta}$, $W^\nu h \le W^{\nu+1} h + \alpha^\nu$.*

Proof. We first demonstrate that $W : C \to C$. W maps C into itself since for any $x \in \operatorname{dom} Wh$, there exists a $u \in \mathbb{R}^n$ such that $x \in \operatorname{dom} c(\cdot, u)$ and $h(Ax + Bu + b) < \infty$, so that

$$|Wh(x)| \le \sup_{(x,u) \in \operatorname{dom} c} |c(x, u)| + \delta \sup_{x \in \operatorname{dom} h} |h(x)| \le K + \frac{\delta K}{1 - \delta} = \frac{K}{1 - \delta}.$$

And additionally,

$$Wh(x) = \inf_u \left\{ c(x, u) + \delta h(Ax + Bu + b) \right\}$$
$$\le \inf_u \left\{ c(x, u) + \delta Q(Ax + Bu + b) \right\} = Q(x).$$

To show that $\{W^\nu h\}_{\nu \in N}$ is almost nondecreasing, we will make use of the fact that for all $\nu \in \mathbb{N}$, for all $x \in \mathbb{R}^s$, $W^\nu h(x) = \min P_h^\nu(x)$. Fix $x \in \mathbb{R}^s$. If there is no feasible solution for $P_h^{\nu+1}(x)$, then $W^{\nu+1} h(x) = +\infty$ and then trivially $W^\nu h(x) \le W^{\nu+1} h(x) + \alpha^\nu$. Suppose there is a feasible solution $(u_1, \ldots, u_{\nu+1})$, with trajectory $(x_0, \ldots, x_{\nu+1})$ for $P_h^{\nu+1}(x)$. then $(u_1, \ldots, u_\nu)$ and $(x_0, \ldots, x_\nu)$ are feasible for $P_h^\nu(x)$, and

$$h(x_\nu) \le \frac{K}{1 - \delta} + \left(c(x_\nu, u_{\nu+1}) + K \right) + \left(\delta h(x_{\nu+1}) + \frac{\delta K}{1 - \delta} \right)$$
$$= c(x_\nu, u_{\nu+1}) + \delta h(x_{\nu+1}) + \frac{2K}{1 - \delta}.$$

This implies that $\sum_{t=1}^{\nu} \delta^{t-1} c(x_{t-1}, u_t) + \delta^{\nu} h(x_{\nu}) \leq \sum_{t=1}^{\nu+1} \delta^{t-1} c(x_{t-1}, u_t) + \delta^{\nu+1} h(x_{\nu+1}) + \alpha^{\nu}$. Since this is true of any feasible point of $P_h^{\nu+1}(x)$, it follows that $\min P_h^{\nu}(x) \leq \min P_h^{\nu+1}(x) + \alpha^{\nu}$, which translates to $W^{\nu} h(x) \leq W^{\nu+1} h(x) + \alpha^{\nu}$ as claimed. Next, observe that

$$\sum_{t=1}^{\infty} \alpha^t = \sum_{t=1}^{\infty} \delta^t \frac{2K}{1-\delta} = \frac{2K}{(1-\delta)^2} < \infty,$$

which implies that $\left\{ W^{\nu} h \right\}_{\nu \in N}$ is almost nondecreasing.   □

Lemma 2. (pointwise convergence of iterates). $W^{\nu} h$ converges to Q pointwise.

Proof. Fix $x \in \mathbb{R}^s$ and let $(u_1, u_2, \dots)$ be any sequence such that $(x_0, x_1, \dots)$ is the corresponding trajectory. By the boundedness of h on its domain we have that

$$\liminf_{\nu} \left\{ \delta^{\nu} h(x_{\nu}) + \sum_{k=1}^{\nu-1} \alpha^k \right\} \geq \sum_{k=1}^{\infty} \alpha^k.$$

This implies that

$$\liminf_{\nu} \left\{ \min P_h^{\nu}(x) + \sum_{k=1}^{\nu-1} \alpha^k \right\} \geq \min P(x) + \sum_{k=1}^{\infty} \alpha^k.$$

Using the facts that $W^{\nu} h(x) = \min P_h^{\nu}(x)$ and $Q(x) = \min P(x)$, and that $W^{\nu} h + \sum_{k=1}^{\nu-1} \alpha^k$ is nondecreasing, yields

$$\lim_{\nu} \left\{ W^{\nu} h(x) + \sum_{k=1}^{\nu-1} \alpha^k \right\} = Q(x) + \sum_{k=1}^{\infty} \alpha^k,$$

whereby $\lim_{\nu} W^{\nu} h(x) = Q(x)$.   □

Lemma 3. (lower semicontinuity via uniform level-boundedness). *If c is level-bounded in u locally uniformly in x, and $h \in C$ is proper and lsc, then $W^{\nu} h$ is proper, lsc, and level-bounded in u locally uniformly in x.*

Proof. This follows from Theorem 4 via the same arguments used in the proof of Theorem 5, applied first to $g(x, u) = c(x, u) + \delta h(Ax + Bu + b)$, and then relying on induction since $Wh \in C$ whenever $h \in C$.   □

Theorem 11. (epi-convergence of iterates). *If $W^{\nu} h$ is lsc for all $\nu \in \mathbb{N}$, in particular when c is uniformly level-bounded and h is proper and lsc, then $W^{\nu} h$ epi-converges to Q and Q is lsc.*

Proof. This applies Theorem 9 (b) about the epi-convergence of almost non-decreasing functions to $W^\nu h$ via Theorem 10 and Lemma 2. Lemma 3 supplies the lower semicontinuity of $W^\nu h$ when the uniform level-boundedness condition is satisfied. That Q is lsc comes from Theorem 7 (a).   □

It is not in general true that if h is lsc then $W^\nu h$ is lsc. The uniform level-boundedness of c is one possible sufficient condition for this. In Sect. 5 we concentrate on a broad class of piecewise linear-quadratic functions for which lower semicontinuity of the iterates holds even when the uniform level-boundedness assumption may fail. We conclude this section by establishing the convergence of optimal policies for P.

Theorem 12. (convergence of optimal policies). *Suppose that $h : \mathbb{R}^s \to \overline{\mathbb{R}}$ lies in C, and that for all $\nu \in \mathbb{N}$, $W^\nu h$ is lsc (e.g. if c is uniformly level-bounded and h is proper and lsc). Suppose B has full row rank. For each $x \in \operatorname{dom} Q$ let $g_x, g_x^\nu : \mathbb{R}^n \to \overline{\mathbb{R}}$ be defined by*

$$g_x^\nu(u) = c(x, u) + \delta W^\nu h(Ax + Bu + b),$$

$$g_x(u) = c(x, u) + \delta Q(Ax + Bu + b).$$

Then $g_x^\nu \xrightarrow{e} g_x$. In particular, the conclusions of Theorem 2 are valid.

Proof. For each $x \in \operatorname{dom} Q$, let

$$f_x^\nu(u) = W^\nu h(Ax + Bu + b),$$

$$f_x(u) = Q(Ax + Bu + b).$$

The pointwise convergence of $W^\nu h$ to Q implies that $f_x^\nu \to f_x$ pointwise. Now let $u \in \mathbb{R}^n$, and let $u^\nu \to u$. Then $Ax + Bu^\nu + b \to Ax + Bu + b$. By Theorem 11 we obtain

$$\liminf_\nu f_x^\nu(u^\nu) = \liminf_\nu W^\nu h(Ax + Bu^\nu + b)$$
$$\geq Q(Ax + Bu + b) = f_x(u).$$

For the lim sup direction, there is a sequence $x^\nu \to Ax + Bu + b$ such that $\limsup_\nu W^\nu h(x^\nu) \leq Q(Ax + Bu + b)$ by Theorem 11. Since B has full row rank we can find a sequence $u^\nu \in \mathbb{R}^n$ that satisfies $Bu^\nu = x^\nu - Ax - b$. Then applying Theorem 11 we obtain

$$\limsup_\nu f_x^\nu(u^\nu) = \limsup_\nu W^\nu h(Ax + Bu^\nu + b)$$
$$= \limsup_\nu W^\nu h(x^\nu)$$
$$\leq Q(Ax + Bu + b) = f_x(u).$$

Thus we have that $f_x^\nu \xrightarrow{e} f_x$. For fixed x, $g_x^\nu = c(x, \cdot) + \delta f_x^\nu$ and $g_x = c(x, \cdot) + \delta f_x$. Theorem 8 may now be applied to obtain that $g_x^\nu \xrightarrow{e} g_x$, which completes the proof.   □

4   Finite Horizon Approximations

This section establishes some possible lower bounds for the infinite horizon problem P, using two different methods. The aim is to come up with a function h that approximates the future of the system, so that P_h^T is a good lower bound of the original problem. The first method sets up an iterative scheme, relying on the approximation framework for Q that has been developed in the previous sections. The second method also concentrates on approximations from below, by combining a terminal cost function with terminal dynamics that in some way average the future, for the case in which the cost function is convex. This adds more information to the problem, but removes the possibility of using iterated methods to obtain better bounds. The idea of this method stems from the work of Grinold [6], and Flåm and Wets [5], and the proof relies on the approximation results in Sect. 3.

We begin with a fairly simple lower bound which is significant in that it satisfies the assumptions which allow iterated convergence to Q, as discussed in the previous section. Such iterations will be especially useful when some explicit representation is available for them. This is the case when c is piecewise linear-quadratic, and we investigate that case in particular in Sect. 5. Recalling the form of the approximating problems $P_h^T(x)$ given in Sect. 3, for each x, given an optimal solution $(u_1^x, \ldots, u_T^x)$, we will call a function $u :$ feas $P_h^T \to I\!\!R^n$ an *optimal first stage policy* for P_h^T if $u(x) = u_1^x$ for each $x \in$ feas P_h^T. The next theorem provides a very intuitive lower bound for Q. See [8] for bounds when dom Q is known a priori.

Theorem 13. (lower bounds when dom Q unknown). *Let $h : I\!\!R^s \to \overline{I\!\!R}$ be defined by $h(x) = \frac{1}{1-\delta} \inf_u c(x, u)$. Then,*

(a) *$h \in C$, $h \leq Q$, $W^\nu h \leq Q$, $\{W^\nu h\}_{\nu \in N}$ is almost nondecreasing, and $W^\nu h \to Q$ pointwise. If $W^\nu h$ is lsc for all $\nu \in I\!\!N$, in particular if c is level-bounded in u locally uniformly in x, then $W^\nu h \xrightarrow{e} Q$ on $I\!\!R^s$ and Q is lsc.*

(b) *If $W^\nu h$ is lsc for all $\nu \in I\!\!N$, in particular if c is uniformly level-bounded, and if the matrix B has full row rank, then for any $T \in I\!\!N$, for fixed $x \in$ feas P, any sequence of optimal first stage policies of $P_{W^\nu h}^T$ converges pointwise in x to an optimal policy of P in the sense of Theorem 2.*

Proof. That $h \in C$ follows from its expansion, $h(x) = \inf_u \sum_{t=1}^\infty \delta^{t-1} c(x, u)$. The remaining facts in (a) come from a direct application of Theorem 10, Lemma 2 and Theorem 11, observing that h is proper and lsc via Theorem 4 in the case that c is uniformly level-bounded. Part (b) is derived from the results of Theorem 12. $\square$

These results are going to be significant for computations only when the terms $W^\nu h$ are easily computable or explicitly available, see Sect. 5. The next situation we deal with may be useful when this is not necessarily the case; i.e., we still want a good approximation of the problem P that does not necessarily rely on "iterations." For a general convex cost function c, we examine a possible lower bound for P in a manner similar to that considered by Grinold [6], Flåm and Wets [5] for infinite horizon optimization problems. The results give lower bounds for P that rely on "averaging" the future. We remain interested in optimal policies, so the epi-convergence results we seek will once again be in u for fixed x. The finite horizon problems we propose have the form

$$\text{minimize} \sum_{t=1}^{T} \delta^{t-1} c(x_{t-1}, u_t) + \frac{\delta^T}{1-\delta} c(x_T, u_{T+1})$$

$$\text{s.t. } x_t = Ax_{t-1} + Bu_t + b_t \text{ for } t = 1, \dots, T-1$$

$$(I - \delta A)x_T - \delta Bu_{T+1} = (1 - \delta)(Ax_{T-1} + Bu_T + b) + \delta b$$

The unintuitive terminal dynamics will be justified shortly. Through the developments of the previous sections, the objective functions in u, (inclusive now of all dynamics and constraints) then have the form, for fixed $x \in \mathbb{R}^s$,

$$g_x^T(u) = c(x, u) + \delta \min P_F^T(Ax + Bu + b),$$

$$g_x(u) = c(x, u) + \delta Q(Ax + Bu + b).$$

We make the assumption throughout that a policy u of P satisfies

$$\sum_{t=1}^{\infty} \delta^{t-1} |u_t| < \infty \text{ and } \sum_{t=1}^{\infty} \delta^{t-1} |x_{t-1}| < \infty, \tag{2}$$

under the identifications given in (1) for a solution given a policy, where $|\cdot|$ is taken componentwise. A policy $u : \text{feas } P \to \mathbb{R}^n$ is called a *feasible policy* for P if for any $x \in \text{feas } P$, the associated objective is finite, i.e. $g_x(u(x)) < \infty$.

Theorem 14. (feasibility). *If $P(x)$ is feasible, then $P_F^T(x)$ is feasible.*

Proof. Suppose $x \in \text{feas } P$, and let u be a feasible policy for P. Then the corresponding solution $(u_1, u_2, \dots)$ and trajectory $(x_0, x_1, \dots)$ is feasible for initial state x. For fixed $T \in \mathbb{N}$, let

$$\bar{x}_T = (1 - \delta) \sum_{t=T}^{\infty} \delta^{t-T} x_t, \quad \bar{u}_{T+1} = (1 - \delta) \sum_{t=T}^{\infty} \delta^{t-T} u_{t+1}, \tag{3}$$

which exist by assumption (2). Next we show that $(u_1, \dots, u_T, \bar{u}_{T+1})$ and $(x_0, \dots, x_{T-1}, \bar{x}_T)$ are feasible for $P_F^T(x)$. For this it suffices to show that the terminal dynamics are satisfied. Observe that

$$\bar{x}_T = (1 - \delta) \sum_{t=T}^{\infty} \delta^{t-T} x_t$$

$$= (1 - \delta) \sum_{t=T}^{\infty} \delta^{t-T} (Ax_{t-1} + Bu_t + b)$$

$$= (1 - \delta) Ax_{T-1} + \delta A\bar{x}_T + (1 - \delta) Bu_T + \delta B\bar{u}_{T+1} + b.$$

This is just a rearrangement of the terminal dynamics given for $P_F^T(x)$, whereby $x \in \text{feas } P_F^T$. $\square$

Next, we show that the objective functions are monotonically increasing, and provide lower bounds for P.

Theorem 15. (monotonically increasing lower bounds). *For any $x \in \mathbb{R}^s$, $T \in \mathbb{N}$, $g_x^T \leq g_x^{T+1} \leq g_x$.*

Proof. It suffices to show from the definition of the objectives that $\min P_F^T \leq \min P_F^{T+1} \leq Q$. Let u be a feasible policy for P. Fix $x \in \mathbb{R}^s$ and let $(u_1, u_2, \dots)$ and $(x_0, x_1, \dots)$ be a corresponding solution with $\bar{x}_T$ and $\bar{u}_{T+1}$ defined by (3). By a convexity argument using the fact that $(1-\delta)\sum_{t=1}^{\infty} \delta^{t-1} = 1$, we obtain

$$\frac{1}{1-\delta}c(\bar{x}_T, \bar{u}_{T+1}) \leq c(x_T, u_{T+1}) + \frac{\delta}{1-\delta}c(\bar{x}_{T+1}, \bar{u}_{T+2}).$$

This immediately implies that $g_x^T \leq g_x^{T+1}$. In the same manner,

$$\frac{1}{1-\delta}c(\bar{x}_T, \bar{u}_{T+1}) \leq \sum_{t=T+1}^{\infty} \delta^{t-T-1}c(x_{t-1}, u_t),$$

which shows that for any $T \in \mathbb{N}$, $\min P_F^T(x) \leq Q(x)$. So we may conclude that $g_x^T \leq g_x^{T+1} \leq g_x$. $\square$

We conclude this section by showing that the objective functions epi-converge as $T \to \infty$, which in turn ensures the convergence of optimal policies to an optimal policy of P. This will follow from the results in Sect. 3.

Theorem 16. (convergence of optimal policies). *If $\min P_F^T(\cdot)$ are lsc for all $T \in \mathbb{N}$, in particular if c is uniformly level-bounded and $\min P_F^1(\cdot)$ is proper and lsc, and the matrix B has full row rank, then $g_x^T \xrightarrow{e} g_x$, and the conclusions of Theorem 2 are satisfied.*

Proof. This applies Theorem 12 through the observation that $\min P_F^1(\cdot)$ is in C, and $\min P_F^T = W \min P_F^{T-1}$. $\square$

5 Piecewise Linear-Quadratic Costs

Let's now consider the case in which the cost function has the form

$$c(x, u) = p \cdot u + \tfrac{1}{2}u \cdot Pu + c \cdot x + \rho_{V,Q}(q - Cx - Du), \qquad (4)$$

where P and Q are (symmetric) positive semidefinite, V is polyhedral convex, and

$$\rho_{V,Q}(u) := \sup_{v \in V}\{v \cdot u - \tfrac{1}{2}v \cdot Qv\}.$$

Note that c is convex, lsc (possibly infinite-valued), and piecewise linear-quadratic on its domain. This is an important class of problems, due to their modeling flexibility and exploitable structure, c.f. [12], [11] and [8]. Our objective is to investigate the form of the functions h and $W^\nu h$ obtained in Sect. 4 when the cost has the piecewise linear-quadratic form of (4). We make use of the following lemma, whose proof relies on extended linear-quadratic programming duality, detailed in [12], [11] and [8].

Lemma 4. *Let V be a convex polyhedral subset of $\mathbb{R}^d$, and let $J : \mathbb{R}^s \times V \to \mathbb{R}$ be defined by*

$$J(u,v) = p \cdot u + \tfrac{1}{2}u \cdot Pu + q \cdot v - \tfrac{1}{2}v \cdot Qv - Du \cdot v,$$

where P and Q are symmetric positive semidefinite matrices. Suppose

$$0 \in V \ \text{ and } \ p \in \operatorname{col} P. \tag{5}$$

Then

$$\inf_u \sup_{v \in V} J(u,v) = \sup_{v \in V} \inf_u J(u,v).$$

Proof. It's always true that

$$\inf_u \sup_{v \in V} J(u,v) \geq \sup_{v \in V} \inf_u J(u,v),$$

so the equality is automatically satisfied when $\sup_{v \in V} \inf_u J(u,v) = \infty$. Observe that when $\sup_{v \in V} \inf_u J(u,v) < \infty$, $\sup_{v \in V} \inf_u J(u,v)$ is finite since for $v = 0$,

$$\inf_u J(u,0) = \inf_u p \cdot u + \tfrac{1}{2}u \cdot Pu = -\tfrac{1}{2}p \cdot P'p > -\infty,$$

by virtue of $p \in \operatorname{col} P$. Applying extended linear-quadratic programming duality (c.f. [12]) to the fact that $\sup_{v \in V} \inf_u J(u,v)$ is finite implies that

$$\inf_u \sup_{v \in V} J(u,v) = \sup_{v \in V} \inf_u J(u,v),$$

which completes the proof. $\square$

This leads us to the first main result, which shows that piecewise linear-quadraticity in the form of (4) is preserved under inf-projection. We will make use of some facts about symmetric positive semidefinite matrices. For a matrix P that is positive semidefinite of rank r, we can write $P = S\Lambda S^*$ where S is orthogonal and Λ is a diagonal matrix of eigenvalues whose first r entries are nonzero (positive). Let S_2 denote the matrix consisting of the last $n - r$ columns of S, and P' the pseudo-inverse of P.

Theorem 17. (preservation of piecewise linear-quadraticity). *Let c be piecewise linear-quadratic as in (4) and* $g(x) = \inf_u c(x, u)$. *If assumption (5) is satisfied for the p, P, and V in the definition of c in (4), then*

$$g(x) = c^0 \cdot x + \rho_{V^0, Q^0}(q^0 - C^0 x) + a^0,$$

where $V^0 = V$, $Q^0 := DP'D^* + Q$ *is symmetric positive semidefinite,* $c^0 = c$, $q^0 = q + DP'p$, $C^0 = C$, *and* $a^0 = -\frac{1}{2}p \cdot P'p$.

Proof. We begin by letting $J_x(u, v) = p \cdot u + \frac{1}{2}u \cdot Pu + v \cdot (q - Cx) - v \cdot Du - \frac{1}{2}v \cdot Qv$. Observe that

$$g(x) = c \cdot x + \inf_u \left\{ p \cdot u + \frac{1}{2}u \cdot Pu + \rho_{V,Q}(q - Cx - Du) \right\}$$
$$= c \cdot x + \inf_u \sup_{v \in V} J_x(u, v).$$

The assumption in (5) along with the conclusions of Lemma 4 allow the limits to be interchanged, so that $g(x) = c \cdot x + \sup_{v \in V} \inf_u J_x(u, v)$.

For fixed $v \in V$, setting $\nabla_u J_x(\cdot, v) = 0$ results in the equation $Pu = D^*v - p$. Since $p \in \text{col } P$, this equation always has a solution when $v = 0$. For P' the pseudoinverse of P (which exists since P is symmetric, hence diagonalizable), whenever $D^*v \in \text{col } P$ (or equivalently $S_2^* D^*v = 0$), we have that $u = P'(D^*v - p)$, is a particular solution of $Pu = D^*v - p$. This yields

$$g(x) = c \cdot x + \rho_{V^0, Q^0}(DP'p + q - Cx) - \frac{1}{2}p \cdot P'p,$$

as claimed. Q^0 is symmetric positive semidefinite since both Q and P' are, the symmetry and positive semidefiniteness of P' coming from that of P. □

The next proposition gives an explicit formula for the lower bound h obtained in Theorem 3.

Proposition 1. (piecewise linear-quadraticity of end term). *Let* $h : \mathbb{R}^s \to \overline{\mathbb{R}}$ *be defined as in Theorem 3 by* $h(x) = \frac{1}{1-\delta} \inf_u c(x, u)$. *With c piecewise linear-quadratic of the form (4), if assumption (5) is satisfied, then*

$$h(x) = c^0 \cdot x + \rho_{V^0, Q^0}(q^0 - C^0 x) + a^0,$$

where $V^0 = V$, $Q^0 := \frac{1}{1-\delta}(DP'D^* + Q)$ *is symmetric positive semidefinite,* $c^0 = \frac{1}{1-\delta}c$, $q^0 = \frac{1}{1-\delta}(q + DP'p)$, $C^0 = \frac{1}{1-\delta}C$, *and* $a^0 = -\frac{1}{2(1-\delta)}p \cdot P'p$.

Proof. This is a direct consequence of Theorem 17, after the observation that for any $\alpha > 0$, $\alpha \rho_{V,Q}(z) = \rho_{V, \alpha Q}(\alpha z)$, which follows directly from the definition of $\rho_{V,Q}$. □

What we have established so far is a rough lower bound for Q that is expressible as a piecewise linear-quadratic function. But this may be improved upon by performing the successive iterations,

$$Wh(x) := \inf_u \{c(x, u) + \delta h(Ax + Bu + b)\},$$

$$W^\nu h(x) := \inf_u \{c(x, u) + \delta W^{\nu-1} h(Ax + Bu + b)\},$$

each a better lower bound approximation to Q than its predecessor, and each expressible explicitly in piecewise linear-quadratic form. The result that follows proves the piecewise linear-quadraticity for the iterates.

Theorem 18. (piecewise linear-quadraticity of iterates). *Let* $c : \mathbb{R}^s \times \mathbb{R}^n \to \overline{\mathbb{R}}$ *be piecewise linear-quadratic as in (4). Suppose* $W^{\nu-1} h(x) = c^{\nu-1} \cdot x + \rho_{V^{\nu-1}, Q^{\nu-1}} (q^{\nu-1} - C^{\nu-1} x) + a^{\nu-1}$, *and that* $0 \in V^{\nu-1}$. *If assumption (5) is satisfied, then* $W^\nu h(x) = c^\nu \cdot x + \rho_{V^\nu, Q^\nu} (q^\nu - C^\nu x) + a^\nu$, *where*

$$V^\nu = V \times V^{\nu-1}, \ \tilde{D} = \begin{pmatrix} D \\ \delta C^{\nu-1} B \end{pmatrix}, \ C^\nu = \begin{pmatrix} C \\ \delta C^{\nu-1} A \end{pmatrix},$$

$$Q^\nu = \tilde{D} P' \tilde{D}^* + \begin{pmatrix} Q \\ & \delta Q^{\nu-1} \end{pmatrix}, \ q^\nu = \tilde{D} P' p + \begin{pmatrix} q \\ \delta (q^{\nu-1} - C^{\nu-1} b) \end{pmatrix},$$

$c^\nu = c + \delta c^{\nu-1}$, *and* $a^\nu = -\frac{1}{2} p \cdot P' p + \delta a^{\nu-1}$. V^ν *is polyhedral convex with* $0 \in V^\nu$, *and* Q^ν *is symmetric positive semi-definite.*

Proof. Observe that

$$W^\nu h(x) = \inf_u \left\{ p \cdot u + \tfrac{1}{2} u \cdot P u + \rho_{V^\nu, \tilde{Q}} (\tilde{q} - C^\nu x - \tilde{D} u) \right\}$$
$$+ (c + \delta c^{\nu-1}) \cdot x + \delta a^{\nu-1},$$

where

$$\tilde{Q} = \begin{pmatrix} Q \\ & \delta Q^{\nu-1} \end{pmatrix} \text{ and } \tilde{q} = \begin{pmatrix} q \\ \delta (q^{\nu-1} - C^{\nu-1} b) \end{pmatrix}.$$

Now apply Theorem 17 in this setting to obtain the result.   □

These results considered in the framework of the approximation theorems and finite horizon approximates in Sect. 3 and Sect. 4, allow one to approximate an infinite horizon problem with piecewise linear-quadratic cost with a sequence of almost nondecreasing finite horizon extended linear-quadratic problems with explicit end term. The structure of such problems is exploited fully in [12], [11] and [8]. Now we turn to an application of the above results, in which extended linear-quadratic penalties arise naturally in an inventory management problem.

6   An Inventory Management Problem

Inventory management problems provide one example of dynamic problems which are commonly handled in the dynamic programming setting. Many such models therefore assume simple (linear) cost structures, a finite state space, and only the simplest constraints. Here, we allow for the possibility of more complex costs, penalty expressions and constraints, rendering the model more realistic and flexible. In addition, the state space will not be assumed discrete.

We consider a store which is supplied by a central warehouse. Inventory is reviewed weekly and if necessary, new inventory is delivered from the central warehouse to the store to maintain the optimal stock levels. Associated with an inventory shipment is a fixed charge p_i per unit of product i, $i = 1, \ldots, k$. In addition, there is a daily storage charge for the inventory kept at the store at the rate of b dollars per unit for up to r units of product and d dollars per unit for quantities in excess of r units due to added storage facility costs. The policy of the store requires that the weekly sales s_i, $i = 1, \ldots, k$, which we assume fixed are supplied by in-store inventory or penalized via the penalty expression

$$l_i(x) = \begin{cases} 0 & \text{if } x \geq s_i \\ a_i(s_i - x)^2 & \text{if } x < s_i \end{cases}$$

as a function of inventory x^i of product i, when demand s_i is not satisfied. In stochastic models, the sales (demand) may be modeled by a random variable. This would fit in the same framework generalized to the stochastic case as in [8]. In that instance, the penalty expression is necessary since we could not ask that the demand be satisfied by the current inventory with certainty.

The store seeks a reordering policy which will minimize the *discounted* cost of inventory replenishment and storage over an infinite time horizon. An infinite horizon is used because the store does not want to use up its inventory at the end of any finite time period. The discounting assures convergence and emphasizes the greater importance of the "here and now" due to more certainty about sales and other paramenters. To formulate this problem as a stationary infinite horizon problem with discounted costs, in the piecewise linear-quadratic framework, we will need the following observations and assumptions.

Cost function. The cost function is piecewise linear-quadratic, given by

$$c_0(x, u) = \sum_{i=1}^{k} p_i u^i + \sum_{i=1}^{k} bx^i + d \Big[\sum_{i=1}^{k} x^i - r \Big]_+ + \sum_{i=1}^{k} l_i(x^i).$$

States and controls. The state of the system keeps track of the current in-store inventory for each product. We denote the state at time t by $x_t =$

$(x_t^1 \cdots x_t^k)^*$, where x_t^i = total amount of product i in store at time t, $i = 1, \ldots, k$. The controls are given by $u_t = (u_t^1 \cdots u_t^k)^*$, where u_t^i = units of product i to ship at time t. Then the dynamics become

$$x_t = x_{t-1} + u_t - s, \quad x_0 = x,$$

for $t = 1, 2, \ldots$, where $s = (s_1 \cdots s_k)^*$ are the expected sales of products $1, \ldots, k$ respectively.

Constraints. In order for the problem to make sense, the shipped inventory must be nonnegative, i.e. $u_t \geq 0$. We also require that the in-store inventory is nonnegative ($x_t \geq 0$), and can never exceed a fixed amount K which amounts to $\sum_{i=1}^k \{x_{t-1}^i + u_t^i\} \leq K$. We give this in matrix form by $\bar{C}x_{t-1} + \bar{D}u_t \geq \bar{q}$ where $\bar{C} = (-e \ 0 \ I)^*$, $\bar{D} = (-e \ I \ 0)^*$, and $\bar{q} = (-K \ 0)^*$, where e is a vector of 1's.

Objective function. The objective is to minimize the total discounted cost over an infinite horizon, which may be written as

$$\min \sum_{t=1}^{\infty} \delta^{t-1} c_0(x_{t-1}, u_t)$$

subject to the aforementioned dynamics and constraints. If we include the constraints in the objective the new cost becomes

$$c(x, u) = c_0(x, u) + \delta_{\{\bar{C}x + \bar{D}u \geq \bar{q}\}}(x, u),$$

where $\delta_C(\cdot)$ is the indicator function for the set C, i.e. $\delta_C(x) = 0$ if $x \in C$ and $+\infty$ otherwise. Note that the objective is bounded on its domain. It may equivalently be written in the desired piecewise linear-quadratic form as

$$c(x, u) = p \cdot u + \tfrac{1}{2}u \cdot Pu + c \cdot x + \rho_{V,Q}(q - Cx - Du),$$

where $p = 0$, $P = 0$, $c = (b \ \cdots \ b)^*$, $V = \Pi_{i=1}^k [0, p_i] \times [-d, 0] \times [0, \infty)^{3k+1}$,

$$Q = \begin{pmatrix} 0 \\ & 0 \\ & & \frac{1}{2a_1} \\ & & & \ddots \\ & & & & \frac{1}{2a_k} \\ & & & & & 0 \end{pmatrix}, \quad q = \begin{pmatrix} 0 \\ r \\ s \\ \bar{q} \end{pmatrix}, \quad C = \begin{pmatrix} 0 \\ e^* \\ I \\ \bar{C} \end{pmatrix}, \quad \text{and } D = \begin{pmatrix} I \\ 0 \\ 0 \\ \bar{D} \end{pmatrix},$$

where e is a vector of 1's. Hence the problem now becomes

$$\text{minimize} \sum_{t=1}^{\infty} \delta^{t-1}\big(c \cdot x_{t-1} + \rho_{V,Q}(q - Cx_{t-1} - Du_t)\big)$$

$$\text{s.t. } x_t = x_{t-1} + u_t - s \text{ for } t = 1, 2, \ldots$$

$$x_0 = x,$$

which is of the form $P(x)$. Note that $0 \in V$ and $p \in \operatorname{col} P$ trivially.

Finite horizon approximations. We are now ready to derive finite horizon bounds for the problem. With $c(x, u) = c \cdot x + \rho_{V,Q}(q - Cx - Du)$ as the starting point, applying the techniques developed in the previous sections, one may obtain the finite horizon problem in piecewise linear-quadratic form,

$$\text{minimize } \sum_{t=1}^{T} \delta^{t-1} c(x_{t-1}, u_t) + \frac{\delta^T}{1 - \delta} c(x_T, u_{T+1})$$

$$\text{s.t. } x_t = x_{t-1} + u_t - s \text{ for } t = 1, \dots, T - 1$$

$$(1 - \delta)x_T - \delta u_{T+1} = (1 - \delta)(x_{T-1} + u_T - s) - \delta s$$

This bound comes from the results of Sect. 4.2. Alternatively, if the assumption in (4) is satisfied, we may rely on the approximation theorems to obtain lower bounds that are almost monotonically increasing to P as $T \to \infty$:

$$\text{minimize } \sum_{t=1}^{T} \delta^{t-1} c(x_{t-1}, u_t) + \delta^T \left(c_T \cdot x_T + \rho_{V_T, Q_T}(q_T - C_T x_T) \right)$$

$$\text{s.t. } x_t = x_{t-1} + u_t - s \text{ for } t = 1, \dots, T,$$

$$x_0 = x,$$

where the parameters c_T, V_T, Q_T, q_T, and C_T in the last term in the sum are derived from Theorem 18. Either of these lower bounds provide good approximations to the original problem (in the sense of epi-convergence), and have a flexible structure that is highly amenable to computational schemes.

References

1. Arrow, K., Blackwell, D., Girschick, M. (1949) Bayes and Minimax Solutions of Sequential Decision Problems. Econometrica **17** 213–244
2. Bellman, R. (1957) Dynamic Programming. Princeton University Press, Princeton
3. Bertsekas, D. (1987) Dynamic Programming: Deterministic and Stochastic Models. Prentice-Hall, Englewood Cliffs
4. Eckstein, J., Ferris, M. (1996) Operator Splitting Methods for Monotone Affine Variational Inequalities, with a Parallel Application to Optimal Control. Manuscript, Rutgers University
5. Flåm, S., Wets, R. (1987) Existence Results and Finite Horizon Approximates for Infinite Horizon Optimization Problems. Econometrica **55** 1187–1209
6. Grinold, R. (1977) Finite Horizon Approximates of Infinite Horizon Linear Programs. Mathematical Programming **12** 1–17
7. King, A., Rockafellar, R., Somlyódy, L., Wets, R. (1988) Lake Eutrophication Management: The Lake Balaton Project. In: Ermoliev Y, Wets R. (Eds.) Numerical Techniques fro Stochastic Optimization. Springer-Verlag, Berlin, 415–424

8. Korf, L. (1998) Approximation and Solution Schemes for Stochastic Dynamic Optimization Problems. Ph.D. Thesis, University of California, Davis
9. Korf, L. (1998) Insurers' Portfolios of Risks: Approximating Infinite Horizon Stochastic Dynamic Optimization Problems. IIASA Interim Report 98–061
10. Massé, P. (1946) Les Réserves et la Régulation de l'Avenir dans la Vie Economique. Herman, Paris, 2 vols
11. Rockafellar, R. (1998) Duality in Stochastic Programming. In: Wets R., Ziemba W. (Eds.) Stochastic Programming: State of the Art 1998. Annals of Operations Research, Baltzer, Amsterdam, to appear
12. Rockafellar, R., Wets, R. (1990) Generalized Linear-Quadratic Problems of Deterministic and stochastic Optimal Control in Discrete Time. SIAM Journal of Control and Optimization **28**, 810–822
13. Rockafellar, R., Wets, R. (1997) Variational Analysis. Grundlehren der mathematischen Wissenschaften 317, Springer-Verlag, Berlin
14. Salinger, D. (1997) A Splitting Algorithm for Multistage Stochastic Programming with Application to Hydropower Scheduling. Ph. D. Thesis, University of Washington, Seattle
15. Wald, A. (1947) Sequential Analysis. Wiley, New York

An Ergodic Theorem
for Stochastic Programming Problems*

Lisa A. Korf[1] and Roger J-B Wets[2]

[1] IBM T.J. Watson Research Center, Yorktown Heights, NY
[2] Department of Mathematics, University of California, Davis

Abstract. To justify the use of sampling to solve stochastic programming problems one usually relies on a law of large numbers for random lsc (lower semicontinuous) functions when the samples come from independent, identical experiments. If the samples come from a stationary process, one can appeal to the ergodic theorem proved here. The proof relies on the 'scalarization' of random lsc functions.

Key words: ergodic theorem, laws of large numbers, stochastic programming, stochastic programs with recourse, stochastic programs with chance constraints, random lsc functions, epi-convergence

1 Introduction

Stochastic programming models can be viewed as extensions of linear and nonlinear programming models to accommodate situations in which only information of a probabilistic nature is available about some of the parameters of the problem. The following formulation includes both the *stochastic programming with recourse models* and the *stochastic programming with chance constraints models* :

$$
\begin{aligned}
\min \quad & E\{f_0(\boldsymbol{\xi}, x)\} && (1)\\
\text{so that} \quad & E\{f_i(\boldsymbol{\xi}, x)\} \le 0, \quad i = 1, \dots, m,\\
& x \in \mathbb{R}^n
\end{aligned}
$$

where
- $\boldsymbol{\xi}$ is a random vector with support $\Xi \subset \mathbb{R}^N$,
- P is a probability distribution function on $\mathbb{R}^N$,
- $f_0 : \mathbb{R}^n \times \Xi \to \overline{\mathbb{R}} = [-\infty, \infty]$,
- $f_i : \mathbb{R}^n \times \Xi \to \mathbb{R}, \quad i = 1, \dots, m,$
- for $i = 0, \dots, m$: $Ef_i(x) := E\{f_i(\boldsymbol{\xi}, x)\} = \int_\Xi f_i(\xi, x)dP(\xi)$ is assumed finite unless $\{\xi \mid f_0(\xi, x) = \infty\}$ has positive probability and then $Ef_0(x) = \infty$.

Let's also assume that the feasibility set

$$
S = \left\{ x \in \mathbb{R}^n \mid Ef_i(x) \le 0, \ i = 1, \dots m \right\} \cap \left\{ x \mid Ef_0(x) < \infty \right\}
$$

* Research supported in part by a grant of the National Science Foundation

is nonempty. We are led to include the possibility that f_0 and Ef_0 take on the value ∞ to allow for the presence of *induced constraints* as will be explained shortly.

The two-stage version of a stochastic program with recourse reads:

$$\min \quad q_1(x) + E\{Q(\xi, x)\} \text{ so that } f_i(x) \leq 0, \quad i = 1, \dots, m,$$

where

$$Q(\xi, x) = \inf_y \left\{ q_2(\xi, y) \mid y \in S_2(\xi, x) \subset \mathbb{R}^{n_2} \right\}.$$

Immediate costs $q_1(x)$ as well as future (recourse) costs $EQ(x) = E\{Q(\xi, x)\}$ must be taken into account in the search for an optimal decision. In terms of our canonical problem, f_0 is simply $q_1(x) + Q(\xi, x)$ and the $Ef_i \equiv f_i$ since these constraints don't depend on ξ. If $S_2(\xi, x) = \emptyset$, i.e., no feasible recourse is available in this situation, then $Q(\xi, x) = \infty$. $P\{\xi \in \Xi \mid Q(\xi, x) = \infty\} > 0$ means that there is a positive probability that no recourse will be available if x is chosen as the first stage decision. The 'induced' constraints restrict the choice of x to those for which, with probability 1, there will be a feasible recourse. Multistage recourse models can be 'reduced' to two-stage problems and consequently also fit our general framework, for example, cf. [1,2].

Reliability considerations lead to the inclusion of chance constraints in the formulation of the stochastic programming problem. Usually, they are expressed in the following probabilistic terms:

$$P\{\xi \in \Xi \mid g_k(\xi, x) \leq 0, \ k = 1, \dots, q\} \geq \alpha$$

with $\alpha \in (0, 1]$ the reliability level, or they may include constraints on the moments of certain quantities such as

$$E\{g_k(\xi, x)\} + \beta[\text{var } g_k(\xi, x)]^{1/2} \leq 0$$

with β a positive constant. To bring the probabilistic constraints in concordance with the canonical form (1), define f_i as follows:

$$f_i(\xi, x) = \begin{cases} \alpha - 1 & \text{if } g_k(\xi, x) \leq 0, \ k = 1, \dots, q \\ \alpha & \text{otherwise .} \end{cases}$$

Similarly, the constraint on the moments involves the sum of two functions that are both expectation functionals, since var $g_k(\xi, x)(x) = E\{g_k(\xi, x) - Eg_k(x)\}^2$.

Let's finally observe that standard nonlinear programming problems are included as special cases of problems of type (1) since one could have $f_i(\xi, x) \equiv g_i(x)$, a function that doesn't depend on ξ, and then $Ef_i(x) = g_i(x)$. At the same time, stochastic programs of the type (1) can also be viewed as a particular class of nonlinear programming problems. Indeed, one can rewrite (1) as follows:

$$(2) \qquad \min Ef_0(x) \text{ so that } Ef_i(x) \leq 0, \qquad i = 1, \dots, m, \ x \in \mathbb{R}^n.$$

The only difference is that one makes explicit the fact that that the evaluation of some, or all, of the functions Ef_i, $i = 0, \ldots, m$, might require the calculation of a (multi-dimensional) integral. This is why our concerns need to go much beyond identifying the properties (linearity, convexity, differentiability) of the *expectation functionals* Ef_i and leaving the task of solving (2) to the appropriate nonlinear programming package. The major obstacle to proceeding in this manner comes precisely from the fact that evaluating Ef_i at any given x, or calculating its (sub)gradient at this x, may be a much more onerous task than solving a typical nonlinear programming problem.

Except for some special cases when the integral $\int_{\Xi} f_i(\xi, x) P(d\xi)$ is one-dimensional, or can be expressed as a sum of one-dimensional integrals, to evaluate this integral one must generally rely on approximation schemes with P replaced by a discrete measure P^ν obtained either from some partitioning of the sample space or as the empirical measure derived from a sample of the random quantities. In this latter instance, one needs to justify that the solution derived with the empirical measure P^ν is, at least in a probabilistic sense, an approximate solution. This paper deals with such a justification without making the usual assumption that the sample points have been obtained from independent experiments. The examples in Sect. 3 and Sect. 4 provide some of the motivation for relaxing the independence assumption, but it is also what is required to obtain the consistency of M-estimates for the parameters of regression models involving constraints coming from *a priori* information, cf. [3, Sect. 2].

2   Ergodic Theorem

A comprehensive and powerful technique to obtain the 'consistency' of the optimal solutions of the approximating problems is to actually prove that the approximating stochastic optimization problems themselves are 'consistent'. And to do this, as explained in Sect. 3 and Sect. 4, one can appeal to a general ergodic theorem for random lsc (lower semicontinuous) functions that can be formulated as follows: Let $\mathcal{B}$ be the Borel field on $\mathbb{R}^n$, $(\Xi, \mathcal{S}, P)$ a probability space with $\mathcal{S}$ P-complete; the P-completeness assumption is harmless for the application we have in mind. A *random lsc (lower semicontinuous) function* is then an extended real-valued function $f : \Xi \times \mathbb{R}^n \to \overline{\mathbb{R}}$ such that

(i) the function $(\xi, x) \mapsto f(\xi, x)$ is $\mathcal{S} \otimes \mathcal{B}$-measurable;
(ii) for every $\xi \in \Xi$, the function $x \mapsto f(\xi, x)$ is lsc.

Theorem 1. (Ergodic Theorem). *Let f be a random lsc function defined on $\Xi \times \mathbb{R}^n$, $\varphi : \Xi \to \Xi$ an ergodic measure preserving transformation. Then, whenever $\xi \mapsto \inf_{\mathbb{R}^n} f(\xi, \cdot)$ is summable,*

$$\frac{1}{\nu} \sum_{k=1}^{\nu} f(\varphi^k(\xi), \cdot) \xrightarrow{e} Ef, \quad P\text{-a.s.,}$$

where $\xrightarrow{e}$ stands for epi-convergence.

The immediate precursors of this theorem are the laws of large numbers for random lsc functions [4–6], that all posit iid (independent identically distributed) sampling cf. also [7,8] for further extensions. Here only stationarity is assumed; the argument relies on a 'scalarization' of random lsc functions developed in Sect. 6. The proof of the ergodic theorem can be found in Sect. 7.

Definition 1. A sequence of functions $\{g^\nu : \mathbb{R}^n \to \overline{\mathbb{R}}, \nu \in \mathbb{N}\}$ *epi-converges to* $g : \mathbb{R}^n \to \overline{\mathbb{R}}$, written $g^\nu \xrightarrow{e} g$, if for all $x \in \mathbb{R}^n$,
  (i) $\liminf_\nu g^\nu(x^\nu) \geq g(x)$ for all $x^\nu \to x$;
  (ii) $\limsup_\nu g^\nu(x^\nu) \leq g(x)$ for some $x^\nu \to x$.

Epi-convergence entails the convergence of the minimizers of the g^ν to those of g as is made precise below; cf. [9–11] for more about epi-convergence, theory and applications. Epi-convergence at a point x can also be characterized in terms of lower and upper epi-limits.

Definition 2. For a sequence of functions $\{g, g^\nu : \mathbb{R}^n \to \overline{\mathbb{R}}, \nu \in \mathbb{N}\}$, the *lower* and *upper epi-limits* are:

$$\text{e-}\liminf g^\nu(x) := \sup_{\rho>0} \liminf_{\nu\to\infty} \inf_{y\in B(x,\rho)} g^\nu(y),$$

$$\text{e-}\limsup g^\nu(x) := \sup_{\rho>0} \limsup_{\nu\to\infty} \inf_{y\in B(x,\rho)} g^\nu(y).$$

If $\text{e-}\limsup g^\nu = \text{e-}\liminf g^\nu = g$, then $g =: \text{e-}\lim g^\nu$ is the *epi-limit* of the sequence $\{g^\nu\}_{\nu\in\mathbb{N}}$.

It follows immediately from Definition 3 that the lower and upper epi-limits are always lsc; epi-convergence of g^ν to g corresponds to the set convergence of epi g^ν to epi g. It is neither implied by, nor does it imply pointwise convergence, but instead can be viewed as a one-sided uniform convergence. But, it's exactly what is needed to ensure the convergence of minimizers of g^ν to the minimizers of g, in the following sense.

Theorem 2. [11, Chapter 7]. *Suppose that $\{g^\nu\}_{\nu\in\mathbb{N}}$ is a sequence of extended real-valued lsc functions such that $g^\nu \xrightarrow{e} g$. Then every cluster point of* $\arg\min g^\nu$ *is an element of* $\arg\min g$*. Moreover, if* $\arg\min g$ *is nonempty, and there exists a compact $K \subset \mathbb{R}^n$ such that* dom $g^\nu \subset K$, *then*

$$\arg\min g = \bigcap_{\varepsilon>0} \liminf_\nu (\varepsilon\text{-}\arg\min g^\nu),$$

where $\varepsilon\text{-}\arg\min g := \{x \in \mathbb{R}^n \mid g(x) \leq \inf g + \varepsilon < \infty\}$.

Here we state only sufficient conditions for convergence of the ε-argmin. For necessary conditions, consult [11, Chapter 7], and one can refer to [6] for some extensions.

In order to obtain our almost sure epi-convergence result for random lsc functions, the following two results based on the separability of $\mathbb{R}^n$ are essential. They tell us that epi-convergence needs only to be verified at the points in a countable dense set.

Lemma 1. [5]. *Let $f, g : \mathbb{R}^n \rightarrow \overline{\mathbb{R}}$ with f lsc. Let $R \subset \mathbb{R}^n$ be the projection on $\mathbb{R}^n$ of a countable dense subset of $\operatorname{epi} g$. If $f \leq g$ on R, then $f \leq g$ on all of $\mathbb{R}^n$.*

Proof. The set R above exists since $\mathbb{R}^n \times \mathbb{R}$ is separable. Suppose $f \leq g$ on R. This is equivalent to $\{(x, \alpha) \mid \alpha \geq g(x), x \in R\} \subset \operatorname{epi} f$. Since f is lsc, $\operatorname{epi} f$ is closed. Taking closures on both sides yields $\operatorname{epi} g \subset \operatorname{epi} f$, which is equivalent to $f \leq g$ on $\mathbb{R}^n$. □

Lemma 2. [5]. *Let $\{g^\nu\}_{\nu \in N}$ be a sequence of extended real-valued functions defined on $\mathbb{R}^n$ and $g : \mathbb{R}^n \rightarrow \overline{\mathbb{R}}$ an lsc function. Let $R \subset \mathbb{R}^n$ be the union of R_1 and R_2, the projections onto $\mathbb{R}^n$ of a countable dense subset of $\operatorname{epi} g$ and a countable dense subset of $\operatorname{e-lim\,inf} g^\nu$ respectively. Then $g = \operatorname{e-lim} g^\nu$ on R implies $g = \operatorname{e-lim} g^\nu$ on $\mathbb{R}^n$.*

Proof. In order to show that $g = \operatorname{e-lim} g^\nu$, the following must hold.

$$\operatorname{e-lim\,sup} g^\nu \leq g \leq \operatorname{e-lim\,inf} g^\nu.$$

Since these are inequalities between lsc functions, we can use Lemma 2.5 to prove each inequality by having the first one satisfied on the countable set, R_1, and the second satisfied on the countable set, R_2. Then both inequalities hold on $\mathbb{R}^n$ if they are satisfied on $R := R_1 \cup R_2$. □

3   Stochastic Programs with Recourse

Consider again the two-stage stochastic program with recourse:

$$\min \quad q_1(x) + E\{Q(\xi, x)\} \text{ so that } f_i(x) \leq 0, \quad i = 1, \ldots, m,$$

with

$$Q(\xi, x) = \inf_y \{q_2(\xi, y) \mid y \in S_2(\xi, x) \subset \mathbb{R}^{n_2}\}.$$

Let's assume that this program has an optimal solution and let's denote it by x^*.

Let $\xi^1, \ldots, \xi^\nu$ be a sample of size ν of the random quantities ξ and let P^ν be the empirical measure obtained by assigning probability $1/\nu$ to each one of these sample points. Replacing P by P^ν leads us to the stochastic program:

$$\min \quad q_1(x) + \frac{1}{\nu} \sum_{k=1}^{\nu} Q(\xi^k, x) \text{ so that } f_i(x) \le 0, \quad i = 1, \ldots, m,$$

where for $k = 1, \ldots, \nu$,

$$Q(\xi^k, x) = \inf_y \left\{ q_2(\xi^k, y) \mid y \in S_2(\xi^k, x) \subset \mathbb{R}^{n_2} \right\}.$$

This can also be written as:

$$\min \quad q_1(x) + \frac{1}{\nu} \sum_{k=1}^{\nu} q_2(\xi^k, y^k)$$
$$\text{so that} \quad f_i(x) \le 0, \quad i = 1, \ldots, m,$$
$$y^k \in S_2(\xi^k, x), \quad k = 1, \ldots, \nu.$$

If ν is not too large, this problem can be solved by an appropriate linear or nonlinear programming package. Let x^ν be the x-component of the solution of this optimization problem. Because P^ν depends on the sample, it actually is a random measure, and consequently x^ν itself is a random variable. Proving consistency consists in showing that x^ν converges to x^* with probability 1.

The answer to this question is provided by Theorem 2 and the Ergodic Theorem 1 if the samples are iid or more generally, are generated from an ergodic process:

$$\{\xi^1, \xi^2, \ldots, \xi^\nu, \ldots\},$$

and the following function f is a random lsc function:

$$f(\xi, x) = \begin{cases} q_1(x) + Q(\xi, x) & \text{if } f_i(x) \le 0, \quad i = 1, \ldots, m, \\ \infty & \text{otherwise}. \end{cases}$$

For example, this will be the case under the following assumptions:
- $q_1, f_i, i = 1, \ldots, m$ are lsc functions;
- $f_2(\xi, x, y) = \begin{cases} q_2(\xi, x, y) & \text{if } y \in S_2(\xi, x), \\ \infty & \text{otherwise}, \end{cases}$ is a random lsc function;
- for all (ξ, x): $y \mapsto f_2(\xi, x, y)$ is inf-compact (lsc with bounded level sets).

A proof could be constructed on the basis of Proposition 1.18 (about epigraphical projections) and Theorem 14.37 (about the measurability of optimal values) in [11].

The need to go beyond the iid case comes from situations when the sample is obtained from a time series. In such situations the samples aren't iid but usually the process is ergodic. This is typically the case when dealing with applications where the uncertainty comes from the environment, cf. [12] for an application dealing with lake eutrophication management and [13] for an application involving the control of water reservoirs to generate hydropower.

4   Stochastic Programs with Chance Constraints

As we have seen in Sect. 1, in a stochastic program with probabilistic constraints the expectation functional appears in the constraints in the following form: $E\{f(\xi,x)\} \leq 0$ where

$$f(\xi,x) = \begin{cases} \alpha - 1 & \text{if } g_k(\xi,x) \leq 0, \ k=1,\dots,q \\ \alpha & \text{otherwise} . \end{cases}$$

For an introduction to stochastic programs with chance constraints one could consult [14], a number of applications are described in [15] and a comprehensive treatment can be found in [16]. To obtain 'consistency' we follow an approach similar to that in [17].

We need the following result about the convergence of level sets of an epi-convergent sequence of functions; the *inner* and *outer limits* of a sequence $\{C^\nu\}_{\nu \in N}$ of subsets of $\mathbb{R}^n$ are defined as follows:

$$\liminf_{\nu \to \infty} C^\nu = \left\{ x = \lim_{\nu \to \infty} x^\nu \,\middle|\, x^\nu \in C^\nu \text{ eventually} \right\}$$

$$\limsup_{\nu \to \infty} C^\nu = \left\{ x = \lim_{k \to \infty} x^{\nu_k} \,\middle|\, x^{\nu_k} \in C^{\nu_k}, \ k \in \mathbb{N} \right\}$$

The *limit* of the sequence exists if the outer and inner limit sets are equal:

$$\lim_{\nu \to \infty} C^\nu := \limsup_{\nu \to \infty} C^\nu = \liminf_{\nu \to \infty} C^\nu .$$

Proposition 1. [11, Proposition 7.7]. *For functions g^ν and g on $\mathbb{R}^n$, one has:*

(a) *$g \leq$ e-$\liminf g^\nu$ if and only if $\limsup_\nu (\text{lev}_{\leq \alpha^\nu} g^\nu) \subset \text{lev}_{\leq \alpha} g$ for all sequences $\alpha^\nu \to \alpha$;*

(b) *$g \geq$ e-$\limsup g^\nu$ if and only if $\liminf_\nu (\text{lev}_{\leq \alpha^\nu} g^\nu) \supset \text{lev}_{\leq \alpha} g$ for some sequence $\alpha^\nu \to \alpha$, in which case such a sequence can be chosen with $\alpha^\nu \downarrow \alpha$.*

(c) *$g =$ e-$\lim g^\nu$ if and only if both conditions hold.*

Theorem 2, about the convergence of the minimizers of epi-convergent functions, and the Ergodic Theorem 1 combined with Proposition 1 yield the following:

Theorem 3. *Let's consider the following stochastic program with chance constraints:*

$$\min f_0(x) \text{ so that } P\{\xi \in \Xi \,|\, g_k(\xi,x) \leq 0, \ k=1,\dots,q\} \geq \alpha \qquad (P)$$

where $f_0 : \mathbb{R}^n \to \mathbb{R}$ is continuous, the functions g_k are lsc on $\Xi \times \mathbb{R}^n$ and $\alpha \in (0,1]$. Let $\xi^1, \xi^2, \dots$ be random samples of ξ, P^ν the empirical measure associated with $\xi^1, \xi^2, \dots, \xi^\nu$, and consider the following stochastic programs:

$$\min f_0(x) \text{ so that } P^\nu\{\xi \in \Xi \,|\, g_k(\xi,x) \leq 0, \ k=1,\dots,q\} \geq \alpha^\nu, \qquad (P^\nu)$$

with $\alpha^\nu \to \alpha$. Suppose that for all ν,

$$x^\nu \in S^\nu = \left\{ x \,\middle|\, P^\nu(\{\xi \,|\, g_k(\xi, x) \leq 0, \ k = 1, \dots, q\}) \geq \alpha^\nu \right\},$$

then almost surely every cluster point of the sequence $\{x^\nu\}_{\nu \in N}$ is a feasible solution of the stochastic program with chance constraints (P).

Moreover there exists a sequence $\alpha^\nu \uparrow \alpha$ such that for x^ν optimal solutions of (P^ν), every cluster point of the sequence $\{x^\nu\}_{\nu \in N}$ is actually an optimal solution of (P).

Proof. The assumptions immediately imply that

$$f(\xi, x) = \begin{cases} -1 & \text{if } g_k(\xi, x) \leq 0, \ k = 1, \dots, q \\ 0 & \text{otherwise}, \end{cases}$$

is a random lsc function; one can also appeal to [11, Theorem 14.31]. Also, $Ef(x) := \int_\Xi f(\xi, x) P(d\xi)$ and $E^\nu f(x) := \int_\Xi f(\xi, x) P^\nu(d\xi)$. The Ergodic Theorem 1 implies that $Ef = \text{e-lim} E^\nu f$ almost surely. In turn, this yields via Proposition 1, that

$$\limsup_\nu (\text{lev}_{\leq -\alpha^\nu} E^\nu f) \subset \text{lev}_{\leq -\alpha} Ef$$

which means that whenever $\bar{x}$ is a cluster point of a sequence of points $\{x^\nu\}_{\nu \in N}$ with $x^\nu \in \text{lev}_{\leq -\alpha^\nu} E^\nu f$, then $\bar{x} \in \text{lev}_{\leq -\alpha} Ef$.

Proposition 1 also guarantees the existence of a sequence $\alpha^\nu \uparrow \alpha$ such that

$$\text{lev}_{\leq -\alpha} Ef =: C = \lim_\nu C^\nu \text{ with } C^\nu := \text{lev}_{\leq -\alpha^\nu} E^\nu f.$$

And thus $\delta_C = \text{e-lim} \, \delta_{C^\nu}$ where δ_C is the indicator function of the set C. It easy to verify that

$$f_0 + \delta_C = \text{e-lim}(f_0 + \delta_{C^\nu}).$$

Finally, Theorem 2 tells us: if $x^\nu \in \text{argmin}(f_0 + \delta_{C^\nu})$ and the x^ν cluster at some point $\bar{x}$, then this cluster point $\bar{x} \in \text{argmin}(f_0 + \delta_C)$, i.e., solves (P).

$\square$

In the case of a constraint involving bounds on moments or on the variance, the argument is similar.

5 Probabilistic Framework

Let lsc-fcns($\mathbb{R}^n$) denote the space of lsc (lower semicontinuous) extended real-valued functions defined on $\mathbb{R}^n$, and $(\Xi, \mathcal{S}, P)$ a probability space; we assume that $\mathcal{S}$ is P-complete. We adapt the standard probabilistic framework to lsc-fcns($\mathbb{R}^n$)-valued random variables.

Here we adopt a slightly different viewpoint of random lsc functions; we think of a *random lsc (lower semicontinuous) function* as a function $f : \Xi \to$ lsc-fcns($I\!\!R^n$) such that the associated bivariate function $(\xi, x) \mapsto f(\xi, x)$ is jointly measurable, i.e., $\mathcal{S} \otimes \mathcal{B}$-measurable where $\mathcal{B}$ is the Borel field on $I\!\!R^n$. It's convenient to identify an lsc function $f(\xi)$ with its bivariate representation so we write $f(\xi, x)$ instead of $f(\xi)(x)$ for the value of $f(\xi)$ at x. This brings us back to the framework introduced in Sect. 2. The concept of a random lsc function goes back to the work of Rockafellar in the Calculus of Variations where it comes up in the form of a 'normal integrand;' see [11, Chapter 14] for a systematic exposition.

To every random lsc function f one associates its *distribution* P_f defined by

$$P_f(\mathcal{A}) := P(\{\xi \in \Xi \mid f(\xi, \cdot) \in \mathcal{A}\}) \quad \text{for } \mathcal{A} \in \mathcal{F};$$

here $\mathcal{F}$ is the σ-field defined on lsc-fcns($I\!\!R^n$). Among the measurable sets in $\mathcal{F}$ are all those of the following form $\{f \in$ lsc-fcns($I\!\!R^n$)$\mid \inf_D f(\xi, \cdot) \leq \alpha\}$ for D any open or closed subset of $I\!\!R^n$; a short argument can be constructed from the joint measurability of f in (ξ, x), the lower semicontinuity in x and the separability of $I\!\!R^n$, or one could rely on the more comprehensive approach found in [11, Chapter 14].

Two random lsc functions, f and g, are *identically distributed* if for all $\mathcal{A} \in \mathcal{F}$, $P_f(\mathcal{A}) = P_g(\mathcal{A})$. The *joint distribution* of a finite collection $\{f^1, \ldots, f^k\}$ of random lsc functions is given, for $\mathcal{A}_1, \ldots, \mathcal{A}_k \in \mathcal{F}$, by

$$P_{\{f^1, \ldots, f^k\}}(\mathcal{A}_1, \ldots, \mathcal{A}_k) := P(\{\xi \in \Xi \mid f^1(\xi, \cdot) \in \mathcal{A}_1, \ldots, f^k(\xi, \cdot) \in \mathcal{A}_k\}).$$

For a sequence $\{f^\nu, \nu \in I\!\!N\}$ of random lsc functions, let's denote by P^∞ the probability measure on the sequence space (lsc-fcns($I\!\!R^n$)$^\infty, \mathcal{F}^\infty$) that is consistent with the joint distribution of the f^ν; that such a measure exists follows from Kolmogorov's Extension Theorem.

Random lsc functions are said to be *independent* if their distributions are independent. A sequence $\{f^1, f^2, \ldots\}$ is said to be *independent* if for any finite subcollection, $\{f^{\nu_1}, \ldots, f^{\nu_k}, k \in I\!\!N\}$,

$$P_{\{f^{\nu_1}, \ldots, f^{\nu_k}\}}(\mathcal{A}_1, \ldots, \mathcal{A}_k) = \Pi_{i=1}^k P_{f^{\nu_i}}(\mathcal{A}_i) \text{ for any sets } \mathcal{A}_1, \ldots, \mathcal{A}_k \in \mathcal{F}.$$

Definition 3. (iid and stationarity). A sequence, $\{f^\nu, \nu \in I\!\!N\}$ of random lsc functions is *iid (independent and identically distributed)* if it is independent and for any $k, l \in I\!\!N$, f^k and f^l are identically distributed. The sequence is *stationary* if its joint distributions are invariant under shifts in the sequence, more precisely, for any finite subcollection $\{f^{\nu_1}, \ldots, f^{\nu_k}\}$, $k \in I\!\!N$, any $l \in I\!\!N$ and any $\mathcal{A}_1, \ldots, \mathcal{A}_k \in \mathcal{F}$, one has

$$P_{\{f^{\nu_1}, \ldots, f^{\nu_k}\}}(\mathcal{A}_1, \ldots, \mathcal{A}_k) = P_{\{f^{\nu_1+l}, \ldots, f^{\nu_k+l}\}}(\mathcal{A}_1, \ldots, \mathcal{A}_k).$$

Stationarity can also be characterized in terms of a measure preserving transformation. Recall that a function $\varphi : \Xi \to \Xi$ is *measure preserving* if for all $A \in \mathcal{S}$, $P(\varphi^{-1}(A)) = P(A)$. If f is a random lsc function, one verifies easily that the sequence $\{f, f \circ \varphi, f \circ \varphi^2, \dots\}$ is stationary. In fact, every stationary sequence of random lsc functions can be redefined in terms of a (single) random lsc function and a measure preserving transformation: Say $\{f^\nu, \nu \in \mathbb{N}\}$ is a stationary sequence of random lsc functions and P^∞ the measure induced on $(\text{lsc-fcns}(\mathbb{R}^n)^\infty, \mathcal{F}^\infty)$. Redefine the f^ν as follows:

$$f^\nu : \text{lsc-fcns}(\mathbb{R}^n)^\infty \to \text{lsc-fcns}(\mathbb{R}^n) \text{ with } f^\nu(\zeta) := \zeta^\nu,$$

i.e., the ν-th element of the sequence $\zeta \in \text{lsc-fcns}(\mathbb{R}^n)^\infty$. The new sequence $\{f^\nu\}_{\nu \in \mathbb{N}}$ is stationary and has the same joint distributions as the original one, but now with respect to the new probability space. Letting $\varphi : \text{lsc-fcns}(\mathbb{R}^n)^\infty \to \text{lsc-fcns}(\mathbb{R}^n)^\infty$ be the shift operator,

$$\varphi(\zeta^1, \zeta^2, \dots) := (\zeta^2, \zeta^3, \dots),$$

and defining $f : \text{lsc-fcns}(\mathbb{R}^n)^\infty \to \text{lsc-fcns}(\mathbb{R}^n)$ as $f(\zeta) = \zeta^1$, one has that $f(\varphi^\nu(\zeta)) = \zeta^{\nu+1}$, so that $f, f \circ \varphi, f \circ \varphi^2, \dots$, defines the same stationary sequence on $\text{lsc-fcns}(\mathbb{R}^n)^\infty$ with respect to the measure preserving shift transformation φ; it is easy to check that φ is measure preserving.

If $\varphi : \Xi \to \Xi$ is measure preserving, then $A \in \mathcal{S}$ is an *invariant event* if $\varphi^{-1}(A) = A$ almost surely, i.e., in terms of the symmetric difference, $P(\varphi^{-1}(A) \triangle A) = 0$.

Definition 4. (ergodicity). Let $\mathcal{I}$ denote the σ-field of invariant events and call it the *invariant σ-field*. A measure preserving map $\varphi : \Xi \to \Xi$ is *ergodic* if $\mathcal{I}$ is trivial, i.e., for all $A \in \mathcal{I}$, $P(A) \in \{0, 1\}$. A sequence $\{f^\nu, \nu \in \mathbb{N}\}$ of random lsc functions is *ergodic* if the associated (measure preserving) shift operator φ on the sequence space $(\text{lsc-fcns}(\mathbb{R}^n)^\infty, \mathcal{F}^\infty, P^\infty)$ is ergodic.

6   Scalarization of Random Lsc Functions

The framework of reference is still that of Sect. 5. In this section, it will be shown that a random lsc function f is completely identified by a countable collection of extended real-valued random variables

$$f \longleftrightarrow \{\pi_{x,\rho} \mid x \in R, \rho \in \mathbb{Q}_+\} \text{ where } R \text{ is a countable dense subset of } \mathbb{R}^n.$$

We refer to such an identification as a *scalarization* of the random lsc function f; results about the scalarizations of random lsc functions with values in $\text{lsc-fcns}(X)$, for X a Polish space, appear in [18]. When $X = \mathbb{R}^n$, it's not necessary to assume $\mathcal{S}$ is P-complete, and this version can also be found in [11, Theorem 14.40]:

Theorem 4. *Theorem 6.1 (scalarization). Let $f : \Xi \to \mathrm{lsc\text{-}fcns}(I\!\!R^n)$,*

$$\textit{and for } D \subset \boldsymbol{R}^n : \quad \textit{let } \pi_D(\xi) := \inf_{x \in D} f(\xi, x).$$

Then f is a random lsc function if and only if for all $D \in \mathcal{D}$, π_D is measurable where $\mathcal{D}$ is any one of the following collection of sets:

(a) $\mathcal{D} =$ *the open sets* $O \subset I\!\!R^n$;
(b) $\mathcal{D} =$ *the closed sets* $C \subset \boldsymbol{R}^n$;
(c) $\mathcal{D} =$ *the closed balls* $\boldsymbol{B}(x, \rho) \subset I\!\!R^n$;
(d) $\mathcal{D} =$ *the closed rational balls* $\boldsymbol{B}(x, \rho) \subset I\!\!R^n$ *with* $x \in R$, R *a countable dense subset of* $I\!\!R^n$ *and* $\rho \in Q_+$;

Corollary 1. *(countable scalarization). Let $f : \Xi \to \mathrm{lsc\text{-}fcns}(I\!\!R^n)$. For $x \in R$, a countable dense subset of $I\!\!R^n$, and $\rho \in Q_+$, define*

$$\pi_{x,\rho}(\xi) := \pi_{B(x,\rho)}(\xi) = \inf_{y \in B(x,\rho)} f(\xi, y).$$

Then f is a random lsc function if and only if the random variables in the countable collection

$$\left\{ \pi_{x,\rho} : \Xi \to \overline{R} \,\middle|\, x \in R, \rho \in Q_+ \right\}$$

are measurable.

Proof. This is just a reformulation of part (d) of the theorem.   □

To each sequence of random lsc functions $\{ f^\nu : \Xi \to \mathrm{lsc\text{-}fcns}(X), \nu \in I\!\!N \}$ we can associate, by scalarization, a sequence of vector-valued random variables

$$\left\{ \pi^\nu_{x,\rho}, \nu \in I\!\!N \,\middle|\, x \in R, \rho \in Q_+ \right\}.$$

Independence, stationarity and ergodicity properties of the sequence of the random lsc functions are inherited by these vectors generated through scalarization. Here we are only interested in the ergodicity properties of these scalarizations.

Proposition 2. *If $\{f \circ \varphi^\nu\}$ is an ergodic sequence of random lsc functions, then $\{\pi_{x,\rho} \circ \varphi^\nu\}$ is an ergodic sequence of random variables for all $x \in R$, $\rho \in Q_+$.*

Proof. The shift operator, $\varphi : \mathrm{lsc\text{-}fcns}(X)^\infty \to \mathrm{lsc\text{-}fcns}(X)^\infty$ is ergodic, and $\pi_{x,\rho}$ defined on $\mathrm{lsc\text{-}fcns}(X)^\infty$ by $\pi_{x,\rho}(\zeta) := \inf_{B(x,\rho)} \zeta_1$ is measurable. Therefore the sequence, $\{\pi_{x,\rho} \circ \varphi^\nu\}$ is ergodic, and equivalent to the original sequence.   □

7   Proof of the Ergodic Theorem

The proof of the Ergodic Theorem relies on the following theorem of independent interest. It says that to verify the almost sure epi-convergence of the empirical means of a sequence of random lsc functions, it suffices to check the almost sure convergence of the empirical means of the corresponding scalarizations.

Theorem 5. Let $\{f, f^\nu, \nu \in \mathbb{N}\}$ be a sequence of random lsc functions on $\mathbb{R}^n$, and let R be a countable dense subset of $\mathbb{R}^n$ that contains the projection onto $\mathbb{R}^n$ of a countable dense subset of epi Ef. For $x \in R$, $\rho \in \mathbb{Q}_+$, let $\pi_{x,\rho}^\nu := \inf_{y \in B(x,\rho)} f^\nu(\cdot, y)$ and $\pi_{x,\rho} := \inf_{y \in B(x,\rho)} f(\cdot, y)$. Suppose that for all $x \in R$, $\rho \in \mathbb{Q}_+$,

$$\frac{1}{\nu} \sum_{k=1}^{\nu} \pi_{x,\rho}^k(\xi) \to E\pi_{x,\rho} \quad P\text{-a.s.}$$

with $E\pi_{x,\rho} := E\{\pi_{x,\rho}(\xi)\}$. Then, whenever $\xi \mapsto \inf_{\mathbb{R}^n} f(\xi, \cdot)$ is summable,

$$\frac{1}{\nu} \sum_{k=1}^{\nu} f^k(\xi, \cdot) \xrightarrow{e} Ef \quad P\text{-a.s.}$$

Proof. Fix $x \in R$, $\rho \in \mathbb{Q}_+$. Let $\Xi_{x,\rho} \subset \Xi$ be such that $P(\Xi_{x,\rho}) = 1$ and

$$\frac{1}{\nu} \sum_{k=1}^{\nu} \pi_{x,\rho}^k(\xi) \to E\pi_{x,\rho}$$

for all $\xi \in \Xi_{x,\rho}$. Let $\Xi_R := \bigcap_{x \in R} \bigcap_{\rho \in \mathbb{Q}_+} \Xi_{x,\rho}$. Then $P(\Xi_R) = 1$, since Ξ_R is obtained from a countable intersection of sets of measure one.

To show that for all $\xi \in \Xi_R$, e-$\liminf \frac{1}{\nu} \sum_{k=1}^{\nu} f^k(\xi, \cdot) \geq Ef$ on $\mathbb{R}^n$, let $\xi \in \Xi_R$, $x \in \mathbb{R}^n$. Then

$$\text{e-}\liminf \frac{1}{\nu} \sum_{k=1}^{\nu} f^k(\xi, x) = \sup_{\rho > 0} \liminf_\nu \inf_{y \in B(x,\rho)} \frac{1}{\nu} \sum_{k=1}^{\nu} f^k(\xi, y)$$

$$\geq \sup_{\rho > 0} \liminf_\nu \frac{1}{\nu} \sum_{k=1}^{\nu} \inf_{y \in B(x,\rho)} f^k(\xi, y)$$

$$\geq \sup_{\ell} \liminf_\nu \frac{1}{\nu} \sum_{k=1}^{\nu} \pi_{x^\ell, \rho^\ell}^k(\xi),$$

where for all $\ell \in \mathbb{N}$, $x^\ell \in R$, $\rho^\ell \in \mathbb{Q}_+$, $x^\ell \to x$, $\rho^\ell \downarrow 0$, $x \in \text{int } B(x^\ell, \rho^\ell)$, and $B(x^{\ell+1}, \rho^{\ell+1}) \subset B(x^\ell, \rho^\ell)$. For each ℓ, $\xi \in \Xi_R$, from the assumptions, one has

$$\liminf_\nu \frac{1}{\nu} \sum_{k=1}^{\nu} \pi_{x^\ell, \rho^\ell}^k(\xi) = E\pi_{x^\ell, \rho^\ell}.$$

Continuing, we obtain

$$\text{e-}\liminf \frac{1}{\nu} \sum_{k=1}^{\nu} f^k(\xi, x) \geq \sup_{\ell} E\pi_{x^\ell, \rho^\ell} = Ef(x)$$

by the Monotone Convergence Theorem and the lower semicontinuity of $x \mapsto f(\xi, x)$ for all ξ. Hence for all $\xi \in \Xi_R$, $\text{e-}\liminf \frac{1}{\nu} \sum_{k=1}^{\nu} f^\nu(\xi, \cdot) \geq Ef$ on $\mathbb{R}^n$.

For the lim sup inequality, observe that for $\xi \in \Xi_R$, $x \in R$, if $x^\nu \equiv x$, then by assumption

$$\limsup_{\nu} \frac{1}{\nu} \sum_{k=1}^{\nu} f^k(\xi, x^\nu) = \limsup_{\nu} \frac{1}{\nu} \sum_{k=1}^{\nu} \pi_{x,0}^k(\xi) = E\pi_{x,0} = Ef.$$

Using the facts that Ef is lsc (by Fatou's Lemma) and R contains the projection on $\mathbb{R}^n$ of a countable dense subset of epi Ef, along with Lemma 1 and the fact that

$$\text{e-}\limsup \frac{1}{\nu} \sum_{k=1}^{\nu} f^k(\xi, \cdot) \leq \limsup_{\nu} \frac{1}{\nu} \sum_{k=1}^{\nu} f^k(\xi, \cdot) \leq Ef \quad \text{on } R,$$

it follows that

$$\text{e-}\limsup \frac{1}{\nu} \sum_{k=1}^{\nu} f^k(\xi, \cdot) \leq Ef \quad \text{on } \mathbb{R}^n.$$

In summary, it has been shown that $P(\Xi_R) = 1$, and for all $\xi \in \Xi_R$,

$$\text{e-}\limsup \frac{1}{\nu} \sum_{k=1}^{\nu} f^k(\xi, \cdot) \leq Ef \leq \text{e-}\liminf \frac{1}{\nu} \sum_{k=1}^{\nu} f^k(\xi, \cdot)$$

on $\mathbb{R}^n$ as claimed. $\qquad\square$

We are now ready to prove the Ergodic Theorem.

Theorem 6. *Let $(\Xi, \mathcal{S}, P)$ be a probability space, $\varphi : \Xi \to \Xi$ an ergodic measure preserving transformation, and f a random lsc function on $\mathbb{R}^n$. Then, whenever $\xi \mapsto \inf_{R^n} f(\xi, \cdot)$ is summable,*

$$\frac{1}{\nu} \sum_{k=1}^{\nu} f(\varphi^k(\xi), \cdot) \xrightarrow{e} Ef \quad P\text{-a.s.}$$

Proof. Let $\{\pi_{x,\rho} \mid x \in R, \rho \in \mathbb{Q}_+\}$ denote the scalarization of f with R the projection of a countable dense subset of epi Ef on $\mathbb{R}^n$. Since φ is measure preserving and ergodic, we obtain that for all $x \in R$, $\rho \in \mathbb{Q}_+$, the sequence, $\{\pi_{x,\rho} \circ \varphi^\nu\}_{\nu \in N}$ is also ergodic by Proposition 6.3. Hence by the classical Birkhoff-Khintchine Ergodic Theorem [19] with a straightforward extension to include functions which take on the value $+\infty$, for all $x \in R$, $\rho \in Q_+$, we obtain,

$$\frac{1}{\nu} \sum_{k=1}^{\nu} \pi_{x,\rho}(\varphi^k(\xi)) \to E\pi_{x,\rho} \quad P\text{-a.s.}$$

Appealing to Theorem 5 it follows immediately that

$$\frac{1}{\nu} \sum_{k=1}^{\nu} f(\varphi^k(\xi), \cdot) \xrightarrow{e} Ef \quad P\text{-a.s.},$$

as claimed.   □

References

1. Wets, R. J-B (1989) Stochastic Programming. Handbook for Operations Research and Management sciences # 1, Elsevier Science, North Holland
2. Birge, J.R., Louveaux, F. (1997) Introduction to Stochastic Programming. Springer, Heidelberg
3. Korf, L.A., Wets, R. J-B (1999) Ergodic limit laws for stochastic optimization problems. Manuscript, University of California, Davis
4. King, A.J., Wets, R. J-B (1990) Epi-consistency of convex stochastic programs. Stochastics and Stochastics Reports 34 83–92
5. Attouch, H., Wets, R. J-B (1990) Epigraphical processes: laws of large numbers for random lsc functions. Séminaire d'Analyse Convexe, Montpellier
6. Artstein, Z., Wets, R. J-B (1995) Consistency of minimizers and the SLLN for stochastic programs. J. of Convex Analysis 2 1–17
7. Hess, C. (1991) Epi-convergence of sequences of normal integrands and strong consistency of the maximum likelihood estimator. Cahiers no. 9121, Cérémade, Paris
8. Castaing, C., Ezzaki, F. (1995) SLLN for convex random sets and random lower semicontinuous integrands. Manuscript, Université de Montpellier
9. Attouch, H. (1984) Variational Convergence for Functions and Operators. Applicable Mathematics Series, Pitman
10. Aubin, J.-P., Frankowska, H. (1990) Set-Valued Analysis. Birkhäuser, Basel
11. Rockafellar, R.T., Wets, R. J-B (1997) Variational Analysis. Grundlehren der mathematischen Wissenschaften 317, Springer-Verlag, Heidelberg
12. King, A.J., Rockafellar R.T., Somlyódy, L., Wets, R. J-B (1988) Lake eutrophication management: the Lake Balaton project. In: Ermoliev Y., Wets R. (Eds.) Numerical Techniques for Stochastic Optimization, Springer-Verlag, Heidelberg, 435–444
13. Salinger, D. (1997) A Splitting Algorithm for Multistage Stochastic Programming with Application to Hydropower Scheduling. Ph. D. Thesis, University of Washington, Seattle
14. Kall, P., Wallace, S. (1994) Stochastic Programming. J. Wiley & Sons, New York
15. Kibzun, A.I., Kan, Y.S. (1996) Stochastic Programming Problems with Probability and Quantile Functions. J. Wiley & Sons, New York

16. Prékopa, A. (1995) Stochastic Programming. Kluwer, Dordrecht
17. Wets, R. J-B (1998) Stochastic programs with chance constraints: generalized convexity and approximation issues. In: Crouzeix J.-P., Martinez-Legaz J.E., Volle M. (Eds.) Generalized Convexity, Generalized Monotonicity: Recent Results. Kluwer, Dordrecht, 61–74
18. Korf, L.A., Wets, R. J-B (1999) Random lsc functions: scalarization. Manuscript, University of California, Davis
19. Durrett, R. (1991) Probability: Theory and Examples, Wadsworth and Brooks

Proximal Methods: Decomposition and Selection

N. Lehdili[1,2] and C. Ould Ahmed Salem[3]

[1] LAMETA CNRS UPRES-A 5474, Faculté des Sciences Economiques, Espace Richter - Avenue de La Mer BP 9606, 34054 Montpellier Cedex 1, France
[2] Department of Mathematics, University of Trier, D-54286 Trier, Germany
E-mail : lehdili@uni-trier.de
[3] Département de Mathématiques, Université de Montpellier II,
Place Eugène Bataillon, 34096 Montpellier, France
E-mail : ould@math.univ-montp2.fr

Abstract. We study the convergence of a general splitting method for finding a particular zero of the sum of two maximal monotone operators A and B, in which a forward-backward splitting iteration is applied to a sequence of approximating operators A_n, B_n converging to A and B. We prove that when these approximating operators converge sufficiently slowly the sequence generated by the method strongly converges towards a particular solution of the given problem, provided that there is one. In the process, we revisited and made a link between the Passty, (PS), and the Barycentric, (BPM), methods (cf. [22], [9]). It is well-known that a weighted average of the iterates of (PS), weighted by stepsizes, weakly converges to a solution of the given problem. However such ergodic convergence does not seem very useful in practice. Here, we combine (PS) iteration with Tikhonov regularization to obtain strong convergence of the generated sequence to the solution of minimal norm. Finally, to motivate the selection concept, we briefly propose an application in the finance area : Portfolio Selection.

Key words: approximation, splitting method, maximal monotone operators, selection, Tikhonov regularization, proximal point algorithm, portfolio selection

AMS subject classification: 65K, 49M07, 90C25

1 Introduction

In this work, we are interested in finding, for given maximal monotone operators $A_i, i = 1, 2, ..., m (m \geq 2)$ on a Hilbert space H, an $x \in H$ satisfying $0 \in Tx := (A_1 + .. + A_m)x$. Many problems that involve convexity can be reduced to the above problem, for example, convex minimization problems, convex-concave saddle-point problems and solutions of games. In particular, when $f : H \to I\!R \cup \{+\infty\}$ is a proper closed convex function, its subdifferential $T = \partial f$ is a maximal monotone operator and a point $x \in H$ minimizes f if and only if $0 \in \partial f(x)$. Another example is $T = F + N_C$, where C is a nonempty closed convex set in H, F is a maximal monotone operator that is single-valued and continuous on C, and N_C is the normal cone to C. Then, $0 \in F(x) + N_C(x)$ if and only if $x \in C$ satisfies the variational inequality $\langle F(x), x' - x \rangle \geq 0 \quad \forall x' \in C$.

A classical method for solving the given problem is the proximal point algorithm (PPA) proposed by Martinet ([13], 1972) and generalized by Rockafellar ([17], 1976):

$$x_{n+1} = J_{\lambda_n}^T(x_n) := (I + \lambda_n T)^{-1}(x_n) \quad \forall n \in I\!N \quad \text{with} \quad \lambda_n \geq 0.$$

This method and its dual version in the context of convex programming, have been extensively studied and are known to yield, as a special cases, decomposition methods (cf. [22,12]). In the case $T = A + B$, when A and B are maximal monotone operators with A singlevalued, the following splitting method

$$x_{n+1} = J_{\lambda_n}^B(I - \lambda_n A)x_n \quad \forall n \in I\!N \quad \text{with} \quad \lambda_n \geq 0 \quad (*)$$

was proposed by Lions-Mercier([12]) and extensively analyzed by Passty([22]).

In this paper, we propose a general splitting method which includes the Passty scheme (PS, [22]) as a particular case. To begin with, we prove that the latter includes, in turn, the barycentric-proximal method (BPM, [9]) as a special case. Then, we investigate the coupling of (*) with approximations through a diagonal process. In the specific case of Tikhonov regularization, we prove the strong convergence of the generated sequence to a solution of the given problem, namely, the solution which minimizes the norm over the closed convex solution set. Our approach is general and allows us to recover, as a special case, iterative methods by Bakushinsky and Goncharsky in ([4], chapter 5) for solving particular variational inequalities. Finally, to motivate the *selection concept*, we briefly propose an application in the finance theory : Portfolio Selection.

The paper is organized as follows. In Section 2 we give our general decomposition methods and convergence results. An analysis of the convergence of our new selection method, called Thikonov-Passty method, is done in Section 3; and we give the connection between the Passty scheme, the barycentric-proximal method. The last section is devoted to an application of the selection concept to portfolio and Markowitz's mean-variance criterion ([18], 1952).

2   Decomposition Methods

All definitions, notations and properties on maximal monotone operators used in this work are the usual ones in convex analysis theory and can be found in the classical books of Brézis and Rockafellar [1,17].

Let H be a real Hilbert space with inner product and norm denoted by $\langle .,. \rangle$ and $\|.\|$, respectively. Let T be a maximal monotone operator on H. $J_\lambda^T := (I + \lambda T)^{-1}$ denotes the resolvent (or proximal mapping) with parameter $\lambda > 0$ associated with T and $T_\lambda := \frac{1}{\lambda}(I - J_\lambda^T)$ the corresponding Yosida approximate. We recall that J_λ^T is a *firmly nonexpansive self mapping* on H, that is,

$$\langle J_\lambda^T x - J_\lambda^T y, x - y \rangle \geq \|J_\lambda^T x - J_\lambda^T y\|^2 \quad \forall x, y \in H, \tag{1}$$

or equivalently,

$$\|J_\lambda^T x - J_\lambda^T y\|^2 \leq \|x - y\|^2 - \|(I - J_\lambda^T)x - (I - J_\lambda^T)y\|^2 \quad \forall x, y \in H,$$

and that T_λ is $\frac{1}{\lambda}$-Lipschitz continuous. Moreover, the set of zeros of T, that is $S := T^{-1}0$ coincides with the set of fixed-points of J_λ^T and the set of zeros of T_λ.

The proximal point algorithm (PPA) is a method for solving the following inclusion

$$\text{find } x \in H \ 0 \in T(x), \tag{P}$$

where T is a monotone operator on a Hilbert space. The algorithm is nothing else but the iteration method for the proximal mapping : from an initial point $x_0 \in H$, it generates a sequence $\{u_n\}$ by

$$u_{n+1} = J_{\lambda_n}^T u_n, \tag{2}$$

where $\{\lambda_n\}$ is a given sequence of positive reals. This method is one of the most powerful and versatile solution techniques for solving variational inequalities, convex programs, and convex-concave mini-max problems and is the basis for a wide variety of decomposition methods called *splitting methods*. Passty and Lehdili & Lemaire especially, have proposed two splitting methods called *Passty scheme* and *barycentric proximal method*, respectively, for finding a zero of the operator $T := A + B$ when A and B are two maximal monotone operators on H with maximal monotone sum $A + B$ and $\{\lambda_n\} \subset [0, +\infty[$. First, we recall that the scheme of Passty consists in a double backward step in the calculation of each iterate

$$\begin{cases} u_0 \in H \\ \\ u_{n+1} = J_{\lambda_n}^A \circ J_{\lambda_n}^B u_n \quad \forall n \geq 0 \end{cases} \tag{SP}$$

and that the associated sequence of weighted averages z_n is defined by

$$z_n := \frac{\sum_{i=1}^n \lambda_i u_i}{\sum_{i=1}^n \lambda_i}. \tag{3}$$

Below, we state some important results of convergence for the scheme (SP)-(3) whose proofs can be found in Passty([22]):

Theorem 1. *(i) Let $\{z_n\}$ be the sequence of weighted averages determined by (PS) and (3), and let $\{\lambda_n\} \in \ell^2/\ell^1$. If $A + B$ is maximal monotone and $(A + B)^{-1}0$ is nonempty, then $\{z_n\}$ converges weakly to some zero of $A + B$.*

(ii) If in addition we assume that either B is strongly monotone or $(A+B)^{-1}0$ has nonempty interior, then $\{u_n\}$ converges strongly to a zero of the operator $A + B$.

In the sequel, we consider problem (P) and we denote by S the set of solutions of problem (P), i.e. $S = (A + B)^{-1}0$. Now, let us replace (P) by the following sequence of approximate problems

$$\text{find } x_n \in H, \quad 0 \in (A_n + B_n)x_n \qquad (P_n),$$

where $\{A_n\}$ denotes a sequence of univoque operators on H converging to A in an appropriate sense and where $\{B_n\}$ is a sequence of maximal monotone operators approximating B. (P_n) can be rewritten as a sequence of fixed-point problems, namely

$$\text{find } x_n \in H, \quad x_n = J_{\lambda_n}^{B_n}[x_n - \lambda_n A_n x_n], \qquad (5)$$

where $\{\lambda_n\}$ denotes, as previously, a sequence of strictly positive reals. This leads to the following diagonal gradient-prox method

$$\begin{cases} u_0 \in H, \\ u_{n+1} = J_{\lambda_n}^{B_n}[u_n - \lambda_n A_n u_n]. \end{cases} \qquad (DGP)$$

It should be noted that this scheme includes the diagonal gradient-projection method when $B_n \equiv \partial \psi$ is the subdifferential of a proper closed convex function ψ. In this case (P) reduces to the following variational inequality problem

$$\text{find } x \in H, \quad \langle Ax, x - y \rangle + \psi(x) - \psi(y) \leq 0, \quad \forall y \in H. \qquad (6)$$

This also includes the diagonal proximal method ([17]) when $A_n \equiv 0$. Now, if $B_n \equiv B$ and $A_n \equiv A_{\lambda_n}$ with A_{λ_n} the Yosida approximation of A with parameter λ_n, then the method (DGP) is nothing else but Passty's method (PS).

The iterative method (DGP) is also studied in Lemaire ([11]) and Ould([15] but the authors assume that the operator A is only single-valued and that the form of operators is $A_n := A + G_n)$ which doesn't include the Passty scheme. In our analysis, we will establish, in particular, the strong convergence of the Passty scheme without assuming the strong monotonicity and we prove that this method converges to a specified solution.

Most of the research works on general forward-backward splitting methods rely on the two following notions.

Definition 1. (i) The operator A is strongly monotone if there exists a constant $L \geq 0$ such that

$$\forall \, x, y \, \in \, X, \ \langle Ax - Ay, x - y \rangle \geq \ L\|x - y\|^2. \tag{7}$$

(ii) The operator A is cocoercive if there exists a constant $L \geq 0$ such that

$$\forall \, x, y \, \in \, X, \ \langle Ax - Ay, x - y \rangle \geq \frac{1}{L}\|Ax - Ay\|^2. \tag{8}$$

If A is L–cocoercive and $L = 0$ then $A = $ cst. If A is L–cocoercive and $L > 0$, it implies that the operator is monotone and Lipschitz with constant L. In general the converse is not true except if A is the gradient of a convex function. Actually it amounts to saying that the set valued inverse of A is $1/L$–strongly monotone or that A/L is 1–firmly nonexpansive.

Lemma 1. ([15]) *Let H be a Hilbert space and A be a L–cocoercive operator on H. If $L \leq 2\alpha$ then $A - \alpha I$ is α–Lipschitizian.*

Let A be a L–cocoercive operator and λ a positive real. Thanks to the lemma above, if $\lambda L \leq 2$, then $I - \lambda A$ is a nonexpansive map.

Lemma 2. ([15]) *Let H be a Hilbert space and A be a maximal monotone operator on H, then the Yosida approximation A_λ is $1/\lambda$–co-coercive.*

The following lemma will be needed in the proof of the next theorem which is the main result of this section.

Lemma 3. ([4]) *If the sequence of positive reals $\{\theta_n\}_{n \in \mathbb{N}}$ satisfies the inequality*

$$\theta_{n+1} \leq (1 - a_n)\theta_n + b_n,$$

where $0 \leq a_n \leq 1$, $\displaystyle\sum_{n=1}^{\infty} a_n = \infty$, $\displaystyle\lim_{n \to \infty} \frac{a_n}{b_n} = 0$,

then $\displaystyle\lim_{n \to \infty} \theta_n = 0$.

In the following theorem, $\{x_n\}$ and $\{u_n\}$ denote respectively the sequence of solutions to problems (P_n) and to (DGP).

Theorem 2. *Let $\{A_n\}$ and $\{B_n\}$ be two sequences of maximal monotone operators such that, for all $n \in \mathbb{N}$, A_n is L_n-cocoercive and B_n is β_n–strongly monotone. Assume that*

(i) $\lambda_n L_n \leq 2$ *and* $\sum_{n=1}^{+\infty} \lambda_n \beta_n = +\infty$,

(ii) $\dfrac{\|x_n - x_{n-1}\|}{\lambda_n \beta_n} \to 0$.

Then

$$\lim_{n \to +\infty} \|u_n - x_n\| = 0. \tag{9}$$

Proof. Since the operator A_n is L_n-co-coercive and B_n is strongly monotone (with parameter β_n), then $(I - \lambda_n A_n)$ is a nonexpansive map if $\lambda_n L_n \leq 2$ (by virtue of Lemma 1) and the resolvent $J_{\lambda_n}^{B_n}$ is a strong contraction with parameter $\dfrac{1}{1 + \lambda_n \beta_n}$. From this we deduce

$$\|u_n - x_n\| \leq \frac{1}{1 + \lambda_n \beta_n} \|u_{n-1} - x_n\|. \tag{10}$$

Setting $\theta_n = \|u_n - x_n\|$ and using the triangular inequality, we obtain

$$\theta_n \leq (1 - a_n)\theta_{n-1} + b_n, \tag{11}$$

where $a_n = \dfrac{\lambda_n \beta_n}{1 + \lambda_n \beta_n}$ and $b_n = \|x_{n-1} - x_n\|$. We get the desired result by applying Lemma 3. $\square$

REMARK 2.1. Since the exact computation is not achievable in computer implementations, we replace the iteration scheme (DGP) by the following weaker relation

$$\|u_{n+1} - J_{\lambda_n}^{B_n}[u_n - \lambda_n A_n u_n]\| \leq \varepsilon_n$$

where $\{\varepsilon_n\}$ is a sequence of positive reals which tends to zero. If the conditions of the theorem above are fulfilled and $\dfrac{\varepsilon_n}{\lambda_n \beta_n} \to 0$, then the conclusion of Theorem 2 still holds true. If the solution x_n of the problem (P_n) converges towards a particular solution, then it will be the same for the sequence $\{u_n\}$ of the iterates. This is a common feature of all viscosity approximation methods ([23,15]) and the Tikhonov regularization for minimization problems ([2,18,4,24]). It is, in particular, checked by the barrier method in linear programming ([7]).

3   Applications

Throughout this paper we assume that the set $S := (A + B)^{-1}0$ is non empty. As shown in the previous section, the two sequences $\{x_n\}$ and $\{u_n\}$ have the same asymptotic behavior. In this section we prove that under a specific kind of approximation we can select a particular solution, for example, the solution of minimal norm.

3.1   Prox-Tikhonov Method

In [8] the authors propose an approximation method which combines the Tikhonov method with the proximal point algorithm. This technique can be seen as a particular case of the approach used in section 2 for deducing the general scheme (SPD). Indeed, in this case, the problem we are interested in is

$$\text{find } x \in H \text{ such that } \quad 0 \in B(x), \tag{12}$$

which is nothing else but (P) with $A \equiv 0$. So, we approximate the operator B by the sequence $\{B_n\}_{n \in \mathbb{N}}$ defined by $B_n \equiv B + r_n I$, $\forall n \in \mathbb{N}$ and $(P_n)_{n \in \mathbb{N}}$ can be rewritten as the following family of fixed-point problems

$$(P_n) \quad \Longleftrightarrow \quad x_n = J_{\lambda_n}^{B+r_n I}(x_n).$$

The diagonal scheme (DGP) takes the following particular form :

$$
\begin{cases}
u_0 \in H, \\[2mm]
u_{n+1} = J_{\lambda_n}^{B+r_n I}(u_n) = J_{\mu_n}^B \left(\frac{u_n}{1+\lambda_n r_n} \right), \\[2mm]
\text{with } \mu_n = \frac{\lambda_n r_n}{1+\lambda_n r_n}.
\end{cases}
\tag{PT}
$$

As a direct consequence of Theorem 2 we get the following convergence result.

Corollary 1. ([8]) *If the parameters λ_n and r_n satisfy the following conditions*

$$\lim_{n \to \infty} r_n = 0, \quad \sum_{n=1}^{\infty} \lambda_n r_n = \infty, \quad \lim_{n \to \infty} \left| \frac{r_{n+1} - r_n}{\lambda_n r_n r_{n+1}} \right| = 0, \tag{13}$$

then the sequence $\{u_n\}_{n \in \mathbb{N}}$ generated by (PT) converges strongly to the norm-minimal solution of (P).

Let us notice that, if the sequence $\{\lambda_k\}_{n \in \mathbb{N}}$ is bounded, then the conditions on the parameters in Corollary 1 amount to

$$\lim_{n \to \infty} r_n = 0, \quad \sum_{n=1}^{\infty} r_n = \infty, \quad \lim_{n \to \infty} \left| \frac{1}{r_n} - \frac{1}{r_{n+1}} \right| = 0, \tag{14}$$

that is, the parameters r_n must go slowly to zero.

3.2   Passty-Tikhonov Method

Let us notice that in the Passty scheme the convergence, in general, is only ergodic, but a strong convergence can be obtained if one assumes that the operator A or B is strongly monotone (a restrictive assumption). The aim of this subsection is to combine the Tikhonov regularization and Passty's scheme to get a strong convergence to the solution of (P) with the minimal norm, without the assumption of strong monotonicity. So, in what follows we focus our attention on the following approximate problems

$$\text{find } x_n \in H, \quad 0 \in (A_{\lambda_n} + B + r_n I)x_n \tag{P_n}$$

namely, we replace in problem (P) the operator A by its Yosida approximate A_{λ_n} and B by $B + r_n I$, respectively $\{r_n\}_{n \in \mathbb{N}}$ is a sequence of positive reals converging to zero. It is clear that the operator $A_{\lambda_n} + B + r_n I$ is maximal and strongly monotone, from which we infer the existence and the uniqueness of the solution x_n of (P_n). Obviously, (P_n) can be rewritten as a fixed-point problem, i.e.

$$\text{find } x_n \in H; \quad x_n = J_{\lambda_n}^{B+r_nI} \circ J_{\lambda_n}^A (x_n) = \left[J_{\frac{\lambda_n r_n}{1+\lambda_n r_n}}^B \right] \circ \left[\frac{1}{1+\lambda_n r_n} J_{\lambda_n}^A \right] (x_n).$$

The following theorem gives convergence results of the sequence $\{x_n\}$ associated with the regularized problems (P_n).

Theorem 3. *If $A + B$ is maximal monotone, A or B is a bounded operator and the parameters satisfy the following conditions*

$$\lim_{n \to \infty} r_n = 0, \quad \lim_{n \to \infty} \frac{\lambda_n}{r_n} = 0, \tag{21}$$

then the sequence generated by the regularized problem (P_n) converges strongly to the solution of minimal norm of (P).

Proof. Consider the auxiliary sequence $\{\tilde{x}_n\}_{n \in \mathbb{N}}$, where $\tilde{x}_n$ is the unique zero of $A + B + r_n I$ and set $\varepsilon_n := r_n^{-1}$. It is easy to verify that

$$\tilde{x}_n = J_{\varepsilon_n}^{A+B}(0), \quad \text{and} \quad x_n = J_{\varepsilon_n}^{A_{\lambda_n}+B}(0). \tag{22}$$

It is well known that the sequence $\tilde{x}_n$ strongly converges to the solution of (P) of minimal norm. We get the desired result if we show that x_n and $\tilde{x}_n$ have the same asymptotic behavior. Indeed, from Brézis ([1], Theorem 2.4), we obtain

$$\|x_n - \tilde{x}_n\| \leq \|z_n\| \sqrt{\frac{\lambda_n}{r_n}} \quad \forall z_n \in (-Ax_n) \cap (B + r_n I)x_n.$$

Taking into account that $\{x_n\}$ is bounded, if A or B is a bounded operator, we see that $\{z_n\}$ is bounded. Now, if conditions (21) are met, then $\|x_n - \tilde{x}_n\| \to 0$ as $n \to \infty$. $\quad\square$

EXAMPLE 3.1 : PENALIZATION-VISCOSITY APPROXIMATION.

We give, as an application, an example concerning the combination of Tikhonov's approximation and penalization : Let $C \subset H$ be a closed convex set, f be a convex and Lipschitz continuous function, and consider the constrained optimization problem

$$\min\{f(x) \ : \ x \in C\}. \tag{R}$$

Assume that the solution set $S := \arg\min_{x \in C} f(x)$ of this minimization problem is nonempty and consider, for each $n \in I\!N$, the approximate minimization problem

$$\min \left\{ f(x) + \frac{1}{2\lambda_n}\operatorname{dist}(x, C)^2 + \frac{r_n}{2}\|x\|^2 \ : \ x \in H \right\}. \tag{R_n}$$

Since this problem is a particular case of (P_n) $(B \equiv \nabla f, A \equiv \mathcal{N}_C)$, then Theorem 3 ensures the convergence of x_n, the unique solution of (R_n) to the element of minimal norm in S. In particular the convergence is assured when we choose $\lambda_n = r_n^\theta$ $(r_n > 0,\ \theta > 1$ this case is analyzed in ([7])).

In this context, the method (DGP) generates a sequence $\{u_n\}$ by the following scheme

$$\begin{cases} u_0 \in H \\[2mm] u_{n+1} = J_{\lambda_n}^{B+r_n I} \circ J_{\lambda_n}^A (u_n) = J_{\mu_n}^B \circ \left(\frac{1}{1+\lambda_n r_n} J_{\lambda_n}^A \right)(u_n) \\[2mm] \text{with } \mu_n = \frac{\lambda_n r_n}{1+\lambda_n r_n}. \end{cases} \tag{PT}$$

This method is nothing else but the Passty method coupled with Tikhonov regularization. We have the following convergence results as a consequence of Theorem 2.

Theorem 4. *Let A, B be two maximal monotone operators such that A or B is bounded and the parameters satisfy the following conditions*

(i) $\sum \lambda_n r_n = +\infty, \ \dfrac{\lambda_n}{r_n} \to 0;$

(ii) $\displaystyle\lim_{n \to \infty} \frac{1}{r_n r_{n-1}} \left| \frac{1}{\lambda_{n-1}} - \frac{1}{\lambda_n} \right| = 0, \quad \lim_{n \to \infty} \left| \frac{1}{\lambda_{n-1} r_{n-1}} - \frac{1}{\lambda_n r_n} \right| = 0.$

Then the sequence $\{u_n\}$ strongly converges to the norm-minimal solution of (P).

Proof. We will just check the assumptions of Theorem 2. Let $y_i \in Bx_i$, $(i = 1, 2)$ such that

$$A_{\lambda_i} x_i + r_i x_i + y_i = 0, \ i = 1, 2. \tag{23}$$

We use the symbol $\mathcal{J}_\lambda^A$ which denote $(I - J_\lambda^A)$. So, condition (23) is written in the following form

$$\mathcal{J}_{\lambda_i}^A x_i + \lambda_i r_i x_i + \lambda_i y_i = 0, \ i = 1, 2.$$

Setting $\varepsilon_i = \lambda_i r_i$ and combining the above equation for i=1 and i=2, we obtain

$$\varepsilon_1 x_1 - \varepsilon_2 x_2 + \mathcal{J}_{\lambda_1}^A x_1 - \mathcal{J}_{\lambda_1}^A x_2 + \mathcal{J}_{\lambda_1}^A x_2 - \mathcal{J}_{\lambda_2}^A x_2 + \lambda_1 (y_1 - y_2) + (\lambda_1 - \lambda_2) y_2 = 0. \tag{24}$$

Multiplying (24) by $(x_1 - x_2)$ and using the monotonocity of $\mathcal{J}_\lambda^A$ and B, we get

$$\langle \varepsilon_1 x_1 - \varepsilon_2 x_2, x_1 - x_2 \rangle + \langle \mathcal{J}_{\lambda_1}^A x_2 - \mathcal{J}_{\lambda_2}^A x_2, x_1 - x_2 \rangle + (\lambda_1 - \lambda_2) \langle y_2, x_1 - x_2 \rangle \leq 0,$$

from which it follows that

$$\varepsilon_1 \|x_1 - x_2\|^2 \leq \|x_1 - x_2\| (|\lambda_1 - \lambda_2| \| \mathcal{J}_{\lambda_1}^A x_2 - \mathcal{J}_{\lambda_1}^A x_2\| + |\lambda_1 - \lambda_2| \|y_2\| + |\varepsilon_1 - \varepsilon_2| \|x_2\|).$$

By the definition of $\mathcal{J}_\lambda^A$ one has

$$\| \mathcal{J}_{\lambda_1}^A x_2 - \mathcal{J}_{\lambda_2}^A x_2 \| = \| J_{\lambda_1}^A x_2 - J_{\lambda_2}^A x_2 \|, \tag{25}$$

on the other hand, the resolvent equation gives

$$J_{\lambda_2}^A x_2 = J_{\lambda_1}^A \left[\frac{\lambda_1}{\lambda_2} x_2 + \left(1 - \frac{\lambda_1}{\lambda_2}\right) J_{\lambda_2}^A x_2 \right]. \tag{26}$$

Combining (26) with nonexpansiveness of J_λ^A, we get

$$\| \mathcal{J}_{\lambda_1}^A x_2 - \mathcal{J}_{\lambda_2}^A x_2 \| \leq |\lambda_1 - \lambda_2| \left\| \frac{x_2 - J_{\lambda_2}^A x}{\lambda_2} \right\| = |\lambda_1 - \lambda_2| \| A_{\lambda_2} x_2 \|.$$

Owing to the fact that the sequence $\{x_n\}$ is bounded, so is Bx_n. Hence

$$\|x_{n-1} - x_n\| \leq \frac{C}{\lambda_{n-1} r_{n-1}} (|\lambda_{n-1} - \lambda_n| + |\lambda_{n-1} r_{n-1} - \lambda_n r_n|).$$

Therefore,

$$\frac{\|x_{n-1} - x_n\|}{\lambda_n r_n} \leq C \left(\frac{1}{r_n r_{n-1}} \left| \frac{1}{\lambda_{n-1}} - \frac{1}{\lambda_n} \right| + \left| \frac{1}{\lambda_{n-1} r_{n-1}} - \frac{1}{\lambda_n r_n} \right| \right). \tag{27}$$

Then the sequences $\{u_n\}$ and $\{x_n\}$ have the same asymptotic behavior (a consequence of Theorem 2) and the desired result follows by applying Theorem 3.   □

REMARK 3.1 : The sequences $\lambda_n := \dfrac{1}{\ln n}$, $r_n := \sqrt{\lambda_n}$ satisfy the conditions of Theorem 4.

3.3   The Tikhonov-Barycentric Proximal Method

Let A_i, $i = 1, 2, ..., m$, be m maximal monotone operators on H. We assume that the sum $T := A_1 + A_2 + ... + A_m$ is maximal monotone and we deal with the following general problem :

$$\text{find } x \in H, \quad 0 \in (A_1 + A_2 + ... + A_m)x. \qquad (P)$$

First, we will show that the barycentric-proximal method is a particular case of the Passty scheme above. Indeed, let us define

$$\mathcal{H} := H \times H \times ... \times H \quad (m \text{ times}),$$

which will be endowed with inner product $\langle\langle x, y \rangle\rangle = \sum_{i=1}^{m} \langle x_i, y_i \rangle$.

$$\mathcal{T} := A_1 \times A_2 \times ... \times A_m,$$

$$\mathcal{C} := \{(x_1, x_2, ..., x_m); \ x_1 = x_2, ... = x_m\},$$

and

$$\mathcal{D} := \{(y_1, y_2, ..., y_m); \ y_1 + y_2, .. + y_m = 0\}.$$

The definition of $\mathcal{T}$ means that $(y_1, y_2, ..., y_m) \in \mathcal{T}(x_1, x_2, ..., x_n)$ if, and only if, $y_i \in A_i x_i$, $i = 1, 2, ..., m$. It is worth mentioning that $\mathcal{T} : \mathcal{H} \to \mathcal{H}$ is maximal monotone, $\mathcal{C}$ and $\mathcal{D}$ are complementary subspaces of $\mathcal{H}$ (i.e., $\mathcal{C}^\perp = \mathcal{D}$ and $\mathcal{D}^\perp = \mathcal{C}$). Set $X = (x_1, ..., x_m)$, a simple calculation involving definition of the resolvent gives

$$Y = J_\lambda^\mathcal{T}(X) = (J_\lambda^{A_1} x_1, J_\lambda^{A_2} x_2, ..., J_\lambda^{A_m} x_m)$$

and

$$J_\lambda^{\mathcal{N}_\mathcal{C}}(X) = \Pi_\mathcal{C}(X) = (y_1, y_2, ..., y_m)$$

with

$$y_1 = y_2 = ... = y_m = \frac{x_1 + x_2 + ... + x_m}{m},$$

where $\mathcal{N}_\mathcal{C}$ is the normal cone to the subset $\mathcal{C}$, and $\Pi_\mathcal{C}$ is the projection operator onto $\mathcal{C}$. So, it is easy to see that the problem (P) is equivalent to finding a zero of the sum of two maximal monotone operators, namely

$$\text{find } X \in \mathcal{H}, \quad 0_\mathcal{H} \in (\mathcal{N}_\mathcal{C} + \mathcal{T})X. \qquad (Q)$$

When we apply the Passty method to this equivalent problem, then we obtain

$$\begin{cases} X_0 \in \mathcal{H}, \ X_0 := (x_0, x_0, ..., x_0), \\[2mm] X_{n+1} = \Pi_C \circ J_{\lambda_n}^{\mathcal{T}}(X_n) \quad \forall n \in I\!N, \\[2mm] X_{n+1} = (x_{n+1}, x_{n+1}, ..., x_{n+1}), \end{cases}$$

or equivalently

$$\begin{cases} x_0 \in H, \\[2mm] x_{n+1} = \frac{1}{m} \sum_{j=1}^m J_{\lambda_n}^{A_i} x_n. \end{cases} \tag{28}$$

This scheme is nothing else but the barycentric-proximal method proposed in [9]. It is natural that assumptions which ensure the convergence of the barycentric-proximal method are the same as for Passty's scheme. Furthermore, the convergence is strong if the operator $\mathcal{T}$ defined above is strongly monotone (this is the case when all operators A_i, $i = 1, 2, ..., m$ are strongly monotone).

When we consider the problem of finding a zero of an operator defined by the sum of m maximal monotone operators A_i, $i = 1, 2, ..., m$, and when we are interested in approximating one of its solutions, the barycentric-proximal method can be used for this objective. Nevertheless, the convergence of this method, in general, is only ergodic. So, in order to improve the convergence property we combine the Tikhonov regularization and the barycentric-proximal method. Indeed, in probem (Q) (see Section 2) we approximate $\mathcal{N}_C$ by $\mathcal{N}_C + r_n \mathcal{I}$ and $\mathcal{T}$ by $\mathcal{T}_{\lambda_n}$, respectively. In such a situation, scheme (DGP) becomes

$$\begin{cases} u_0 \in H \\[2mm] u_{n+1} = \frac{1}{1+\lambda_n r_n} \frac{1}{m} \sum_{j=1}^m J_{\lambda_n}^{A_j} u_n. \end{cases} \tag{29}$$

The convergence results are assured by applying Theorem 4.

3.4 Variational Inequalities

A number of authors ([4,8]) have extensively studied this idea of coupling Tikhonov regularization and other iterative schemes in the context of variational inequalities and ill-posed problems (see for example [4]). In this subsection, we show that some iterative methods of these works are a particular case of the (DGP)-scheme proposed in Section 2. The problem of solving a variational inequality is formulated as follows. Let A be some mapping defined on a Hilbert space H and ψ is a proper closed convex function. Solving a variational inequality means finding an element $x^* \in H$ such that

$$\langle A(x^*), x^* - z \rangle + \psi(x^*) - \psi(z) \leq 0, \quad z \in H.$$

In particular if $\partial \psi = \mathcal{N}_C$, then this variational inequality can naturally be rewritten as a fixed point problem (a particular case of problem (P), see Section 2), i.e.

$$\text{find } x^\star \quad \text{such that} \quad x^\star = \Pi_C \circ (I - \lambda A)(x^\star), \tag{P_f}$$

for some $\lambda > 0$. Here Π_C is the projection operator into C. We shall approximate the solutions of (P_f) by the solutions of approximate problems given as follows ($B_n \equiv \mathcal{N}_C + r_n I$ and $A_n \equiv A$) :

$$\text{find } x_n \quad \text{such that} \quad x_n = J_{\lambda_n}^{\mathcal{N}_C + r_n I} \circ (I - \lambda_n A)(x_n). \tag{P_{fn}}$$

With a simple calculation, we show that problem (P_{fn}) is equivalent to

$$\text{find } x_n \text{ such that } x_n = \Pi_C(x_n - \alpha_n(Ax_n + r_n x_n)), \quad \alpha_n = \frac{\lambda_n}{1 + \lambda_n r_n},$$

which is nothing else but the sequence of the following approximate variational inequalities proposed in ([4]) for ill-posed problems

$$\text{find } x_n \in C \quad \text{such that} \quad \langle (A + r_n I)x_n, x_n - z \rangle \leq 0, \quad z \in C. \tag{30}$$

In this context, the (DGP)-algorithm can be reduced to the iterative regularization proposed by Bakushinsky & Goncharsky in ([4], chapter 5)

$$\begin{cases} u_0 \in C, \\ u_{n+1} = \Pi_C(u_n - \alpha_n[A(u_n) + r_n u_n]). \end{cases} \tag{31}$$

In particular, under the same type of assumptions as in Theorem 3 and Corollary 1, they obtain convergence to the solution of the variational problem with minimal norm. More exactly, they use the following assumptions on the parameters α_n, r_n related to the iterative regularization :

$$\lim_{n \to \infty} \frac{\alpha_n}{r_n} = 0, \quad \lim_{n \to \infty} \frac{|r_n - r_{n-1}|}{\alpha_n r_n^2} = 0, \quad \sum_{n=1}^{\infty} \alpha_n r_n = \infty. \tag{32}$$

Obviously, they also assume, that $\alpha_n > 0$, $r_n > 0$ and $\lim_{n \to \infty} r_n = 0$.

4   Portfolio Selection

Currently, we dispose of increasingly efficient methods for solving optimization problems. These methods provide algorithms which enable to calculate an unspecified solution of the considered problem. Nevertheless, such algorithms are unable to select a particular solution when there is uniqueness. The selection concept is of particular interest in many fields of mathematics and other areas (mechanics, economy,...etc.). Game theory and its applications in economy usually use the equilibrium Nash concept, but in a conflict situation the multiplicity of Nash equilibria, in an non-cooperative game, complicate the choice of an optimal solution. It is well-known that this difficulty permitted to evolve a selection theory called refinement Theory. In what follows, we give an application of the selection concept developed in the finance theory. We consider a single period model and an investor who faces a capital market with n riskly financial assets. Let $\tilde{R}_i$, $i = 1, 2, 3, ..., n$ the random variable which represents the stochastic return of capital asset i. The problem of choosing optimal portfolio for an investor is formulated as follows :

• MODEL : EXPECTED UTILITY CRITERION

We assume that there is an investor whose prefered structure is caracterized by its utility function $U : I\!R \to I\!R$. In particular, suppose that U is a differentiable, concave, and strictly increasing function. If we denote by W_0 the investor's initial wealth and by x_i, $i = 1, 2, ..., n$ the proportion of this wealth invested in the financial asset i, then the terminal wealth is a random variable given by

$$\tilde{W}(X) := W_0 \left(\sum_{i=1}^{n} x_i \tilde{R}_i \right), \tag{33}$$

with

$$X = (x_1, x_2, ..., x_n), \quad \sum_{i=1}^{n} x_i = 1, \text{ and } x_i \geq 0, \ i = 1, 2, ..., n.$$

Hence, the optimization problem will be expressed as maximizing the expected utility of terminal wealth, that is

$$\max \left\{ E \left[U \left(\tilde{W}(X) \right) \right]; \ X \in \mathcal{C} \right\} \tag{S}$$

where

$$\mathcal{C} := \left\{ (x_1, x_2, ..., x_n), \quad \sum_{i=1}^{n} x_i = 1, \text{ and } x_i \geq 0, \ i = 1, 2, ..., n \right\}. \tag{34}$$

The investor chooses his best trading strategy by solving problem (S), but if there are several optimal solutions the "good" portfolio must be selected amongst all optimal portfolios, for example, the optimal portfolio which minimizes the risk ! When one uses the variance as a measure of risk, we will show that the problem is to select the optimal solution which minimizes some norm. Indeed, set

$$\Omega := [\sigma_{i,j}]_{1 \leq i,j \leq n}, \quad \sigma_{i,j} := \text{cov}(\tilde{R}_i, \tilde{R}_j), \ i, j = 1, 2, ..., n,$$

i.e. Ω is the $n \times n$ variance-covariance matrix. We easily verify that :

$$\mathrm{var}\left(\frac{\tilde{W}}{W_0}\right) = \sum_{i=1}^{n}\sum_{j=1}^{n}\sigma_{i,j}x_i x_j = \langle \Omega X, X \rangle = \|X\|_{\Omega}^2.$$

Consequently, instead of solving directly problem (S), we consider the sequence of regularized problems (S_n) given by

$$\max_{X \in \mathbf{R}^n}\left\{ E\left[U\left(\tilde{W}(X)\right)\right] - \frac{1}{2\lambda_n}\mathrm{dist}(X, C)^2 - \frac{r_n}{2}\|X\|_{\Omega}^2\right\}, \qquad (S_n)$$

and, applying the analysis of Section 3, we directly select the unique optimal portfolio which presents the minimal risk.

EXEMPLE 4.1 : MARKOWITZ'S MEAN VARIANCE CRITERION([18])

Besides, In this example, we suppose that there is a riskless asset and denote R_0 by its deterministic return. The Markowtz's mean-variance criterion provides the optimal portfolio by solving the regularized problem given as follows (Tikhonov regularization)

$$\max_{X \in \mathbf{R}^n}\left\{ E\left[\tilde{W}(X)\right] - \frac{\varepsilon}{2}\mathrm{var}\left[\tilde{W}(X)\right]\right\}, \qquad (S_\varepsilon)$$

where

$$\tilde{W}(X) := W_0\left(R_0 + \sum_{i=1}^{n}x_i(\tilde{R}_i - R_0)\right).$$

Therefore, this criterion is founded on the Tikhonov regularization of some ill-posed problems.

Acknowledgments. The authors are thankful to Professors B. Lemaire and R. Tichatschke for several helpful discussions and remarks. We would like to thank Professor Abdellatif Moudafi and the reviewers for their thorough work. Their comments and suggestions have greatly contributed to our exposition.

References

1. Attouch, H., Baillon, J.B., Théra, M. (1994) Variational sum of monotone operators. Journal of Convex Analysis **1**, 1, 1–29
2. Attouch, H. (1996) Viscosity solution of minimization problems. SIAM J. Optimization **6**, 3, 769–806
3. Attouch, H., Cominetti, R. (1996) A dynamical approach to convex minimization coupling approximation with the steepset descent method. Journal of Differential Equations **128**, 519–540
4. Bakushinsky, A.B., Goncharsky, A. (1989) Ill posed Problems : Theory and Applications. Mathematics and Applications, Kluwer Academie Publishers

5. Brézis, H. (1973) Opérateurs maximaux monotones et semi-groupes de contractions dans les espaces de Hilbert. North-Holland, Amsterdam, Holland
6. Cominetti, R. Coupling the proximal point algorithm with approximation methods. To appear in JOTA
7. Kaplan, A., Tichatschke, R. (1994) Stable Methods for Ill-Posed Variational Problems - Prox-regularization of Elliptical Variational Inequalities and Semi-infinite optimization problems. Akademie Verlag, Berlin
8. Lehdili, N., Moudafi, A. (1996) Combining the proximal algorithm and Tikhonov regularization. Optimization **37**, 239–252
9. Lehdili, N., Lemaire, B. The barycentric proximal method. To appear in Communication in Applied Nonlinear Analysis
10. Lemaire, B. (1997) Which fixed-point does the iteration method select ? In: Gritzmann P., Horst R., Sachs E., Tichatschke R. (Eds.) Recent Advances in Optimization. Lectures Notes in Economics and Mathematical Systems **452**, Spinger-Verlag, 154–167
11. Lemaire, B. (1996) Stability of the iteration method for non expansive mappings. Serdica Math. J. **22**, 331–340
12. Lions, P.L., Mercier, B. (1979) Splitting algorithms for the sum of two nonlinear operators. SIAM J. Numer. Anal. **6**, 964–978
13. Martinet, B. (1972) Algorithmes pour la résolution de problèmes d'optimisation et de minimax. Thèse d'Etat, Université de Grenoble, France
14. Markowitz, H. (1952) Portfolio Selection. Journal of Finance **VII**, 1, 77–91
15. Ould Ahmed Salem, C. (1998) Approximation de points fixes d'une contraction. Thèse de Doctorat, Université Montpellier II, France
16. Passty, G.B. (1978) Ergodic convergence to a zero of the sum of monotone operators in Hilbert space. Nonlinear Analysis **10**, 215–227
17. Rockafellar, R.T. (1976) Monotone operators and the proximal point algorithm. SIAM J.Control and Optimization **14**, 5, 877–898
18. Tikhonov, A., Arsenine, V. (1976) Méthodes de résolution de problèmes mal posés. Edition MIR de Moscou, traduction française
19. Tossings, P. (1990) Sur les zéros des opérateurs maximaux monotones et applications. Thèse de Doctorat, Université de Liège, Belgium
20. Tossings, P. (1994) The perturbed Tikhonov's algorithm and some of its applications. Modélisation Mathématique et Analyse Numérique **28**, 2, 189–221

Optimization of Plastic Stepped Plates Subjected to Impulsive Loading*

Jaan Lellep

Institute of Applied Mathematics, Tartu University, 51014 Tartu, Estonia

Abstract. An optimization technique is developed for rigid-plastic axisymmetric plates subjected to impulsive loading. The thickness of the plate is assumed to be piecewise constant. Both, ideal homogeneous and fiber reinforced materials are considered. Necessary optimality conditions are established with the aid of variational methods of the theory of optimal control whereas the method of mode form motions is used. Numerical results are presented for an annular plate clamped at the inner edge and free at the outer edge.

Key words: plasticity, axisymmetric plate, optimization

1 Introduction

Dynamic loading of rigid-plastic beams, plates and cylindrical shells of piecewise constant thickness has been studied Ü. Lepik [5], Ü. Lepik and Z. Mróz [6]. It was shown by Ü. Lepik and Z. Mróz [6], also by N. Jones [1], W.J. Stronge and T.X. Yu [8] that the approximate method of mode form motions leads to results quite close to the exact ones. It is assumed in the mentioned papers that the material of a plate or shell is an homogeneous material which obeys the Tresca yield condition and associated flow law.

Cylindrical shells of piecewise constant thickness have been studied by J. Lellep and E. Sakkov [4] assuming the material is a fiber reinforced composite. Annular plates of composite material were studied by J. Lellep and A. Mürk [3].

In the present paper, axisymmetric plates of piecewise constant thickness made of a fiber reinforced composite material will be considered. The problem will be treated as a particular problem of optimal control. Necessary optimality conditions will be derived with the aid of the variational methods of the theory of optimal control.

2 Formulation of the Problem

Let us consider a circular or annular plate of radius R and of internal radius a_0 (in the case of the circular plate $a_0 = 0$). The thickness h of the plate is

* This research was partly supported by the Estonian Science Foundation. The support through Grants No 2426 and 3380 is highly appreciated

piecewise constant with one step in the thickness. Thus (**Fig.1**)

$$h = \begin{cases} h_0, & r \in (a_0, a), \\ h_1, & r \in (a, R), \end{cases} \tag{1}$$

where r is the current radius and h_0, h_1 and a are considered as preliminarily unknown constant parameters. However, the material volume of the plate

$$V = \pi[h_0(a^2 - a_0^2) + h_1(R^2 - a^2)] \tag{2}$$

is considered to be given.

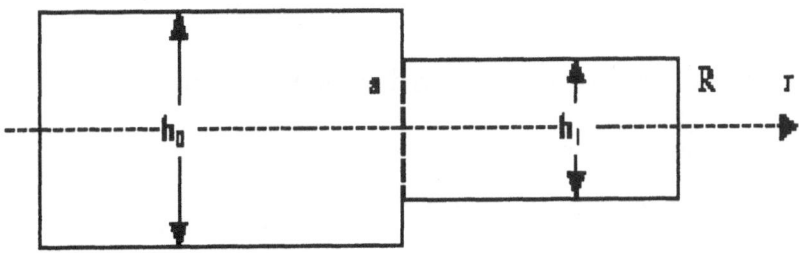

Fig. 1. Stepped plate

Material of the plate is an homogeneous rigid-plastic material obeying the Tresca yield condition or a fiber reinforced composite obeying a piecewise linear yield surface.

The plate under consideration is subjected to initial impulsive loading. The load intensity is assumed to be high enough to cause plastic deformations. It is assumed that the loading of high intensity influences the plate during a short period so that at the initial moment of time each point of the plate has non-zero initial velocity. However, displacements of the plate at $t = 0$ are negligible. Thus the deformation of the plate can be conceived as a motion due to inertia. The initial velocity field can be unspecified whereas the initial kinetic energy K_0 of the plate is considered to be given.

We are looking for the design of the plate for which residual deflections are as small as possible for given total mass (material volume V) of the plate. As the measures of smallness of residual deflections the mean deflection and maximal deflection, respectively are used. Thus the mean deflection and the maximal deflection as the criteria to be minimized are introduced. The mean deflection can be presented as

$$J = \int_0^R W(r, t_0)\, dr \tag{3}$$

whereas the maximal residual deflection is

$$J = \max_{r \in [0,R]} W(r, t_0). \tag{4}$$

It is assumed in (3), (4) that the motion ceases at the time moment $t = t_0$. It is reasonable to expect that the maximum with respect to the coordinate r in (4) is achieved at the center of the plate in the case of a circular plate. The latter means that (4) can be replaced by the criterion

$$J = W(0, t_0). \tag{5}$$

In the case of an annular plate maximum with respect to r is achieved at the free edge.

3   Governing Equations and Basic Assumptions

The equilibrium equations for a shell element can be presented as

$$\frac{\partial}{\partial r} \left(\frac{\partial}{\partial r}(rM) - N \right) = \left(-P + \mu \frac{\partial^2 W}{\partial t^2} h \right) r. \tag{6}$$

In (6) $M = M(r,t)$ and $N = N(r,t)$ stand for the principal bending moments in the radial and circumferential direction, respectively. Here P is the intensity of the distributed pressure loading in the transverse direction and μ stands for the density of the material. Evidently, in the case of impulsive loading one can neglect P in (6), e.g. $P = 0$.

The initial kinetic energy is given by the formulae

$$K_* = 2\pi \int_{a_0}^{a} \frac{\mu h_0}{2} \left(\frac{\partial W}{\partial t}(r,0) \right)^2 r \, dr + 2\pi \int_{a}^{R} \frac{\mu h_1}{2} \left(\frac{\partial W}{\partial t}(r,0) \right)^2 r \, dr, \tag{7}$$

where K_* is a fixed constant.

Curvature rates for an axisymmetric plate are defined as

$$\frac{\partial \kappa_1}{\partial t} = -\frac{\partial^3 W}{\partial r^2 \partial t}, \quad \frac{\partial \kappa_2}{\partial t} = -\frac{1}{r} \frac{\partial^2 W}{\partial r \partial t}. \tag{8}$$

Material of the plate is assumed to be an ideal rigid-plastic material. The yield condition on the plane of principal moments (M, N) is formed by the intersection of curves

$$\Phi_j(M, N, h, \sigma_1, \sigma_2) = 0; \quad j = 1, ..., s \tag{9}$$

where σ_1, σ_2 stand for the yield stresses in the radial and circumferential direction, respectively. Let us denote $\sigma_1 = \nu \sigma_0$. It is worth while mention that the set of points surrounded by (9) is a convex set on the plane (M, N) which includes the origin of coordinates.

An approximate method of mode form motions will be used in the present paper. The method was suggested by J. Martin and P. Symonds [7] and efficiently used by N. Jones [1], Ü. Lepik and Z. Mróz [6], Stronge and T. Yu [8] and others.

According to the concept of mode form motions the deformation shape (mode) retains its form during dynamic plastic deformation of a structure. In this case the displacement rate can presented as

$$\frac{\partial W}{\partial t} = f_*(t)g_*(r). \tag{10}$$

As the plate is considered to be not deformed at the initial moment of time it immediately follows from (10) that

$$W(r, t) = \int_0^t f_*(t)\, dt \cdot g_*(r). \tag{11}$$

On the other hand, the acceleration of the plate takes according to (10) the form

$$\frac{\partial^2 W}{\partial t^2} = \frac{\partial f_*}{\partial t} \cdot g_*(r). \tag{12}$$

In (10)-(12) the function $g_*(r)$ is to be considered as a given (shape) function. It is defined by the suitable flow regime.

Assume that the stress state of the shell corresponds to a side of the yield polygon (10) so that one can determine

$$N = \Phi_*(M, h), \tag{13}$$

provided the material constants σ_0, σ_1 are fixed earlier.

Note that the stress state of the plate can be more complicated so that the plate should be divided into different parts, each part corresponding to a special side of (10). However, the solution procedure in this case is similar to that corresponding to the assumption (13).

Substituting (12) and (13) in (6) and taking (1) into account one can present the equations of equilibrium as

$$\frac{\partial}{\partial r}(rM) - N = rQ;$$

$$\frac{\partial}{\partial r}(rQ) = \mu r g_*(r)\frac{\partial f_*}{\partial t} \cdot \begin{cases} h_0; & r \in (a_0, a); \\ h_1; & r \in (a, R); \end{cases} \tag{14}$$

where Q stands for the shearing force.

It seems to be reasonable to introduce the following non-dimensional quantities

$$\varrho = \frac{r}{R}, \quad \alpha = \frac{a}{R}, \quad \gamma_0 = \frac{h_0}{h_*}, \quad \gamma_1 = \frac{h_1}{h_*}, \quad m = \frac{M}{M_*},$$

$$n = \frac{N}{M_*}, \quad w = \frac{W}{h_*}, \quad q = \frac{QR}{M_*}, \quad \tau = \sqrt{\frac{M_*}{\mu h_* R^2}} \cdot t, \qquad (15)$$

$$v = \frac{V}{\pi h_* R^2} \quad K_0 = \frac{K_*}{\pi h_*^2 M_*}.$$

Here h_* stands for the thickness of the reference plate of constant thickness, M_* being the yield moment, e.g. $M_* = \sigma_0 h_*^2/4$.

Making use of the non-dimensional variables (15) one can present the equations of motion (14) as well as (10), (12) in the form of a set of equations

$$\begin{cases} q'(\varrho, \tau) = -\dfrac{q}{\varrho} + ug \cdot \begin{cases} \gamma_0, & \varrho \in (\alpha_0, \alpha), \\ \gamma_1, & \varrho \in (\alpha, 1), \end{cases} \\[2mm] m'(\varrho, \tau) = q - \dfrac{m}{\varrho} + \dfrac{1}{\varrho} \cdot \begin{cases} \Phi(m, \gamma_0), & \varrho \in (\alpha_0, \alpha), \\ \Phi(m, \gamma_1), & \varrho \in (\alpha, 1), \end{cases} \\[2mm] \dot{w}(\varrho, \tau) = f(\tau) \cdot g(\varrho), \\[1mm] \dot{f}(\tau) = u. \end{cases} \qquad (16)$$

Here and henceforth primes and dots denote the differentiation with respect to ϱ and τ, respectively. The functions $f(\tau)$ and $g(\varrho)$ are introduced so that

$$\frac{1}{h_*}\sqrt{\frac{\mu h_* R^2}{M_*}} f_* \left(\tau \sqrt{\frac{\mu h_* R^2}{M_*}}\right) \cdot g_*(R\varrho) = f(\tau)g(\varrho), \qquad (17)$$

whereas $\Phi(m, \gamma)$ presents the non-dimensional form of (13).

The quantity u can be considered as an auxiliary variable, later it will be treated as a control.

The plates with different support conditions will be considered from the unique point of view. It is assumed that the boundary and intermediate conditions are given at $\varrho = \alpha_0$; $\varrho = 1$ and at $\varrho = \alpha$ as follows

$$\begin{aligned} m(\alpha_0, \tau) &= m_0; \quad q(\alpha_0, \tau) = q_0, \\ m(1, \tau) &= m_1; \quad q(1, \tau) = q_1, \\ m(\alpha, \tau) &= m_\alpha. \end{aligned} \qquad (18)$$

If, for instance, the internal edge of the plate is free, one has $m_0 = 0$, $q_0 = 0$. In the case of the clamped inner edge plastic hinge circle appears at $\varrho = \alpha_0$ and thus $m_0 = -\gamma_0^2$, whereas q_0 is an unknown constant (in the case of the Tresca yield condition).

Similarily, if the outer edge of the plate is free then $m_1 = 0$ and $q_1 = 0$. However, in the case of a simply supported outer edge $m_1 = 0$ and q_1 is preliminarily unknown.

Consider the case of a clamped circular plate. Now $\alpha_0 = 0$ and there are hinges at the edge and at the center of the plate. Thus $m_0 = \gamma_0^2$, $q_0 = 0$, $m_1 = -\gamma_1^2$ whereas q_1 is unknown. The quantity $m_\alpha = \pm\gamma_i^2$, where $\gamma_i = \min(\gamma_0, \gamma_1)$.

According to (7) and (15), (16) the initial kinetic energy can be rewritten as

$$\int_{\alpha_0}^{\alpha} \gamma_0 f^2(0) g^2(\varrho)\varrho \, d\varrho + \int_{\alpha}^{1} \gamma_1 f^2(0) g^2(\varrho)\varrho \, d\varrho - K_0 = 0 \qquad (19)$$

and (2) can be put into the form

$$\gamma_0(\alpha^2 - \alpha_0^2) + \gamma_1(1 - \alpha^2) - v = 0. \qquad (20)$$

It is assumed that the motion stops at the moment $\tau = \tau_0$. Thus according to (10) and (16) at the final moment $f(\tau_0) = 0$.

4   Optimality Conditions

Let us consider the problem of minimization of the residual mean deflection. Evidently, the optimality criterion (3) can be replaced by

$$J = \int_0^1 w(\varrho, \tau_0) \, d\varrho \qquad (21)$$

where τ_0 stands for the time when the motion ceases.

We are looking for the minimum of (21) so that the differential constraints (16) with boundary and intermediate conditions (18) and additional constraints (19), (20) are satisfied. The problem (16)-(21) will be considered as a particular problem of optimal control with distributed parameters. The variables q, m and w will be treated as state variables and u as a control. The quantity f will be handled as a one dimensional state variable and α, γ_0, γ_1 as constant parameters.

In order to establish the necessary conditions of optimality for the problem set up, let us introduce the augmented functional J_* defined by

$$\int_{\alpha_0}^1 w(\varrho, \tau_0)\, d\varrho + \int_{\alpha_0}^\alpha \int_0^{\tau_0} \{\psi_1(q' + \frac{q}{\varrho} - ug\gamma_0) + \psi_2(m' + \frac{m}{\varrho} - \frac{1}{\varrho}\Phi(m, \gamma_0 - q))$$

$$+\psi_3(\dot{w} - f \cdot g)\}\, d\varrho d\tau + \int_0^{\tau_0} \psi_4(\dot{f} - u)\, d\tau + \int_\alpha^1 \int_0^{\tau_0} \psi_1\{(q' + \frac{q}{\varrho} - ug\gamma_1)$$

$$+\psi_2(m' + \frac{m}{\varrho} - \frac{1}{\varrho}\Phi(m, \gamma_1)) + \psi_3(\dot{w} - f \cdot g)\}\, d\varrho d\tau + \lambda\{\int_{\alpha_0}^\alpha \gamma_0 f^2(0)g^2\varrho\, d\varrho$$

$$+\int_\alpha^1 \gamma_1 f^2(0)g^2\varrho\, d\varrho - K_0\} + \int_0^{\tau_0} \{\nu_0(m(o, \tau) - m_0) + \mu_0(q(0, \tau) - q_0)$$

$$+\nu_1(m(1, \tau) - m_1) + \mu_1(q(1, \tau) - q_1) + \nu(m(\alpha, \tau) - m_\alpha)\}\, dt$$

$$+\varphi(\gamma_0(\alpha^2 - \alpha_0^2) + \gamma_1(1 - \alpha^2) - v) + \varphi_0 f(\tau_0). \tag{22}$$

In (22), ψ_1, ψ_2, ψ_3, ψ_4 stand for adjoint variables and λ, ν_0, ν_1, ν_2, μ_0, μ_1, φ_0, φ_1 for certain Lagrangian multipliers. It is expected that ψ_1, ψ_2, ψ_3 are the functions of both, the coordinate and time whereas λ and φ_0, φ_1 are assumed to be constants. However, ψ_4, μ_0, μ_1 and ν_0, ν_1, ν_2 are unknown functions of the time τ.

The total variation ΔJ_* of the functional (22) can be presented as

$$\int_0^1 (\delta w(\varrho, \tau_0) + \psi_3(\varrho, \tau_0)\delta w(\varrho, \tau_0))\, d\varrho + \int_0^{\tau_0} \int_{\alpha_0}^1 \{-\psi_1'\delta q + \frac{\psi_1}{\varrho}\delta q - \psi_2\delta q -$$

$$-\psi_2'\delta m + \frac{\psi_2}{\varrho}\delta m - \frac{\psi_2}{\varrho}\frac{\partial\psi}{\partial m}\delta m - \dot{\psi}_3\delta w - \psi_3 g\delta f\}\, d\varrho d\tau$$

$$+\int_0^{\tau_0} \{\int_{\alpha_0}^\alpha (-\psi_1(g\gamma_0\delta u + ug\Delta\gamma_0) - \frac{\psi_2}{\varrho}\frac{\partial\Phi}{\partial\gamma_0}\Delta\gamma_0)\, d\varrho + \int_\alpha^1 (-\psi_1(g\gamma_1\delta u$$

$$+ug\Delta\gamma_1) - \frac{\psi_2}{\varrho}\frac{\partial\Phi}{\partial\gamma_1}\Delta\gamma_1)\, d\varrho\}\, d\tau$$

$$+\int_0^{\tau_0} \{(\psi_1\delta q + \psi_2\delta m)\Big|_{\varrho=\alpha_0}^{\varrho=\alpha^-} + (\psi_1\delta q + \psi_2\delta m)\Big|_{\varrho=\alpha_+}^{\varrho=1}\}\, d\tau$$

$$+\lambda \int_\alpha^1 (f^2(0)g^2\varrho\Delta\gamma_1 + 2\gamma_1 f(0)g^2\varrho\delta f(0))\, d\varrho + \int_0^{\tau_0} (-\dot{\psi}_4 df - \psi_4\delta u)\, d\tau$$

$$+\int_0^{\tau_0} \{\nu_0(\Delta m(0, \tau) - \Delta m_0) + \nu_1(\Delta m(1, \tau) - \Delta m_1) + \nu_3(\Delta m(\alpha, \tau) - \Delta m_\alpha)$$

$$+\mu_0(\Delta q(0, \tau) - \Delta q_0) + \mu_1(\Delta q(1, \tau) - \Delta q_1)\}\, d\tau + \psi_4(\tau_0)\delta f(\tau_0)$$

$$-\psi_4(0)\delta f(0) + \varphi\left((\alpha^2 - \alpha_0^2)\Delta\gamma_0 + 2\alpha\gamma_0\Delta\alpha - (1 - \alpha^2)\Delta\gamma_1 - 2\alpha\gamma_1\Delta\alpha\right)$$

$$+\varphi_0\Delta f(\tau_0) + \lambda f^2(0)(\gamma_0 - \gamma_1)g(\alpha)\alpha\Delta\alpha. \tag{23}$$

In (23) and henceforth, two types of variations are distinguished. Namely, variation of type δy stands for a weak variation of the variable y. Thus $\delta y \in C^1$. However, $\Delta y(\varrho, \tau_0)$ or $\Delta y(\alpha, \tau)$ denote the total variations of y at the moment $\tau = \tau_0$ and at the point $\varrho = \alpha$, respectively. Evidently,

$$\Delta y(\varrho, \tau_0) = \delta y(\varrho, \tau_0) + \dot{y}(\varrho, \tau_0) \cdot \Delta \tau \tag{24}$$

and

$$\Delta y(\alpha, \tau) = \delta y(\alpha \pm, \tau) + y'(\alpha \pm, \tau) \cdot \Delta \alpha. \tag{25}$$

Here

$$\delta y(\alpha \pm, \tau) = \lim_{\varrho \to \alpha \pm 0} \delta y(\varrho, \tau). \tag{26}$$

Due to physical reasons all the state variables are considered to be continuous for $\varrho \in [0, 1]$ and $\tau \in [0, \tau_0]$. However, it follows from (14) and (16) that the derivatives of q and m with respect to ϱ are discontinuous at $\varrho = \alpha$. Therefore one has to distinguish the left and right hand limits in (25) at the point $\varrho = \alpha$.

From (23) one easily obtains the adjoint equations

$$\begin{aligned} \psi_1' &= \frac{\psi_1}{\varrho} - \psi_2, \\ \psi_2' &= \frac{\psi_2}{\varrho} - \frac{\psi_2}{\varrho} \frac{\partial \Phi}{\partial m}, \\ \dot{\psi}_3 &= 0, \\ \dot{\psi}_4 &= -\int_{\alpha_0}^1 \psi_3 g \, d\varrho. \end{aligned} \tag{27}$$

Taking into account that $\delta u = \delta u(t)$ on the grounds of (23) one can establish

$$\int_{\alpha_0}^\alpha \gamma_0 \psi_1 g \, d\varrho + \int_\alpha^1 \gamma_1 \psi_1 g \, d\varrho + \psi_4 = 0. \tag{28}$$

Evidently $\delta w(\varrho, \tau_0) = \Delta w(\varrho, \tau_0)$ and thus

$$\psi_3(\varrho, \tau_0) = -1. \tag{29}$$

Similarily, $\delta f(0) = \Delta f(0)$ which leads to the equation

$$2\lambda f(0) \left(\int_{\alpha_0}^\alpha \gamma_0 g^2 \varrho \, d\varrho + \int_\alpha^1 \gamma_1 g^2 \varrho \, d\varrho \right) - \psi_4(0) = 0. \tag{30}$$

Substituting $\delta m(\alpha\pm, \tau)$, $\delta q(\alpha\pm, \tau)$ according to (25) and taking (27)-(30) into account one can put the equation (23) into the form

$$\int_0^{\tau_0} \left\{ \int_{\alpha_0}^{\alpha} \left(-\psi_1 u g - \frac{\psi_2}{\varrho} \frac{\partial \Phi}{\partial \gamma_0} \right) \Delta\gamma_0 d\varrho + \int_{\alpha}^{1} \left(-\psi_1 u g - \frac{\psi_2}{\varrho} \frac{\partial \Phi}{\partial \gamma_1} \right) \Delta\gamma_1 \, d\varrho \right\}$$

$$d\tau + \lambda f^2(0) \left\{ \int_{\alpha_0}^{\alpha} g^2 \varrho \Delta\gamma_0 \, d\varrho + \int_{\alpha}^{1} g^2 \varrho \Delta\gamma_1 \, d\varrho \right\} + \lambda \alpha f^2(0)(\gamma_0 - \gamma_1) g(\alpha)$$

$$\Delta\alpha + \int_0^{\tau_0} \left\{ \psi_1(\alpha-, \tau)(\Delta q(\alpha-, \tau) - q'(\alpha-, \tau)\Delta\alpha) - \psi_1(\alpha_0, \tau)\Delta q(\alpha_0, \tau) \right.$$

$$+\psi_1(1, \tau)\Delta q(1, \tau) - \psi_1(\alpha+, \tau)(\Delta q(\alpha+, \tau) - q'(\alpha+, \tau)\Delta\alpha)$$

$$+\psi_2(\alpha-, \tau)(\Delta m(\alpha-, \tau) - m'(\alpha-, \tau)\Delta\alpha) - \psi_2(\alpha_0, \tau)\Delta m(\alpha_0, \tau)$$

$$+\psi_2(1, \tau)\Delta m(1, \tau) - \psi_2(\alpha+, \tau)(\Delta m(\alpha+, \tau) - m'(\alpha+, \tau)\Delta\alpha)$$

$$+ \int_0^{\tau_0} \left\{ \nu_0 \left(\Delta m(\alpha_0, \tau) - \frac{\partial m_0}{\partial \gamma_0} \Delta\gamma_0 \right) + \nu_1 \left(\Delta m(1, \tau) - \frac{\partial m_1}{\partial \gamma_1} \Delta\gamma_1 \right) \right.$$

$$+\nu_2 \left(\Delta m(\alpha, \tau) - \frac{\partial m_\alpha}{\partial \gamma_0} \Delta\gamma_0 - \frac{\partial m_\alpha}{\partial \gamma_1} \Delta\gamma_1 \right) + \mu_0 (\Delta q(\alpha_0, \tau) - \Delta q_0)$$

$$+\mu_1 (\Delta q(1, \tau) - \Delta q_1) \left. \right\} \, d\tau + \psi_4(\tau_0)(\Delta f(\tau_0) - \dot{f}(\tau_0)\Delta\tau_0) + \varphi_0 \Delta f(\tau_0)$$

$$+\varphi((\alpha^2 - \alpha_0^2)\Delta\gamma_0 + 2\alpha\gamma_0\Delta\alpha + (1 - \alpha^2)\Delta\gamma_1 - 2\alpha\gamma_1\Delta\alpha) = 0. \tag{31}$$

Since $\Delta\gamma_0 = const$ and $\Delta\gamma_1 = const$, it follows from the equation (31) that

$$\int_0^{\tau_0} \int_{\alpha_0}^{\alpha} \left(-\psi_1 u g - \frac{\psi_2}{\varrho} \frac{\partial \Phi}{\partial \gamma_0} \right) d\varrho d\tau + \lambda f^2(0) \int_{\alpha_0}^{\alpha} g^2 \varrho \, d\varrho$$

$$+ \int_0^{\tau_0} \left(-\nu_0 \frac{\partial m_0}{\partial g_0} - \nu_2 \frac{\partial m_\alpha}{\partial \gamma_0} \right) d\tau + \varphi(\alpha^2 - \alpha_0^2) = 0 \tag{32}$$

and

$$\int_0^{\tau_0} \int_{\alpha}^{1} \left(-\psi_1 u g - \frac{\psi_2}{\varrho} \frac{\partial \Phi}{\partial \gamma_1} \right) d\varrho d\tau + \lambda f^2(0) \int_{\alpha}^{1} g^2 \varrho \, d\varrho$$

$$+ \int_0^{\tau_0} \left(-\nu_1 \frac{\partial m_1}{\partial g_1} + \nu_2 \frac{\partial m_\alpha}{\partial \gamma_1} \right) d\tau + \varphi(1 - \alpha^2) = 0. \tag{33}$$

It can be seen from (31) that the transversality conditions have the form at $\varrho = \alpha_0$

$$\psi_1(\alpha_0, \tau) = \mu_0, \quad \psi_2(\alpha_0, \tau) = \nu_0 \tag{34}$$

and at $\varrho = 1$

$$\psi_1(1, \tau) = -\mu_1 \quad \psi_2(1, \tau) = -\nu_1, \tag{35}$$

whereas

$$\mu_0 \Delta q_0 = 0, \quad \mu_1 \Delta q_1 = 0. \tag{36}$$

Equations (36) have the following meaning. If, for instance, q_0 is fixed then $\Delta q_0 = 0$, $\mu_0 \neq 0$. However, in the case when $q(\alpha_0, \tau)$ is unknown one has $\Delta q_0 \neq 0$ and thus $\mu_0 = 0$.

The state variables m and q are assumed to be continuous at $\varrho = \alpha$. Therefore $\Delta m(\alpha-, \tau) = \Delta m(\alpha+, \tau)$, $\Delta q(\alpha-, \tau) = \Delta q(\alpha+, \tau)$ and it follows from (31) that

$$\begin{aligned} \psi_1(\alpha-, \tau) - \psi_1(\alpha+, \tau) &= 0, \\ \psi_2(\alpha-, \tau) - \psi_2(\alpha+, \tau) + \nu_2 &= 0. \end{aligned} \tag{37}$$

Evidently, $\Delta f(\tau_0)$ and $\Delta \tau_0$ are arbitrary increments in (31). Thus

$$\psi_4(\tau_0) = \varphi_0 = 0. \tag{38}$$

Due to arbitrariness of $\Delta \alpha$ in (32) one has

$$\lambda f^2(0)(\gamma_0 - \gamma_1)g(\alpha)\alpha + \int_0^{\tau_0} \{-\psi_1(\alpha-, \tau)q'(\alpha-, \tau) + \psi_1(\alpha+, \tau)q'(\alpha+, \tau)$$
$$-\psi_2(\alpha-, \tau)m'(\alpha-, \tau) + \psi_2(\alpha+, \tau)m'(\alpha+, \tau)\} \, d\tau + 2\varphi\alpha(\gamma_0 - \gamma_1) = 0. \tag{39}$$

Equations (27)-(30) and (32)-(39) present the set of necessary conditions of optimality. These equations can be solved together with the state equations (16), boundary conditions (18) and constraints (19), (20).

Note that the last two equations in (27) together with (29) yield

$$\psi_3 = -1 \tag{40}$$

and

$$\dot{\psi}_4 = \int_0^1 g \, d\varrho$$

which in turn with (38) gives

$$\psi_4 = \int_{\alpha_0}^1 g \, d\varrho(\tau - \tau_0). \tag{41}$$

Making use of (41) one can eliminate ψ_4 from the system of optimality conditions. Thus (28) can be presented as

$$\gamma_0 \int_{\alpha_0}^\alpha \psi_1 g \, d\varrho + \gamma_1 \int_\alpha^1 \psi_1 g \, d\varrho + \int_{\alpha_0}^1 g \, d\varrho(\tau - \tau_0) = 0. \tag{42}$$

Equations (30) and (41) admit to define

$$\lambda = \frac{-\tau_0 \int_{\alpha_0}^1 g \, d\varrho}{2f(0)\left(\int_{\alpha_0}^\alpha \gamma_0 g^2 \varrho \, d\varrho + \int_\alpha^1 \gamma_1 g^2 \varrho \, d\varrho\right)}. \tag{43}$$

Substituting the derivatives of state variables from (16) in (39), one eventually obtains

$$\lambda f^2(0)(\gamma_0 - \gamma_1)\alpha g(\alpha) + \int_0^{\tau_0} \{\psi_1(\alpha,\tau)ug(\alpha)(\gamma_1 - \gamma_0) - \psi_2(\alpha-,\tau)(q(\alpha,\tau) -$$

$$\frac{m(\alpha,\tau)}{\alpha} + \frac{1}{\alpha}\Phi(m(\alpha,\tau),\gamma_0)) + \psi_2(\alpha+,\tau)(q(\alpha,\tau) - \frac{m(\alpha,\tau)}{\alpha} + \frac{1}{\alpha}\Phi(m(\alpha,\tau),$$

$$\gamma_1))\} \, d\tau + 2\varphi\alpha(\gamma_0 - \gamma_1) = 0. \tag{44}$$

Making use of (43), (44) one can determine the Lagrangian multiplier

$$\varphi = \frac{\tau_0 f(0)g(\alpha)\int_{\alpha_0}^1 g\,d\varrho}{4\left(\int_{\alpha_0}^\alpha \gamma_0 g^2 \varrho\,d\varrho + \int_\alpha^1 \gamma_1 g^2 \varrho\,d\varrho\right)} - \frac{1}{2\alpha(\gamma_0 - \gamma_1)}\int_0^{\tau_0} \{\psi_1(\alpha,\tau)ug(\alpha)$$

$$(\gamma_1 - \gamma_0) + \nu_2\left(q(\alpha,\tau) - \frac{1}{\alpha}m(\alpha,\tau)\right) - \psi_2(\alpha-,\tau)\frac{1}{\alpha}\Phi(m(\alpha,\tau),\gamma_0)$$

$$+ \psi_2(\alpha+,\tau)\frac{1}{\alpha}\Phi(m(\alpha,\tau),\gamma_1)\} \, d\tau. \tag{45}$$

5   Annular Plate

Let us consider an annular plate clamped at the inner edge and free at the outer edge. The plate is subjected to the initial impulsive loading and has the piecewise constant thickness (1).

The plate is made of a fiber reinforced composite material. Different combinations of fibers and matrix material are considered. It is assumed that the material obeys the yield condition suggested by Lance and Robinson [2]. Let us confine our attention to the case of circumferential orientation of fibers. In this case the yield condition is formed by the intersection of the straight lines

$$|M| = M_0, \quad |M - N| = kM_0, \quad |N| = kM_0$$

on the plane of principal moments M and N (**Fig.2**). Here k stands for the single material parameter.

Assume that the stress state of the plate corresponds to the side DE ofthe yield hexagon (**Fig.2**). It means that equation (13) takes in the present case the form

$$N = -k\sigma_0\frac{h^2}{4}, \tag{46}$$

σ_0 being the yield stress for the matrix material. Making use of non-dimensional quantities (15) and the relation (46) one can state that

$$\Phi(m,\gamma_0) = -k\gamma_0^2, \quad \Phi(m,\gamma_1) = -k\gamma_1^2 \tag{47}$$

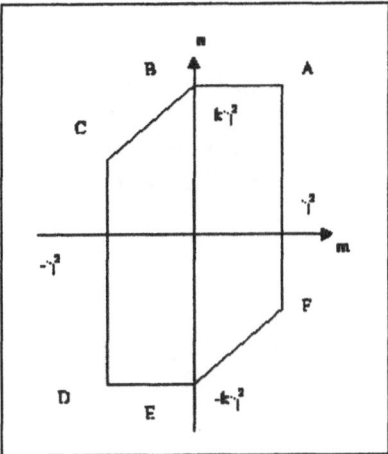

Fig. 2. Yield condition

in the present case. Evidently, $\partial \Phi / \partial m = 0$.

According to the associated flow law the vector with components (8) is to be normal to the side DE in **Fig.2**. Therefore $\dot{w}'' = 0$ and

$$\dot{w} = f \frac{\varrho - \alpha_0}{1 - \alpha_0}. \tag{48}$$

Comparing (15)-(17) with (48) one can see that the mode of deformation can be presented as

$$g(\varrho) = \frac{\varrho - \alpha_0}{1 - \alpha_0}. \tag{49}$$

The boundary conditions for stress components are the following ones:

$$m(\alpha_0, \tau) = -\gamma_0^2, \quad m(1, \tau) = 0, \quad q(1, \tau) = 0, \quad m(\alpha, \tau) = -\gamma_1^2, \tag{50}$$

provided $\gamma_1 < \gamma_0$ and $m \leq 0$ for $\varrho \in (\alpha_0, 1)$. Since $q(\alpha_0, \tau)$ is not fixed it follows from (36), (50) that $\Delta q_0 \neq 0$, $\Delta q_1 = 0$ and

$$\mu_0 = 0, \quad \mu_1 \neq 0. \tag{51}$$

Making use of (18) and (50), it is easy to recheck that

$$\frac{\partial m_0}{\partial \gamma_0} = -2\gamma_0, \quad \frac{\partial m_\alpha}{\partial \gamma_0} = 0, \quad \frac{\partial m_\alpha}{\partial \gamma_1} = -2\gamma_1, \quad \frac{\partial m_1}{\partial \gamma_1} = 0. \tag{52}$$

Substituting the mode form (49) in (19), one can obtain

$$f(0) = \sqrt{12 k_0 (1 - \alpha_0)^2 / \Theta}, \tag{53}$$

where

$$\Theta = \gamma_0(3\alpha^4 - 8\alpha_0\alpha^3 + 6\alpha_0^2\alpha^2 - 5\alpha_0^4) + \gamma_1(3 - 8\alpha_0 + 6\alpha_0^2 - 3\alpha^4 + 8\alpha_0\alpha^3 - 6\alpha_0^2\alpha^2).$$

It can be seen from (46) or (47) that $\partial\Phi/\partial m = 0$ in the present case. Thus the adjoint system (27) takes the form

$$\psi_1' = \frac{\psi_1}{\varrho} - \psi_2,$$

$$\psi_2' = \frac{\psi_2}{\varrho}, \tag{54}$$

whereas (40) and (41) give

$$\psi_3 = -1, \quad \psi_4 = \frac{1}{2}(1 - \alpha_0)(\tau - \tau_0). \tag{55}$$

Boundary and intermediate conditions for (53) are given by (34), (35), (51) and (37). An appropriate solution of (54) can be presented as

$$\psi_1 = \begin{cases} C_0\varrho(\alpha_0 - \varrho), & \varrho \in (\alpha_0, \alpha), \\ -C_1\varrho^2 + (C_0\alpha_0 + (C_1 - C_0)\alpha), & \varrho \in (\alpha, 1), \end{cases}$$

$$\psi_2 = \begin{cases} C_0\varrho, & \varrho \in (\alpha_0, \alpha), \\ C_1\varrho, & \varrho \in (\alpha, 1) \end{cases} \tag{56}$$

where C_0 and C_1 are arbitrary functions of time.

Making use of (19), (49) and (43) one can present the multiplier λ as

$$\lambda = -\frac{\tau_0 f(0)(\alpha_0 - 1)}{4k_0} \tag{57}$$

and the multiplier φ according to (45) as

$$\varphi = \frac{\tau_0 f^3(0) \cdot (\alpha - \alpha_0)}{8k_0} - \frac{1}{2(\gamma_0 - \gamma_1)} \int_0^{\tau_0} \left\{ C_0 \left[\frac{u}{1 - \alpha_0}(\gamma_0 - \gamma_1)(\alpha - \alpha_0)^2 \right. \right.$$
$$\left. - q(\alpha, \tau) - \frac{\gamma_0^2}{\alpha} + \frac{k\gamma_0^2}{\alpha} \right] + C_1 \left[q(\alpha, \tau) + \frac{\gamma_1^2}{\alpha} - \frac{k\gamma_1^2}{\alpha} \right] \right\} d\tau. \tag{58}$$

Let us introduce the notation

$$\int_0^{\tau_0} C_0 d\tau = B_0, \quad \int_0^{\tau_0} C_1 d\tau = B_1. \tag{59}$$

Integrating (42) with respect to τ and substituting (49), (56) one obtains the algebraic equation

$$
B_0 \left[\gamma_0 (6\alpha_0^2(\alpha^2 - \alpha_0^2) - 8\alpha_0(\alpha^3 - \alpha_0^3) + 3\alpha^4 - 3\alpha_0^4) + \gamma_1(\alpha_0 - \alpha) \right.
$$
$$
(6\alpha_0(1 - \alpha^2) + 4\alpha^3 - 4) \right] + B_1\alpha\gamma_1 \left[6\alpha_0(1 - \alpha^2) - 4 + 4\alpha^3 - \frac{4\alpha_0}{\alpha}(1 - \alpha^3) \right.
$$
$$
\left. -\frac{3}{\alpha}(1 - \alpha^4) \right] - 3\tau_0^2(\alpha_0 - 1)^2 = 0, \tag{60}
$$

where the notation (59) is used.

In a similar way, equations (32) and (33) take the form

$$
B_0 \left[-\frac{u}{\alpha_0 - 1} (6\alpha_0^2(\alpha^2 - \alpha_0^2) - 8\alpha_0(\alpha^3 - \alpha_0^3) + 3\alpha^4 - 3\alpha_0^4) + 24\gamma_0 \right.
$$
$$
(k(\alpha - \alpha_0) + \alpha_0)] + 12(\alpha^2 - \alpha_0^2)\varphi + \frac{\lambda f^2(0)}{(\alpha_0 - 1)^2} (6\alpha_0^2(\alpha^2 - \alpha_0^2) - 8\alpha_0 + 3\alpha^4
$$
$$
-8\alpha_0(\alpha^3 - \alpha_0^3) - 3\alpha_0^4) = 0 \tag{61}
$$

and

$$
B_0 \left[(\alpha_0 - \alpha) (6\alpha_0(1 - \alpha^2) - 4(1 - \alpha^3)) \frac{u}{1 - \alpha_0} - 24\alpha\gamma_1 \right] + B_1 \left[\frac{u}{\alpha_0 - 1} \right.
$$
$$
(4\alpha_0 (1 - \alpha^3) + 3\alpha^4 - 3 - 6\alpha\alpha_0(1 - \alpha^2) + 4\alpha(1 - \alpha^3) + 24k\gamma_1(1 - \alpha)
$$
$$
+24\alpha\gamma_1] - \frac{\lambda f^2(0)}{(\alpha_0 - 1)^2} (6\alpha_0^2(1 - \alpha^2) - 8\alpha_0(1 - \alpha^3) + 3(1 - \alpha^4))
$$
$$
+12\varphi((1 - \alpha^2) = 0. \tag{62}
$$

When deriving (61), (62) it was taken into account that according to the concept of mode form motions the acceleration u is a constant.

In order to solve the optimization problem up to the end, one has to integrate the state equations (16) making use of (47). Satisfying the boundary conditions (50) and the continuity requirements $q(\alpha-, \tau) = q(\alpha+, \tau,)$ $m(\alpha-, \tau) = m(\alpha+, \tau)$ one can determine 4 integration constants without using two of these requirements. One of the rest boundary or intermediate conditions is to be employed for determination of the acceleration

$$
u = \frac{12(1 - \alpha_0)(-\alpha + k(\alpha - 1))\gamma_1}{\alpha^4 - 2\alpha_0\alpha^3 + 6\alpha_0\alpha - 4\alpha_0 - 4\alpha + 3}. \tag{63}
$$

The rest boundary condition with the system (60)-(62) and (20) is to be used for determination of unknown parameters. Here we have five equations and five unknown quantities B_0, B_1, α, h_0, h_1, provided $f(0)$, λ and φ are substituted from (53), (57) and (58), respectively.

Note that τ_0 is also unknown but according to (16) and (63) it can be defined from the equation $f(\tau_0) = 0$. Thus

$$f(\tau) = u \cdot \tau + f(0), \tag{64}$$

where $f(0)$ is given by (53). It immediately follows from (64) that

$$\tau_0 = -\frac{f(0)}{u}. \tag{65}$$

6   Numerical Example

Consider the problem of minimization the maximal residual deflection. In the case of the annular plate introduced in the previous section, the deflection is maximal at the free edge, e.g., we are interested in the minimization of the quantity $w(1, \tau_0)$, provided the volume of the plate is given by (20). Evidently, now

$$J = w(1, \tau) = \int_0^{\tau_0} f(\tau)\, d\tau, \tag{66}$$

as $w(1, 0) = 0$.

In the case of the optimality criterion (66), the necessary conditions established in Section 4 have to be modified slightly. First of all, the last equations in (27) gives $\dot{\psi}_4 = 1$. Thus

$$\psi_3 = 0, \quad \psi_4 = \tau - \tau_0. \tag{67}$$

In the present case, $\psi_3(\varrho, \tau_0) = 0$ and $\psi_4(\tau_0) = 0$, whereas τ_0 is defined by (65).

The set of algebraic equations has been solved numerically in the case when $h_0 = h_*$, e.g. $\gamma_0 = 1$ and $\Delta\gamma_0 = 0$. The results of calculations are presented in **Table 1** in the case $k = 2$ and $\alpha_0 = 0.2$.

Table 1. Optimal values of parameters

v	α	γ_1	w_0	w_*	e
0.40	0.24	0.83	0.31	0.32	0.96
0.35	0.28	0.72	0.39	0.42	0.93
0.30	0.33	0.59	0.52	0.57	0.92

Here w_0 stands for the maximal residual deflection of the stepped plate with optimal parameters and w_* is the maximal deflection of the reference plate of constant thickness. Both the stepped plate and the reference plate have common weight (volume) v. Economy of the design is assessed by the ratio $e = w_0/w_*$. It can be seen from Table 1 that the maximal deflection can be reduced up to 8% in the case when $\alpha_0 = 0.2$ and $k = 2$.

7 Concluding Remarks

A method of optimization for axisymmetric plates of piecewise constant thickness has been developed in the present paper. The method is applicable to homogeneous plates material which obeys a piecewise linear yield condition as well as to composite materials obeying the yield condition suggested by R. Lance and D. Robinson [2].

Numerical results have been presented for annular plates clamped at the inner edge and free at the outer edge. Calculations carried out show that the design with one step in the thickness allows to reduce the maximal residual deflection up to 8% (in the case when $\alpha_0 = 0.2$ and $k = 2$). Employing the design with two steps allows to get additionally 3-4 % in the decrease of deflections.

Note that the similar approach could be developed for shallow shells of revolution. However, this is the task for future work.

References

1. Jones, N. (1989) Structural impact. Cambridge University Press, Cambridge
2. Lance, R.H., Robinson, D.N. (1972) Limit analysis of ductile fiber reinforced structures. Proc. ASCE. J. Eng. Mech. Div. **98**, 195–209
3. Lellep, J., Mürk, A. (1997) Optimization of annular plates of composite materials. In: Andersen, S.I. et al. (Eds.) Proc. 18th Riso Symp. Polymeric Composites-Expanding the Limits. Roskilde, Denmark, 405–410
4. Lellep, J., Sakkov, E. (1996) Optimization of cylindrical shells of fiber reinforced composite materials. Mech. Comp. Mater. **32**, 65–71
5. Lepik, Ü. (1982) Optimal design on non-elastic structures in the case of dynamic loading (in Russian), Tallinn, Valgus
6. Lepik, Ü., Mróz, Z. (1977) Optimal design of plastic structures under impulsive and dynamic pressure loading. Int. J. Solids and Struct. **13**, 7, 657–674
7. Martin, J.B., Symonds, P.S. (1966) Mode approximation for impulsively loading rigid-plastic structures. Proc. ASCE. J. Eng. Mech. Div. **92**, 5, 43–66
8. Stronge, W.J., Yu, I.X. (1993) Dynamic models for structural plasticity. Springer, Berlin, Heidelberg

From Fixed Point Regularization to Constraint Decomposition in Variational Inequalities

Bernard Lemaire and Cheikh Ould Ahmed Salem

Université Montpellier II, Place E. Bataillon, 34 060 Montpellier Cedex 05, France

Abstract. Extending the Tikhonov regularization method to the fixed point problem for a nonexpansive self mapping P on a real Hilbert space H, generates a family of fixed points u_r of strongly nonexpansive self mappings P_r on H with positive parameter r tending to 0. If the fixed point set C of P is nonempty, then u_r converges strongly to u^* the unique solution to some monotone variational inequality on the (closed convex) subset C. The iteration method suitably combined with this regularization generates a sequence that converges strongly to u^*. When C is *a priori* defined by finitely many convex inequality constraints, expressing C as the fixed point set of a suitable nonexpansive mapping and applying the above method lead to an iterative scheme in which each step is decomposed in finitely many successive or parallel projections or proximal computations.

Key words: constraint decomposition, fixed point, iteration, nonexpansive, regularization, variational inequalities.

1 Introduction

The purpose of *à la* Tikhonov regularization techniques is to approximate, by a suitable perturbation, a possibly Tikhonov ill-posed problem by a family of Tikhonov well-posed problems of same type (minimization, inclusion or fixed point in our case). This is well known for minimization and inclusion ([24,13,2,9,23]). Extensions to fixed point problems has yet been considered ([4–6,1,16,21]). Here we present such a technique with a more general perturbation than the one of the standard Tikhonov regularization. This is done in section 2 in which we prove the main feature, that is, a selection principle. In section 3 we prove that the selection principle remains true for the iteration method suitably combined, by a staircase technique, with regularization, akin to recent results ([7,12,20,14,15,21]. Applications to constraints decomposition in monotone variational inequalities is done in section 4.

2 From variational regularization to fixed point regularization

Let us recall the regularization (or viscosity) method for convex minimization. Let φ be a closed proper convex function on the real Hilbert space H, g be a continuous real valued strongly convex function on H and $r > 0$. The regularized minimization problem for φ is the minimization problem for

$$\varphi_{g,r} := \varphi + rg \ .$$

Tikhonov regularization corresponds to the particular case where g is the square norm. As $\varphi_{g,r}$ is closed proper, strongly convex, it has a unique minimizer u_r. It is known (see for instance [13,2]) that when r tends to 0 then $\{u_r\}$ is minimizing for φ and if $C := \text{Argmin } \varphi$ is nonempty then u_r norm converges to the unique minimizer of g over C. Since u_r is characterized as the unique zero of the maximal and strongly monotone operator $\partial \varphi + r \partial g$ we can extend the regularization method to the inclusion problem as follows.

Let T be a maximal monotone operator on H, R a maximal and strongly monotone operator on H everywhere defined ($D(R) = H$) and $r > 0$. The regularized inclusion problem for T is the inclusion problem for

$$T_{R,r} := T + rR \ .$$

Tikhonov regularization corresponds to $R := I - \hat{u}$ for some $\hat{u} \in H$. As $T_{R,r}$ is maximal and strongly monotone, it has a unique zero u_r.

Let $J_\lambda^T := (I + \lambda T)^{-1}$ be the resolvent of T with positive parameter λ and let us consider the three equivalent problems

$$0 \in Tu \ \Leftrightarrow \ u = J_\lambda^T u \ \Leftrightarrow \ 0 = (I - J_\lambda^T)u$$

with common solution set C. Let $\hat{u} \in H$ and $r > 0$ be given. Let us consider the Tikhonov regularizations respectively of the equation in the right handside above with regularizing parameter r and of the inclusion in the left handside with regularizing parameter $\dfrac{r}{\lambda(1+r)}$:

$$0 = \left(r(I - \hat{u}) + \underbrace{I - J_\lambda^T}\right) v_r, \quad 0 \in \left(\frac{r}{\lambda(1+r)}(I - \hat{u}) + \underbrace{T}\right) w_r \ .$$

A simple calculation shows that these are respectively equivalent to

$$v_r = J_r^R \circ J_\lambda^T \, v_r \ , \quad w_r = J_\lambda^T \circ J_r^R \, w_r \ , \tag{1}$$

where $R := I - \hat{u}$, because in this case, $J_r^R = \dfrac{I}{1+r} + \dfrac{r}{1+r}\hat{u}$. As it is known ([23]), if C is nonempty then, when r tends to 0, v_r and w_r norm converge to $\text{proj}_C \hat{u}$ the projection of $\hat{u}$ onto the (closed convex set) C, i.e. the unique solution to the variational inequality $0 \in (R + N_C)u$, where N_C denotes the normal cone to C.

Now we can extend the regularization method to fixed point problems in a natural way replacing in (1) J_λ^T by a general nonexpansive mapping P and $I - \hat{u}$ by a more general maximal and strongly monotone operator R.

More precisely, let P be a nonexpansive self mapping on H, R a maximal and α-strongly monotone operator on H and $r > 0$. Let us define the external regularized fixed point problem for P as the fixed point problem for

$$P_{R,r}^e := J_r^R \circ P , \qquad (2)$$

and the internal regularized fixed point problem for P as the fixed point problem for

$$P_{R,r}^i := P \circ J_r^R . \qquad (3)$$

As J_r^R is $\dfrac{1}{1 + \alpha r}$-strongly nonexpansive, $P_{R,r}^e$ (resp. $P_{R,r}^i$) is strongly nonexpansive and so has a unique fixed-point v_r (resp. w_r), and both regularized mappings are fixed point Tikhonov well-posed ([16]). It should be noticed that the external regularized problem is nothing but the regularized inclusion problem for the maximal monotone operator $I - P$. Obviously, we have $w_r = Pv_r$ and $v_r = J_r^R w_r$. For $R := I - \hat{u}$ with $\hat{u} \in H$ see [5,6,16,21].

In the following, we show that the selection principle recalled above for convex minimization and Tikhonov regularization for maximal monotone inclusions holds true for the external and internal fixed point regularization methods. Actually this will be obtained as a corollary of a more general result connected with Theorem 1 of [5].

Theorem 1. *Let T be a maximal monotone operator on H, R be an everywhere defined maximal and α-strongly monotone operator on H, u_r be the zero of the regularized operator $T_{R,r} := T + rR$. Let us assume that R is the subdifferential of a continuous real valued strongly convex function on H or is a bounded operator.*

If $C := T^{-1}0$ is nonempty then $\lim_{r \to 0} u_r = u^\star$ the unique solution to the variational inequality

$$0 \in (R + N_C)u . \qquad (4)$$

If C is empty then $\|u_r\| \to +\infty$.

Proof. First we note that, C being a closed convex subset and $D(T) = H$, if C is nonempty then $R + N_C$ is maximal and strongly monotone with nonempty domain C and so $u^\star$ is well defined.

Let $\{r_n\}$ be a sequence of positive reals converging to 0 and let $u_n := u_{r_n}$. The proof is in four parts.

1. $\{u_n\}$ IS BOUNDED.

By definition of u_n, there exists $z_n \in Ru_n$ such that $-rz_n \in Tu_n$. So, thanks to the monotonicity of T, we get

$$\forall u \in C, \quad \langle z_n, u - u_n \rangle \geq 0 . \qquad (5)$$

Let $u \in C$ and $z \in Ru$. Thanks to the strong monotonicity of R, we get

$$\langle z - z_n, u - u_n \rangle \geq \alpha \|u - u_n\|^2 .$$

So, by (5),

$$\forall u \in C, \forall z \in Ru, \ \langle z, u - u_n \rangle \geq \alpha \|u - u_n\|^2 . \tag{6}$$

Therefore,

$$\forall u \in C, \forall z \in Ru, \ \|u_n - u\| \leq \|z\|/\alpha .$$

2. ANY WEAK LIMIT POINT OF $\{u_n\}$ IS IN C.

Let us consider a subsequence $\{u_{n'}\}$ that converges weakly to u_∞. Let $(u, v) \in T$ and $z \in Ru$. Set $v_{n'} := v + r_{n'} z$. Thanks to the monotonicity of $T + r_n R$, we get $\langle v_{n'}, u - u_{n'} \rangle \geq 0$. Passing to the limit as $n' \to +\infty$, we get

$$\forall (u, v) \in T, \ \langle v, u - u_\infty \rangle \geq 0 ,$$

that is, thanks to the maximality of T, $0 \in T(u_\infty)$.

3. $\{u_n\}$ HAS u^* AS UNIQUE WEAK LIMIT POINT.

3.1. Case where R is the subdifferential of a real valued strongly convex function g. So,

$$\forall u \in H, \ g(u) \geq g(u_n) + \langle z_n, u - u_n \rangle .$$

Using (5) we get $\forall u \in C, \ g(u) \geq g(u_n)$. Therefore, thanks to the weak lower semi continuity of g, any weak limit point u_∞ of $\{u_n\}$ satisfies

$$\forall u \in C, \ g(u_\infty) \leq g(u) ,$$

that is $u_\infty = u^*$.

3.2. Case where R is a bounded operator.

Let $z_n \in Ru_n$. $\{z_n\}$ being bounded, it has a weak limit point z_∞. Let us consider a subsequence $\{z_{n'}\}$ that converges weakly to z_∞, and such that $u_{n'}$ converges weakly to some $u_\infty (\in C)$. From (5) we have

$$\langle z_{n'}, u_\infty \rangle \geq \langle z_{n'}, u_{n'} \rangle ,$$

and so, $\limsup_{n' \to +\infty} \langle z_{n'}, u_{n'} \rangle \leq \langle z_\infty, u_\infty \rangle$. Therefore ([1], proposition 2.5), we get

$$z_\infty \in Ru_\infty \text{ and } \langle z_{n'}, u_{n'} \rangle \to \langle z_\infty, u_\infty \rangle .$$

Passing to the limit in (5) for the considered subsequence, we get

$$\forall u \in C, \ \langle z_\infty, u - u_\infty \rangle \leq 0 ,$$

that is, $-z_\infty \in N_C(u_\infty)$, so $u_\infty = u^*$.

4. STRONG CONVERGENCE.
From above, u_n converges weakly to u^*. Strong convergence results from (6).

If C is empty, assuming that $\|u_r\| \not\to +\infty$, some sequence $\{u_{r_n}\}$ with $0 < r_n \to 0$ would be bounded. From step 2 above, $\{u_{r_n}\}$ would have a weak limit point in C, a contradiction. $\qquad \square$

Corollary 1. *Let P, R, v_r (resp. w_r) defined as above. Let us assume that R is the subdifferential of a continuous real valued strongly convex function on H or is an everywhere defined maximal and strongly monotone bounded operator on H.*

If C, the fixed point set of P, is nonempty, then $\lim_{r \to 0} v_r = \lim_{r \to 0} w_r = u^*$ the unique solution to the variational inequality $0 \in (R + N_C) u$.

If C is empty, then $\lim_{r \to 0} \|v_r\| = \lim_{r \to 0} \|w_r\| = +\infty$.

Proof. First we note that, thanks to the nonexpansiveness of P and J_r^R, $\{v_r\}$ and $\{w_r\}$ are simultaneously bounded or unbounded. As noted above, v_r is nothing but the zero of the regularized operator $rR + I - P$, and as $I - P$ is maximal monotone and as $C = (I - P)^{-1}0$, the results follows immediately from Theorem 1 for v_r and then for w_r using the relationship $w_r = Pv_r$ and the nonexpansiveness of P. $\qquad \square$

Remark 1. (i) A more general result in which P is not necessarily defined everywhere and involving a sequence of approximations P_n to P is given in [17].

(ii) When P is the projector onto a given nonempty closed convex subset C (so its fixed point set is C), the external regularization coincides with the partial Yosida approximation of the variational inequality (4): $0 \in (R + (N_C)_r) u$, where, for a maximal monotone operator A, A_r denotes its Yosida approximation $A_r := \dfrac{I - J_r^A}{r}$. As mentioned in ([19]), a direct consequence of ([1], Theorem 2.4) gives the estimate

$$\forall z \in (-Ru^*) \cap N_C u^*, \ \ \|v_r - u^*\| \leq \|z\| \sqrt{r/\alpha} \, .$$

3 Staircase iteration

Let us consider some problem $\mathcal{P}$ with solution set C. The following is a standard situation in Numerical Analysis. There exists some sequence of approximate problems $\mathcal{P}_n$ of same type as $\mathcal{P}$ with a notion of quasi solution $\tilde{u}_n$ controlled by some positive ε_n such that, when ε_n tends to 0, $\tilde{u}_n$ converges in a suitable sense to some solution $u^* \in C$. Moreover, there exists an iterative method adapted to the considered type problem. A usual approach is to apply this iterative method successively to $\mathcal{P}_n$ for $n = 1, 2, ...$, but for each n stopping after a finite number k_n of iterates and restarting with a new approximate problem from the last iterate, giving so a "staircase iterative process":

$$\begin{cases} u_1^0 \text{ given}, \quad n = 1, 2, ... \\ k = 1, ..., k_n, \quad u_n^k = Q_n\, u_n^{k-1}\,; \\ u_{n+1}^0 := u_n^{k_n}\,. \end{cases}$$

where the iteration mapping Q_n is defined by the method.

Now, the main question is to deal with a "good" stopping rule that is, how defining k_n such that $u_n^{k_n}$ be an ε_n-solution of $\mathcal{P}_n$. Of course a possibility is to fix k_n in advance, which amounts to setting $k_n := 1$ for al n (considering k_n copies of $\mathcal{P}_n$), leading to what we can call a "diagonal iterative process". But, this is nonflexible and in general needs more restrictive assumptions than an adaptative choice to be a "good" stopping rule.

The purpose of this section is to show how fixed point regularization allows to design a simple adaptative stopping rule and get convergence results under mild assumptions. Let us first present this stopping rule in a general framework.

Let X be a real Banach space, $\{Q_n\}$ be a sequence of σ_n-strongly nonexpansive self mappings on X, for all n, $\{\delta_n^k\}$ be a sequence of nonnegative reals such that $\lim_{k \to +\infty} \delta_n^k = 0$, $\{\varepsilon_n\}$ be a sequence of positive reals. Let us consider the following staircase iterative process in which each iterate u_n^{k+1} is performed in an approximate way within the tolerance δ_n^k.

$$\begin{cases} u_1^0 \in X, \quad n = 1, 2, ... \\ k = 1, ..., k_n, \quad \|u_n^k - Q_n\, u_n^{k-1}\| \le \delta_n^k\,, \\ k_n := \text{first } k : \ \|u_n^k - u_n^{k-1}\| + \delta_n^k \le \varepsilon_n\,; \\ u_{n+1}^0 := u_n^{k_n}\,. \end{cases}$$

It should be noticed that the stopping rule is effective because, thanks to the strong nonexpansiveness of Q_n and to the Hardy-Cesaro's theorem on convolution of real sequences, $\lim_{k \to +\infty} u_n^k = u_n$ the unique fixed point of Q_n, and so

$$\lim_{k \to +\infty} (\|u_n^k - u_n^{k-1}\| + \delta_n^k) = 0\,.$$

Proposition 1. *(Comparison result). Let $\{u_n^k\}$ be generated by the above staircase process. Then*

(i) $\forall n,\ \|u_n^{k_n} - u_n\| \le \varepsilon_n/(1 - \sigma_n)$.

(ii) $\forall n,\ k = 0, ..., k_n,\ \|u_n^k - u_n\| \le \dfrac{\varepsilon_{n-1}}{1 - \sigma_{n-1}} + \|u_{n-1} - u_n\| + \dfrac{\sup_k \delta_n^k}{1 - \sigma_n}$.

Proof. (i) $\|u_n^{k_n} - u_n\| \le \delta_n^{k_n} + \sigma_n\|u_n^{k_n-1} - u_n\| \le \delta_n^{k_n} + \|u_n^{k_n-1} - u_n^{k_n}\| + \sigma_n\|u_n^{k_n} - u_n\|$.

(ii) $\|u_n^k - u_n\| \le \|u_n^0 - u_n\| + \sup_k \delta_n^k \sum_{i=0}^k \sigma_n^i \le \|u_{n-1}^{k_{n-1}} - u_{n-1}\| + \|u_{n-1} - u_n\| + \sup_k \delta_n^k/(1 - \sigma_n)$. $\qquad\square$

Remark 2. If $\forall n, k_n = 1$, case of diagonal iteration, an analogue result is given in ([15], Proposition 5.1) under more restrictive assumptions.

In the following we specialize the basic iterative method (i.e. defined by Q_n) by means of fixed point regularization. Let H be a real Hilbert space, P be a nonexpansive self mapping on H, $\{r_n\}$ be a sequence of positive reals. Let us define Q_n as follows:

$$\forall n, \quad Q_n := \begin{cases} J^R_{r_n} \circ P \\ \text{or} \\ P \circ J^R_{r_n} \end{cases}$$

where R satifies assumptions of Corollary 1. We have

$$u_n := v_n \ \text{ or } \ w_n, \quad \sigma_n = 1/(1 + \alpha\, r_n)\,.$$

As the sequence $\{u_n\}$ is a mixing of the two sequences of regularized fixed points, in the context of Corollary 1, it converges strongly to $u^\star$ the unique solution to the variational inequality (4). Therefore, the following proposition is a direct consequence of Corollary 1 and Proposition 1.

Proposition 2. *(i)* $\varepsilon_n/r_n \to 0 \ \Rightarrow\ u_n^{k_n} \to u^\star$.
(ii) $(\varepsilon_n + \sup_k \delta_n^k)/r_n \to 0 \ \Rightarrow\ u_n^k \to u^\star$.

Remark 3. If $\forall n, k_n = 1$, case of diagonal iteration, $u_n^1 \to u^\star$ holds true under more restrictive assumptions: $\{r_n\} \notin \ell^1$, $|\frac{1}{r_n} - \frac{1}{r_{n-1}}| \to 0$ (with standard Tikhonov fixed point regularization see [16,21], see also [10] for a similar result). If moreover $P = \text{proj}_C$, our diagonal iterative process coincides with the Passty's scheme based upon partial Yosida approximation of the variational inequality (4). In this case, the same result holds true under the assumption $\{r_n\} \in \ell^2\backslash\ell^1$ ([22], Theorem 5).

4   Constraint decomposition in variational inequalities

In this section, we adopt the opposite viewpoint. Now the data are a real Hilbert space H, a nonempty closed convex subset C of H, an everywhere defined maximal and α-strongly monotone operator R on H and we look for the unique solution $u^\star$ to the variational inequality

$$0 \in (R + N_C)u\,. \tag{7}$$

The goal is to express C as the fixed point set of a suitable nonexpansive self mapping P on H in order to apply the results of the previous sections. We consider the standard situation where C is defined as the intersection of finitely many closed convex subsets: $C := \cap_{i=1}^m C_i$. Let us assume that each

C_i is the fixed point set of some firmly nonexpansive self mapping P_i on H, i.e.

$$C_i := \{u \in H \ ; \ u = P_i u\} ,$$

with

$$\forall u, v \in H, \ \|P_i u - P_i v\|^2 \leq \|u - v\|^2 - \|(I - P_i)u - (I - P_i)v\|^2 .$$

For example, If C_i is "simple" (a box if H is finite dimensional, or a half-space) we can take as P_i the projector onto C_i, in which case the application of P_i can be explicitly performed; if C_i is described by a convex inequality constraint $C_i := \{u \in H \ ; \ f_i(u) \leq 0\}$, with f_i a closed proper convex function on H, we can take $P_i := J_\lambda^{g_i}$ with $\lambda > 0$ and $g_i := f_i^+$ or $g_i := (f_i^+)^2$, in which case we can use cutting plane techniques like in bundle methods ([8]).

The previous discussion is motivated by the following results.

Proposition 3. *([18,11]). Let us assume that C is the nonempty intersection of finitely many fixed point sets of firmly nonexpansive self mappings P_i on H. Then C is the fixed point set of the nonexpansive self mapping P on H where*

$$P := P_m \circ ... \circ P_1 \quad or \quad P := \sum_{i=1}^{m} \theta_i P_i, \quad \theta_i > 0, \quad \sum_{i=1}^{m} \theta_i = 1 .$$

Applying Corollary 1 and Proposition 3 gives immediately the following selection principle.

Corollary 2. *$R, C, P_i, i = 1, ..., m$ being as above in this section, let us consider the four families of regularized fixed points defined by*

$$v_r^1 = J_r^R \circ P_m \circ ... \circ P_1 \, v_r^1, \quad w_r^1 = P_m \circ ... \circ P_1 \circ J_r^R \, w_r^1 ,$$

$$v_r^2 = J_r^R \circ (\sum_{i=1}^{m} \theta_i P_i) \, v_r^2, \quad w_r^2 = (\sum_{i=1}^{m} \theta_i P_i) \circ J_r^R \, w_r^2 .$$

If R is bounded or is the subdifferential of a continuous real valued strongly convex function on H, then

$$\lim_{r \to 0} v_r^k = \lim_{r \to 0} w_r^k = u^*, \quad k = 1, 2 .$$

Remark 4. There is no need of a qualification condition contrarily to decomposition based on partial Yosida approximation of (7) using the standard product space setup and leading to the fixed point problem for

$$\frac{1}{m} \sum_{i=1}^{m} J_r^{R + \partial \psi_{C_i}} \quad or \quad \frac{1}{m+1} (J_r^R + \sum_{i=1}^{m} proj_{C_i}) .$$

Applying the general staircase process of section 3 leads to the following iterative decomposition methods.

(Sequential decomposition):

let $\varepsilon_n > 0$, $0 < r_n \to 0$, $\dfrac{\varepsilon_n}{r_n} \to 0$, $\forall n \in I\!N\backslash\{0\}$, $i = 0, ..., m+1$, $0 \le \delta_n^{i,k} \xrightarrow{k} 0$.

$$
\begin{cases}
u_1^0 \in H, \quad n = 1, 2, ... \\
k = 1, ..., k_n, \quad x(0) := u_n^{k-1} \quad [\|x(0) - J_{r_n}^R u_n^{k-1}\| \le \delta_n^{0,k}], \\
i = 1, ..., m, \quad \|x(i) - P_i\, x(i-1)\| \le \delta_n^{i,k} ; \\
\|u_n^k - J_{r_n}^R x(m)\| \le \delta_n^{m+1,k} \quad [u_n^k := x(m)], \\
k_n := \text{first } k : \|u_n^k - u_n^{k-1}\| + \sum_{i=1[0]}^{m+1[m]} \delta_n^{i,k} \le \varepsilon_n ; \\
u_{n+1}^0 := u_n^{k_n}.
\end{cases}
$$

(Parallel-sequential decomposition):

let $\varepsilon_n > 0$, $0 < r_n \to 0$, $\dfrac{\varepsilon_n}{r_n} \to 0$, $\forall n \in I\!N\backslash\{0\}$, $i = 0, ..., m+1$, $0 \le \delta_n^{i,k} \xrightarrow{k} 0$, $\theta_i > 0$, $\sum_{i=1}^m \theta_i = 1$, $\theta_0 := \theta_{m+1} := 1$.

$$
\begin{cases}
u_1^0 \in H, \quad n = 1, 2, ... \\
k = 1, ..., k_n, \quad x(0) := u_n^{k-1} \quad [\|x(0) - J_{r_n}^R u_n^{k-1}\| \le \delta_n^{0,k}], \\
i = 1, ..., m, \quad \|x(i) - P_i\, u_n^k\| \le \delta_n^{i,k}, \quad \tilde{x} := \sum_{i=1}^m \theta_i\, x(i) ; \\
\|u_n^k - J_{r_n}^R \tilde{x}\| \le \delta_n^{m+1,k} \quad [u_n^k := \tilde{x}], \\
k_n := \text{first } k : \|u_n^k - u_n^{k-1}\| + \sum_{i=1[0]}^{m+1[m]} \theta_i\, \delta_n^{i,k} \le \varepsilon_n ; \\
u_{n+1}^0 := u_n^{k_n}.
\end{cases}
$$

In both cases, for each n, the terms in [ ] correspond to internal regularization of P, while otherwise external regularization is involved. Moreover, due to composition of nonexpansive mappings, the tolerance δ_n^k in the general scheme is given here by $\sum_{i=1[0]}^{m+1[m]} \delta_n^{i,k}$, explaining this sum in the stopping rule. Applying Proposition 2 we get $u_n^{k_n} \to u^*$.

References

1. Alber, Ya. I., Guerre-Delabrire, S. (1994) Problems of fixed point theory in Hilbert and Banach spaces. Funct. Diff. Equations. Proceedings of the Israel Seminar **2**, 5–10
2. Attouch, H. (1996) Viscosity solutions of minimization problems. SIAM J. Optimization **6**, 3, 769–806
3. Brezis, H. (1973) Opérateurs maximaux monotones et semi-groupes de contractions dans les espaces de Hilbert. Mathematics Studies, **5**, North Holland Publishing Company, Amsterdam
4. Browder, F.E. (1966) Fixed point theorems for nonlinear semicontractive mappings in Banach spaces. Arch. for Rational Mech. and Anal. **21**, 259–269

5. Browder, F.E. (1966) Existence and approximation of solutions of nonlinear variational inequalities. Proc. Nat. Acad. Sci. U.S.A. **56**, 1080–1086

6. Browder, F.E. (1967) Convergence of approximants to fixed points of nonexpansive nonlinear mappings in Banach spaces. Arch. for Rational Mech. and Anal. **24**, 82–90

7. Cominetti, R. Coupling the proximal point algorithm with approximation methods. To appear in JOTA

8. Correa, R., Lemaréchal, C. (1993) Convergence of some algorithms for convex minimization. Math. Programming **62**, 261–276

9. Kaplan, A., Tichatschke, R. (1994) Stable methods for ill-posed problems. Akademie Verlag, Berlin

10. Halpern, B. (1967) Fixed points of nonexpanding maps. Bull. AMS **73**, 957–961

11. Lehdili, N., Lemaire, B. (1999) The barycentric proximal method. To appear in Communications in Applied Nonlinear Analysis **6**, 2

12. Lehdili N., Moudafi, A. (1996) Combining the proximal algorithm and Tikhonov regularization. Optimization **37**, 3, 239–252

13. Lemaire, B. (1971) Régularisation et pénalisation en optimisation convexe. Séminaire d'Analyse Convexe **1**, 17, Université Montpellier II

14. Lemaire, B. (1995) On the convergence of some iterative methods for convex minimization. In: Durier R., Michelot Ch. (Eds.) Recent Developments in Optimization, Lecture Notes in Economics and Mathematical Systems, **429**, Springer-Verlag, Berlin Heidelberg, 252–268

15. Lemaire, B. (1997) Which fixed-point does the iteration method select? In: Gritzmann P., Horst R., Sachs E., Tichatschke R. (Eds.) Recent Advances in Optimization, Lecture Notes in Economics and Mathematical Systems **452**, Springer, Berlin Heidelberg, 154–167

16. Lemaire, B. (1999) Well-posedness, conditioning and regularization of minimization, inclusion and fixed point problems. To appear in Pliska Studia Mathematica Bulgarian Journal

17. Lemaire, B. (1999) Regularization of fixed point problems and staircase iteration. To appear in the proceedings of the workshop on ill-posed variational problems and regularization techniques, Trier, 1998

18. Martinet, B. (1972) Algorithmes pour la résolution de problèmes d'optimisation et de minimax. Thèse de Doctorat d'Etat, Université de Grenoble

19. Mahey, P., Pham Dinh Tao (1995) Partial regularization of the sum of two maximal monotone operators. Math. Model. and Num. Anal. (M^2AN) **92**, 375–392

20. Moudafi, A. (1994) Coupling proximal algorithm and Tikhonov method. Nonlinear Times and Digest **1**, 203–210

21. Ould Ahmed Salem, C. (1998) Approximation de points fixes d'une contraction. Thèse de Doctorat, Université Montpellier II

22. Passty, G.B. (1979) Ergodic convergence to a zero of the sum of monotone operators in Hilbert spaces. J. of Math. Anal. and Appl. **72**, 383–390

23. Tossings, P. (1994) The perturbed Tikhonov's algorithm and some of its applications. Math. Model. and Num. Anal. (M^2AN) **28**, 189–221

24. Tikhonov, A., Arsenine, V. (1974) Méthodes de résolution de problèmes mal posés. MIR, Moscow

Large Scale Molecular Conformation via the Exact Distance Geometry Problem

Le Thi Hoai An and Pham Dinh Tao

Mathematical Modelling and Applied Optimization Group,
Laboratory of Mathematics - CNRS UPRES-A 60 85,
National Institute for Applied Sciences-Rouen
BP 8, F-76 131 Mont Saint Aignan Cedex, France.

Abstract. We develop in this paper a method based on a d.c. (difference of convex functions) optimization approach called DCA for solving large-scale exact distance geometry problem. Requiring only matrix-vector products and one Cholesky factorization, the DCA seems to be robust and efficient in large scale problems. Moreover it allows exploiting sparsity of the given distance matrix. A technique using the triangle inequality to generate a complete approximate distance matrix was investigated in order to compute a good starting point for the DCA. Numerical simulations of the molecular conformation problems with up to 12288 variables are reported which prove the robustness, the efficiency, and the globality of our algorithms.

Key words: d.c. programming, d.c. algorithm (DCA), exact distance geometry problem, molecular conformation

1 Introduction

In this paper we are interested in the large-scale molecular conformation from the distance geometry problem which can be formulated as follows: find positions of n atoms $x^1, \ldots, x^n$ in $\mathbb{R}^3$ such that

$$\|x^i - x^j\| = \delta_{ij}, \ (i, j) \in \mathcal{S}, \tag{1}$$

where $\mathcal{S}$ is a subset of the atom pairs, δ_{ij} with $(i, j) \in \mathcal{S}$ is the given distance between atoms i and j, and $\|\cdot\|$ denotes the Euclidean norm. Usually a small subset of pairwise distances is known, i.e., $\mathcal{S}$ is sparse.

The above formulation corresponds to the exact distance geometry problem. By the error in the theoretical or experimental data, there may not exist any solution to this problem, for example, when the triangle inequality

$$\delta_{ij} \le \delta_{ik} + \delta_{kj}$$

is violated for atoms i, j, k. Then an ε-optimal solution of (1), namely a configuration $x^1, \ldots, x^n$ satysfying

$$\mid \|x^i - x^j\| - \delta_{ij} \mid \le \varepsilon, \ (i, j) \in \mathcal{S}, \tag{2}$$

is useful in practice.

A more general case, called the general distance geometry problem, is to find positions $x^1, \ldots, x^n$ in $\mathbb{R}^3$ verifying

$$l_{ij} \leq \|x^i - x^j\| \leq u_{ij}, \ (i,j) \in \mathcal{S}, \tag{3}$$

where l_{ij} and u_{ij} are lower and upper bounds on the distance constraints, respectively.

In all the sequel $\mathcal{M}_{n,p}(\mathbb{R})$ denotes the space of real matrices of order $n \times p$ and for $X \in \mathcal{M}_{n,p}(\mathbb{R})$, X_i (resp. X^i) is its ith row (resp. i^{th} column). By identifying a set of positions $x^1, \ldots, x^n$ with the matrix X (i.e. $(X^T)^i = (X_i)^T = x^i$ for i = 1,..., n), we can advantageously express the exact and/or general distance geometry problems in the matrix space $\mathcal{M}_{n,p}(\mathbb{R})$:

$$\text{(EDP)} \qquad \min\left\{\sigma(X) := \frac{1}{2} \sum_{(i,j)\in S, i<j} w_{ij}\theta_{ij}(X) : \quad X \in \mathcal{M}_{n,p}(\mathbb{R})\right\},$$

where $w_{ij} > 0$ for $i \neq j$ and the pairwise potential $\theta_{ij} : \mathcal{M}_{n,p}(\mathbb{R}) \longrightarrow \mathbb{R}$ is defined for problem (1) by either

$$\theta_{ij}(X) = \left(\delta_{i,j}^2 - \|X_i^T - X_j^T\|^2\right)^2 \tag{4}$$

or

$$\theta_{ij}(X) = \left(\delta_{ij} - \|X_i^T - X_j^T\|\right)^2, \tag{5}$$

and for problem (2) by

$$\theta_{ij}(X) = \min^2\left\{\frac{\|X_i^T - X_j^T\|^2 - l_{ij}^2}{l_{ij}^2}, 0\right\} + \max^2\left\{\frac{\|X_i^T - X_j^T\|^2 - u_{ij}^2}{u_{ij}^2}, 0\right\}. \tag{6}$$

Note that for simplicity $(X_j)^T$ is written through the paper as X_j^T. In the molecular optimization problem p is equal to 3. It is easy to see that except for θ_{ij} defined by (4), (EDP) with θ_{ij} given by (5) and (6) are nondifferentiable optimization problems. However they are all d.c. programs.

Observe that X is a solution of the distance geometry problem if and only if it is a global minimizer of Problem (EDP) and $\sigma(X) = 0$.

When all pairwise distances are available and a solution exists, the exact distance geometry problem (1) can be solved by a polynomial time algorithm (Blumenthal [1], Crippen and Havel [2]). However, in practice, one knows only a subset of the distances, and it is well known (Saxe [17]) that $p-$ dimensional distance geometry problems are strongly NP-complete with $p = 1$ and strongly NP-hard for all $p > 1$.

Several methods have been proposed for solving the distance geometry problems (1) and/or (2). For a short survey and a comprehensive collection of references on this topic, we refer to recent works [7], [10].

We present in this paper an efficient method to solve the exact distance geometry problem (EDP) with θ defined in (5) because this formulation seems to be advantageous to DCA. It is interesting to note that most existing algorithms (see e.g. [6], [9], [10]) have been developed to solve (EDP) with θ defined by (4) rather than (5), since one is faced with a smooth optimization problem in such a case. Our method is based on a d.c. optimization approach called DCA. The DCA was first introduced by Pham Dinh Tao [11], [12] and later developed in our work, see, e.g. Le Thi H. An [7], Le Thi H. An and Pham Dinh Tao [8], Pham Dinh Tao and Le Thi H. An [14], [15] and references therein. The DCA is a primal-dual subdifferential method for solving a general d.c. program of the form

$$(P_{dc}) \qquad \alpha = \inf\{f(x) := g(x) - h(x) : \quad x \in E\},$$

with g, $h \in \Gamma_o(E)$, and its dual given by

$$(D_{dc}) \qquad \alpha = \inf\{h^*(y) - g^*(y) : \quad y \in E^*\}.$$

Here $\Gamma_o(E)$ denotes the set of all proper lower semi-continuous convex functions on $E = \mathbb{R}^n$ which is equipped with the canonical inner product $\langle \cdot, \cdot \rangle$, E^* is the dual space of E which can be identified with E itself, g^* is the conjugate function belonging to $\Gamma_o(E^*)$ and is defined by

$$g^*(y) = \sup\{\langle x, y \rangle - g(x) : x \in \mathbb{R}^n\}.$$

One says that $g - h$ is a d.c. decomposition (or d.c. representation) of f, and g, h are its d.c. components.

Based on the d.c. duality and the local optimality, the DCA consists in the construction of two sequences $\{x^k\}$ and $\{y^k\}$ such that x^{k+1} (resp. y^k) is a solution to the convex program (P_k) (resp. (D_k)) defined by

$$(P_k) \qquad \inf\{g(x) - [h(x^k) + \langle x - x^k, y^k \rangle] : \quad x \in \mathbb{R}^n\},$$

$$(D_k) \qquad \inf\{h^*(y) - [g^*(y^{k-1}) + \langle x^k, y - y^{k-1} \rangle] : \quad y \in \mathbb{R}^n\}.$$

In view of the relation: (P_k) (resp. (D_k)) is obtained from (P_{dc}) (resp. (D_{dc})) by replacing h (resp. g^*) with its affine minorization defined by $y^k \in \partial h(x^k)$ (resp. $x^k \in \partial g^*(y^{k-1})$), the DCA yields the next scheme:

$$y^k \in \partial h(x^k); \quad x^{k+1} \in \partial g^*(y^k). \tag{7}$$

It is proved in Pham Dinh Tao [12], Pham Dinh Tao and Le Thi H. An [14], [15] that

(i) The sequences $\{g(x^k) - h(x^k)\}$ and $\{h^*(y^k) - g^*(y^k)\}$ are decreasing and

- $g(x^{k+1}) - h(x^{k+1}) = g(x^k) - h(x^k)$ if and only if $y^k \in \partial g(x^k) \cap \partial h(x^k)$, $y^k \in \partial g(x^{k+1}) \cap \partial h(x^{k+1})$ and $[\rho(g) + \rho(h)]\|x^{k+1} - x^k\| = 0$.
- $h^*(y^{k+1}) - g^*(y^{k+1}) = h^*(y^k) - g^*(y^k)$ if and only if $x^{k+1} \in \partial g^*(y^k) \cap \partial h^*(y^k)$, $x^{k+1} \in \partial g^*(y^{k+1}) \cap \partial h^*(y^{k+1})$ and $[\rho(g^*) + \rho(h^*)]\|y^{k+1} - y^k\| = 0$.

($\rho(g)$ denotes the modulus of strong convexity of g). In such a case the DCA terminates at the kth iteration.

(ii) Every limit point x^* (resp. y^*) of the sequence $\{x^k\}$ (resp. $\{y^k\}$) is a critical point of $g - h$ (resp. $h^* - g^*$), say $\partial g(x^*) \cap \partial h(x^*) \neq \emptyset$ (resp. $\partial h^*(y^*) \cap \partial g^*(y^*) \neq \emptyset$).

At the present time the DCA is one of few algorithms in the local approach which has been successfully applied to many large-scale d.c. optimization problems and proved to be more robust and efficient than related standard methods. Due to its local character it cannot guarantee the globality of computed solutions for general d.c. programs. However, we observe that with a suitable starting point it converges quite often to a global one (see e.g. [7], [8], [14] and references therein). This property motivates us to investigate a technique for computing a "good" starting point for the DCA in the solution of (EDP). Our main algorithm is then composed of two phases. In the first phase we complete the matrix of distances by using the triangle inequality and then apply the DCA to the new problem where all pairwise "distances" are known. In the second phase we solve the original problem (EDP) by applying the DCA from the point obtained by Phase 1. We work then with both (*dense* and *sparse*) sets of constraints. In contrast, the existing methods work only on either a full set of constraints (see e.g. [2]) or a sparse set of constraints ([10], [19]). As in the embed algorithm ([2]) which serves to solve (2), we use the triangle inequality to compute a complete estimating distance matrix. But the main difference between the two approaches is that we do generate only one complete "distance" matrix while the embed algorithm may require many trial choices of the distance matrix before a solution of (1) is obtained. Another advantage of our method is that we can exploit sparsity of the given distance matrix. This is important because only a small subset of constraints is known in practice.

Our algorithms are quite simple and easy to implement. They only require matrix-vector products and solving a linear system with constant symmetric positive definite matrix. We have tested our codes on the artificial distance geometry problems (Moré & Wu [9]) with up to 12288 variables (the molecule contains 4096 atoms).

2 The main algorithm for solving the exact distance geometry problem

We develop in this section our main algorithm based on the DCA for solving Problem (1) with θ_{ij} defined in (5), i.e.

$$(EDP_1) \quad \min \sigma(X) := \frac{1}{2} \sum_{(i,j)\in\mathcal{S},i<j} w_{ij} \left(\|X_i^T - X_j^T\| - \delta_{ij} \right)^2 : X \in \mathcal{M}_{n,p}(\mathbb{R})$$

(recall that the molecular problem correspond to case $p = 3$). By identifying an $n \times p$ matrix X with a $p.n$-vector, in what follow, we use either $\mathcal{M}_{n,p}(\mathbb{R})$ or $\mathbb{R}^{p.n}$ for indicating the same notation. We can identify by rows (resp. columns) each matrix $X \in \mathcal{M}_{n,p}(\mathbb{R})$ with a row-vector (resp. column-vector) in $(\mathbb{R}^p)^n$ (resp. $(\mathbb{R}^n)^p$) by writing respectively

$$X \longleftrightarrow \mathcal{X} = (X_1,\dots,X_n),\ X_i^T \in \mathbb{R}^p,\ \mathcal{X} \in (\mathbb{R}^p)^n, \tag{8}$$

and

$$X \longleftrightarrow \overline{\mathcal{X}} \in (\mathbb{R}^n)^p : \overline{\mathcal{X}} = \begin{pmatrix} X^1 \\ \ddots \\ X^p \end{pmatrix},\ X^j \in \mathbb{R}^n,\ \overline{\mathcal{X}} \in (\mathbb{R}^n)^p. \tag{9}$$

The inner product in $\mathcal{M}_{n,p}(\mathbb{R})$ is defined as the inner product in $(\mathbb{R}^p)^n$ or $(\mathbb{R}^n)^p$. In the sequel, for simplicity, we shall suppress, if no possible ambiguity, the indices for the inner product and denote by $\|.\|$ the corresponding Euclidean norm on $\mathcal{M}_{n,p}(\mathbb{R})$. Evidently we must choose either representation in a convenient way.

We define the weight matrix $W = (w_{ij})$ of (EDP_1) by taking the value zero for w_{ij} when $(i,j) \notin \mathcal{S}$. Thourought this paper, we assume that W is irreducible which can be viewed as the associated graph $G(N,\mathcal{S})$ with $N = \{1,...,n\}$ is connected. This assumption is not restrictive for Problem (EDP) since it can be decomposed into a number of smaller problems otherwise. Then we work under the next assumptions for the two symmetric matrices $\Delta = (\delta_{ij})$ (the distance matrix) and $W = (w_{ij})$:

(a1) for $i \neq j$, $\delta_{ij} > 0$ when $(i,j) \in \mathcal{S}$ (i.e., two different atoms are not in the same position),

(a2) for $i \neq j$, $w_{ij} = 0$ if and only if δ_{ij} is unknown, say $(i,j) \notin \mathcal{S}$,

(a3) the weight matrix W is irreducible.

Remark that if we set $\delta_{ij} = 0$ for $(i,j) \notin \mathcal{S}$, then $G(N,\mathcal{S})$ is the graph associated with the distance matrix Δ too.

The case where $w_{ij} = c$ (c is a given positive number) for all $i \neq j$ is called *the normal case*. Clearly that this case can occur if and only if the distance matrix Δ is completely defined, say all pairwise distances are known.

We first formulate Problem (EDP$_1$) in the form of d.c. program. The objective function of (EDP$_1$) can be written as $\sigma(X) =$

$$\frac{1}{2} \sum_{(i,j)\in\mathcal{S},\, i<j} w_{ij} d_{ij}^2(X) - \sum_{(i,j)\in\mathcal{S},\, i<j} w_{ij}\delta_{ij}d_{ij}(X) + \frac{1}{2} \sum_{(i,j)\in\mathcal{S},\, i<j} w_{ij}\delta_{ij}^2, \tag{10}$$

with $d_{ij}(X) = \|X_i^T - X_j^T\|$.

Under assumption (a$_2$), althought δ_{ij} is unknown for any $(i,j) \notin \mathcal{S}$, in (10) the summation over pairs $(i,j) \in \mathcal{S}$ can be extended to that over all pairs (i,j). Then (EDP$_1$) is equivalent to the following problem:

$$\min\left\{F(X) := \frac{1}{2}\eta^2(X) - \xi(X): \quad X \in \mathcal{M}_{n,p}(\mathbb{R})\right\}, \tag{11}$$

where η and ξ are the functions defined on $\mathcal{M}_{n,p}(\mathbb{R})$ by

$$\eta(X) = \left[\sum_{i<j} w_{ij}d_{ij}^2(X)\right]^{1/2}, \quad \xi(X) = \sum_{i<j} w_{ij}\delta_{ij}d_{ij}(X). \tag{12}$$

It is not difficult to verify that η and ξ are two seminorms in $\mathcal{M}_{n,p}(\mathbb{R})$ and then (11) is a d.c. program to which the DCA can be applied.

Performing the scheme DCA for solving (11) is reduced to calculating subdifferentials of the functions ξ and $((1/2)\eta^2)^*$.

2.1 Solving (11) by the DCA

Under assumption (a2) and (a3), we can restrict the working space to an appropriate set which is, as will be seen in the next, favourable to our calculations. Indeed, let $\mathcal{A}$ denote the set of matrices in $\mathcal{M}_{n,p}(\mathbb{R})$ whose rows are identical, i.e.,

$$\mathcal{A} := \{X \in \mathcal{M}_{n,p}(\mathbb{R}): X_1 = \cdots = X_n\}$$

and let $Proj_\mathcal{A}$ be the orthogonal projection on $\mathcal{A}$, we have

Lemma 1. *(i)* $\mathcal{A}^\perp = \{Y \in \mathcal{M}_{n,p}(\mathbb{R}): \sum_{i=1}^n Y_i = 0\}$.
(ii) $\mathcal{A} \subset \xi^{-1}(0); \quad \mathcal{A} \subset \eta^{-1}(0)$.
(iii) $Proj_\mathcal{A} = (1/n)ee^T; \quad Proj_{\mathcal{A}^\perp} = I - (1/n)ee^T$ *(e is the vector of ones in $\mathbb{R}^n$).*
(iv) *If the weight matrix W is irreducible, (resp. W is irreducible and $w_{ij}\delta_{ij} > 0$ whenever $w_{ij} > 0$), then $\mathcal{A} = \eta^{-1}(0)$ (resp. $\mathcal{A} = \xi^{-1}(0)$).*
(v) *If the conditions (a2) and (a3) hold, then Problems (11) is equivalent to*

$$\min\left\{\frac{1}{2}\eta^2(X) - \xi(X): \quad X \in \mathcal{A}^\perp\right\}, \tag{13}$$

in the sense that X^* is an optimal solution of (13) if and only if $X^* + Z$ is an optimal solution of (11) for all $Z \in \mathcal{A}$.

Proof. (i) and (ii) are straightforward from the definition of $\mathcal{A}$. The proof of (iii) is easy.

Let $X \in \mathcal{M}_{n,p}(\mathbb{R})$ such that $\eta(X) = 0$ and $(i,j) \in \{1,\ldots,n\}^2$ with $i \neq j$. Since the matrix W is irreducible there is a finite sequence $\{i_1,\ldots,i_r\} \subset \{1,\ldots,n\}$ verifying $w_{ii_1} > 0, w_{i_k i_{k+1}} > 0$ for $k = 1,\ldots,r-1$, and $w_{i_r j} > 0$. It follows that $X_i = X_{i_1} = \cdots = X_{i_r} = X_j$. Hence $\mathcal{A} = \eta^{-1}(0)$. By the same reasoning we get the rest of (iv).

Under assumtions (a2) and (a3) we have, in virtue of (iv), $\eta^{-1}(0) = \xi^{-1}(0) = \mathcal{A}$. Since $E = \mathcal{A} + \mathcal{A}^{\perp}$, there is, for every $T \in \mathcal{M}_{n,p}(\mathbb{R})$, one and only one pair $(X, U) \in \mathcal{A}^{\perp} \times \mathcal{A}$ such that $T = X + U$. So

$$\eta(T) = \eta(X + U) \leq \eta(X) + \eta(U) = \eta(X),$$

$$\xi(T) = \xi(X + U) \leq \xi(X) + \xi(U) = \xi(X).$$

On the other hand we can write

$$\eta(X) = \eta(T - U) \leq \eta(T) + \eta(-U) = \eta(T),$$

$$\xi(X) = \xi(T - U) \leq \xi(T) + \xi(-U) = \xi(T).$$

Consequently $\eta(T) = \eta(X + U) = \eta(X)$ and $\xi(T) = \xi(X + U) = \xi(X)$. The proof is then complete. $\qquad\square$

Subdifferential of ξ By the definition of ξ, say

$$\xi(X) = \sum_{i<j} w_{ij} \delta_{ij} d_{ij}(X),$$

we have

$$\partial \xi(X) = \sum_{i<j} w_{ij} \delta_{ij} \partial d_{ij}(X).$$

d_{ij} can be expressed as (using the row-representation of $X \in \mathcal{M}_{n,p}(\mathbb{R})$):

$$d_{ij} = \|\cdot\| \circ L_{ij} : (\mathbb{R}^p)^n \longrightarrow \mathbb{R}^p \longrightarrow \mathbb{R}$$

$$X \longmapsto L_{ij}(X) = X_i^T - X_j^T \longmapsto \|X_i^T - X_j^T\|.$$

It follows that ([32])

$$\partial d_{ij}(X) = L_{ij}^T \partial (\|\cdot\|)(L_{ij}(X)).$$

Hence

$$Y(i,j) \in \partial d_{ij}(X) \Leftrightarrow Y(i,j) = L_{ij}^T y, \quad y \in \partial(\|\cdot\|)(X_i^T - X_j^T)$$

which implies

$$Y(i,j)_k = 0 \text{ if } k \notin \{i,j\} \text{ and } Y(i,j)_i^T = -Y(i,j)_j^T \in \partial(\|\cdot\|)(X_i^T - X_j^T).$$

$$\tag{14}$$

So ξ is not differentiable on $\Omega = \{X \in \mathcal{M}_{n,p}(\mathbb{R}) : X$ has two identical rows $\}$, but on the complement of Ω in $\mathcal{M}_{n,p}(\mathbb{R})$. We can choose the subgradient $Y(i,j) \in \partial d_{ij}(X)$ defined by

$$Y(i,j)_i = -Y(i,j)_j = \begin{cases} \frac{X_i - X_j}{\|X_i^T - X_j^T\|} & \text{if } X_i \neq X_j, \\ 0 & \text{if } X_i = X_j. \end{cases} \tag{15}$$

In this case, each $Y \in \partial \xi(X)$ is called special subgradient of $\xi(X)$ and defined explicitly by

$$\begin{aligned} Y_k &= \sum_{i<j} w_{ij} \delta_{ij} Y(i,j)_k = \sum_{i<k} w_{ik} \delta_{ik} Y(i,k)_k + \sum_{j>k} w_{kj} \delta_{kj} Y(k,j)_k \\ &= \sum_{i<k} w_{ki} \delta_{ki} s_{ki}(X)(X_k - X_i) + \sum_{j>k} w_{kj} \delta_{kj} s_{kj}(X)(X_k - X_j) \\ &= \left[\sum_{i=1}^{n} w_{ki} \delta_{ki} s_{ki}(X) \right] X_k - \sum_{i=1}^{n} w_{ki} \delta_{ki} s_{ki}(X) X_i, \end{aligned}$$

where

$$s_{ij}(X) = \begin{cases} 1/(\|X_i^T - X_j^T\|) & \text{if } X_i \neq X_j, \\ 0 & \text{otherwise.} \end{cases}$$

Let $B(X) = (b_{ij}(X))$ be the $n \times n$ matrix given as

$$b_{ij}(X) = \begin{cases} -w_{ij} \delta_{ij} s_{ij}(X) & \text{if } i \neq j, \\ -\sum_{k=1, k\neq i}^{n} b_{ik} & \text{if } i = j. \end{cases} \tag{16}$$

It follows that $Y = B(X)X$.

In all the sequel $\mathcal{N}(A)$ and $\text{Im}A$ denote the null space and the range of the matrix A respectively.

Proposition 1. Let $B(X)$ be the matrix defined by (16). Then

(i) $\mathcal{N}(B(X)) \supset \mathcal{A}$, $\text{Im}\,(B(X)) \subset \mathcal{A}^\perp$ for all $X \in \mathcal{M}_{n,p}(\mathbb{R})$.
(ii) $B(X)$ is symmetric positive semidefinite for all $X \in \mathcal{M}_{n,p}(\mathbb{R})$.
(iii) $\xi(X) = \langle X, B(X)X \rangle$.

Proof. (i) is immediate from the fact that $\mathcal{A} = \{ev^T : v \in \mathbb{R}^p\}$ and $B(X)e = 0$ for all $X \in \mathcal{M}_{n,p}(\mathbb{R})$.
(ii) $B(X)$ is symmetric, diagonally dominant whose diagonal entries are non-negative. Thus it is positive semidefinite ([18]).
(iii) is the generalized Euler's relation for functions convex nondifferentiable and positively homogeneous of degree 1 ([32]).   □

Subdifferential of $((1/2)\eta^2)^*$ First we state some fundamental properties of the function $(1/2)\eta^2$. From the definition of η, say

$$\eta^2(X) = \sum_{i<j} w_{ij}\|X^i - X^j\|^2 = \sum_{i<j} w_{ij}d_{ij}^2(X),$$

we have

$$\partial\left(\frac{1}{2}\eta^2\right)(X) = \sum_{i<j} w_{ij}d_{ij}(X)\partial d_{ij}(X).$$

Thus

$$Y \in \partial\left(\frac{1}{2}\eta^2\right)(X) \Leftrightarrow Y = \sum_{i<j} w_{ij}d_{ij}(X)Y(i,j)$$

with $Y(i,j)$ defined by (14). It follows that η^2 is differentiable on $\mathcal{M}_{n,p}(\mathbb{R})$ and $Y = \nabla(\frac{1}{2}\eta^2)(X)$ is defined as

$$Y_k = \sum_{i<k} w_{ki}(X_k - X_i) + \sum_{j>k} w_{kj}(X_k - X_j) = \left(\sum_{i=1}^n w_{ki}\right)X_k - \sum_{i=1}^n w_{ki}X_i.$$

Hence $Y = VX$ where $V = (v_{ij})$ given by

$$v_{ij} = \begin{cases} -w_{ij} & \text{if } i \neq j, \\ \sum_{k=1}^n w_{ik} & \text{if } i = j. \end{cases} \tag{17}$$

Like Proposition 1 for the function ξ, one has

Proposition 2. *Let V be the matrix defined by (17). Then*

(i) V is positive semidefinite, $\nabla(\frac{1}{2})\eta^2(X) = VX$ and $\eta^2(X) = \langle X, VX \rangle$.
(ii) If the weight matrix W is irreducible, then

$$\mathcal{A} = \eta^{-1}(0) = \{X \in \mathcal{M}_{n,p}(\mathbb{R}) : VX = 0\}, \tag{18}$$

rank $V = n - 1$ and

$$\mathcal{A}^\perp = \{Y = VX : X \in \mathcal{M}_{n,p}(\mathbb{R})\} = \{Y = V^+X : X \in \mathcal{M}_{n,p}(\mathbb{R})\}. \tag{19}$$

(iii) $(\frac{1}{2}\eta^2)^*(Y) = \frac{1}{2}\langle Y, V^+Y \rangle$ if $Y \in \mathcal{A}^\perp$, $+\infty$ otherwise (V^+ denotes the pseudo-inverse of V).

Proof. (i) The positive semidefiniteness of V comes from [18] as in Proposition 1. Since $\nabla(\frac{1}{2}\eta^2)(X) = VX$, then from the generalized Euler's relation ([32]) follows $\eta^2(X) = \langle X, VX \rangle$.
(ii) The first equality of (18) is immediate from Lemma 1. If $VX = 0$, then $\eta(X) = \sqrt{\langle X, VX \rangle} = 0$, so $\{X \in \mathcal{M}_{n,p}(\mathbb{R}) : VX = 0\} \subset \eta^{-1}(0)$. Conversely, if $\eta(X) = 0$, then $\langle X, VX \rangle = 0$ and therefore $VX = 0$ because V is semidefinite positive. Then the second equality of (18) holds.

Let $\mathcal{L}$ be the linear transformation from $\mathcal{M}_{n,n}(\mathbb{R})$ to $\mathcal{M}_{np,np}(\mathbb{R})$ defined by: for $A \in \mathcal{M}_{n,n}(\mathbb{R})$, $\mathcal{L}(A)$ is the matrix having p diagonal blocks and each block is A. For $X, Y \in \mathcal{M}_{n,p}(\mathbb{R})$, according to the column-representation (9), one has

$$\mathcal{L}(A)\overline{\mathcal{X}} = 0 \Leftrightarrow AX^k = 0, \quad k = 1, \ldots, p \quad \Leftrightarrow AX = 0, \tag{20}$$

$$\overline{\mathcal{Y}} = \mathcal{L}(A)\overline{\mathcal{X}} \Leftrightarrow Y^k = AX^k, \quad k = 1, \ldots, p \quad \Leftrightarrow Y = AX. \tag{21}$$

It follows that if A is symmetric then we have

$$\{Y = AX : X \in \mathcal{M}_{n,p}(\mathbb{R})\} = \{X \in \mathcal{M}_{n,p}(\mathbb{R}) : AX = 0\}^{\perp}.$$

Hence (19).

As for rank V, we constate first that rank $V \leq n - 1$ because $Ve = 0$. On the other hand, if rank $V < n - 1$, then there exists $u \notin \mathbb{R}e$ such that $Vu = 0$. Let $X = uv^T$ with $v \in \mathbb{R}^p \setminus \{0\}$. Clearly $VX = 0$, and therefore $X \in \mathcal{A}$ following (18). By the definition of $\mathcal{A}$, all row of X are identical which implies that $u \in \mathbb{R}e$. This contradiction proves that rank $V = n - 1$.

(iii) It is well-known that if A is an $n \times n$ real symmetric and positive semidefinite matrix and $f(x) = \frac{1}{2}\langle x, Ax \rangle$, then ([32])

$$f^*(y) = \begin{cases} \frac{1}{2}\langle y, A^+y \rangle & \text{if } y \in \text{Im } A, \\ +\infty & \text{otherwise.} \end{cases}$$

On the other hand, by using the column-representation (9), it is clear from (20) and (21) that A is positive semidefinite if and only if $\mathcal{L}(A)$ is positive semidefinite. Moreover the pseudo-inverse of $\mathcal{L}(A)$ is $\mathcal{L}(A^+)$. The proof (iii) is then straightforward. $\qquad\square$

It follows that

$$(\tfrac{1}{2}\eta^2)^*(Y) = \frac{1}{2}\langle V^+Y, Y \rangle + \chi_{\mathcal{A}^\perp}(Y) \text{ for } Y \in \mathcal{M}_{n,p}(\mathbb{R}).$$

Since V^+ is symmetric positive semidefinite, we have $\partial(\tfrac{1}{2}\eta^2)^*(Y) = V^+Y + \mathcal{A}$ for $Y \in \mathcal{A}^\perp$.

Hence, determine a subgradient of $(\tfrac{1}{2}\eta^2)^*(Y)$ with $Y \in \mathcal{A}^\perp$ amounts to compute the pseudo-inverse of V. The next result permits to calculate V^+:

Proposition 3. *Let Λ be an $n \times n$ symmetrix matrix and $u \in \mathbb{R}^n, u \neq 0$ such that $\Lambda u = 0$. Then $\Lambda + \frac{1}{\|u\|^2}uu^T$ is nonsingular if and only if rank $\Lambda = n - 1$. In this case for every $y \in \mathbb{R}^n$ there exists $x \in \text{Im } \Lambda$ satisfying $\Lambda x = \text{Proj}_{\text{Im } \Lambda} y$ and*

$$(\Lambda + \frac{1}{\|u\|^2}uu^T)(x + \frac{u^Ty}{\|u\|^2}u) = y \quad i.e., \quad \Lambda^+ = (\Lambda + \frac{1}{\|u\|^2}uu^T)^{-1} - \frac{1}{\|u\|^2}uu^T.$$

Proof. If $(\Lambda + \frac{1}{\|u\|^2}uu^T)$ is nonsingular, then $\text{Im}\left(\Lambda + \frac{1}{\|u\|^2}uu^T\right) = \mathbb{R}^n$. Thus

$$\text{Im}\,\Lambda + \text{Im}\,\left(\frac{1}{\|u\|^2}uu^T\right) = \mathbb{R}^n = \text{Im}\,\Lambda + \mathcal{N}(\Lambda).$$

This implies $\mathcal{N}(\Lambda) = \text{Im}\,(\frac{1}{\|u\|^2}uu^T) = \mathbb{R}u$. Thus rank Λ = n-1.
Conversely, if rank $\Lambda = n - 1$, then $\mathcal{N}(\Lambda) = \mathbb{R}u$. Let $x \in \mathbb{R}^n$ such that

$$(\Lambda + \frac{1}{\|u\|^2}uu^T)x = 0, \text{ i.e., } \Lambda x = -\frac{1}{\|u\|^2}uu^Tx.$$

This implies $\Lambda x = 0$ and $u^Tx = 0$. Hence $x = 0$. Thus we can deduce that $\mathcal{N}(\Lambda + \frac{1}{\|u\|^2}uu^T) = \{0\}$, i.e., $\Lambda + \frac{1}{\|u\|^2}uu^T$ is nonsingular. In this case the projection on $\mathcal{N}(\Lambda) = \mathbb{R}u$ is given by $Proj_{\mathcal{N}(\Lambda)} = (1/\|u\|^2)uu^T$ and

$$Proj_{\text{Im}\,\Lambda} = I - \frac{1}{\|u\|^2}uu^T. \tag{22}$$

Let y be an arbitrary vector in $\mathbb{R}^n$. Since $Proj_{\text{Im}\,\Lambda}(y)$ is an element in Im Λ, there exists $\bar{x} \in \mathbb{R}^n$ such that $\Lambda\bar{x} = Proj_{\text{Im}\,\Lambda}(y)$. The decomposition $\mathbb{R}^n = \text{Im}\,\Lambda + \mathcal{N}(\Lambda)$ ensures the existence of $x \in \text{Im}\,\Lambda, x_1 \in \mathcal{N}(\Lambda)$ such that $\bar{x} = x + x_1$ and $\Lambda\bar{x} = \Lambda x = Proj_{\text{Im}\,\Lambda}(y)$. Observing $\Lambda u = 0, u^Tx = 0$ (since $x \in \text{Im}\,\Lambda$) we have from (22)

$$\begin{aligned}(\Lambda + \frac{1}{\|u\|^2}uu^T)(x + \frac{u^Ty}{\|u\|^2}u) &= \Lambda x + \frac{1}{\|u\|^2}uu^T\frac{u^Ty}{\|u\|^2}u \\ &= Proj_{\text{Im}\,\Lambda}y + \frac{u^Ty}{\|u\|^2}u = y.\end{aligned} \tag{23}$$

This implies

$$x = (\Lambda + \frac{1}{\|u\|^2}uu^T)^{-1}y - \frac{u^Ty}{\|u\|^2}u.$$

Therefore

$$\Lambda^+ = (\Lambda + \frac{1}{\|u\|^2}uu^T)^{-1} - \frac{1}{\|u\|^2}uu^T. \tag{24}$$

$\square$

The matrix V satisfies the assumptions of Proposition 3, then using (24) we have for $Y \in \mathcal{M}_{n,p}(\mathbb{R})$

$$V^+Y = \left(V + \frac{1}{n}ee^T\right)^{-1}Y - \frac{1}{n}ee^TY.$$

That implies, for $Y \in \mathcal{A}^\perp$,

$$X = V^+Y = (V + \frac{1}{n}ee^T)^{-1}Y, \tag{25}$$

i.e.,

$$(V + \frac{1}{n}ee^T)X = Y. \tag{26}$$

In particular we have in the normal case where $V = ncI - cee^T$: $ncX = Y$, i.e., $X = Y/(nc)$.

The description of the DCA for solving (11)

Algorithm 1 *(DCA applied to (11))*
Let $\varepsilon > 0$, and $0 \neq X^{(0)} \in \mathcal{A}^\perp$ be given.
For $k = 0, 1, \dots$
until either $\|X^{(k+1)} - X^{(k)}\| \leq \varepsilon$ or $F(X^{(k)}) - F(X^{(k+1)}) \leq \varepsilon$
take $Y^{(k)} = B(X^{(k)})X^{(k)}$ and

$$X^{(k+1)} = V^+ Y^{(k)}. \tag{27}$$

Algorithm 1bis *(DCA applied to (11) in the normal case).*
Replace in Algorithm 1 (27) by $X^{(k+1)} = \frac{1}{nc}Y^{(k)}$.

As indicated above, one of interesting features of the DCA is the nice effect of a *good* starting point. This property motivates us to investigate a technique for computing a "good" starting point $X^{(0)}$ to Algorithm 1.

How to find a good starting point for Algorithm 1 ?

First, with the help of the triangle inequality we determine a full "distance matrix" $\tilde{\Delta} = (\tilde{\delta}_{ij})$ by either deducing bounds (l_{ij}), (u_{ij}) from the given distances or imposing additional bounded constraints and taking $\tilde{\delta}_{ij} \in [l_{ij}, u_{ij}]$ for $(i, j) \notin \mathcal{S}$. We solve then the exact distance geometry problem (EDP_1) where all pairwise "distances" $\tilde{\delta}_{ij}$ are known and take its solution as $X^{(0)}$. The idea of this technique comes from two facts:

- Algorithm 1bis to solve (11) when all pairwise distances are known is very simple, it is in explicit form which requires only matrix-vector products, and works very well in practice.
- In the general case where only a small subset of distances are known one can *approximate* a solution of (2) by using a dense set of constraints which is extended of the given distances, and then working with this set.

There are many ways to complete the "distance matrix". In our experiments we use the following procedure:

Procedure CP:
    If $(i, j) \in \mathcal{S}$, then $\tilde{\delta}_{ij} = \delta_{ij}$.
    If $(i, j) \notin \mathcal{S}$, then $\tilde{\delta}_{ij} = +\infty$.
    While $\tilde{\delta}_{ij} = +\infty$ do $\tilde{\delta}_{ij} = \min_{k=1,\dots,n}\{\tilde{\delta}_{ij}, \tilde{\delta}_{ik} + \tilde{\delta}_{kj}\}$.

2.2 The main algorithm for solving (EDP$_1$)

We now describe our main algorithm for solving the distance geometry problem (EDP$_1$).

Algorithm EDCA:

   Phase 1. Find an initial point.
   Step 1. *Determine an approximate distance matrix* $\tilde{\Delta}$ by Procedure CP.
   Step 2. *Solve Problem*

$$\min\left\{\frac{1}{2}\sum_{i<j} c(\tilde{\delta}_{ij} - \|X_i^T - X_j^T\|)^2 : X \in \mathcal{M}_{n,p}(\mathbb{R})\right\}$$

by applying Algorithm 1bis to problem (11), where w_{ij} and δ_{ij} are replaced by c and $\tilde{\delta}_{ij}$ respectively, to obtain the point noted $\tilde{X}$.

   Phase 2. Solve the original problem (EDP$_1$) by applying Algorithm 1 to problem (11) from the point $\tilde{X}$.

3   Computational experiments

Our algorithm is coded in FORTRAN and runs on a SGI Origin 2000 multiprocessor with IRIX system. We have tested our algorithm on two model problems from Moré-Wu [9] where the molecule has $n = s^3$ atoms located in the three-dimensional lattice

$$\{(i_1, i_2, i_3) : 0 \leq i_1 < s, \ 0 \leq i_2 < s, \ 0 \leq i_3 < s\}$$

for some integer $s \geq 1$.

   In the first model propblem the ordering for the atoms is specified by letting i be the atom at position (i_1, i_2, i_3)

$$i = 1 + i_1 + si_2 + s^2 i_3,$$

and distance data is generated for all pairs of atoms in

$$\mathcal{S} = \{(i,j) : |i - j| \leq r\}, \tag{28}$$

where r is an integer between 1 and n.

   In the second model problem the set $\mathcal{S}$ is specified by ($X_i^T = (i_1, i_2, i_3)$)

$$\mathcal{S} = \{(i,j) : \|X_i^T - X_j^T\| \leq \sqrt{r}\}. \tag{29}$$

   We considered $w_{ij} = 1$ for all $i \neq j$ in Phase 1, and $w_{ij} = 1$ for $(i,j) \in \mathcal{S}$, $i \neq j$ in Phase 2. For starting Algorithm 1bis, we first took a random point

X in $(0, s-1)$ and then set $X^{(0)} = Proj_{A^\perp}(X)$. We terminated Algorithm 1bis when

$$|F(X^{(k+1)}) - F(X^{(k)})| \le 10^{-7}|F(X^{(k+1)})|.$$

In our computational results a matrix $X^* \in \mathcal{M}_{n,p}$ solves Problem (EDP$_1$) if

$$|\|X^{*T}_i - X^{*T}_j\| - \delta_{ij}| \le \varepsilon, \ \forall(i,j) \in \mathcal{S}$$

for some telerance ε. We took $\varepsilon = 0.005$ when $n < 1331$, $\varepsilon = 0.01$ when $1331 \le n < 2744$ and $\varepsilon = 0.03$ when $n \ge 2744$. For solving the linear system

$$\left(V + \frac{1}{n}ee^T\right)X = Y^{(k)} \tag{30}$$

in Phase 2 we first decomposed the matrix $V + \frac{1}{n}ee^T = R^T R$ by the Cholesky factorization, and then at each iteration we solved two systems $RU = Y^{(k)}$ and $R^T X = U$.

The main aim of our computational experiments is to show that Algorithm ECDA is an efficient approach to solve (EDP$_1$) with large dimensions. We consider molecules containing at most 2744 atoms (then (EDP$_1$) has 8232 variables) in the first model problem, and 4096 atoms (12288 variables) in the second one.

The second aim of our computational experiments is to study the effect of the number of given distances on the performance of the algorithm. We have tested our algorithm on different values of r ($r = 1$, $r = 2$, and $r = s^2$) that vary the cardinality of $\mathcal{S}$.

In the tables 1 and 2 we indicate the following informations:

t0: CPU time for determining the matrix $\tilde{\Delta}$;

it1 and *time1*: respectively, the number of iterations and CPU time, of Algorithm 1bis;

it2 and *time2*: respectively, the number of iteration and CPU time of Phase 2. *time2* is composed two parts: the first part is CPU time for the Cholesky factorization of matrix $V + \frac{1}{n}ee^T$ and the second one is CPU time of Algorithm 1.

ttotal: the total CPU time of the main algorithm EDCA;

$\sigma(X^*)$: the optimal value of (EDP$_1$);

data: the number of given distances, i.e., $\frac{1}{2}|\mathcal{S}|$ where $|\mathcal{S}|$ denotes the cardinality of $\mathcal{S}$. Note that, in the complete distance matrix (i.e., $\mathcal{S} = \{(i,j) \in N \times N : i \ne j\}$ we have $|\mathcal{S}| = n(n-1)/2$.

All CPU times were computed in seconds.

In the first experiment (Table 1) we study the performance of Algorithm EDCA in two model problems (28) and (29) where the parameter r was set to $r = s^2$. As indicated in Moré-Wu [9], a difference between both definitions of $\mathcal{S}$ is that (29) includes all nearby atoms, while (28) includes some of nearby atoms and some relatively far away atoms. Then these model problems may capture various features in distance data from applications. We note that when $r = s^2$ the set defined by (28) is included in the set defined by (29).

In the second experiment (Table 2) we compare the behavior of our method when the number of given distance varies with a same model problems. We consider the second model problems with $r = 1$, $r = 2$, and $r = s^2$.

Table 1. The performance of EDCA for the first model problem, $r = s^2$

n	data	t0	iter1	time1	iter2	time2	ttotal	$\sigma(X^*)$
27	198	0.000	60	0.015	37	0.000 + 0.012	0.028	0.00007
64	888	0.003	64	0.092	86	0.001 + 0.120	0.216	0.00005
125	2800	0.023	102	0.553	148	0.006 + 0.687	1.268	0.00008
216	7110	0.119	75	1.212	240	0.027 + 2.975	4.334	0.00010
343	15582	0.497	92	3.879	364	0.110 + 10.414	14.899	0.00014
512	30688	2.160	57	6.445	517	0.355 + 38.358	47.318	0.00032
729	55728	9.164	99	27.341	661	1.024 + 132.875	170.405	0.00038
1000	94950	28.697	101	58.837	856	2.607 + 383.240	473.381	0.00063
1331	153670	128.528	99	164.098	870	6.498 + 1355.730	1654.850	0.00321
1728	238392	293.446	91	258.790	1070	14.312 + 2924.534	3491.080	0.00473
2197	356928	623.202	127	596.273	1277	33.138 + 5761.941	7014.554	0.00729
2744	518518	1410.874	102	894.039	755	90.402 + 6795.002	9190.320	0.01322

Comments

A direct comparison with the efficiency of the promising existing methods such as Moré-Wu and Zou et al. [19] is difficult, mainly because we do not work with the same objective function, and another also because they did not give costs of their results in seconds, and solutions computed by the two approaches are not of the same accurracy. For this type of problem, we considered the molecules with a very large number of atoms (up to 4096 atoms), while the results presented in the just mentioned papers deal only with 216 atoms ([9]) and 343 atoms ([19]). Hendrickson [6] and Zou et al. [19] solved another test model problem with at most 777 atoms. Our main concern in this paper are both ability to treat large-scale problems, and the cost of algorithms. The most important fact is that in all experiments Algorithm EDCA gives a global solution of (EDP$_1$).

We observe that the rate of convergence of the DCA in Phase 1 (Algorihm 1bis) does not seem to depend on data (the number and the length of given distances, say distance data is determined between *nearby* or *far away* atoms). Generally, the sequences $F(X^{(k)})$ and $\{\|X^{(k+1)} - X^{(k)}\|\}$ in Algorithm 1bis decrease rapidly and remain practically unchanged after at most 113 iterations (except for $n = 2197$ in the first model). So the DCA is quite stable in the normal case. On the contrary, the DCA in Phase 2 (Algorithm 1) is sensitive to data. In the first experiment where given distances were determined between both nearby and far away atoms, the number of

Table 2. The performance of EDCA for the second model problem, $r = 1$, $r = 2$ and $r = s^2$

n	r	data	t0	iter1	time1	iter2	time2	ttotal	$\sigma(X^*)$
27	$r = 1$	54	0.000	37	0.010	1	0.000+0.000	0.011	0.00026
	$r = 2$	126	0.000	44	0.012	2	0.000 + 0.000	0.012	0.00017
	$r = s^2$	347	0.000	49	0.013	0	0.000 + 0.000	0.013	0.00000
64	$r = 1$	144	0.004	50	0.070	7	0.001 + 0.005	0.080	0.00035
	$r = 2$	360	0.004	50	0.070	4	0.001 + 0.004	0.078	0.00019
	$r = s^2$	1880	0.000	57	0.080	8	0.001 + 0.018	0.100	0.00088
125	$r = 1$	300	0.027	55	0.292	17	0.006 + 0.036	0.363	0.00059
	$r = 2$	780	0.029	57	0.304	9	0.005 + 0.025	0.363	0.00059
	$r = s^2$	7192	0.003	67	0.355	15	0.005 + 0.130	0.494	0.00095
216	$r = 1$	540	0.142	61	0.966	27	0.028 + 0.069	1.191	0.00079
	$r = 2$	1440	0.147	60	0.947	10	0.027 + 0.069	1.190	0.00079
	$r = s^2$	21672	0.015	59	0.932	19	0.027 + 0.496	1.471	0.00135
343	$r = 1$	882	0.575	86	3.565	35	0.119 + 0.513	4.772	0.00119
	$r = 2$	2394	0.608	51	2.186	17	0.116 + 0.277	3.188	0.00038
	$r = s^2$	53799	0.075	113	4.624	3	0.110 + 0.202	5.012	0.00199
512	$r = 1$	1344	2.620	71	7.844	46	0.357 + 2.125	12.948	0.00139
	$r = 2$	3696	2.589	91	9.988	12	0.377 +0.573	13.528	0.00180
	$r = s^2$	119692	0.308	86	9.452	1	0.355 + 0.172	10.288	0.02086
729	$r = 1$	1944	11.169	67	20.211	58	1.050 + 8.684	41.115	0.00167
	$r = 2$	5400	11.361	57	15.851	18	1.029 + 2.770	31.012	0.00074
	$r = s^2$	243858	1.107	87	23.615	9	1.025 + 3.859	29.606	0.00117
1000	$r = 1$	2700	35.458	67	40.681	81	2.688 + 30.552	109.379	0.00110
	$r = 2$	7560	35.635	88	53.607	13	2.647 + 5.069	96.958	0.00409
	$r = s^2$	456872	3.460	75	43.195	18	2.604 + 16.485	65.744	0.00696
1331	$r = 1$	3630	159.438	94	163.382	55	6.695 + 82.352	411.866	0.01289
	$r = 2$	10230	159.259	112	193.147	10	6.753 + 15.564	374.724	0.02052
	$r = s^2$	809763	14.595	67	111.128	29	6.525 + 75.837	208.080	0.01051
1728	$r = 1$	4752	342.327	101	285.686	64	14.312 + 154.711	797.036	0.01514
	$r = 2$	13464	341.081	87	246.426	11	14.311 + 26.720	628.537	0.02238
	$r = s^2$	1359216	33.879	77	218.960	18	4.378 + 81.363	338.580	0.01185
2197	$r = 1$	6084	708.901	89	411.171	73	32.846 + 291.364	1444.283	0.01783
	$r = 2$	17316	706.695	67	310.502	25	32.853 + 01.005	1151.055	0.00796
	$r = s^2$	2014666	68.777	90	414.914	17	32.874 + 125.490	642.055	0.02998
2744	$r = 1$	7644	1619.529	83	723.645	27	81.289 + 218.304	2642.768	0.33067
	$r = 2$	21840	1614.741	73	637.749	10	81.929 + 80.757	2415.177	0.07326
	$r = s^2$	3436528	156.124	110	958.317	1	81.995 + 13.571	1210.008	0.14925
3375	$r = 1$	9450	3536.546	81	1161.486	30	163.451 + 401.991	5263.470	0.37332
	$r = 2$	27090	3526.546	72	1026.892	9	157.118 + 119.524	4830.081	0.12410
	$r = s^2$	5196129	334.742	69	982.825	21	155.842 + 463.341	1936.751	0.30644
4096	$r = s^2$	7640952	1885.085	81	3632.0010	21	547.952 + 1226.400	7291.440	0.3008

iterations of Algorithm 1 is much greater than in the second experiment. A simple explanation is that when the given distances are determined between relatively far way atoms, the approximate distance matrix $\tilde{\Delta}$ seems to be not good which implies that $\tilde{X}$ is not relatively *near* a solution of (EDP$_1$). Then Algorithms 1 need more iterations to yield a solution. In general, the cost of Algorithm EDCA in the first model problem is much more expensive than in the second model problem.

Consider now the second experiment with different numbers of distance data. When $n \leq 343$ EDCA is fastest in case $r = 2$ while when $n \geq 729$ the more the number of given distances increases, the faster EDCA is. Moreover, with $n \geq 1000$, we observe that when the number of given distances is small ($r = 1$ and $r = 2$) the cost for determining $\tilde{\Delta}$ is the most important in Algorithm EDCA (35% to 73% of the total cost). However this step is necessary, because Phase 1 is indispensable for EDCA to obtain a global solution of (EDP$_1$) with $n \geq 64$. On the other hand, although the sequences $\{X^{(k)}\}$ in Algorithm is not in explicit form, the cost of one iteration of this algorithm (which contains the cost for determining matrix $B(X^{(k)})$, the product $B(X^{(k)})X^{(k)}$ and the solution of two triangular systems) is not more expensive than the cost of Algorithms 1bis which requires only matrix-vector products. This shows that Algorithm 1 exploits well sparsity of S (in the determination of matrix $B(X^{(k)})$ and the product $B(X^{(k)})X^{(k)}$).

Interesting problems for future research arise from our results. Firstly, what is the "best" DCA for solving (EDP$_1$). Various regularization techniques are to be studied in order to obtain different d.c. decomposition of (EDP$_1$) which can improve the qualities (robustness, stability, convergence rate and globality of computed solutions) of the DCA (see [7], [15]). Lagrangian duality without gap presented in [7], [13] permits also to state interesting equivalent forms of (EDP$_1$).

Secondly, in the case of large-scale problems, it could be desirable to find in a less expensive way, if possible, a good approximate distance matrix (Phase 1).

These issues are currently being studied.

References

1. Blumenthal, L.M. (1953) Theory and Applications of Distance Geometry. Oxford University Press
2. Crippen, G.M., Havel, T.F. (1998) Distance Geometry and Molecular Conformation. John Wiley & Sons
3. De Leeuw, J. (1977) Applications of convex analysis to multidimensional scaling. In: Barra J.R. *et al.* (Eds.) Recent developments in statistics, North-Holland Publishing Company, 133–145
4. Glun, W., Hayden, T.L., Raydan, M.(1993) Molecular Conformation from distance matrices. J. Comp. Chem. **14**, 114–120

5. Havel, T.F. (1991) An evaluation of computational strategies for use in the determination of protein structure from distance geometry constraints obtained by nuclear magnetic resonance. Prog. Biophys. Mol. Biol. **56** , 43–78
6. Hendrickson, B.A. (1995) The molecule problem: Exploiting structure in global optimization. SIAM J. Optimization **5**, 835–857.
7. Le Thi Hoai An (1997) Contribution à l'optimisation non convexe et l'optimisation globale: Théorie, Algorithmes et Applications. Habilitation à Diriger des Recherches, Université de Rouen
8. Le Thi Hoai An, Pham Dinh Tao (1997) Solving a class of linearly constrained indefinite quadratic problems by D.c. algorithms. Journal of Global Optimization **11**, 253–285
9. Moré, J.J., Wu, Z. (1997) Global continuation for distance geometry problems. SIAM J. Optimization **8**, 814–836
10. Moré, J.J., Wu, Z. (1996) Issues in large-scale Molecular Optimization. Preprint MCS-P539-1095, Argonne National Laboratory, Argonne, Illinois 60439
11. Pham Dinh Tao, Souad El. B. (1986) Algorithms for solving a class of non convex optimization problems. Methods of subgradients. In: Hiriart Urruty J.B. (Ed.) Fermat days 85. Mathematics for Optimization. Elsevier Science Publishers B.V., North-Holland, 249–271
12. Pham Dinh Tao, Souad El. B. (1988) Duality in d.c. (difference of convex functions) optimization. Subgradient methods. Trends in Mathematical Optimization, International Series of Numer Math. **84**, Birkhauser, 277–293
13. Pham Dinh Tao, Le Thi Hoai An (1994) Stabilité de la dualité lagrangienne en optimisation d.c. (différence de deux fonctions convexes) C.R. Acad. Paris **318**, Série I, 79–384
14. Pham Dinh Tao, Le Thi Hoai An (1998) D.c. optimization algorithms for trust region problem. SIAM J. Optimization **8**, No 2, 476–505
15. Pham Dinh Tao, Le Thi Hoai An (1997) Convex analysis approach to d.c. programming: Theory, Algorithms and Applications (dedicated to Professor Hoang Tuy on the occasion of his 70th birthday). Acta Mathematica Vietnamica **22**, No 1, 289–355
16. Rockafellar, R.T. (1970) Convex Analysis. Princeton University, Princeton
17. Saxe, J. B. (1979) Embeddability of Weighted Graphs in k-space is Strongly NP-hard. Proc. 17 Allerton Conference in Communications, Control and Computing, 480–489
18. Varga, R. (1962) Matrix iterative analysis. Prentice Hall
19. Zou, Z., Richard H. Bird, Schnabel, R.B. (1997) A Stochastic/Pertubation Global Optimization Algorithm for Distance Geometry Problems. J. of Global Optimization **11**, 91–105

Adaptive Scaling and Convergence Rates of a Separable Augmented Lagrangian Algorithm

Philippe Mahey[1], Jean-Pierre Dussault[2], Abdelhamid Benchakroun[2], and Abdelouahed Hamdi[3]

[1] Laboratoire d'Informatique, de Modélisation et d'Optimisation des Systèmes, (L.I.M.O.S), I.S.I.M.A - Université Blaise-Pascal, Clermont-Ferrand. France
e-mail: Philippe.Mahey@isima.fr
[2] Département de Mathématique et Informatique, Université de Sherbrooke, Québec, Canada
[3] Fachbereit Mathematik, University of Trier, Germany

Abstract. We analyze the numerical behaviour of a separable Augmented Lagrangian algorithm. This algorithm which is equivalent to the Proximal Decomposition algorithm in the convex case, uses a dual sequence which convergence is associated with the inverse of the primal penalty parameter. As a consequence, an optimal value for that parameter, which is thus more like a scaling parameter than like a penalty one, is expected, confirming former theoretical results in the strongly convex case.

We propose an implementable algorithm where the scaling parameter is adjusted at each iteration in order to keep the same rate of convergence for both primal and dual sequences. The correcting effect of that parameter update is illustrated on small quadratic problems. The autoscaled decomposition algorithm is then tested on larger block-angular problems with convex or non convex separable cost functions.

Key words: augmented Lagrangian, decomposition, proximal regularization

1 Introduction

In [4], a separable augmented lagrangian algorithm, called (SALA), was proposed to solve separable programs with coupling constraints. This algorithm can be interpreted as a mixed primal- and dual-decomposition method and it presents strong connections with the Partial Inverse method of Spingarn [7], the Alternate Direction of Multiplier method ([2], [3]) and the Proximal Decomposition on the graph of monotone operators [5]. It has been successfully applied to some network optimization problems with convex cost functions, showing nice properties for distributed computations (see [6]). In [5], the basic algorithm was shown to be very sensitive to the scaling parameter which multiplies the quadratic penalty term in the subproblems. Partial answers to the role of that parameter were proposed but very poor bounds for the convergence rates of the algorithm were derived from these results. We aim in the present paper at improving these results and deriving sharper bounds for these convergence rates when applied to the separable augmented algorithm (SALA).

2   A Separable Augmented Lagrangian Algorithm

We present first the algorithm (SALA), a decomposition method based on separable Augmented Lagrangians for convex and nonconvex large scale programs (see [4]) .

Let consider the following problem defined on a product space $X_1 \times X_2 \cdots \times X_p$ $(X_i = \mathbb{R}^{n_i})$.

$$(PS) \quad \begin{array}{ll} \text{minimize} & \sum_{i=1}^{p} f_i(x_i) \\ x_i \in S_i, i = 1, \ldots, p \\ \sum_{i=1}^{p} g_i(x_i) = 0 \end{array}$$

where $x_i \in \mathbb{R}^{n_i}, i = 1, \ldots, p$, $f_i, \mathbb{R}^{n_i} \to \mathbb{R}$, $i = 1, \ldots, p$ and $g_i, \mathbb{R}^{n_i} \to \mathbb{R}^m, i = 1, \ldots, p$. All the functions are assumed to be $\mathbb{C}^2$. The nonempty sets $S_i, i = 1, \ldots, p$ are supposed to be closed and bounded subsets of $\mathbb{R}^{n_i}$. Introducing a resource allocation vector $(y_1, \cdots, y_p)$, we get an equivalent embedded formulation with a distributed coupling :

$$\begin{array}{lll} \text{minimize} & & \sum_{i=1}^{p} f_i(x_i) \\ x_1, \ldots, x_p & & x_i \in S_i, \ i = 1, \ldots, p \\ y_1, \ldots, y_p & & g_i(x_i) + y_i = 0, \ i = 1, \ldots, p \\ & & \sum_{i=1}^{p} y_i = 0 \end{array}$$

This formulation is equivalent to (PS) for each $(y_1, \ldots, y_p)$ with $\sum_i y_i = 0$ such that there exist $x_i \in S_i$ with $g_i(x_i) + y_i = 0 \ \forall i$.

Algorithm (SALA) generates subproblems which correspond to minimizing Augmented Lagrangians functions associated, at iteration k, with the resource allocation constraints in each block, i.e., $g_i(x_i) = -y_i^k$. These allocation vectors y_i and their dual counterpart u_i, $i = 1, \ldots, p$, are updated and projected on their respective mutually orthogonal feasibility subspaces :

$$A = \{(u_1, \ldots, u_p) \mid u_1 = u_2 = \cdots = u_p\}$$

$$A^\perp = \{(y_1, \ldots, y_p) \mid \sum_{i=1}^{p} y_i = 0\}$$

The main step of (SALA) is sketched below with a fixed proximal parameter $\lambda > 0$. Hereafter, the primal residual is denoted by $r = \sum_{i=1}^{p} g_i(x_i)$.

Separable Augmented Lagrangian Algorithm (SALA)

1. Initialize : x^0, ε, , $y^0 \in A^{\perp}$, $i = 1, \ldots, p$, $k = 0$

2. Determine :
$$\forall i = 1, \ldots, p \;\; x_i^{k+1} := \arg \min_{x_i \in S_i} \; f_i(x_i) + u^k(g_i(x_i) + y_i^k) + \frac{\lambda}{2}\|g_i(x_i) + y_i^k\|^2$$

3. Compute the residual $r^{k+1} = \sum_{i=1}^{p} g_i(x_i^{k+1})$

$\qquad$ **If :** $\|r^{k+1}\| < \varepsilon$ stop.
$\qquad$ **Else :** go to step 4

4. Update the multipliers and the allocations :

$$\begin{cases} u^{k+1} = u^k + \dfrac{\lambda}{p} r^{k+1} \\ y_i^{k+1} = -g_i(x_i^{k+1}) + \dfrac{1}{p} r^{k+1}, \; i = 1, \ldots, p \end{cases}$$

and return to step 2

A general convergence result in the nonconvex twice differentiable case has been given by Hamdi et al in [4]. It relies on the following classical second-order hypothesis.

$$\begin{cases} \text{There exists an isolated local optimal solution } x^* \\ \text{the optimal multipliers set is nonempty and bounded} \\ x^* \text{ is a regular point w.r.t. coupling surface} \\ x^* \text{ satisfies the second-order sufficient optimality conditions.} \end{cases}$$

We will show now that (SALA) is a special instance of the Proximal Decomposition method which applies to the solution of the following concave constrained maximization problem :

(P) $\qquad\qquad\qquad$ Maximize $H(u)$ such that $u \in A$

Indeed, (P) is equivalent to the Lagrangian dual problem of the original problem (PS) by taking :

$$u = (u_1, \ldots, u_p), u \in X = (\mathbb{R}^m)^p, H(u) = \sum_i H_i(u_i)$$

where H_i is the i-th part of the dual function associated to the coupling constraints of (PS), i.e. :

$$H_i(u_i) = \inf_{x_i \in S_i} f_i(x_i) + \langle u_i, g_i(x_i) \rangle \, .$$

and A is the subspace of X made of the copies of the original dual variables, i.e. :

$$A = \{(u_1, \cdots, u_p) \in (I\!\!R^m)^p \mid u_1 = \cdots = u_p\}$$

The Proximal Decomposition algorithm (see [5]) generates a primal-dual sequence in $X \times X$, (u^k, y^k), such that $(u^k, y^k) \in A \times A^{\perp}$, which converges to (u^*, y^*) where u^* are the p copies of the optimal multipliers associated to the coupling constraints in (PS) and $y^* = (y_1^*, \dots, y_p^*)$ are the p optimal resource allocation vectors defined above (Observe that these allocations are feasible as $A^{\perp} = \{(y_1, \dots, y_p) \in X \mid \sum_i y_i = 0\}$).

The Proximal Decomposition iteration is based on the fact that the optimality condition for (P) can be stated formally as :

$$(u^*, y^*) \in A \times A^{\perp} \cap \mathrm{Gr}(T)$$

where T is the subdifferential operator of the concave function H and $\mathrm{Gr}(T)$ is the graph of that operator, i.e. $\mathrm{Gr}(T) = \{(u, y) \mid y \in Tu\}$. T is indeed maximal monotone and the iterates are given in two steps, beginning with the **proximal** step for a given primal-dual pair (u^k, y^k) at iteration k :

$$u'^k = (I + T)^{-1}(u^k + y^k) \tag{1}$$
$$y'^k = (I + T^{-1})^{-1}(u^k + y^k)$$

Of course, only one proximal calculus is needed as $(I + T^{-1})^{-1} = I - (I + T)^{-1}$. The second step is the **projection** step where (u'^k, y'^k) are projected on their respective subspaces A and $A^{\perp}$.

The proximal step is equivalent to applying a proximal iteration on the dual function H which leads naturally to separable augmented lagrangian subproblems in the original x_i variables. These subproblems can be shown after straightforward transformations equivalent to step 2 of algorithm (SALA) described above (with $\lambda = 1$).

Convergence results in the convex case are presented in [5] and we will now introduce the scaling parameter λ to study the speed of convergence in the strongly convex situation.

3   New Bounds for the Total Convergence Rate of Proximal Decomposition

In [5], an upper bound for the convergence rate of the Proximal Decomposition method is given which is based on the properties (strong monotonicity and Lipschitz constant) of the operator T attached to the primal sequence.

Ignoring the iteration superscript k to simplify notations, we will analyze the proximal step (1) which inherits the contraction properties of the resolvent operator $(I + T)^{-1}$. We will suppose that T is strongly monotone with radius ρ and Lipschitz with constant L to get the following bound in $X \times X$, denoting $z = u + y$ (see [5]) :

$$\|(u - u', y - y')\|^2 \leq (1 - \frac{2\rho}{(1 + L)^2})\|z - z'\|^2$$

But, performing the same reasoning on the dual sequence which is based on operator T^{-1} (with radius L^{-1} and Lipschitz constant ρ^{-1}), we get the following bounds :

$$\|(u - u', y - y')\|^2 \leq (1 - \frac{2\rho^2}{L(1 + \rho)^2})\|z - z'\|^2$$

Introducing a scaling parameter $\lambda > 0$ like in [5] (thus substituting T by λT and T^{-1} by $\lambda^{-1}T^{-1}$), we obtain two distinct rates, one for each sequences, i.e. r_u for the primal sequence, which already appeared in [5], and r_y for the dual sequence :

$$r_u(\lambda) = \sqrt{1 - \frac{2\lambda\rho}{(1 + \lambda L)^2}}$$

with an optimal value $\lambda_u = 1/L$ associated with the optimal rate $r_u(\lambda_u) = \sqrt{1 - \frac{\rho}{2L}}$, and

$$r_y(\lambda) = \sqrt{1 - \frac{2\lambda\rho^2}{L(1 + \lambda\rho)^2}}$$

with an optimal value $\lambda_y = 1/\rho$ associated with the optimal rate $r_y(\lambda_y) = \sqrt{1 - \frac{\rho}{2L}}$.

Observe that the optimal rates are equal but for different values of λ. Figure 12 shows the two rates as functions of λ. The best compromise should be the result of minimizing $\sup (r_u(\lambda), r_y(\lambda))$ which yields :

$$\tilde{\lambda} = \frac{1}{\sqrt{\rho L}}$$

Even if these bounds seem to give poor estimates of the real rates, they show how the primal and dual sequences have opposite behaviour with respect to the scaling parameter. In fact, as early noticed in [5], one sequence depends on λ and the other one depends on λ^{-1}. In consequence, increasing the parameter will accelerate one sequence but stop its dual counterpart, with a serious risk to converge to a nonoptimal solution.

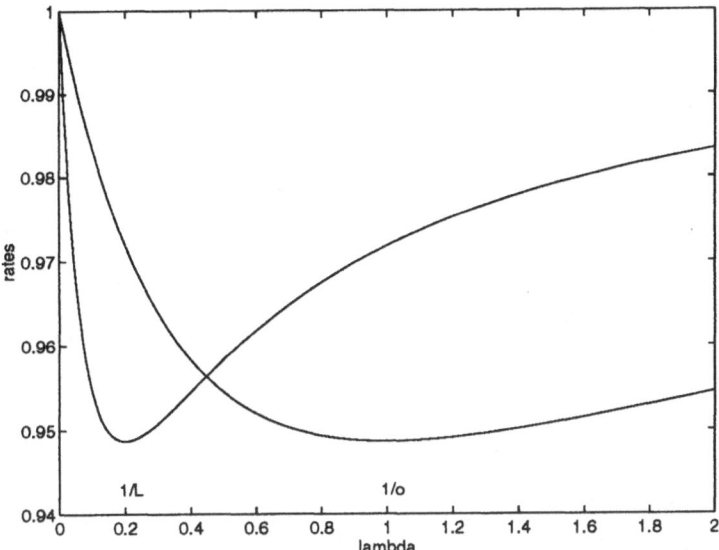

Fig. 1. Primal and dual total rates

4   Individual Convergence Rates for Primal and Dual Sequences

We will now derive direct bounds on the convergence of each primal and dual sequences using properties of T and T^{-1} :

$$\frac{1}{1+\lambda L}\|z - z'\| \le \|u - u'\| \le \frac{1}{1+\lambda\rho}\|z - z'\|$$

$$\frac{1}{1+\frac{1}{\lambda\rho}}\|z - z'\| \le \|y - y'\| \le \frac{1}{1+\frac{1}{\lambda L}}\|z - z'\|$$

We will denote by LB_u, UB_u, LB_y, UB_y, the four bounds above for the u and y sequences respectively.

We can observe in figure 2 the four curves representing these bounds in function of λ. Again, setting $\lambda = \tilde{\lambda} = 1/\sqrt{\rho L}$, we get the same bounds for both sequences and the best compromise for convergence; these common bounds are given by :

$$\tilde{LB} = \frac{1}{1+\sqrt{\frac{L}{\rho}}} \le 0.5$$

$$\tilde{UB} = \frac{1}{1+\sqrt{\frac{\rho}{L}}} \ge 0.5$$

Ideally, one would expect a convergence rate of 0.5 for both sequences when λ is tuned to keep them at the same speed. This point has been confirmed by numerical experiments on small quadratic problems.

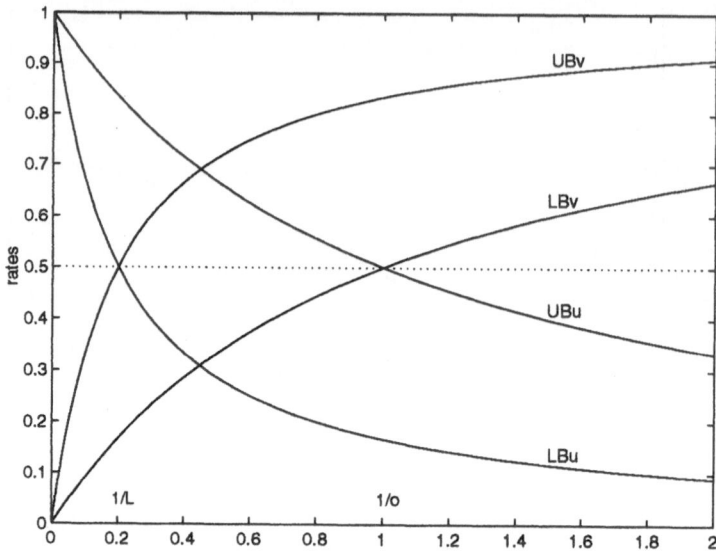

Fig. 2. Primal and dual individual rates bounds

The figures 3 and 4 show the evolution of the number of iterations as a function of the parameter λ (plotted on a logarithmic axis to get a symmetric curve). We have considered here a very simple quadratic problem with two variables :

$$\text{minimize} \quad \frac{1}{2}(ax_1^2 + x_2^2)$$
$$\text{subject to} \quad b(x_1 + x_2 - 2) = 0$$

It is easy to see that the corresponding dual operator T possesses the desired properties with $\rho = b^2$ and $L = \frac{b^2}{a}$. The sensitivity with respect to the parameter λ and the optimal intervals for its values are clearly illustrated on these curves (all tests have been made using Matlab 4.0).

Figure 5 shows that the best rate of convergence of 0.5 is attained in an ideal situation (a=1, b=1 and $\lambda = 1$) when plotting the two residual sequences given by $\tau_x = \frac{\|r^{k+1}\|}{\|r^k\|}$ and $\tau_y = \frac{\|\nabla L^{k+1}\|}{\|\nabla L^k\|}$, where r^k is the residual of the coupling constraints and L^k is the ordinary Lagrangian function of problem (PS) computed at iteration k.

5   Scaling Parameter Update

The figure 2 in the previous section suggests that λ should be update in order to keep both sequences with closer speed of convergence.

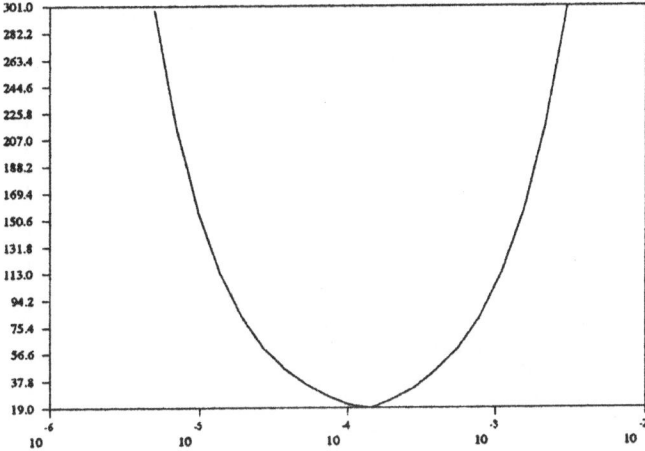

Fig. 3. Iteration number versus λ; $a = 1, b = 0.01$

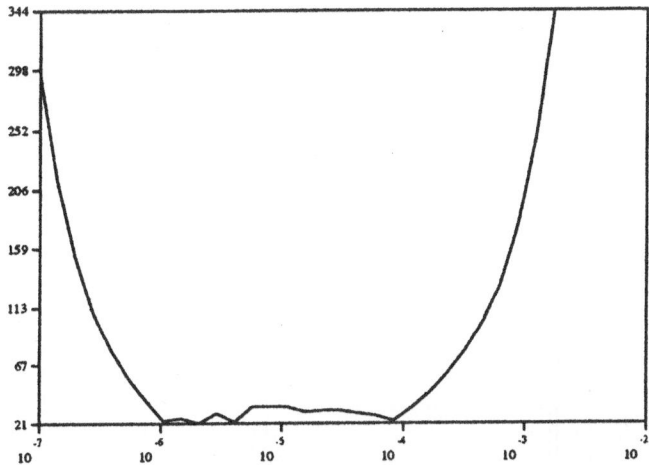

Fig. 4. Iteration number versus λ; $a = 100, b = 0.01$

Let τ_x and τ_y be the current estimate of the primal and dual convergence rates. From figure 2, we deduce the following empirical rule :

If $\tau_y < \tau_x$, then λ_k should be increased

If $\tau_y > \tau_x$, then λ_k should be decreased

This could be achieved with the following update rule :

$$\lambda_{k+1} = (\frac{\tau_x}{\tau_y})^\alpha \lambda_k$$

where α is a user defined parameter such that $0 < \alpha < 1$.

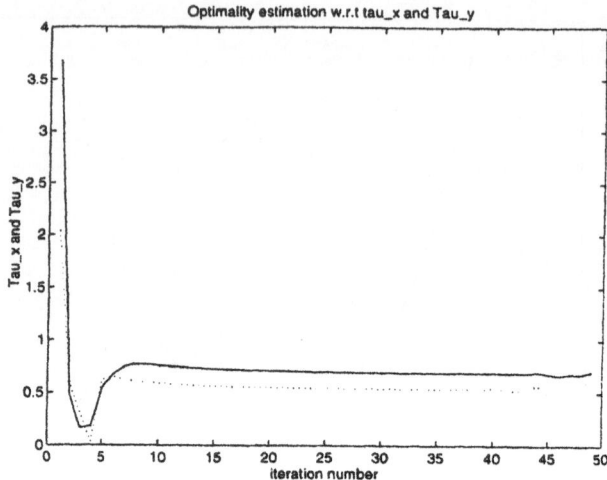

Fig. 5. Primal and dual rates versus Iterations; fixed λ

Figure 6 shows the evolution of the convergence rates on the same example as before with this updating rule.

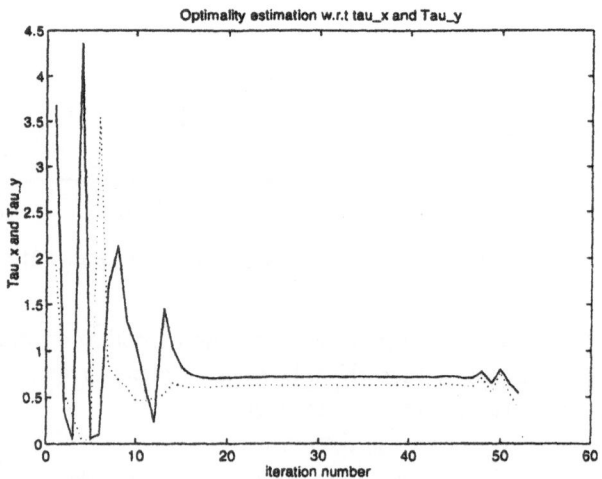

Fig. 6. Primal and dual rates versus Iterations; varying λ

6   Conclusion

The numerical analysis of Augmented Lagrangian methods is very sensitive to the penalty parameter. To be efficient, these algorithms must be taylored so that that parameter diverges to infinity in order to approach superlinear convergence. Numerical instabilities due to large parameters can be overcome (see [1]) yielding so-called augmented penalty algorithms but they cannot apply to decomposition as the penalized terms loose the separability property. When dealing with separable augmented lagrangian based on primal-dual proximal steps, the parameter has no more the role of a penalty parameter, but of a scaling parameter.

We haveshown here the trade-off between the convergence rates of the primal and dual sequences which leads toward an optimal value for the parameter. As that value depends on the second-order properties of the problem functions, it is in general very hard even if impossible to estimate. We have shown however how to update the parameter to keep both primal and dual sequences at the same relative speed. The global rate of convergence is thus expected to be linear and we conjecture that it can tend to an ideal compromise value of 0.5.

References

1. Dussault, J. P. (1997) Augmented penalty algorithms. IMA J. of Numer. Anal. **14**, 1–24
2. Eckstein, J., Bertsekas, D. P. (1992) On the Douglas-Rachford splitting method and the proximal point algorithm for maximal monotone operators. Math. Programming **55**, 293–318
3. Gabay, D., Mercier, B. (1976) A dual algorithm for the solution of nonlinear variational inequalities via finite element approximation. Comput. Math. Applic. **2**, 17–40
4. Hamdi, A., Mahey P., Dussault, J. P. (1997) A new decomposition method in nonconvex programming via a separable augmented Lagrangian. In: Gritzmann P., Horst R., Sachs E., Tischatschke R. (Eds.) Recent Advances in Optimization, Lecture Notes in Economics and Mathematical Systems **452**, 90–104
5. Mahey, P., Oualibouch, S., Pham, D. T. (1995) Proximal decomposition on the graph of a maximal monotone operator. SIAM J. Optimization **5**, 454–466
6. Mahey, P., Ouorou, A., Leblanc, L. J., Chifflet, J. An efficient implementation of the proximal decomposition algorithm for routing in telecommunications networks. Networks **31**, 227–238
7. Spingarn, J. E. (1985) Applications of the method of partial inverse to convex programming : decomposition. Math. Programming **32**, 199–223

Numerical Considerations on Decomposition and Augmented Lagrangians

Jean-Baptiste Malézieux[1,2] and Jean-Christophe Culioli[1,2]

[1] Centre Automatique et Systèmes, Ecole des Mines de Paris,
35 rue Saint-Honoré, F-77305 Fontainebleau, France
[2] Service Recherche Opérationnelle, Air France DI.IZ,
1, av du Maréchal Devaux, F-91550 Paray Vieille Poste, France

Abstract. We consider a general large scale constrained minimization problem with separable and differentiable objective function and constraints. The augmented Lagrangian method cannot be directly applied to such a problem because the resulting penalty function is not separable. We first recall two existing approaches (a method based on a linearization of the quadratic term in the augmented Lagrangian) and SALA (for Separable Augmented Lagrangian Algorithm) which is a generalization of Proximal Decomposition. We also present a new approach based on the decomposition of the line search in primal-dual algorithms but which requires the convexity of the augmented Lagrangian function. We finally compare the three methods on small and large scale test examples, insisting on robustness in parameter tuning.

Key words: augmented Lagrangians, decomposition, numerical experiments

1 Introduction

It is well known to any user of primal-dual methods that parameter tuning is a very time consuming task. Several primal-dual algorithms have excellent theoretical convergence behaviour while in practice they often fail due to the inability of the user to find the "right" parameters. This research is oriented toward finding robust algorithms, i.e. algorithms that will almost always converge to the solution in a reasonnable time and for a large domain of the tuning parameters. In the following, we present several primal-dual algorithms and test them for a wide range of each one of their parameters on some sample problems in order to validate their robustness with respect to this criterion. We successively present the problem, describe several existing approaches and propose a new method based on the decomposition of a line search strategy. We finally give our tests results and discuss them. As a brief conclusion, we propose further investigations to be done.

Let X be a finite dimensional space which decomposes into an orthogonal set of subspaces :

$$X = X_1 + X_2 + \cdots + X_N \, , \, \forall j \neq l \, : \, X_j \perp X_l \, .$$

We will assume for simplicity that the components $d_1 \in X_1$, $d_2 \in X_2,\ldots,$ $d_N \in X_N$ of a vector $d \in X$ are themselves vectors of X. Thus : $d = d_1 + \cdots + d_N$. The norm and scalar product over $X_1,\ldots,X_N$ and X will be the Euclidean norm and scalar product. Since d_j vectors are in X, we can write $\|d_j\|_X = \|d_j\|_{X_j}$. We consider the following problem :

$$
\begin{aligned}
\min\ & f(x) \\
\text{s.t.}\ & g(x) \leq 0 \\
& h(x) = 0 ,
\end{aligned}
\tag{1}
$$

where f, g and h are additive with respect to the decomposition of X :

$$
x \in X\ ,\ f(x) \in \mathbb{R}\ ,\ f(x) = \sum_{j=1}^{N} f_j(x_j)\ ,
$$
$$
g(x) = (g_1(x), g_2(x), \ldots, g_G(x))^T\ ,
$$
$$
h(x) = (h_1(x), h_2(x), \ldots, h_H(x))^T\ .
$$

The constraints of problem (1) can thus be written as :

$$
\forall i = 1,\ldots,G\ :\ g_i(x) = \sum_{j=1}^{N} g_{i,j}(x_j) \leq 0\ ,
$$
$$
\forall i = 1,\ldots,H\ :\ h_i(x) = \sum_{j=1}^{N} h_{i,j}(x_j) = 0\ .
$$

All functions $f_j, g_{i,j}, h_{i,j}$ are assumed to be sufficiently smooth. The feasible set of the problem is supposed non empty.

Standard primal-dual Lagrangian methods cannot be applied efficiently to such a problem, due to a potential lack of differentiability of the dual function. The idea is then natural of using augmented Lagrangian methods which regularize the dual function. The augmented Lagrangian function for problem (1) is :

$$
L_c(x, p, q) = f(x) + \frac{1}{2c} \sum_{i=1}^{G} \left[\max(0, c\, g_i(x) + p_i)^2 - (p_i)^2 \right]
$$
$$
+ q^T h(x) + \frac{c}{2} \|h(x)\|^2
$$
$$
= \sum_{j=1}^{N} J_j(x_j) + \frac{1}{2c} \sum_{i=1}^{G} \left[\max\left(0, c \sum_{j=1}^{N} g_{i,j}(x_j) + p_i\right)^2 - (p_i)^2 \right]
$$
$$
+ \sum_{i=1}^{H} q_i \sum_{j=1}^{N} h_{i,j}(x_j) + \frac{c}{2} \sum_{i=1}^{H} \left[\sum_{j=1}^{N} h_{i,j}(x_j) \right]^2 .
$$

This function which is to be minimized in x and maximized in p and q is unfortunately not separable with respect to the decomposition of X. To overcome this drawback, we will present several possible methods, all of them being primal-dual methods.

2   Existing Approaches

We will not survey all existing methods some of which seem to have never been fully tested in practice, however many interesting works can be cited : see page 113 of Hamdi's PhD thesis [5] for another recent bibliography. In the convex case, Eckstein [4] proposed the *epigraphic projection method* which seems efficient on quadratic block-separable convex problems with affine inequality constraints. For general separable convex problems, his method requires accurate epigraphic projections at every step which makes the practicality of the method an open question. Spingarn [9] did an interesting work on partial inverses which was extended by Hamdi, Mahey and Dussault [6] to nonconvex cases. Auslender [1] devised a cutting plane method in the convex case. Cohen and Zhu [3] generalized the work of Stephanopoulos and Westerberg to some nondifferentiable cases. Tanikawa and Mukai [11] added a separable convexifying term to the Lagrangian function of a nonconvex separable problem. Their method requires two levels of iterative optimization and introduces two extra parameters to be tuned. Previously Bertsekas [2] had published a similar method but which needed three levels of imbedded optimizations. More recently Shin-Yeu Lin [7] proposed a greedy method whose iteration solves a block-diagonal convex quadratic program which approximates the nonconvex problem. We focus here on the method of Hamdi [5,6] and on a linearization scheme from Cohen and Zhu [3].

2.1   Separated Augmented Lagrangian Algorithm (SALA)

SALA (for Separated Augmented Lagrangian Algorithm) was devised by Hamdi and Mahey [6]. See Hamdi [5] for a complete bibliography and presentation. This algorithm can be seen either as a generalization of proximal decomposition algorithm for diffentiable nonconvex problems, or as an application of an augmented Lagrangian algorithm to a resource allocation version of the original problem.

The standard version of SALA (the one we implemented) only works with equality constrained problems i.e. when $G = 0$.

Let us define allocations $y_j^0 \in \mathbb{R}^H$ such that $\sum_{j=1}^N y_j^0 = 0$. For $j \in \{1, \ldots, N\}$ let $x_j(y_j)$ be a solution if any of the following problem (2) :

$$\begin{aligned} &\min f_j(x_j) \\ &\text{s.t.} \quad x_j \in X_j \\ &\qquad h_j(x_j) + y_j = 0 \, , \end{aligned} \qquad (2)$$

Let $v_j(y_j) = \begin{cases} f_j(x_j(y_j)) & \text{if } x_j(y_j) \text{ exists,} \\ +\infty & \text{otherwise.} \end{cases}$

SALA is the application of an augmented Lagrangian method to the problem :

$$\min v(y) = \sum_{j=1}^{N} v_j(y_j)$$

$$\text{s.t. } \sum_{j=1}^{N} y_j = 0 .$$

The algorithm we implemented is :

1) For every $j = 1, \ldots, N$, solve :

$$\min_{x_j} f_j(x_j) + \sum_{i=1}^{H} q_i^k \left(h_{i,j}(x_j) + y_{i,j} \right) + \frac{c^k}{2} \sum_{i=1}^{H} (h_{i,j}(x_j) + y_{i,j})^2 \rightsquigarrow x_j^{k+1}$$

2) multiplier and allocations update :

$$r^{k+1} = \sum_{j=1}^{N} h_j(x_j^{k+1}) \in \mathbb{R}^H ,$$

$$q^{k+1} = q^k + \frac{c^k}{N} r^{k+1} ,$$

$$y_j^{k+1} = -h_j(x_j^{k+1}) + \frac{1}{N} r^{k+1} ,$$

$$c^{k+1} = \beta\, c^k .$$

3) increment k and go to step 1.

This version of SALA is the most simple in [5]. It's main advantage is the low number of parameters that are to be tuned, Those parameters are the following : c^k measures the quadratic penalization term, β measures the increase of c^k through iterations and a third parameter n_{LS} in our tests gives the maximum number of iterations in the primal minimization of step 1.

2.2   Linearized Augmented Lagrangian (LAL)

We recall here an algorithm presented in Cohen and Zhu [3] whose results apply to additive convex but non strongly convex problems. However the same ideas have also been studied by Stephanopoulos and Westerberg [10] for example and some convergence results exist which apply to nonconvex cases. The algorithm exposed here derives from the Auxiliary Problem Principle of Cohen. Convergence results using constraint qualification, mild convexity and Lipschitz properties of objective and constraints functions are presented in Cohen and Zhu [3].

Algorithm

Wishing to solve the minimization of f under a general constraint $\Theta(x) \in -C$ (where C typically has to be a closed convex cone). We start from Algorithm 14 of Cohen and Zhu [3] (pp.230 to 238) which for each k is :

Algorithm 14

$$\min_{x \in X} f(x) + \frac{1}{\varepsilon} \left[K_k(x) - \langle K_k'(x^k), x \rangle \right] \quad \leadsto \quad x^{k+1}$$
$$+ \langle \Theta(x), \varphi_\Theta'(\Theta(x^k), p^k) \rangle$$
$$p^{k+1} = p^k + \rho \, \varphi_p'(\Theta(x^{k+1}), p^k) \,,$$

with :

$$\varphi(\Theta, p) = \frac{1}{2c} \left[\|\Pi(p + c\Theta)\|^2 - \|p\|^2 \right]$$
$$\varphi_\Theta'(\Theta, p) = \Pi(p + c\Theta) \,,$$
$$\varphi_p'(\Theta, p) = \frac{1}{c} \left[\Pi(p + c\Theta) - p \right] \,.$$

- We can then choose as K_k :

$$K_k(x) = \frac{1}{2} \|x\|^2 \,. \tag{3}$$

Each iteration then consists in the resolution of N independant subproblems $(^1)$. The minimized criteria for each j are (up to an additive constant $\|x^k\|^2/2$) :

$$\min_{x_j \in X_j} f_j(x_j) + \frac{1}{2\varepsilon} \|x_j - x_j^k\|^2$$
$$+ \sum_{i=1}^{G+H} \Theta_{i,j}(x_j) \, \Pi_i(p^k + c\,\Theta(x^k))$$
$$\leadsto \quad x_j^{k+1} \,.$$

With our notations :

$$\min_{x_j \in X_j} f_j(x_j) + \frac{1}{2\varepsilon} \|x_j - x_j^k\|^2$$
$$+ \sum_{i=1}^{G} g_{i,j}(x_j) \max\left(0, p_i^k + c \sum_{l=1}^{N} g_{i,l}(x_l^k)\right)$$
$$+ \sum_{i=1}^{H} h_{i,j}(x_j)\left(p_i^k + c \sum_{l=1}^{N} h_{i,l}(x_l^k)\right) \quad \leadsto \quad x_j^{k+1} \,.$$

---

^{1} for $j = 1, \ldots, N$

- We can also choose :

$$
K_k(x) = \frac{1}{2}\|x\|^2 + \frac{\varepsilon c}{2} \sum_{j=1}^{N} \sum_{i=1}^{G+H} \Theta_{i,j}(x_j)^2
$$

$$
- \varepsilon c \sum_{j=1}^{N} \sum_{i=1}^{G+H} \Theta_{i,j}(x_j)\, \Theta_{i,j}(x_j^k) \ . \tag{4}
$$

Each iteration then consists in the resolution of N independant subproblems (^{2}). The minimized criteria for each j are (up to an additive constant $\|x^k\|^2/2$) :

$$
\min_{x_j \in X_j}\ f_j(x_j) + \frac{1}{2\varepsilon}\|x_j - x_j^k\|^2 + \sum_{i=1}^{G+H} \Theta_{i,j}(x_j)\Big[c(\tfrac{1}{2}\Theta_{i,j}(x_j)
$$
$$
- \Theta_{i,j}(x_j^k)) + \Pi_i(p^k + c\,\Theta(x^k))\Big] \quad \rightsquigarrow x_j^{k+1} \ .
$$

With our notations :

$$
\min_{x_j \in X_j}\ f_j(x_j) + \frac{1}{2\varepsilon}\|x_j - x_j^k\|^2
$$
$$
+ \sum_{i=1}^{G} g_{i,j}(x_j)\left[c(\tfrac{1}{2}g_{i,j}(x_j) - g_{i,j}(x_j^k)) + \max\left(0, p_i^k + c\sum_{l=1}^{N} g_{i,j}(x_l^k)\right)\right]
$$
$$
+ \sum_{i=1}^{H} h_{i,j}(x_j)\left[p_i^k + c\left(\tfrac{1}{2}h_{i,j}(x_j) - h_{i,j}(x_j^k) + \sum_{l=1}^{N} h_{i,l}(x_l^k)\right)\right]
$$
$$
\rightsquigarrow x_j^{k+1} \ ,
$$

or :

$$
\min_{x_j \in X_j}\ f_j(x_j) + \frac{1}{2\varepsilon}\|x_j - x_j^k\|^2
$$
$$
+ \sum_{i=1}^{G} g_{i,j}(x_j)\left[c(\tfrac{1}{2}g_{i,j}(x_j) - g_{i,j}(x_j^k)) + \max\left(0, p_i^k + c\sum_{l=1}^{N} g_{i,j}(x_l^k)\right)\right]
$$
$$
+ \sum_{i=1}^{H} h_{i,j}(x_j)\left[p_i^k + c\left(\tfrac{1}{2}h_{i,j}(x_j) + \sum_{l \neq j} h_{i,l}(x_l^k)\right)\right] \quad \rightsquigarrow x_j^{k+1} \ .
$$

---

^{2} for $j = 1, \ldots, N$

In both cases the multiplier and c^k updates are :

$$\forall i = 1, \ldots, G \; : \; p_i^{k+1} = (1 - \varrho)p_i^k + \varrho\, c^k \, \max\left(0, p_i^k + c^k \sum_{j=1}^{N} g_{i,j}(x_j^{k+1})\right),$$

$$\forall i = 1, \ldots, H \; : \; q_i^{k+1} = (1 - \varrho)q_i^k + \varrho\, c^k \sum_{j=1}^{N} h_{i,j}(x_j^{k+1}),$$

$$c^{k+1} = \beta\, c^k .$$

Implementation

In practice, we implemented both versions of K_k but the first choice (3) presented such a bad behaviour that we did not include it into our numerical results. Moreover, having the quadratic term $\Theta_{i,j}(x_j)^2$ in K_k can prevent successive x_k to diverge when $c \to +\infty$, thus bringing more stability at a low cost. The $1/(2\varepsilon)\|x_j - x_j^k\|^2$ term also brings more numerical stability but at the cost of one more parameter. We will further see that this parameter is hard to tune in practice.

The parameters used in our implementation LAL algorithm are the following : we have just seen ε ; as in SALA, c^k measures the quadratic penalization term and β measures the increase of c^k through iterations ; finally, ϱ rules the relaxation in lagrange multiplier update and a fifth parameter n_{LS} in our tests gives the maximum number of iterations in the primal minimization of step 1.

2.3   Decomposed Line Search (DLS)

This approach allows to straightforwardly use any Armijo line search based method as primal minimizations, thus allowing a global convergent strategy when the augmented Lagrangian function is convex w.r.t. x. Our method is a parallel implementation of line search based primal minimization methods for augmented Lagrangians.

Presentation of DLS

Let us consider an Armijo line search on $L_c(\,.\,,p,q)$ starting from $x \in X$ in direction $d = d_1 + \cdots + d_N$. This expensive line search on L_c in the *global* direction d will here be decomposed into N independant line searches in respective directions d_j for $j = 1, \ldots, N$.

Let $0 < m_L < 1$. A search from x in direction d_j ensures that one gets a positive step α_j such that a sufficient decrease condition is met :

$$L_c(x + \alpha_j\, d_j, p, q) \le L_c(x, p, q) - m_L\, \alpha_j\|d_j\|^2 \tag{5}$$

We will further see how such independant searches can be easily executed.

Proposition 1. *Let $J : X \longrightarrow \mathbb{R}$ be a convex function, let $m_L \in \,]0,1[$ be a fixed parameter. If there exists vectors $x_j, d_j \in X_j$, $j = 1, \ldots, N$ (d_j being non zero for all j) and numbers $\alpha_j > 0$, $j = 1, \ldots, N$ such that*

$$d_1 + \cdots + d_N = d \in X \,,$$

$$x_1 + \cdots + x_N = x \in X \,,$$

$$\forall\, j = 1, \ldots, N \,:\; J(x + \alpha_j\, d_j) \le J(x) - m_L\, \alpha_j\, \|d_j\|^2 \,,$$

then for $\bar{\alpha} = 1/(\sum_{j=1}^{N}(1/\alpha_j))$ one gets :

$$J(x + \bar{\alpha}\, d) \;\le\; J(x) - m_L\, \bar{\alpha}\, \|d\|^2 \,. \tag{6}$$

Proof. We look for a step $\bar{\alpha} > 0$ in direction d such that $\bar{\alpha}\, d = \bar{\alpha} \sum_{j=1}^{N} d_j$ and such that $\bar{\alpha}\, d$ is a convex combination of the vectors $\alpha_j\, d_j$ taken as vectors of X for $j = 1, \ldots, N$, with respective weights ε_j. We would then have :

$$\bar{\alpha}\, d = \sum_{j=1}^{N} \varepsilon_j\, \alpha_j\, d_j \,. \tag{7}$$

Multiplying both sides of this equation by d_j for every j, we obtain for all j": $\bar{\alpha} = \varepsilon_j\, \alpha_j$ or :

but since $\sum_{j=1}^{N} \varepsilon_j = 1$, we finally get :

$$\bar{\alpha} = 1/\sum_{j=1}^{N} \frac{1}{\alpha_j} \,.$$

Then by convexity of J :

$$J(x + \bar{\alpha}\, d) \;=\; J\!\left(x + \sum_{j=1}^{N} \varepsilon_j\, \alpha_j\, d_j\right) \le \sum_{j=1}^{N} \varepsilon_j\, J(x + \alpha_j\, d_j) \,.$$

Using the sufficient decrease assumption for each direction d_j, we further have :

$$J(x + \bar{\alpha}\, d) \le \sum_{j=1}^{N} \varepsilon_j\, \left(J(x) - m_L\, \alpha_j\, \|d_j\|^2\right) \,,$$

then considering that $\sum_{j=1}^{N} \varepsilon_j = 1$, and $\sum_{j=1}^{N} \|d_j\|^2 = \|d\|^2$, we get :

$$J(x + \bar{\alpha}\, d) \le J(x) - m_L \sum_{j=1}^{N} \varepsilon_j\, \alpha_j\, \|d_j\|^2$$

$$= J(x) - m_L \sum_{j=1}^{N} \bar{\alpha}\, \|d_j\|^2 \;=\; J(x) - m_L\, \bar{\alpha}\, \|d\|^2 \,,$$

which completes the proof. $\square$

Remark 1. If any vector d_j is zero, Proposition 1 can still be applied, by summing over the set of indices $\{1,\ldots,N\}\backslash\{j\}$.

Application

When L_c is convex w.r.t. x, Proposition 1 can be applied : sufficiently decreasing $\alpha_j\,d_j$ line searches thus provide a sufficient decrease conditions for L_c in global direction d :

$$L_c(x + (1/\sum_{j=1}^{N} 1/\alpha_j)d, p, q) = L_c(x, p, q) - m_L(1/\sum_{j=1}^{N} 1/\alpha_j)\|d\|^2 .$$

Implementation

Our implementation of Armijo line searches could be replaced by more sophisticated line searches such a Wolfe line search, the theoretical interest is to be ensured of a sufficient decrease of the global function L_c. The computations of values of L_c are as follows :

$$
\begin{aligned}
L_c(x + \alpha_j\,d_j, p, q) = & f(x) - f_j(x_j) + \underline{f_j(x_j + \alpha_j\,d_j)} \\
& + \frac{1}{2c}\sum_{i=1}^{G}\Big\{ \max\big[0, p_i + c\,(g_i(x) - g_{i,j}(x_j) + \underline{g_{i,j}(x_j + \alpha_j\,d_j)})\big]^2 - p_i^2\Big\} \\
& + \sum_{i=1}^{H}\Big\{ \big(h_i(x) - h_{i,j}(x_j) + \underline{h_{i,j}(x_j + \alpha_j\,d_j)}\big)\big[q_i + \frac{c}{2}(h_i(x) \\
& - h_{i,j}(x_j) + \underline{h_{i,j}(x_j + \alpha_j\,d_j)})\big]\Big\}
\end{aligned}
$$

This requires only the evaluation of underlined expressions (which are $f_j(x_j + \alpha_j\,d_j)$, $g_{i,j}(x_j + \alpha_j\,d_j)$ and $h_{i,j}(x_j + \alpha_j\,d_j)$), and a few elementary operations.

This turns out the same for the computation of a part of the gradient in x_j of L_c :

$$\nabla_{x_j} L_c(x, p, q) = \underline{\nabla_{x_j} f_j(x_j)} + \frac{1}{c}\sum_{i=1}^{G}\Big[\max(0, p_i + c\,g_i(x))\underline{\nabla_{x_j} g_{i,j}(x_j)}\Big] \quad (8)$$

$$+ \sum_{i=1}^{H}\Big[(q_i + c\,h_i(x))\underline{\nabla_{x_j} h_{i,j}(x_j)}\Big] \quad (9)$$

This requires only the evaluation of underlined expressions in (8) ($\nabla f_j(x_j)$, $\nabla_{x_j} g_{i,j}(x_j)$, $\nabla_{x_j} h_{i,j}(x_j)$) and a few elementary operations.

DLS Algorithm

Set $x_0, p^0, q^0, c^0, k = 0$, compute for all i and j : $f_j(x_j^k)$, $g_{i,j}(x_j^k)$, $h_{i,j}(x_j^k)$.
Do

- Do
    - In each block : compute $d_j = -\nabla_{x_j} L_c(x^k, p^k, q^k)$ Line search : find $\alpha_j > 0$ such that Armijo condition (5) is met in direction d_j.
    - Once for all blocks : (coordination step) compute $\bar{\alpha} = 1/\sum_{j=1}^{N}(1/\alpha_j)$ and $L_c(x^k + \bar{\alpha} d, p^k, q^k)$.
      Choose the best step s among $\{\alpha_j d_j, \bar{\alpha} d\}$, and set $x^k + s$ as the new current point x^k.

  Until a maximum number of successive line searches has been reached, OR until a stopping criterion is met.
- set x^{k+1} as the new current point.
- Multiplier and c^k update :

$$p_i^{k+1} = (1 - \varrho)p_i^k + \varrho c^k \max\left(0, p_i^k + c^k \sum_{j=1}^{N} g_{i,j}(x_j^{k+1})\right),$$

$$q_i^{k+1} = (1 - \varrho)q_i^k + \varrho c^k \sum_{j=1}^{N} h_{i,j}(x_j^{k+1}),$$

$$c^{k+1} = \beta c^k.$$

Until convergence.

A proof of convergence can be drawn for this algorithm in the case where exact primal minimizations are obtained. This is unfortunately never the case in practice.

Parameters

The parameters used in our DLS algorithm are the following : As in SALA, c^k measures the quadratic penalization term and β measures the increase of c^k through iterations ; ϱ rules the relaxation in lagrange multiplier update and a fourth parameter n_{LS} in our tests gives the maximum number of iterations in the primal minimization of step 1.

Interest of the Approach

Compared to the global augmented Lagrangian function, time is saved if some of the line searches in sub-directions d_j take lower time than others. Moreover, since the step which gave the least value for L_c is chosen, we take advantage of simultaneous searches in N directions.

However our method is not designed for heterogeneous problems but rather for separable problems just too big to be handled at once.

3   Numerical Results

All algorithms were implemented using basic steepest descent algorithm with Armijo line search ([3]). At most n_{LS} Armijo gradient steps were performed before multiplier update. Stopping criterions for line searches as well as convergence were common to all algorithms.

These tests cannot pretend to be fully objective since the method used for unconstrained primal minimizations is rather simplistic. Further tests should be executed using an efficient unconstrained minimization method and a modern line search.

3.1   Computational Remarks

The computation of a step in a subproblem j of LAL (namely a new linearized augmented Lagrangian funciton in $x_j + \alpha_j d_j$ and its gradient) requires the same function evaluations as part j of the original L_c augmented Lagrangian function of DLS and of its gradient, then if function and gradients evaluations are the most time consuming computations, DLS and LAL are almost equivalent for the computation of a line search step.

A similar remark can be formulated between DLS and SALA, the advantage of LAL and SALA being that primal subproblems are really independant. Moreover, memory requirements for DLS, LAL and SALA are also quite similar.

3.2   Parameters

Since the important factor was robustness of the algorithms regarding the tuning of their parameters, for each algorithm and each test example, we gave to each of the parameters of the algorithm 5 different values :

$$\varepsilon \in \{0.01, 0.1, 1, 10, 100\} \quad \text{for LAL,}$$
$$c^0 \in \{0.001, 0.1, 1, 10, 100\} \quad \text{for DLS, LAL and SALA,}$$
$$\beta \in \{1, 1.005, 1.05, 1.1, 2\} \quad \text{for DLS, LAL, and SALA,}$$
$$\varrho \in \{0.2, 0.4, 0.6, 0.8, 1\} \quad \text{for DLS and LAL.}$$
$$n_{LS} \in \{1, 2, 9, 64, 625\} \quad \text{for DLS, LAL and SALA.}$$

So that DLS had 4 parameters (c^0, β, ϱ and n_{LS}), LAL had 5 parameters (ε, c^0, β, ϱ and n_{LS}), SALA had 3 parameters (c^0, β and n_{LS}). This gave for each test example :

3125   tests for LAL,

625   tests for DLS,

125   tests for SALA.

---

[3] starting from α_0, find the first $\alpha_0\mu^m$ which satisfies the sufficient decrease condition (6) where $0 < \mu < 1$ and $m = 1, 2, \ldots$

We considered that convergence dit not occur when the algorithm needed more than 500 iterations or more than 10^6 calls to any part of the objective function. What interested us in the results was for every algorithm/test example pair :

- number of convergent runs / proportion of convergent runs for a given algorithm and a given test example,
- average overall number of calls to any part of the objective function over all tests,
- least overall number of calls to any part of the objective function among all tests,
- average overall number of calls to any part of the gradient of the objective function over all tests,
- overall number of calls to any part of the gradient of the objective function for the test which led to the least number of calls to the objective function.

We also included in our tables :

- the average number of iterations over all tests for an algorithm/test example pair, We notice that this number of iterations is not a good comparison criterion between algorithms since iterations in different algorithms are of different nature.
- the number of iteration of the test which led to the least number of calls to the objective function.
- the number of convergent runs of which the number of calls to the objective function parts is less than 1.5 times the best number of calls met. In other words, the number of runs close to what we considered as the best run. We associated this number of runs with the equivalent proportion amongst all runs for this algorithm/test example pair.

3.3   Test Examples

Hamdi's PhD, Problem 16a

Taken from Hamdi's PhD thesis [5](p.191) :

$$\min f(x) = a_1(x_1 - 0.5)^2 + \sum_{i=2}^{m} a_i(x_i + 1)^2 + \sum_{i=m+1}^{M} a_i(x_i - 1)^2 ,$$

$$\text{s.t.}\ \ h(x) = b_1(1 - x_1) + \sum_{P} i = 2^m b_i x_i^2 + \sum_{i=m+1}^{M} b_i(x_i - 1)^2 = 0 ,$$

where for $i = 1, \ldots, M$:

$$a_i = \psi - \frac{\psi^2 - 1}{\psi(M-1)}(i-1) , \quad b_i = 1 - \frac{\psi^2 - 1}{M-1}(i-1) ,$$

with $\psi = 0.8$, $m = 25$, $M = 50$; number of subdomains : 10 ; number of variables : 50 ; sizes of the respective subdomains : 2, 1, 17, 7, 2, 1, 12, 1, 2, 5 ; number of constraints : 1 ; starting point : $x^0 = (1, 1, ..., 1)$. See Table 1 for results.

Table 1. Problem Hamdi 16a tested with DLS, LAL and SALA

Method	DLS	LAL	SALA
Number of parameters	4	5	3
Total number of runs	625	3125	125
Convergent runs / proportion	358 / 0.5728	86 / 0.02572	78 / 0.624
CV runs 50% from best	16 / 0.0256	11 / 0.00352	2 / 0.016
Average iter.	100	115	69
Iter. best run	44	48	61
Average fct/grad calls	45951 / 3809	146102 / 5509	44912 / 4063
Fct/grad calls of best run	4724 / 680	744 / 1072	1135 / 914

Hamdi's PhD, Problem 16d

Taken from Hamdi's PhD thesis [5] (p.191) :

$$\min f(x) = a_1(x_1 - 0.5)^2 + \sum_{i=2}^{m} a_i(x_i + 1)^2 + \sum_{i=m+1}^{M} a_i(x_i - 1)^2 ,$$

$$\text{s.t.} \ \ h(x) = b_1(1 - x_1) + \sum_{P} i = 2^m b_i x_i^2 + \sum_{i=m+1}^{M} b_i(x_i - 1)^2 = 0 ,$$

where for $i = 1, \ldots, M$:

$$a_i = \psi - \frac{\psi^2 - 1}{\psi(M - 1)}(i - 1) , \ \ b_i = 1 - \frac{\psi^2 - 1}{M - 1}(i - 1) ,$$

with $\psi = 0.8$, $m = 500$, $M = 1000$; number of subdomains : 10 ; number of variables : 1000 ; sizes of the respective subdomains : 40, 20, 340, 140, 40, 20, 240, 20, 40, 100 ; number of constraints : 1 ; starting point : $x^0 = (1, 1, ..., 1)$. See Table 2 for results.

Table 2. Problem Hamdi 16d tested with DLS and SALA

Method	DLS	SALA
Number of parameters	4	3
Total number of runs	625	125
Convergent runs / proportion	299 / 0.4784	73 / 0.584
CV runs 50% from best	24 / 0.0384	4 / 0.032
Average iter.	121	72
Iter. best run	36	31
Average fct/grad calls	80846 / 6862	70616 / 4895
Fct/grad calls of best run	10887 / 1050	3367 / 600

Schittkowski's Book, Problem 269

Taken from [8] :

$$\min f(x) = (x_1 - x_2)^2 + (x_2 + x_3 - 2)^2 + (x_4 - 1)^2 + (x_5 - 1)^2 ,$$
$$\text{s.t.} \quad h_1(x) = x_1 + 3\,x_2 = 0$$
$$h_2(x) = x_3 + x_4 - 2\,x_5 = 0$$
$$h_3(x) = x_2 - x_5 = 0 .$$

Number of subdomains : 3 ; number of variables : 5 ; sizes of the respective subdomains : 3,1,1 ; number of constraints : 3 ; starting point : $x^0 = (2, 2, 2, 2, 2)$. See Table 3 for results.

Rosen Suzuki

$$\min f(x) = x_1^2 - 5\,x_1 + x_2^2 - 5\,x_2 + 2\,x_3^2 - 21\,x_3 + x_4^2 + 7\,x_4 ,$$
$$\text{s.t.} \quad g_1(x) = 2\,x_1^2 + 2\,x_1 + x_2^2 - x_2 + x_3^2 - x_4 - 5 \leq 0$$
$$g_2(x) = x_1^2 + x_1 + x_2^2 - x_2 + x_3^2 + x_3 + x_4^2 - x_4 - 8 \leq 0$$
$$g_3(x) = x_1^2 - x_1 + 2\,x_1^2 + x_3^2 + 2\,x_4^2 - x_4 - 10 .$$

Number of subdomains : 4 ; number of variables : 4 ; sizes of the respective subdomains : 1,1,1,1 ; number of constraints : 3 (inequalities) ; starting point : $x^0 = (12, -0.3, 54, 0.002)$. See Table 4 for results.

3.4 Comments

We give here our impression on the tests presented above.

Table 3. Problem Schittkowski 269 tested with DLS, LAL and SALA

Method	DLS	LAL	SALA
Number of parameters	4	5	3
Total number of runs	625	3125	125
Convergent runs / proportion	187 / 0.2992	123 / 0.03936	35 / 0.28
CV runs 50% from best	8 / 0.0128	9 / 0.00288	2 / 0.016
Average iter.	161	227	115
Iter. best run	78	59	85
Average fct/grad calls	16775 / 1755	106869 / 4166	72629 / 7270
Fct/grad calls of best run	1189 / 234	749 / 510	1046 / 504

Table 4. Problem Rosen-Suzuki tested with DLS and LAL

Method	DLS	LAL
Number of parameters	4	5
Total number of runs	625	3125
Convergent runs / proportion	215 / 0.344	185 / 0.0592
CV runs 50% from best	4 / 0.0064	5 / 0.0016
Average iter.	159	263
Iter. best run	28	126
Average fct/grad calls	25454 / 1641	217530 / 7210
Fct/grad calls of best run	3030 / 304	5037 / 1585

Linearized Augmented Lagrangian

On most problems LAL gets the lowest number of function calls for its best run but the average number of calls over all convergent runs is usually an order greater than that of other algorithms. Moreover ε is very difficult to tune, and most of the time convergence does not occur if primal minimizations are not pushed far which means that only a high n_{LS} is acceptable in practice, although this implies more calls to the objective function.

Separated Augmented Lagrangian Algorithm

This rather simple version of the algorithm presented the advantage of requiring the tuning of only 3 parameters, i.e. we did not over-relax neither underrelax the multiplier updates.

SALA needs a good stopping criterion for primal minimizations although behaves better than LAL when primal sub minimizations are not very accurate (i.e. n_{LS} low). Moreover the performances are much higher when β is not close to 1,

Decomposed Line Search

DLS usually requires less calls to the gradients than SALA, behaves as well when $\varrho = 1$ and when $\beta = 1$.

DLS sticks closer to the original Augmented Lagrangian method but the result in Proposition 1 given only stands when the Augmented Lagrangian function is convex with respect to its primal variable. However we can apply this method to any problem, considering that N directions are permanently being sought. The method as presented here is stuck with Armijo line search based methods but one could easily implement more sophisticated line searches such as Wolfe.

A final remark on DLS is that it requires a synchronization step after every line search, it is thus not designed for heterogeneous problems. At almost no extra cost compared to our method, we could have implemented the original augmented Lagrangian as a parallel method by keeping on each node about the same amount of information as is done in our scheme. The only requirement would be to be able to keep in memory several vectors of X on the same processor (as much as possible). This would simply result in an even more synchronized algorithm but which could take advantage of all existing results on augmented Lagrangian methods. Moreover, such a direct implementation of augmented Lagrangian would be compatible with almost any unconstrained minimization algorithm that would only need first order derivatives (for instance any quasi-Newton method).

4   Conclusion

We have tested three strategies of decomposition of augmented Lagrangians, one of which is new (DLS). This last method tested with 4 parameters and SALA tested with only 3 parameters had comparable results, but DLS does not enjoy as desirable theoretical convergence results as SALA does. The linearized augmented Lagrangian methods did not look really competitive compared to SALA, although LAL often did the best run. However more practical testing should be done on real test examples and with efficient minimization methods in order to fully validate those remarks.

References

1. Auslender, A. (1985) Two general methods for computing saddle points with applications for decomposing convex programming problems. Appl. Math. Optim. **13**, 79–95
2. Bertsekas, D. P. (1979) Convexification procedures and decomposition methods for nonconvex optimization problems. JOTA **29**, 169–197
3. Cohen, G., Zhu, D. L. (1984) Decomposition coordination methods in large-scale optimization problems. The nondifferentiable case and the use of augmented lagrangian. In: Cruz J. (Ed.) Advances in large scale systems theory and applications, vol. I, JAI Press, Greenwitch, Connecticut, 203–266
4. Eckstein, J. (1994) Parallel alternating direction multiplier decomposition of convex programs. JOTA **80**, 39–61
5. Hamdi, A. (1997) Méthode de Lagrangiens augmentés en programmation convexe et non convexe et applications à la décomposition. PhD Thesis, Blaise Pascal University, Clermont-Ferrand, France
6. Hamdi, A., Mahey, P., Dussault, J. P. (1997) A new decomposition method in nonconvex programming via separable augmented Lagrangians. In: Gritzmann P., Horst R., Sachs E., Tichatschke R. (Eds.) Recent Advances in Optimization. Lecture Notes in Economics and Mathematical Systems **452**, 90–104
7. Lin, S.-Y. (1992) Complete decomposition algorithm for nonconvex separable optimization problems. Automatica **28-6**, 1249–1254
8. Schittkowski, K. (1987) More test examples for nonlinear programming codes, Springer-Verlag, Lecture Notes in Economics and Mathematical Systems, **212**
9. Spingarn, J. E. (1985) Application of the method of partial inverses to convex programming, decomposition. Math. Prog. **32**, 199–223
10. Stephanopoulos, G., Westerberg, W. (1975) The use of Hestenes' method of multipliers to resolve dual gaps in engineering system optimization. JOTA **15**, 285–309
11. Tanikawa, A., Mukai, H. (1987) New Lagrangian function for nonconvex primal-dual decomposition. Comput. Math. Applic. **13-8**, 661–676

Bicriteria Linear Fractional Optimization

Christian Malivert[1] and Nicolae Popovici[2]

[1] LACO, ESA-CNRS 6090, University of Limoges, France.
[2] Department of Mathematics, "Babeş-Bolyai" University of Cluj, Romania.

Abstract. In multiple objective programming it is generally more convenient to study the efficient outcome instead of the efficient solution set. In general an approximation of this efficient outcome is obtained by solving a sequence of optimization problems. In this work we consider a special class of bicriteria optimization problems with linear fractional objectives and linear constraints. It is shown that the efficient outcome is the graph of a piecewise linear fractional curve in the plane, which can be easily computed. A finite algorithm is presented which generates this curve by simply considering some particular edges of the constraint polyhedron.

Key words: bicriteria optimization, fractional programming, efficiency

1 Introduction

Many studies concerned with numerical aspects of multiple objective programming consist to develop procedures for finding all the solutions in the decision set. These solutions form a subset of the feasible region D in $\mathbb{R}^n$ which is called the efficient set. Most of algorithms presented in the literature consider only the linear case (see [1],[14] and references therein), or have been written for problems having a special structure as location problems [11] , [12]. These restrictions are justified by the fact that in many nonlinear situations the complexity of the efficient set grows rapidly with the size of the problems and the determination of all efficient solutions becomes rapidly unmanageable. For that reason it may be interesting to turn out our attention to the set of efficient values instead of efficient solutions, mainly when the number of objectives is small compared with the number of variables. It is typically the case in bicriteria optimization where the efficient outcome lies in $\mathbb{R}^2$. Theoretical results about this approach are developped for instance in [5] and [3] and applications of this idea to industrial problems can be found in [2], [8] and [9]. The numerical applicability of the method depends strongly on the class of functions involved in the problem and in general only an approximation of the efficient outcome is obtained. The aim of this work is to show that for a special class of nonconvex bicriteria problems it is possible to determine exactly the set of efficient values which can be represented by a curve in $\mathbb{R}^2$. In this class of problems, the objectives are linear fractional functions and the constraint set is given by a polyhedron of $\mathbb{R}^n$. Although these functions are not convex, they enjoy good properties and it is proved for example in [5] or [12] that all the efficient values can be obtained by simply considering the edges of the constraint polyhedron. The method proposed in this work find iteratively these efficient edges and provides the decision maker with an explicit equation of the curve in $\mathbb{R}^2$ representing the set of efficient values.

The paper is organized as follows. In the second section we introduce the notations and the preliminary results. The third section studies the efficient subset contained in a given edge. It is shown that it is a line segment which can be directly obtained from the simplex tableau corresponding to one of the ends of the edge that we consider. Then the main algorithm is developped and in the last section an illustrative example is given.

2   Preliminary results

We consider the problem of minimizing two linear fractional functions over a convex polyhedron :

$$(\mathcal{P}) \qquad \min\{f(x) = (f_1(x), f_2(x)) : x \in D\}$$

where

$$f_i(x) = \frac{c_i x + s_i}{d_i x + t_i}, \ c_i \in \mathbb{R}^n, \ d_i \in \mathbb{R}^n, \ s_i \in \mathbb{R}, \ t_i \in \mathbb{R}, \ i = 1, 2$$

and

$$D = \{x \in \mathbb{R}^n : Ax \leq b, x \geq 0\}$$

with A a $m \times n$ matrix and $b \in \mathbb{R}^m_+$. No restriction occurs from this formulation of D as it is well-known in linear programming that every nonempty polyhedron can be written under this form. We suppose also that D is bounded and that the denominators $d_i x + t_i$ are strictly positive for $i = 1, 2$ and $x \in D$.

Recall that a level surface of a linear fractional function g defined on D given by

$$S_\alpha = \{x \in D : g(x) = \alpha\}$$

can be written $S_\alpha = H_\alpha \cap D$ where H_α is an hyperplane of $\mathbb{R}^n$. Using this property many algorithms have been developped, see for instance [4] or [7], in order to minimize a linear fractional objective over a polyhedron. All these algorithms adapt the simplex method to the fractional case and provide an optimal vertex in a finite number of iterations.

If $x, y \in D$, the notation $f(x) \leq f(y)$ (resp. $<$) means $f_i(x) \leq f_i(y)$ (resp. $<$) for $i = 1, 2$. A point $x \in D$ is called an efficient solution of $(\mathcal{P})$ (notation : $x \in E$) and then $f(x) = (f_1(x), f_2(x))$ is called an efficient value, if it does not exist $y \in D$ such that $f(y) \leq f(x), \ f(y) \neq f(x)$.

Proposition 1. *The set E of the efficient solutions of $(\mathcal{P})$ is closed.*

Proof. Suppose to the contrary that (x_n) is a sequence of efficient solutions in D such that $\lim x_n = \bar{x}$ with $\bar{x} \notin E$. From the definition of efficiency there is $y \in D$ such that $f(y) \leq f(\bar{x})$, $f(y) \neq f(\bar{x})$. If $f(y) < f(\bar{x})$ the continuity of f_1 and f_2 implies that $f(y) < f(x_n)$ for n large enough, contradicting the efficiency of x_n. Thus we may suppose that for instance $f_1(y) = f_1(\bar{x})$ and $f_2(y) < f_2(\bar{x})$. In that case we have

$$f_1(z) \geq f_1(y) \quad \forall z \in D. \tag{1}$$

Indeed, if we suppose that $f_1(z_0) < f_1(y)$ with $z_0 \in D$, taking $y' = y + \varepsilon(z_0 - y) \in D$ with $\varepsilon > 0$ small enough, we have $f_1(y') < f_1(y) = f_1(\bar{x})$ and $f_2(y') < f_2(\bar{x})$. The first inequality is a property of linear fractional functions and the second one follows from the continuity of f_2. Thus $f(y') < f(\bar{x})$ and we get a contradiction as in the first case. Now choosing $z = x_n$ in (1) it follows that $f_1(x_n) \geq f_1(y)$ and using the continuity of f_2, we get $f_2(x_n) > f_2(y)$ for n large enough, contradicting the efficiency of x_n.   □

Note that the above result fails to be true for a general bicriteria problem. For instance the efficient set of

$$\min\{(4 - (1 - x)^2, x + 4) : 0 \leq x \leq 3\}$$

is $\{0\} \cup]2, 3]$ which is not closed in $\mathbb{R}$.

Notice also that the result of Proposition 1 cannot be extended to three or more linear fractional objectives. If we consider for example the problem

$$\min\left\{\left(\frac{x_1 - 4}{-x_2 + 3}, \frac{-x_1 + 4}{x_2 + 1}, -x_2\right) : -x_1 + 4x_2 \leq 0, \; x_1 - \frac{x_2}{2} \leq 4, \; x_1, x_2 \geq 0\right\}$$

we have $\{(4, y) : 0 \leq y < 1\} \in \operatorname{cl}(E) \setminus E$.

Let us introduce the notations

$$\underline{f}_i = \min\{f_i(x) : x \in D\}$$

$$\bar{f}_1 = \min\{f_1(x) \; : \; x \in D, f_2(x) = \underline{f}_2\} \tag{2}$$

$$\bar{f}_2 = \min\{f_2(x) \; : \; x \in D, f_1(x) = \underline{f}_1\} \tag{3}$$

Note that these optimization problems consists to minimize a linear fractional function over a polyhedron and it is easy to prove that $(\underline{f}_1, \bar{f}_2)$ as well as $(\bar{f}_1, \underline{f}_2)$ are two efficient values for the problem $(\mathcal{P})$.

Under general assumptions of semistrict quasiconvexity on f_1 and f_2, Schaible [13] has proved the following result :

Lemma 1. [13] *If $x \in E$, then $\underline{f}_i \leq f_i(x) \leq \bar{f}_i$ $i = 1, 2$. Conversely, if t satisfies $\underline{f}_i \leq t \leq \bar{f}_i$ for $i = 1$ or 2 then there exists $x \in E$ such that $f_i(x) = t$.*

Based on this result a new characterization of the set of efficient values is proposed in [6].

Lemma 2. [6] *There exists a continuous, strictly decreasing function φ, from $[\underline{f}_1, \bar{f}_1]$ onto $[\underline{f}_2, \bar{f}_2]$ such that*

$$f(E) = \{(t, \varphi(t)) \ : \ t \in [\underline{f}_1, \bar{f}_1]\}$$

where

$$\varphi(t) = \min\{f_2(x) : x \in D, f_1(x) = t\}. \tag{4}$$

Observe that the closedness of E (Proposition 1) follows directly from Lemma 2. Indeed the graph G_φ of the continuous function φ being closed, $E = f^{-1}(G_\varphi)$ is also a closed subset.

If we denote by D_t the polyhedron $D \cap \{x \in \mathbb{R}^n : f_1(x) = t\}$ and by s_t a vertex of D_t which is a solution of

$$(\mathcal{P}_t) \qquad \min\{f_2(x) : x \in D_t\},$$

it follows from Lemma 2 that

$$f(E) = \{(f_1, f_2)(s_t) : t \in [\underline{f}_1, \bar{f}_1]\}.$$

Observe that s_t being a vertex of D, it is either a vertex of D or an interior point for some edge of D. Thus every efficient value can be obtained by simply considering the image of the efficient points located on the edges of D. As mentioned in [5] and [12] this property justifies the interest for the set of efficient values which is generally simpler than the efficient set.

In the next section we consider a situation in which we have a solution $s_t \in]\bar{x}, \bar{y}[\cap E$ of $(\mathcal{P})$ belonging to the relative interior of an edge $\mathcal{E} = [\bar{x}, \bar{y}]$ of D and we want to determine entirely the efficient subset $]\bar{x}, \bar{y}[\cap E$.

3   Efficient subset of an edge

In [5] it is shown that the efficient part of a face of D is convex where a face is defined using two subsets $I \subset \{1, ..., n\}$ and $J \subset \{1, ..., m\}$ by the formula

$$F = \{x \in D : A_i x < b_i \ i \in I, A_i x = b_i \ i \notin I, \ x_i > 0 \ i \in J, x_i = 0 \ i \notin J\}.$$

It follows that in the particular case of an edge $[\bar{x}, \bar{y}]$ of D, its relative interior is a face in the previous sense and then $]\bar{x}, \bar{y}[\cap E$ is an interval. The purpose of this section is to find explicitly this interval.

As usual in linear programming we denote by B a nonsingular $m \times m$-matrix formed by choosing m columns of the matrix $[A, Id_m]$ such that $B^{-1}b \geq 0$ and by N the $m \times n$-matrix formed with the remaining columns taken in their order. B is called a feasible basis for the system of inequalities $Ax \leq b$, $x \geq 0$, this basis corresponding to a basic solution (vertex) denoted by $\bar{x}$. It is well-known that each element x of the polyhedron D is characterized by its basic and nonbasic components (x_B, x_N) which are nonnegative and satisfy the formula

$$x_B = B^{-1}b - B^{-1}Nx_N. \tag{5}$$

In particular $(\bar{x}_B, \bar{x}_N) = (B^{-1}b, 0)$ characterizes the vertex $\bar{x}$. We say that $\bar{x}$ is a non degenerate vertex if $B^{-1}b > 0$, which means that exactly n of the inequalities $A\bar{x} \leq b$, $\bar{x} \geq 0$ are equalities. With respect to the new variables x_N the affine functionals $P_i(x) = c_i x + s_i$ and $Q_i(x) = d_i x + t_i$ for $i = 1, 2$ can be written $P_i(x) = \bar{c}_i x_N + \bar{s}_i$ and $Q_i(x) = \bar{d}_i x_N + \bar{t}_i$ for $i = 1, 2$ where $\bar{c}_i, \bar{d}_i$ are vectors in $\mathbb{R}^n$ and $\bar{s}_i, \bar{t}_i$ are real numbers.

For $x \in D$ we denote by

$$B(x) = \{i \in \{1 \ldots m\} : (x_B)_i = 0\} \tag{6}$$

and by

$$N(x) = \{i \in \{1 \ldots n\} : (x_N)_i = 0\}. \tag{7}$$

Theorem 1. *A point $x \in D$ is efficient for $(\mathcal{P})$ if and only if there exist $\lambda_1 > 0$ and $\lambda_2 > 0$ such that for all h_N satisfying $(h_N)_i \geq 0$, $i \in N(x)$ and $(B^{-1}Nh_N)_i \leq 0, i \in B(x)$ we have*

$$\sum_{i=1}^{2} \lambda_i \left[(\bar{d}_i x_N + \bar{t}_i)\bar{c}_i - (\bar{c}_i x_N + \bar{s}_i)\bar{d}_i \right] h_N \geq 0 \tag{8}$$

Proof. From [10], $x \in D$ is efficient if and only if there exists $\lambda_1 > 0$ and $\lambda_2 > 0$ such that for all $h \in T_x D$

$$\sum_{i=1}^{2} \lambda_i [(d_i x + t_i)c_i - (c_i x + s_i)d_i] h \geq 0 \tag{9}$$

where $T_x D = \mathbb{R}_+(D - x)$ is the tangent cone to D at the point x. As D is a polyhedron, $h \in T_x D$ means that $x + th \in D$ for t sufficiently small or equivalently that there is $t_0 > 0$ such that $(x_N + th_N) \geq 0$ and $B^{-1}b - B^{-1}N(x_N + th_N) \geq 0$ as soon as $t \in [0, t_0]$. Using (5), these conditions are equivalent to $(h_N)_i \geq 0$, $i \in N(x)$ and $(B^{-1}Nh_N)_i \leq 0$, $i \in B(x)$. $\qquad\square$

Consider an edge $[\bar{x}, \bar{y}]$ of D and a point $x \in]\bar{x}, \bar{y}[$. We suppose that B is a basis associated to the vertex $\bar{x}$ and that we describe the edge $[\bar{x}, \bar{y}]$ by increasing the nonbasic variable x_{j_0} from 0 to $\bar{y}_{j_0}$. In other words x belongs to the edge $[\bar{x}, \bar{y}]$ if and only if

$$x_N = (0, \ldots, x_{j_0}, \ldots, 0), \quad x_{j_0} \in [0, \bar{y}_{j_0}] \tag{10}$$

where $\bar{y}_{j_0}$ is given by $\bar{y}_N = (0, \ldots, \bar{y}_{j_0}, \ldots, 0)$.

Theorem 2. *Let be given an edge $[\bar{x}, \bar{y}]$. One has*

$$]\bar{x}, \bar{y}[\cap E =]\bar{x}, \bar{y}[\cap [\bar{\bar{x}}, \bar{\bar{y}}]$$

where the points $\bar{\bar{x}}$ and $\bar{\bar{y}}$ are obtained by solving two linear problems.

Proof. For every $x \in]\bar{x}, \bar{y}[$, we have $x_N = (0, \ldots, x_{j_0}, \ldots, 0)$ and $N(x) = N(\bar{x}) \setminus \{j_0\}$. Let us show that $B(x)$ is also constant for all $x \in]\bar{x}, \bar{y}[$. Indeed for $y \in]\bar{x}, \bar{y}[$ such that $y_N = (0, \ldots, \beta_{j_0}, \ldots, 0)$ we have

$$y_B = (B^{-1}b)_i - (B^{-1}N)_{i,j_0}\beta_{j_0} \geq 0, \quad i = 1, .., m, \; \forall \beta_{j_0} \in]0, \bar{y}_{j_0}[\tag{11}$$

and for $x \in]\bar{x}, \bar{y}[$ such that $x_N = (0, \ldots, \alpha_{j_0}, \ldots, 0)$, $\alpha_{j_0} \in]0, \bar{y}_{j_0}[$ we get from the definition of $B(x)$

$$(B^{-1}b)_i - (B^{-1}N)_{i,j_0}\alpha_{j_0} = 0, \quad \forall i \in B(x). \tag{12}$$

Since $\bar{x}_B = B^{-1}b \geq 0$ we have from (11) and (12)

$$(B^{-1}b)_i = 0 \text{ and } (B^{-1}N)_{i,j_0} = 0, \quad \forall i \in B(x). \tag{13}$$

Then for all $y_{j_0} \in]0, \bar{y}_{j_0}[$ we have $(B^{-1}b)_i - (B^{-1}N)_{i,j_0}\beta_{j_0} = 0$. We conclude that $i \in B(y)$ and then $B(x) \subset B(y)$ for all $y \in]\bar{x}, \bar{y}[$. As this result holds for all $x \in]\bar{x}, \bar{y}[$ the set $B(x)$ is constant for $x \in]\bar{x}, \bar{y}[$ and we adopt the notation $B(x) = B(j_0)$.

Using Theorem 1, $x \in]\bar{x}, \bar{y}[$ is efficient if and only if there exist $\lambda_1 > 0$ and $\lambda_2 > 0$ such that

$$(h_N)_i \geq 0, \; i \in N(\bar{x}) \setminus \{j_0\} \text{ and } (B^{-1}Nh_N)_i \leq 0, \; i \in B(j_0)$$

implies

$$\sum_{i=1}^{2} \lambda_i [(\bar{d}_i x_N + \bar{t}_i)\bar{c}_i - (\bar{c}_i x_N + \bar{s}_i)\bar{d}_i] h_N \geq 0.$$

Here $x_N = (0, \ldots, x_{j_0}, \ldots, 0)$ so that the previous formula becomes

$$\sum_{i=1}^{2} \lambda_i [(\bar{d}_{ij_0} x_{j_0} + \bar{t}_i)\bar{c}_i - (\bar{c}_{ij_0} x_{j_0} + \bar{s}_i)\bar{d}_i] h_N \geq 0.$$

¿From the Farkas' Lemma it follows that $x \in]\bar{x}, \bar{y}[\cap E$ is equivalent to the existence of $\lambda_1 > 0$, $\lambda_2 > 0$, $\mu_j \geq 0$, $j \in N \setminus \{j_0\}$, $\gamma_i \geq 0$, $i \in B(j_0)$ such that

$$\sum_{i=1}^{2} \lambda_i [(\bar{d}_{ij_0} x_{j_0} + \bar{t}_i) \bar{c}_i - (\bar{c}_{ij_0} x_{j_0} + \bar{s}_i) \bar{d}_i] = \sum_{\substack{j \in N \\ j \neq j_0}} \mu_j e_j - \sum_{i \in B(j_0)} \gamma_i (B^{-1}N)_i \tag{14}$$

where $e_j = (e_{j_1}, \ldots, e_{j_n}) \in \mathbb{R}^n$ with $e_{j_i} = 1$ if $j = i$ and 0 otherwise. Dividing (14) by $\lambda_1 > 0$ and setting

$$P_i = \bar{c}_{ij_0} x_{j_0} + \bar{s}_i, \quad Q_i = \bar{d}_{ij_0} x_{j_0} + \bar{t}_i, \quad i = 1, 2 \tag{15}$$

the efficiency of x is equivalent to the existence of $\lambda > 0$ and $\gamma_i' \geq 0$, $i \in B(j_0)$ such that

$$Q_1 \bar{c}_{1j} - P_1 \bar{d}_{1j} + \lambda(Q_2 \bar{c}_{2j} - P_2 \bar{d}_{2j}) + \sum_{i \in B(j_0)} \gamma_i'(B^{-1}N)_{ij} \geq 0, \quad j \neq j_0 \tag{16}$$

and

$$Q_1 \bar{c}_{1j_0} - P_1 \bar{d}_{1j_0} + \lambda(Q_2 \bar{c}_{2j_0} - P_2 \bar{d}_{2j_0}) + \sum_{i \in B(j_0)} \gamma_i'(B^{-1}N)_{ij_0} = 0. \tag{17}$$

As from (13), $(B^{-1}N)_{i,j_0} = 0$ for all $i \in B(j_0)$, (17) is equivalent to

$$Q_1 \bar{c}_{1j_0} - P_1 \bar{d}_{1j_0} + \lambda(Q_2 \bar{c}_{2j_0} - P_2 \bar{d}_{2j_0}) = 0. \tag{18}$$

If

$$Q_2 \bar{c}_{2j_0} - P_2 \bar{d}_{2j_0} = 0 \tag{19}$$

then from (18)

$$Q_1 \bar{c}_{1j_0} - P_1 \bar{d}_{1j_0} = 0 \tag{20}$$

and it is easy to see that these equalities imply that f_1 and f_2 are constant on $[\bar{x}, \bar{y}]$. In that case, if $]\bar{x}, \bar{y}[$ is nonempty then $]\bar{x}, \bar{y}[\cap E =]\bar{x}, \bar{y}[$. If (19) does not hold then λ must be equal to

$$\lambda^* = -\frac{Q_1 \bar{c}_{1j_0} - P_1 \bar{d}_{1j_0}}{Q_2 \bar{c}_{2j_0} - P_2 \bar{d}_{2j_0}} = -\frac{\bar{t}_1 \bar{c}_{1j_0} - \bar{s}_1 \bar{d}_{1j_0}}{\bar{t}_2 \bar{c}_{2j_0} - \bar{s}_2 \bar{d}_{2j_0}}. \tag{21}$$

Replacing λ by λ^* in (16) and using (15) we get that $x \in]\bar{x}, \bar{y}[$ is efficient if and only if there exist $\gamma_i' \geq 0$, $i \in B(j_0)$ such that

$$\alpha_j x_{j_0} + \beta_j + \sum_{i \in B(j_0)} \gamma_i'(B^{-1}N)_{ij} \geq 0, \quad j \neq j_0 \tag{22}$$

where

$$\alpha_j = \bar{d}_{1j_0}\bar{c}_{1j} - \bar{c}_{1j_0}\bar{d}_{1j} + \lambda^*(\bar{d}_{2j_0}\bar{c}_{2j} - \bar{c}_{2j_0}d_{2j}) \tag{23}$$

and

$$\beta_j = \bar{t}_1\bar{c}_{1j} - \bar{s}_1\bar{d}_{1j} + \lambda^*(\bar{t}_2\bar{c}_{2j} - \bar{s}_2\bar{d}_{2j}). \tag{24}$$

These terms α_j, β_j do not depend on $x \in]\bar{x}, \bar{y}[$ and can be easily obtained from the simplex tableau corresponding to the basis B. The set of all x_{j_0} such that there exists $\gamma_i', i \in B(j_0)$ satisfying the linear inequalities (22) is an interval. Now in order to find this interval we introduce the linear problems

$$(LP_1) \qquad \bar{\bar{x}}_{j_0} = \min(z)$$
$$\alpha_j z + \beta_j + \sum_{i \in B(j_0)} \gamma_i'(B^{-1}N)_{ij} \geq 0, \quad j \neq j_0$$
$$0 \leq z \leq \bar{y}_{j_0}$$
$$\gamma_i' \geq 0, \quad i \in B(j_0)$$

and

$$(LP_2) \qquad \bar{\bar{y}}_{j_0} = \max(z)$$
$$\alpha_j z + \beta_j + \sum_{i \in B(j_0)} \gamma_i'(B^{-1}N)_{ij} \geq 0, \quad j \neq j_0$$
$$0 \leq z \leq \bar{y}_{j_0}$$
$$\gamma_i' \geq 0, \quad i \in B(j_0).$$

The optimal values of (PL_1) and (PL_2) define $\bar{\bar{x}}$ and $\bar{\bar{y}}$ in $[\bar{x}, \bar{y}]$ such that $\bar{\bar{x}}_N = (0, \ldots, \bar{\bar{x}}_{j_0}, \ldots, 0)$ and $\bar{\bar{y}}_N = (0, \ldots, \bar{\bar{y}}_{j_0}, \ldots, 0)$. Then it follows from (22) that $x \in]\bar{x}, \bar{y}[$ is efficient if and only if $\bar{\bar{x}} \leq x_{j_0} \leq \bar{\bar{y}}$ and we conclude that

$$]\bar{x}, \bar{y}[\cap E =]\bar{x}, \bar{y}[\cap [\bar{\bar{x}}, \bar{\bar{y}}]. \tag{25}$$

$\square$

We say that an edge $[\bar{x}, \bar{y}]$ is efficient if $]\bar{x}, \bar{y}[\cap E \neq \emptyset$. Notice that the data of problems (PL_1) and (PL_2) derive easily from the simplex tableau associated to the basis B. It follows from the closedness of E and (25) that

$$\mathrm{cl}(]\bar{x}, \bar{y}[\cap [\bar{\bar{x}}, \bar{\bar{y}}]) = [\hat{x}, \hat{y}] \subset E.$$

The next proposition shows that $f([\hat{x}, \hat{y}])$ can be interpreted as the graph of a linear fractional function. Recall the notations

$$f_i(x) = \frac{c_i x + s_i}{d_i x + t_i} = \frac{P_i(x)}{Q_i(x)}, \quad i = 1, 2$$

Proposition 2. *Let x and y be two points in D such that $f_1(x) < f_1(y)$. Then*

$$f([x, y]) = \{(X, Y) : Y = \varphi(X), \quad X \in [f_1(x), f_1(y)]\}$$

where

$$\varphi(X) = \frac{[P_2(x)Q_1(y) - Q_1(x)P_2(y)]X + [P_1(x)P_2(y) - P_2(x)P_1(y)]}{[Q_2(x)Q_1(y) - Q_1(x)Q_2(y)]X + [P_1(x)Q_2(y) - Q_2(x)P_1(y)]}.$$

Proof. For $x_\mu = (1 - \mu)x + \mu y$ with $\mu \in [0, 1]$ one has

$$f_i(x_\mu) = \frac{\mu c_i(y - x) + P_i(x)}{\mu d_i(y - x) + Q_i(x)} \quad i = 1, 2. \tag{26}$$

On the other hand

$$\nabla f_1(x_\mu) = \frac{c_1 - f_1(x_\mu)d_1}{d_1 x_\mu + t_1}. \tag{27}$$

The assumption $f_1(x) \neq f_1(y)$ entails $\nabla f_1(x_\mu)(y - x) \neq 0$ which is equivalent to

$$c_1(x - y) - f_1(x_\mu)d_1(x - y) \neq 0 \tag{28}$$

It follows from (26) for $i = 1$, and (28) that

$$\mu = \frac{P_1(x) - f_1(x_\mu)Q_1(x)}{c_1(x - y) - f_1(x_\mu)d_1(x - y)} \tag{29}$$

Now reporting this value in (26) for $i = 2$ and using the formulae $c_i(x - y) = P_i(x) - P_i(y)$, $d_i(x - y) = Q_i(x) - Q_i(y)$ we get

$$f_2(x_\mu) = \frac{f_1(x_\mu)[P_2(x)Q_1(y) - Q_1(x)P_2(y)] + [P_1(x)P_2(y) - P_2(x)P_1(y)]}{f_1(x_\mu)[Q_2(x)Q_1(y) - Q_1(x)Q_2(y)] + [P_1(x)Q_2(y) - Q_2(x)P_1(y)]}.$$

$\square$

4   The efficient outcome

We summarize Sections 2 and 3 in the following algorithm which gives a curve in $\mathbb{R}^2$ representing the set of all optimal values.

The Algorithm

Step 0. Compute

$$\underline{f}_i = \min\{f_i(x) : x \in D\}, \quad i = 1, 2$$

$$\bar{f}_2 = \min\{f_2(x) : x \in D, \ f_1(x) = \underline{f}_1\}$$
$$\bar{f}_1 = \min\{f_1(x) : x \in D, \ f_2(x) = \underline{f}_2\}.$$

As $D \cap \{x : f_2(x) = \underline{f}_2\}$ is a face of D the last minimization problem admits a vertex $\bar{x}$ of D as a solution. This point $\bar{x}$ is an efficient solution. We set $\mathcal{E} = \{\bar{x}\}$ and $F_1 = \{f_1(\bar{x})\}$.

Step 1. If $[\underline{f}_1, \bar{f}_1] \setminus F_1 = \emptyset$ then the algorithm stops, else choose $t \in [\underline{f}_1, \bar{f}_1] \setminus F_1$ and solve

$$(\mathcal{P}_t) \qquad \min\{f_2(x) : x \in D, \ f_1(x) = t\}.$$

Let s_t be a vertex of $D_t = \{x \in D : f_1(x) = t\}$ which is a solution of $(\mathcal{P}_t)$.

Step 2. If s_t is a vertex of D then we set $\mathcal{E} = \mathcal{E} \cup \{s_t\}$ $F_1 = F_1 \cup \{f_1(s_t)\}$ and go to the Step 1. Otherwise s_t belongs to the interior $]\bar{x}, \bar{y}[$ of an edge of D then we solve the linear problems (PL_1) and (PL_2) of the Section 3 in order to obtain the points $\hat{x}$ and $\hat{y}$. Then we set $\mathcal{E} = \mathcal{E} \cup [\hat{x}, \hat{y}]$, $F_1 = F_1 \cup f_1([\hat{x}, \hat{y}])$ and go to the Step 1.

Proposition 3. *The algorithm stops after a finite number of iterations.*

Proof. In Step 1, s_t satisfies $f_1(s_t) = t \notin F_1$, where F_1 contains the image by f_1 of the efficient solutions previously obtained. Thus in Step 2 we cannot discover twice the same efficient vertex or the same efficient edge of D. Then the number of iterations is bounded by the total number of vertices and edges of the polyhedron. □

The algorithm provides a subset $\mathcal{E} \subset E$ of efficient solutions located on the edges of the polyhedron D.

Proposition 4. *One has*

$$f(\mathcal{E}) = f(E).$$

Proof. It is sufficient to prove that $f(E) \subset f(\mathcal{E})$. If $x \in E$ we have from Lemma 1, $f_1(x) \in [\underline{f}_1, \bar{f}_1]$. As the algorithm stops when $[\underline{f}_1, \bar{f}_1] = F_1$, we have $f_1(x) \in F_1$. It follows that there exists $x' \in \mathcal{E}$ such that $f_1(x) = f_1(x')$. The efficiency of x and x' implies that $f_2(x) = f_2(x')$ and we conclude $f(x) = f(x') \in f(\mathcal{E})$. □

Notice that at the end of the algorithm we have a familly of efficient vertices $(v^k)_{k \in K}$ and a family of efficient intervals $([\hat{x}^k, \hat{y}^k])_{k \in L}$ such that

$$f(E) = (\cup_{k \in K}\{f(v^k)\}) \cup (\cup_{k \in L}f([\hat{x}^k, \hat{y}^k])). \tag{30}$$

¿From Lemma 2 $f(E)$ is the graph of a continuous function and $\cup_{k \in K}\{f(v^k)\})$ being a finite subset it can be removed from (30). Therefore we have

$$f(E) = \cup_{k \in L}f([\hat{x}^k, \hat{y}^k]) \tag{31}$$

Proposition 5. *The set $f(E)$ of the efficient values of $(\mathcal{P})$ is a piecewise linear fractional curve in $\mathbb{R}^2$.*

Proof. It follows directly from (31) and Proposition 2. □

The end of this section concerns an improvement of our algorithm. Indeed in Step 2, when a new efficient vertex v^k of D is discovered it is generally easy to select an efficient edge emanating from v^k. More precisely let $\bar{x}$ be an efficient vertex and denote by $x^1, \ldots, x^r$ the adjacent vertices to $\bar{x}$. Let us consider for every $k = 1, \ldots, r$ the unit vectors in $\mathbb{R}^n$, $\varepsilon^k = \dfrac{x^k - \bar{x}}{\|x^k - \bar{x}\|}$ and introduce the sets

$$K_1 = \{k \in \{1, .., r\} : \nabla f_1(\bar{x})\varepsilon^k > 0, \nabla f_2(\bar{x})\varepsilon^k < 0\}$$

$$K_2 = \{k \in \{1, .., r\} : \nabla f_1(\bar{x})\varepsilon^k < 0, \nabla f_2(\bar{x})\varepsilon^k > 0\}$$

$$K_3 = \{k \in \{1, .., r\} : \nabla f_1(\bar{x})\varepsilon^k = \nabla f_2(\bar{x})\varepsilon^k = 0\}$$

Notice that in the nondegenerate case all these subsets are easily available from the simplex tableau characterizing $\bar{x}$.

Proposition 6. *If $]\bar{x}, x^k[\cap E \neq \emptyset$ then $k \in K_1 \cup K_2 \cup K_3$.*

Proof. If $k \notin K_1 \cup K_2 \cup K_3$ then one of the following situation occurs :

$$(S_1) \qquad \nabla f_1(\bar{x})\varepsilon^k > 0 \text{ and } \nabla f_2(\bar{x})\varepsilon^k \geq 0$$

$$(S_2) \qquad \nabla f_1(\bar{x})\varepsilon^k < 0 \text{ and } \nabla f_2(\bar{x})\varepsilon^k \leq 0$$

$$(S_3) \qquad \nabla f_1(\bar{x})\varepsilon^k = 0 \text{ and } \nabla f_2(\bar{x})\varepsilon^k \neq 0$$

On the other hand it is proved in [10] that $f_i, i = 1, 2$ satisfy for all x, y in D

$$f_i(y) - f_i(x) = \frac{d_i x + t_i}{d_i y + t_i} \nabla f_i(x)(y - x) \tag{32}$$

where $\dfrac{d_i x + t_i}{d_i y + t_i} > 0$ for $x, y \in D$.

Suppose that $y^k \in]\bar{x}, x^k[\cap E$. In the situation (S_1) we get from (32), that for all $y^k \in]\bar{x}, x^k[$ we have $f_1(y^k) > f_1(\bar{x})$ and $f_2(y^k) \geq f_2(\bar{x})$, which contradicts the efficiency of y^k. In the situation (S_2), it comes $f_1(y^k) < f_1(\bar{x})$ and $f_2(y^k) \leq f_2(\bar{x})$ which contradicts the efficiency of $\bar{x}$. In the situation (S_3) we have $f_1(y^k) = f_1(\bar{x})$ and either $f_2(y^k) > f_2(\bar{x})$ or $f_2(y^k) < f_2(\bar{x})$. In any case we get a contradiction with the efficiency of y^k or $\bar{x}$. □

If there exists an adjacent vertex x^k to $\bar{x}$ such that $k \in K_3$ then it is obvious that $[\bar{x}, x^k]$ is entirely efficient. Indeed in that case the functions f_i, $i = 1, 2$ are constant on $[\bar{x}, x^k]$ and $f([\bar{x}, x^k]) = f(\bar{x})$.
For $k \in K_1 \cup K_2$ we introduce the ratio

$$R(k) = -\frac{\nabla f_1(\bar{x})\varepsilon^k}{\nabla f_2(\bar{x})\varepsilon^k} > 0$$

Proposition 7. *Let $\bar{x}$ be an efficient vertex such that $K_3 = \emptyset$. If there is an adjacent vertex x^{k_0} with $k_0 \in K_1 \cup K_2$ such that*

$$\max_{k \in K_2, k \neq k_0} R(k) < R(k_0) < \min_{k \in K_1, k \neq k_0} R(k)$$

then $[\bar{x}, y^{k_0}] \subset E$ for some $y^{k_0} \in]\bar{x}, x^{k_0}]$.

Proof. Choose a decomposition $[B, N]$ such that B is a basis corresponding to the vertex $\bar{x}$ and suppose that we describe the edge $[\bar{x}, x^{k_0}]$ by increasing the nonbasic variable x_{j_0}. From (22), a point $x \in]\bar{x}, x^{k_0}]$ with $x_N = (0, \ldots, x_{j_0}, \ldots, 0)$ is efficient if and only if there exists $\gamma_i' \geq 0, i \in B(j_0)$ such that

$$\alpha_j x_{j_0} + \beta_j + \sum_{i \in B(j_0)} \gamma_i'(B^{-1}N)_{ij} \geq 0, \quad j \neq j_0, \tag{33}$$

where

$$\alpha_j = \bar{d}_{1j_0}\bar{c}_{1j} - \bar{c}_{1j_0}\bar{d}_{1j} + \lambda^*(\bar{d}_{2j_0}\bar{c}_{2j} - \bar{c}_{2j_0}d_{2j}),$$
$$\beta_j = \bar{t}_1\bar{c}_{1j} - \bar{s}_1\bar{d}_{1j} + \lambda^*(\bar{t}_2\bar{c}_{2j} - \bar{s}_2\bar{d}_{2j}),$$

and

$$\lambda^* = -\frac{\bar{t}_1\bar{c}_{1j_0} - \bar{s}_1\bar{d}_{1j_0}}{\bar{t}_2\bar{c}_{2j_0} - \bar{s}_2 d_{2j_0}}.$$

Using $\nabla f_i(\bar{x})\varepsilon_{k_0} = \dfrac{\bar{t}_i\bar{c}_{ij_0} - \bar{s}_i\bar{d}_{ij_0}}{\bar{t}_i^2}$, observe that $\lambda^* = \left(\dfrac{\bar{t}_1}{\bar{t}_2}\right)^2 R(k_0)$. The end of the proof consists to show that $\beta_j > 0$ for $j \neq j_0$, this condition implying that (33) holds for x_{j_0} small enough with $\gamma_i' = 0, i \in B(j_0)$. Consider $j \neq j_0$. When we increase the nonbasic variable x_j we describe an adjacent edge say $[\bar{x}, x^{k_j}]$. As $\bar{t}_1\bar{c}_{1j} - \bar{s}_1\bar{d}_{1j} = \bar{t}_1^2\nabla f_1(\bar{x})\varepsilon^{k_j}$ and $\bar{t}_2\bar{c}_{2j} - \bar{s}_2\bar{d}_{2j} = \bar{t}_2^2\nabla f_2(\bar{x})\varepsilon^{k_j}$ we have

$$\beta_j = \bar{t}_1^2[\nabla f_1(\bar{x})\varepsilon^{k_j} + R(k_0)\nabla f_2(\bar{x})\varepsilon_{k_j}].$$

If $\nabla f_2(\bar{x})\varepsilon^{k_j} = 0$ we deduce from $\bar{x} \in E$ and $K_3 = \emptyset$ that $\nabla f_1(\bar{x})\varepsilon^{k_j} > 0$, thus $\beta_j > 0$.
If $\nabla f_2(\bar{x})\varepsilon^{k_j} < 0$ we deduce from $\bar{x} \in E$ that $\nabla f_1(\bar{x})\varepsilon^{k_j} > 0$ and $k_j \in K_1$. Then $R(k_0) < R(k_j)$ entails $\beta_j > 0$.
If $\nabla f_2(\bar{x})\varepsilon^{k_j} > 0$ and $\nabla f_1(\bar{x})\varepsilon^{k_j} \geq 0$ then $\beta_j > 0$.
If $\nabla f_2(\bar{x})\varepsilon^{k_j} > 0$ and $\nabla f_1(\bar{x})\varepsilon^{k_j} < 0$ then $k_j \in K_2$ and $R(k_0) > R(k_j)$ entails $\beta_j > 0$. $\qquad\square$

5   An illustrative example

Consider the problem

$$\min(\frac{-x_1 - 1}{x_2 + 1}, \frac{-x_3}{-x_2 + 10})$$

$$x_1 + x_2 \leq 16$$
$$x_2 + x_3 \leq 10$$
$$x_1 + 2x_3 \leq 15$$
$$x_1 + x_2 + 2x_3 \leq 20$$
$$0 \leq x_1 \leq 11$$
$$0 \leq x_2 \leq 8$$
$$0 \leq x_3 \leq 5$$

Step 0.
$$\underline{f_1} = \min\{f_1(x) : x \in D\} = -12$$
$$\underline{f_2} = \min\{f_2(x) : x \in D\} = -1$$

$$\bar{f_1} = \min\{f_1(x) : x \in D \;\; f_2(x) = -1\} = -1, \text{ and } f_1(E) = [\underline{f_1}, \bar{f_1}] = [-12, -1]$$

$$\bar{f_2} = \min\{f_2(x) : x \in D, \;\; f_1(x) = -12\} = 0.2 \text{ with } \bar{x} = (11, 0, 2)$$

$$\mathcal{E} = \{\bar{x}\}, F_1 = \{-12\}.$$

Step 1. Choose $t = -12 + 0.1 \in [-12, -1] \setminus \{-12\}$

$\quad\quad x^t = (10.9, 0, 2.05)$ is a vertex of D_t solution of $(\mathcal{P}_t)$.

Step 2. From the simplex tableau corresponding to x^t we deduce that x^t is not a vertex of D because only two constraints are equalities. Thus x^t belongs to the relative interior of some edge, we find $x^t \in \text{ri}[(11, 0, 2), (5, 0, 5)]$.

$$\bar{\bar{x}} = \hat{x} = (11, 0, 2) \text{ solves } (PL_1) \text{ and } f_1(\hat{x}) = -12$$

$$\bar{\bar{y}} = \hat{y} = (5, 0, 5) \text{ solves } (PL_2) \text{ and } f_1(\hat{y}) = -6$$

$$\mathcal{E} = \{\bar{x}\} \cup [(11, 0, 2), (5, 0, 5)], F_1 = \{-12\} \cup [-12, -6].$$

Step 1. Choose $t = -6 + 0.1 \in [-12, -1] \setminus F_1$

$\quad\quad x^t = (5, 0.02, 5)$ is a vertex of D_t solution of $(\mathcal{P}_t)$.

Step 2. $x^t \in \text{ri}[(5, 0, 5), (5, 5, 5)]$.

$$\bar{\bar{x}} = \hat{x}_2 = (5, 0, 5) \text{ solves } (PL_1) \text{ and } f_1(\hat{x}) = -6$$

$$\bar{\bar{y}} = \hat{y}_2 = (5, 3.1, 5) \text{ solves } (PL_2) \text{ and } f_1(\hat{y}) = -1.45$$

$$\mathcal{E} = \{\bar{x}\} \cup [(11, 0, 2), (5, 0, 5)] \cup [(5, 0, 5), (5, 3.1, 5)]$$

$$F_1 = \{-12\} \cup [-12, -6] \cup [-6, -1.45].$$

Step 1. Choose $t = -1.45 + 0.1 \in [-12, -1] \setminus F_1$

$$x^t = (7.1272, 5, 3.9364) \text{ is a vertex of } D_t \text{ solution of } (\mathcal{P}_t).$$

Step 2. $x^t \in \mathrm{ri}[(11, 5, 2), (5, 5, 5)]$.

$$\bar{\bar{x}} = \hat{x} = (7.7273, 5, 3.6364) \text{ solves } (PL_1) \text{ and } f_1(\hat{x}) = -1.45$$

$$\bar{\bar{y}} = \hat{y} = (5, 5, 5) \text{ solves } (PL_2) \text{ and } f_1(\hat{y}) = -1$$

$$\mathcal{E} = \{\bar{x}\} \cup [(11, 0, 2), (5, 0, 5)] \cup [(5, 0, 5), (5, 3.1, 5)] \cup [(7.7273, 5, 3.6364), (5, 5, 5)]$$

$$F_1 = \{-12\} \cup [-12, -6] \cup [-6, -1.45] \cup [-1.45, -1].$$

A new Step 1 stops the algorithm as $F_1 = [\underline{f_1}, \bar{f}_1] = [-12, -1]$. Then from Proposition 2 applied to the intervals $[-12, -6]$, $[-6, -1.45]$ and $[-1.45, -1]$ we get a graphic representation of $f(E)$.

References

1. Armand, P., Malivert, C. (1991) Determination of the efficient set in multiobjective linear programming. Journal of Optimization Theory and Application **70**, 467–489
2. Armand, P., Malivert, C., Valentin, P. (1997) Une méthode d'optimisation bicritère adaptée à la constitution de mélanges. Rapport de recherche, Elf Aquitaine (Direction Raffinage Distribution) Solaize
3. Benson, H.P. (1997) Generating the efficient outcome set in multiple objective linear programs: The bicriteria case. Acta Matematica Vietnamica **22**, 29–51
4. Charnes, A., Cooper, W. W. (1962) Programming with linear fractional functionals. Naval Reseach Logistics Quarterly **9**, 181–196
5. Choo, E.U., Atkins, D.R. (1982) Bicriteria linear fractional programing. JOTA **36**, 2, 203–220
6. Daniilidis, A., Hadjisavvas, N., Schaible, S. (1996) Connectedness of the Efficient Set for Three-Objective Quasiconcave Maximization Problems. The A. Gary Anderson Graduate School of Management, University of California, Riverside Working Paper Series No. 96-18
7. Gillmore, P.C., Gomory, R.E. (1963) A linear programming approach to the cutting stock problem. Part II. Operation Research **11**, 863–888
8. Jahn, J. (1991) On the application of a method of reference point approximation to bicriterial optimization problems in chemical engineering. Lecture Notes in Econom. and Math. Systems (Advances in optimization), 478–491
9. Jahn, J. (1997) A bicriteria optimization problem of antenna design. Computational Optimization and Applications **7**, 3, 261–276

10. Malivert, C. (1995) Multicriteria fractional optimization. In: Sofonea M., Corvellec J.N. (Eds.) Proc. of the 2nd Catalan days on applied mathematics, 189–198
11. Michelot, C. (1987) Localization in multifacility location theory. European Journal of Operations Research **31**, 177–184
12. Plastria, F. (1995) Continuous location problems in Facility location: A survey of Applications and Methods. Springer-Verlag, New York
13. Schaible, S. (1983) Bicriteria Quasiconcave programs. Cahiers du Centre d'Etudes de Recherche Oprationnelle **25**, 93–101
14. Steuer, R. (1986) Multiple criteria optimization : Theory, Computation, Application, Wiley

Tikhonov Fixed–Point Regularization [*]

Abdellatif Moudafi

Université des Antilles et de la Guyane,
Département de Mathématiques et d'Informatique,
97159 Pointe-à-Pitre, Guadeloupe, France

Abstract. The main purpose of this note is to propose viscosity approximation methods which amount to selecting a particular fixed-point of a given nonexpansive self mapping in a general Hilbert space. The connection with the selection principles of Attouch, in the context of convex minimization and monotone inclusion problems, is made and an application to a semi-coercive elliptic problem is then stated.

Key words: nonexpansive mappings, fixed-points, monotone operators, iterative methods, convex optimization, selection, penalty methods, monotone inclusions

1 Introduction

In this paper we are concerned with the problem of finding a fixed-point of a nonexpansive self mapping P defined on a closed convex set C of a real Hilbert space X. When P is strongly nonexpansive, that is, for all $x, y \in C$ one has $|Px - Py| \leq \theta |x - y|$ with $0 \leq \theta < 1$, the well-known Banach principle asserts that a fixed point x^* exists and is unique. Moreover, the sequence generated by the method of successive approximations strongly converges to x^* and the convergence is stable with respect to perturbations of P. As θ goes to 1, the problem is a priori unstable and it is necessary to apply some regularizing procedures. At the same time the existence of x^* is asserted on bounded closed convex set C but the convergence can only be weak. Hereafter we are interested in a less standard situation (i.e., $\theta = 1$) and in covering the case where (1) has multiple solutions. Assuming that the fixed points set, S, is nonempty, we will propose viscosity approximation methods which generate sequences that strongly converge to particular fixed-points of P. To this end we associate to the initial problem, namely

$$\text{find} \quad \bar{x} \in C \quad \text{such that} \quad \bar{x} = P(\bar{x}), \tag{1}$$

the following approximate well-posed problem

$$\text{find } x_k \in C \quad \text{such that} \quad x_k = \frac{1}{1 + \varepsilon_k} P(x_k) + \frac{\varepsilon_k}{1 + \varepsilon_k} \pi(x_k), \tag{2}$$

---

[*] The author is greatly indebted to Professor B. Lemaire, Université de Montpellier II, for his helpful comments and suggestions.

where $\{\varepsilon_k\}$ is a sequence of positive real numbers having to go to zero and $\pi : X \rightarrow C$ is a strongly nonexpansive mapping (with constant θ).

Note that Banach's theorem ensures the existence and uniqueness of x_k. Moreover by taking $\pi(x) = x_0$ for all $x \in X$ we recover the well-known continuous regularization method as a special case (see for example, [1], [2], [6] and [8]).

All definitions and notations used throughout this work are the usual ones in convex and nonlinear analysis and they can be found in the Brézis book [5].

2   The main results

Theorem 1. *The sequence $\{x_k\}$ generated by the proposed method strongly converges to the unique solution of the variational inequality*

$$find \ \tilde{x} \in S \quad such \ that \quad \langle(I - \pi)\tilde{x}, \tilde{x} - x\rangle \leq 0 \quad \forall x \in S, \tag{3}$$

in other words the unique fixed-point of the operator $proj_S \circ \pi$.

Proof. We have $-\varepsilon_k(I - \pi)x_k = (I - P)(x_k)$. By invoking the monotonocity of $(I - P)$, we get

$$\langle(I - \pi)x_k, x - x_k\rangle \geq 0 \quad \forall x \in S. \tag{4}$$

On the other hand non expansiveness of $(I - \pi)$ yields

$$\langle(I - \pi)x_k - (I - \pi)x, x_k - x\rangle \geq (1 - \theta)|x_k - x|^2. \tag{5}$$

Combining the last inequalities we obtain

$$\langle(I - \pi)x, x - x_k\rangle \geq (1 - \theta)|x_k - x|^2 \quad \forall x \in S. \tag{6}$$

Thus

$$|x_k - x| \leq (1 - \theta)^{-1}|(I - \pi)x| \quad \forall x \in S, \tag{7}$$

which implies the boundedness of $\{x_k\}$. Let $\bar{x}$ be any weak-cluster point of $\{x_k\}$. Then there exists a subsequence $\{x_{k_\nu}\}$ that weakly converges to $\bar{x}$. ¿From (2), we can write

$$0 = ((I - \pi) + \frac{1}{\varepsilon_{k_\nu}}(I - P))x_{k_\nu}. \tag{8}$$

Thanks to a result of P. L. Lions ([12], Proposition 2), we have that $\{\varepsilon_k^{-1}(I - P)\}$ graph converges to N_S, the normal cone to the solution set. This combined with a result of Brézis ([5]) gives the graph convergence of $\{(I - \pi) + \varepsilon_k^{-1}(I - P)\}$ to $(I - \pi) + N_S$. Now passing to the limit in relation (8) and taking into account the fact that the graph of a maximal monotone operator is weakly-strongly closed, we infer

$$0 \in (I - \pi)\bar{x} + N_S(\bar{x}), \quad \text{that is,} \quad \bar{x} = \text{proj}_S(\pi(\bar{x})). \tag{9}$$

Thus $\bar{x} = \tilde{x}$ and, the weak cluster point being unique, the whole sequence weakly converges to $\tilde{x}$. Then the desired result follows by setting $x = \tilde{x}$ in relation (6) and passing to the limit. $\qquad\square$

When $\pi(x) = x_0$ for all $x \in X$, we recover the convergence result of the continuous regularization method to the element of minimal norm, i.e., $\tilde{x} = \text{proj}_S(x_0)$, (see for example [1], [6], [8] or [2], Theorem 3.2.).

Let us now consider the following iterative method which generates from an initial point $z_0 \in C$, a sequence $\{z_k\}$ in C such that

$$z_k = \frac{1}{1 + \varepsilon_k} P(z_{k-1}) + \frac{\varepsilon_k}{1 + \varepsilon_k} \pi(z_{k-1}), \tag{10}$$

with P, π and ε_k as above. We will show that $\{z_k\}$ and $\{x_k\}$ have the same asymptotical behavior.

Theorem 2. *Suppose* $\displaystyle\sum_{k=1}^{+\infty} \varepsilon_k = +\infty$ *and* $\displaystyle\lim_{k \to +\infty} |\varepsilon_k^{-1} - \varepsilon_{k-1}^{-1}| = 0$.
Then, for all z_0, *the sequence* $\{z_k\}$ *strongly converges to* $\tilde{x}$.

Proof. From (2) and (10), we have

$$|z_k - x_k| \le \frac{1 + \theta\varepsilon_k}{1 + \varepsilon_k}(|z_{k-1} - x_{k-1}| + |x_k - x_{k-1}|). \tag{11}$$

Since $-\varepsilon_k(I - \pi)x_k = (I - P)x_k$ and $-\varepsilon_{k-1}(I - \pi)x_{k-1} = (I - P)x_{k-1}$ and thanks to the monotonicity of $(I - P)$, we get

$$|x_k - x_{k-1}|^2 \le \langle \pi(x_k) - \pi(x_{k-1}), x_k - x_{k-1} \rangle + \left(1 - \frac{\varepsilon_k}{\varepsilon_{k-1}}\right)$$
$$(\langle x_k, x_k - x_{k-1} \rangle - \langle \pi(x_k), \pi(x_k) - \pi(x_{k-1}) \rangle).$$

Since π is strongly nonexpansive and $\{x_k\}$ is bounded, there exists a constant γ such that

$$|x_k - x_{k-1}| \le \gamma \left|1 - \frac{\varepsilon_k}{\varepsilon_{k-1}}\right|. \tag{12}$$

We conclude by invoking the following result (see for example [15]): Let $\mu_k \ge 0$ and $\gamma_k \ge 0$ such that $\displaystyle\sum_{k=1}^{+\infty} \gamma_k = +\infty$ and $\mu_k/\gamma_k \to 0$; if the sequence $\{a_k\}$ satisfies $0 \le a_k \le (1 - \gamma_k)a_{k-1} + \mu_k$ then $a_k \to 0$. $\qquad\square$

When $\pi(x) = x_0$ for all $x \in X$, we recover as a special case a result given in [1], [8] (which is closely related to a result in [2]).

3 Link with other selection methods

3.1 Convex optimization

Let f be a convex lower semicontinuous function and consider the problem of finding a minimizer of f on X. A simple calculation shows that

$$\tilde{x} \in \text{Argmin} f \Leftrightarrow 0 \in \partial f(\tilde{x}) \Leftrightarrow \tilde{x} = \text{prox}_{\lambda f}(\tilde{x}), \quad \forall \lambda > 0, \qquad (13)$$

where ∂f denotes the convex subdifferential of f and $\text{prox}_{\lambda f}$ the unique minimizer of f_λ, the Moreau-Yosida approximate of f defined by

$$f_\lambda(x) = \inf_{y \in X} \{ f(y) + \frac{1}{2\lambda} |x - y|^2 \}. \qquad (14)$$

The function f_λ is differentiable and its gradient is given by $\nabla f_\lambda = (\partial f)_\lambda :=$ $\lambda^{-1}(I - \text{prox}_{\lambda f})$ and it is well known that the proximal mapping, $\text{prox}_{\lambda f} :$ $x \rightarrow \text{prox}_{\lambda f} x$, is nonexpansive.

Let $\lambda > 0$. If we take $P = \text{prox}_{\lambda f}$ and $\pi = \text{prox}_{\lambda g}$, where $g : X \rightarrow \mathbb{R}^+ \cup \{+\infty\}$ is a strongly convex (with modulus $\alpha/2$) and lower semicontinuous function, then S is nothing else than $\text{Argmin} f$, the viscosity approximation method corresponds to

$$x_{\lambda,k} = \text{Argmin}\{ f_\lambda(x) + \varepsilon_k g_\lambda(x) ; x \in X \}, \quad \forall k \in \mathbb{N} \qquad (15)$$

and the point $\tilde{x}_\lambda$ is characterized by

$$\tilde{x}_\lambda = \text{Argmin}\{ g_\lambda(x) ; x \in \text{Argmin} f \}. \qquad (16)$$

In what follows we will show that by letting λ go to 0, we obtain at the limit a viscosity principle proposed by Attouch in [3].

Proposition 1. *The sequence $\{x_{\lambda,k}, \tilde{x}_\lambda\}$ defined by (15)-(16) strongly converges to $\{x_k, \tilde{x}\}$ given by*

$$\begin{cases} x_k = Argmin\{ f(x) + \varepsilon_k g(x) ; x \in X \}, \\ \tilde{x} = Argmin\{ g(x) ; x \in Argmin f \}. \end{cases}$$

Proof. The optimality condition of (16) gives

$$0 \in \nabla g_\lambda(\tilde{x}_\lambda) + N_S(\tilde{x}_\lambda) \quad \text{or in other words} \quad \tilde{x}_\lambda = \text{proj}_S(\text{prox}_{\lambda g} \tilde{x}_\lambda). \qquad (17)$$

By taking $x \in S \cap \text{dom} \partial g$ and according to the fact that $\text{prox}_{\lambda g}$ is a contraction with modulus $(1 + \lambda \alpha)^{-1}$, we have

$$|\tilde{x}_\lambda - \text{proj}_S(\text{prox}_{\lambda g} x)| \leq \frac{1}{1 + \lambda \alpha} |\tilde{x}_\lambda - x|. \qquad (18)$$

This implies

$$|\tilde{x}_\lambda - x| \le \frac{1 + \lambda\alpha}{\lambda\alpha}|x - \text{prox}_{\lambda g}x| = (\lambda + \alpha^{-1})|(\partial g)_\lambda(x)|$$
$$\le (\lambda + \alpha^{-1})|(\partial g)^\circ(x)|,$$

where $(\partial g)^\circ x$ stands for the element of minimal norm of the convex set $\partial g(x)$. ¿From this we infer that $\{\tilde{x}_\lambda\}$ is bounded. Let $\tilde{x}$ be a weak-cluster point of $\{\tilde{x}_\lambda\}$. Then there exists a subsequence $\{\tilde{x}_{\lambda_\nu}\}$ which weakly converges to $\tilde{x}$. Since $\forall \lambda > 0$, $g_\lambda \le g$ and according to (16), we can write

$$g_\lambda(\tilde{x}_\lambda) \le g(x) \quad \forall x \in \text{Argmin} f. \tag{19}$$

Passing to the limit in the last inequality and taking into account the fact that g_λ Mosco converges to g (for the definition and properties of the Mosco convergence, we refer to Mosco [14]), we obtain

$$g(\tilde{x}) \le \liminf_{\nu \to +\infty} g_{\lambda_\nu}(\tilde{x}_{\lambda_\nu}) \le g(x) \quad \forall x \in \text{Argmin} f. \tag{20}$$

Since g is strongly convex, $\tilde{x}$ is unique and hence the whole sequence $\{\tilde{x}_\lambda\}$ weakly converges to $\tilde{x}$.

On the other hand, the optimality condition of (15) gives

$$0 \in \partial(f_\lambda + \varepsilon_k g_\lambda)x_{\lambda,k} = \nabla f_\lambda(x_{\lambda,k}) + \varepsilon_k \nabla g_\lambda(x_{\lambda,k})$$

which can be rewritten as

$$x_{\lambda,k} = \frac{1}{1 + \varepsilon_k}\text{prox}_{\lambda f}(x_{\lambda,k}) + \frac{\varepsilon_k}{1 + \varepsilon_k}\text{prox}_{\lambda g}(x_{\lambda,k}).$$

Let $x \in \text{dom}\partial f \cap \text{dom}\partial g$. We have

$$|x_{\lambda,k} - \frac{1}{1 + \varepsilon_k}\text{prox}_{\lambda f}(x) - \frac{\varepsilon_k}{1 + \varepsilon_k}\text{prox}_{\lambda g}(x)| \le \frac{1}{1 + \varepsilon_k}(1 + \frac{\varepsilon_k}{1 + \lambda\alpha})|x_{\lambda,k} - x|,$$

from which we infer

$$|x_{\lambda,k} - x| \le \frac{1 + \lambda\alpha}{\varepsilon_k\alpha}(|(\partial f)_\lambda(x)| + \varepsilon_k|(\partial g)_\lambda(x)|)$$

$$\le \frac{1 + \lambda\alpha}{\varepsilon_k\alpha}(|(\partial f)^\circ(x)| + \varepsilon_k|(\partial g)^\circ(x)|).$$

This implies the boundedness of $\{x_{\lambda,k}\}$.

Now let x_k be a weak-cluster point of $\{x_{\lambda,k}\}$. There exists a subsequence $\{x_{\lambda_\nu,k}\}$ which weakly converges to x_k. Since for all $\lambda > 0$, $f_\lambda \le f$ and $g_\lambda \le g$, we can write

$$f_\lambda(x_{\lambda,k} + \varepsilon_k g_\lambda(x_{\lambda,k} \le f(x) + \varepsilon_k g(x) \quad \forall x \in X. \tag{21}$$

Passing to the limit in the last inequality and using the Mosco convergence of $f_\lambda + \varepsilon_k g_\lambda$ to $f + \varepsilon_k g$, we get

$$f(x_k) + \varepsilon_k g(x_k) \leq \liminf_{\nu \to +\infty} (f_{\lambda_\nu} + \varepsilon_k g_{\lambda_\nu})(x_{\lambda_\nu, k})$$
$$\leq f(x) + \varepsilon_k g(x) \quad \forall x \in X.$$

As $f + \varepsilon_k g$ is also strongly convex, this implies that x_k is unique and thus that the whole sequence $\{x_{\lambda, k}\}$ weakly converges to x_k. The convergence of $\{x_{\lambda, k}, \tilde{x}_\lambda\}$ to $\{x_k, \tilde{x}\}$ is in fact strong. Indeed, since the strong convexity of g_λ is equivalent to strong monotonicity of ∇g_λ, we have

$$\langle \nabla g_\lambda(\tilde{x}_\lambda) - \nabla g_\lambda(\tilde{x}), \tilde{x}_\lambda - \tilde{x} \rangle \geq \alpha_\lambda |\tilde{x}_\lambda - \tilde{x}|^2. \tag{22}$$

On the other hand, from (17), we get

$$\langle \nabla g_\lambda(\tilde{x}_\lambda), x - \tilde{x}_\lambda \rangle \geq 0 \quad \forall x \in S \tag{23}$$

which, combined with (22), implies

$$\alpha_\lambda |\tilde{x}_\lambda - \tilde{x}|^2 \leq \langle \nabla g_\lambda(\tilde{x}), \tilde{x} - \tilde{x}_\lambda \rangle. \tag{24}$$

It is easy to check that the modulus of strong convexity for ∇g_λ is given by $\alpha_\lambda = \alpha(1 + \alpha\lambda)^{-1}$. Now, by passing to the limit in the last inequality and by taking into account the fact that $\nabla g_\lambda(\tilde{x}) = (\partial g)_\lambda(\tilde{x})$ strongly converges to $(\partial g)^\circ(\tilde{x})$, we obtain

$$\lim_{\lambda \to 0} |\tilde{x}_\lambda - \tilde{x}| = 0. \tag{25}$$

As $f + \varepsilon_k g$ is also strongly convex, mimicking the proof above, we get the strong convergence of $x_{\lambda, k}$ to x_k. $\qquad\square$

This selection principle has already been given in Attouch [3]. Here we obtain it as a consequence of Theorem 1. It should be noticed that applications of this result have been given in some specific cases like the log-barrier and exponential penalty functions for linear programming (see for example [4] and [7]). More precisely, when considering an inequality constrained program of the form

$$\min_{x \in \mathbb{R}^n} \{c^t x; \ Ax \leq b\} \tag{26}$$

which is assumed to have a nonempty and bounded feasible set $\{x \in \mathbb{R}^n; Ax \leq b\}$, the corresponding log-barrier approximation is given by

$$\min_{x \in \mathbb{R}^n} \{c^t x - \varepsilon \sum_{i=1}^{i=m} \ln(b_i - a_i^t x)\} \tag{27}$$

where $a_i, i = 1, \cdots, m$ denote the rows of A. An alternative penalty approach is to consider

$$\min_{x \in \mathbf{R}^n} \{c^t x - \varepsilon \sum_{i=1}^{i=m} \exp(-(b_i - a_i^t x)/\varepsilon)\}. \qquad (28)$$

Problem (27) (resp. (28)) has a unique solution x_ε (resp. $\tilde{x}_\varepsilon$) which converges when $\varepsilon \to 0$ to the analytic center of the optimal set S, that is, the unique solution of

$$\min_{x \in S} \{-\sum_{i \notin I_0} \ln(b_i - a_i^t x)\} \text{ where } I_0 = \{i; \ a_i^t x = b_i \quad \forall x \in S\} \qquad (29)$$

(resp. to $\tilde{x} \in S$ called the centroid). See [7] for more details.

3.2   Monotone inclusions

Let $A : X \to X$ be a multivalued maximal monotone operator and consider the problem of finding a zero of A. A short calculation gives

$$0 \in A(\tilde{x}) \Leftrightarrow \tilde{x} = J_\lambda^A(\tilde{x}) \quad \forall \lambda > 0, \qquad (30)$$

where $J_\lambda^A := (I + \lambda A)^{-1}$ denotes the resolvent of A. It should be noticed that J_λ^A is a nonexpansive mapping. Now let $\lambda > 0$. If we set $P = J_\lambda^A$ and $\pi = J_\lambda^B$, where B is a maximal monotone operator which is strongly monotone (with modulus α), then S is nothing else than $A^{-1}(0)$, the viscosity approximation method generates the points $x_{\lambda,k}, k \in \mathbb{N}$ defined by

$$0 = A_\lambda(x_{\lambda,k}) + \varepsilon_k B_\lambda(x_{\lambda,k}) \qquad (31)$$

and the point $\tilde{x}_\lambda$ is characterized by

$$0 \in B_\lambda(\tilde{x}_\lambda) + N_S(\tilde{x}_\lambda),$$

where $A_\lambda := \lambda^{-1}(I - J_\lambda^A)$ and $B_\lambda := \lambda^{-1}(I - J_\lambda^B)$ denote the Yosida approximates of A and B.

To make the connection with a method proposed by Attouch in [3], let us recall that B_λ graph converges to B and that for all $x \in \text{dom} B$ we have $|B_\lambda(x)| \le |B^\circ x|$, where $B^\circ x$ stands for the element of minimal norm of the convex set Bx.

First, by taking an $x \in \text{dom} A \cap \text{dom} B$ and using the same arguments as in the convex optimization case, we infer

$$|x_{\lambda,k} - x| \le \frac{1 + \alpha\lambda}{\varepsilon_k \alpha}(\varepsilon_k |B^\circ x| + |A^\circ x|). \qquad (32)$$

On the other hand, by setting $x \in S \cap \operatorname{dom} B$, we get

$$|\tilde{x}_\lambda - x| \leq (\alpha^{-1} + \lambda)|B^\circ x|. \tag{33}$$

In the same way, we establish that $\{x_{\lambda,k}, \tilde{x}_\lambda\}$ weakly converges to $\{x_k, \tilde{x}\}$ satisfying

$$\begin{cases} 0 \in Ax_k + \varepsilon_k Bx_k, \\ 0 \in B(\tilde{x}) + N_S(\tilde{x}). \end{cases} \tag{34}$$

Finally, as in the convex optimization case, the strong convergence can be obtain by using the strong monotonicity of B_λ.

Now, we apply the viscosity approximation method to the following semi-coercive elliptic problem: given Ω an open bounded subset of $R^N, N \in \mathbb{N}^*$, with a regular boundary $\partial\Omega$ and given a function $f \in L^2(\Omega)$, find $u \in H^1(\Omega)$, such that

$$\begin{cases} -\Delta u = f \quad \text{on } \Omega, \\ \partial u/\partial n = 0 \text{ on } \partial\Omega. \end{cases} \tag{35}$$

Here, $\partial/\partial n$ denotes the exterior normal derivative. Setting $\bar{f} = |\Omega|^{-1} \operatorname{int} f(x) \, dx$, it is well known that (35) admits a solution (unique upon a constant) if, and only if, $\bar{f} = 0$. The sequence $\{u_\varepsilon\}$ generated by the viscosity approximation method, namely,

$$\begin{cases} \varepsilon u_\varepsilon - \Delta u_\varepsilon = f \quad \text{on } \Omega, \\ \partial u_\varepsilon/\partial n = 0 \quad \text{on } \partial\Omega, \end{cases} \tag{36}$$

converges in $H^1(\Omega)$ to $\tilde{u}$ the unique solution of minimal norm for (35) characterized by

$$\begin{cases} -\Delta\tilde{u} = f \quad \text{on } \Omega, \\ \partial\tilde{u}/\partial n = 0 \quad \text{on } \partial\Omega, \\ \operatorname{int} \tilde{u}(x)\,dx = 0. \end{cases} \tag{37}$$

For more details see, for example, Lemaire [11].

To conclude, we would like to emphasize that the extension of selection methods to the problem of finding fixed points of nonexpansive mappings is justified by the fact that there exist nonexpansive mappings which are not proximal mappings nor resolvent operators. Indeed, if we consider the following periodic problem: Given $c \in X$, $T > 0$ and $f : [0,T] \times X \to \mathbb{R} \cup \{+\infty\}$, find an absolutely continuous function $u : [0,T] \to X$ satisfying

$$\begin{cases} -du(t)/dt \in \partial f(t, u(t)), \quad t \in [0, T], \\ u(0) = u(T) = c. \end{cases} \tag{38}$$

Then, assuming that, for any $c \in X$, the solution x_c exists (the uniqueness being assured by the monotonicity of $\partial f(t, \cdot)$, see Moreau [13] or Lemaire [9]), we define $P : X \to X$ by $P(c) = x_c(T)$. It is clear that the problem of finding fixed points of P and (38) are equivalent problems. Moreover, thanks to the same argument which ensures the uniqueness in (38), we get that P is a nonexpansive self mapping (for more details see Moreau [13] or Lemaire [9]).

References

1. Ould Ahmed Salem, C. (1998) Régularisation des problèmes de points fixes. Thèse de Doctorat, Université Montpellier II
2. Albert, Y., Guerre-Delabriere, S. (1995) Problems of fixed point theory in Hilbert and Banach spaces. Lecture delivered 2 January 1995, Technion University, Haifa
3. Attouch, H. (1996) Viscosity approximation methods for minimization problems. SIAM Journal on Optimization, **6**, 769–806
4. Attouch, H., Cominetti, R. (1996) A dynamical approach to convex minimization coupling approximation with steepest method. Journal of Differenial Equations, **128**, 519–540
5. Brézis, H. (1974) Opérateurs maximaux monotones et semi-groupes de contractions dans les espaces de Hilbert. North Holland, Amsterdam
6. Browder, F. E. (1967) Convergence of approximates to fixed points of nonexpansive nonlinear mappings in Banach spaces. Arch. for Rational Mech. and Anal. **24**, 82–90
7. Cominetti, R., San Martin, J. (1994) Asymptotical analysis of the exponential penalty trajectory in linear programming. Math. prog. **67**, 169–187
8. Halpern, B. (1967) Fixed points of nonexpanding maps. Bull. A.M.S. **73**, 957–961
9. Lemaire, B. (1977) Approximations successives pour certains problèmes périodiques. Séminaire d'Analyse Convexe de Montpellier, Exposé 21, 1–19
10. Lemaire, B. (1997) Which fixed-point does the iteration method select? Recent Advances in Optimization, Lecture Notes in Economic and Mathematical Systems, **452**, Springer, Berlin Heidelberg, 154–167
11. Lemaire, B. (1998) Regularization of fixed-point problems and staircase iteration. Workshop on ill-posed variational problems and regularization techniques, Trier
12. Lions, P. L. (1978) Two remarks on the convergence of the convex functions and monotone operators. Nonlinear Anal. Theo. Meth. and Appli. **2**, 553–562
13. Moreau, J. J. (1971) Rafle par un convexe variable. Séminaire d'Analyse Convexe, Montpellier, Exposé 15
14. Mosco, U. (1969) Convergence of convex set and of solution of variational inequalities. Adv. Math. **3**, 510–585
15. Vasil'ev, F.V. (1988) Numerical methods for solving extremum problems. Nauka, Moscow

Augmented Lagrangians for General Side Constraints

Werner Oettli[1], Dirk Schläger[1], and Michel Théra[2]

[1] Universität Mannheim, Lehrstuhl für Mathematik VII,
D-68131 Mannheim, Germany (oettli@math.uni-mannheim.de)
[2] Université de Limoges, LACO, 123 avenue Albert Thomas,
F-87060 Limoges Cedex, France (thera@unilim.fr)

Abstract. We consider augmented Lagrangians for the general (nonconvex) optimization problem $\inf\{F(x) \mid x \in X, G(x) \in M\}$, where X is an arbitrary set, M is a closed subset of some real topological vector space Z, and $F \colon X \to I\!R$, $G \colon X \to Z$.

1 Introduction

Augmented Lagrangian methods are a combination of primal-dual and penalty methods. They provide a convenient framework for studying optimization problems both in theory and in practice. Moreover, because of the easy implementation they have been extensively studied since the pioneering work of Hestenes, Powell and Rockafellar in the seventies. The reader is referred to the survey paper on augmented Lagrangian methods and their variants till 1976 by Bertsekas [1].

In this contribution, we consider augmented Lagrangians for the general (nonconvex) optimization problem

$$(P) \qquad \inf\{F(x) \mid x \in X, G(x) \in M\},$$

where X is an arbitrary set, M is a closed subset of some real topological vector space Z, and $F \colon X \to I\!R$, $G \colon X \to Z$. Augmented Lagrangians (at least in the convex case) are mainly used for algorithmic purposes (e. g. [2], [5], [10]), and this use restricts possible choices for such functions. We obtain greater freedom in limiting ourselves to the "static" aspects of problem (P), like duality, stability, regularization, which we discuss in terms of augmented Lagrangians. In so doing we follow the pattern set out by Rockafellar [9]. Compare also [6].

By an ordinary Lagrangian for problem (P) one understands usually a function

$$L(x,r) = F(x) + \Phi(x,r), \quad x \in X, \ r \in R,$$

where R is a space of multipliers, and Φ satisfies

$$\sup_{r \in R} \Phi(x,r) = \begin{cases} 0, & \text{if } G(x) \in M, \\ +\infty & \text{else.} \end{cases}$$

Accordingly, by an augmented Lagrangian we understand a function

$$L(x, r, z) = F(x) + \varphi(x, r, z), \quad x \in X, \ r \in R, \ z \in \mathcal{Z},$$

where $\mathcal{Z} \subseteq Z$, and φ satisfies

$$\sup_{(r,z) \in R \times \mathcal{Z}} \varphi(x, r, z) = \begin{cases} 0, & \text{if } G(x) \in M, \\ +\infty & \text{else.} \end{cases}$$

We consider one particular realization of such a function, which is still general enough not to require a norm or distance in the space Z.

2   Augmented Lagrangians

Let X be a set, Z a real topological vector space. Let M be a nonempty, closed subset of Z. Let $F\colon X \to \mathbb{R}$ and $G\colon X \to Z$ be given. We consider the problem

(P) $$\inf\{F(x) \,|\, x \in X, \ G(x) \in M\}.$$

Without loss of generality we may assume $0 \in M$. Let a function $\varphi\colon Z \to \mathbb{R}$ be given such that $\varphi(0) = 0$ and $\varphi(z) > 0$ for all $z \neq 0$. We assume that, for every net $\{z^n\}$ in Z,

$$\varphi(z^n) \to 0 \implies z^n \to 0. \tag{1}$$

We define

$$\varphi^M(z) := \inf_{m \in M} \varphi(z - m), \quad z \in Z.$$

Obviously $0 \leq \varphi^M(z) \leq \varphi(z)$ for all $z \in Z$, since $m = 0 \in M$. Moreover, from (1) and the closedness of M,

$$\varphi^M(z) = 0 \iff z \in M.$$

Proposition 1. (i) *If M is convex, and φ is convex, then $\varphi^M(\cdot)$ is convex.*
(ii) *If φ is upper semicontinuous, then φ^M is upper semicontinuous.*
(iii) *If M is convex, and φ is convex and upper semicontinuous, then, for every $\bar{z} \in Z$, there exists an affine, continuous $\theta\colon Z \to \mathbb{R}$ such that*

$$\begin{aligned} \theta(z) &\leq \varphi(z) & \forall z \in Z, \\ \theta(z) &\geq \varphi^M(\bar{z}) & \forall z \in \bar{z} - M, \\ \theta(z) - \theta(\bar{z}) &\leq \varphi^M(z) - \varphi^M(\bar{z}) & \forall z \in Z. \end{aligned} \tag{2}$$

Proof. (i) Let $z_1, z_2 \in Z$ and $\alpha \in [0,1]$. Then, for all $m_1, m_2 \in M$ and $m := \alpha m_1 + (1 - \alpha)m_2 \in M$,

$$\begin{aligned} \varphi^M(\alpha z_1 + (1 - \alpha)z_2) &\leq \varphi(\alpha z_1 + (1 - \alpha)z_2 - m) \\ &\leq \alpha \varphi(z_1 - m_1) + (1 - \alpha)\varphi(z_2 - m_2), \end{aligned}$$

hence $\varphi^M(\alpha z_1 + (1-\alpha)z_2) \leq \alpha \varphi^M(z_1) + (1-\alpha)\varphi^M(z_2)$.

(ii) This follows from the fact that the pointwise infimum of a family of upper semicontinuous functions is upper semicontinuous.

(iii) Let $\psi(z) := \varphi^M(\bar{z})$ for all $z \in \bar{z} - M$, $\psi(z) := -\infty$ for all $z \notin \bar{z} - M$; this defines a concave function ψ on Z, and $\psi \leq \varphi$. The sandwich theorem [4] yields an affine, continuous function θ such that $\psi \leq \theta \leq \varphi$, i.e., θ satisfies the first two inequalities of (2). The third can be deduced as follows: Given $z \in Z$, we have $\theta(z-m) \leq \varphi(z-m)$ and $\theta(\bar{z}-m) \geq \varphi^M(\bar{z})$ for all $m \in M$. Since θ is affine, we obtain $\theta(z) - \theta(\bar{z}) = \theta(z-m) - \theta(\bar{z}-m) \leq \varphi(z-m) - \varphi^M(\bar{z})$ for all $m \in M$.

$\square$

We denote by p_0 the optimal value of (P), i.e.,

$$p_0 := \inf\{F(x) \mid x \in X, \ G(x) \in M\}.$$

In general, an augmented Lagrangian for problem (P) may be written in the form

$$L(x, r, z) := F(x) + \varphi(x, r, z), \quad x \in X, \ r \in R, \ z \in Z,$$

where R is a set of multipliers, Z is a subset of Z with $0 \in Z$, and $\varphi(x, r, z)$ satisfies

$$\sup_{(r,z) \in R \times Z} \varphi(x, r, z) = \begin{cases} 0, & \text{if } G(x) \in M, \\ +\infty & \text{else.} \end{cases} \tag{3}$$

The case $Z = \{0\}$ includes ordinary Lagrangians. In the following we restrict ourselves to $R := \mathbb{R}_+$. Observe that $G(x) \in M \iff \varphi^M(G(x)) = 0$. Let

$$L(x, r, z) := F(x) + r\varphi^M(G(x) + z) - 2r\varphi(z), \quad x \in X, \ r \in \mathbb{R}_+, \ z \in Z.$$

This is in fact an augmented Lagrangian in the above sense:

Lemma 1. *For every $x \in X$,*

$$\sup_{(r,z) \in \mathbb{R}_+ \times Z} (r\varphi^M(G(x) + z) - 2r\varphi(z)) = \begin{cases} 0, & \text{if } G(x) \in M, \\ +\infty & \text{else.} \end{cases}$$

Proof. It suffices to show that $\sup_{z \in Z}(\varphi^M(G(x) + z) - 2\varphi(z)) \leq 0$ if, and only if, $G(x) \in M$. This is seen as follows: If $G(x) \in M$, then for all $z \in Z$,

$$\varphi^M(G(x) + z) = \inf_{m \in M} \varphi(G(x) + z - m) \leq \varphi(G(x) + z - G(x)) = \varphi(z),$$

hence $\varphi^M(G(x) + z) - 2\varphi(z) \leq -\varphi(z) \leq 0$. And if $G(x) \notin M$, then for $z = 0 \in Z$ we have $\varphi^M(G(x) + z) - 2\varphi(z) = \varphi^M(G(x)) > 0$.

$\square$

We consider the perturbation function $p(u)$ $(u \in Z)$ of (P), namely

$$p(u) := \inf\{F(x) \mid x \in X, \ G(x) \in u + M\}.$$

Then $p_0 = p(0)$, and

$$
\begin{aligned}
\inf_{x \in X} L(x,r,z) &= \inf_{x \in X} \inf_{m \in M} (F(x) + r\varphi(G(x) + z - m) - 2r\varphi(z)) \\
&= \inf_{x \in X} \inf_{u \in Z, u \in G(x) - M} (F(x) + r\varphi(u + z) - 2r\varphi(z)) \\
&= \inf_{u \in Z} \inf_{x \in X, G(x) \in u + M} (F(x) + r\varphi(u + z) - 2r\varphi(z)) \qquad (4) \\
&= \inf_{u \in Z} (p(u) + r\varphi(u + z) - 2r\varphi(z)).
\end{aligned}
$$

At this point let us briefly compare our choice of $\varphi(x,r,z)$, namely

$$\varphi(x,r,z) = \inf_{u \in G(x) - M} (r\varphi(u + z) - 2r\varphi(z)),$$

with a possible alternative if M is convex; it was first introduced for a particular case in [9]:

Assume that $M \subseteq \mathbb{R}^m$ is convex and closed, and set

$$\varphi(x,r,z) := \inf_{u \in G(x) - M} \left(\langle z, u \rangle + \sum_i r_i |u_i|^2 \right), \qquad (5)$$

with $x \in X$, $r \in \mathbb{R}^m_+$, $z \in \mathbb{R}^m$. For this choice of φ, (3) is satisfied. Indeed, if $0 \in G(x) - M$, then for all $r \in \mathbb{R}^m_+$, $z \in \mathbb{R}^m$, $\varphi(x,r,z) \le 0$, the value obtained by choosing $u = 0$ in (5). If $0 \notin G(x) - M$, then from the separation theorem for convex sets [4] there exist $\bar{z} \in \mathbb{R}^m$ and $\delta > 0$ such that $\inf_{u \in G(x) - M} \langle \bar{z}, u \rangle \ge \delta$, hence for arbitrary $\bar{r} \in \mathbb{R}^m_+$, $\varphi(x, \bar{r}, \bar{z}) \ge \delta$ and $\varphi(x, \lambda\bar{r}, \lambda\bar{z}) \ge \lambda\delta \to \infty$ for $\lambda \to \infty$. So (3) holds.

3   Duality

We define the dual (D) of (P) as

(D) $\qquad\qquad\qquad \sup\{H(r,z) \mid (r,z) \in \mathbb{R}_+ \times \mathcal{Z}\},$

where $H(r,z) := \inf_{x \in X} L(x,r,z)$. Let d_0 denote the optimal value of (D),

$$d_0 := \sup_{(r,z) \in \mathbb{R}_+ \times \mathcal{Z}} H(r,z).$$

Thanks to Lemma 1 we have

$$\sup_{(r,z) \in \mathbb{R}_+ \times \mathcal{Z}} L(x,r,z) = \begin{cases} F(x), & \text{if } G(x) \in M, \\ +\infty & \text{else,} \end{cases}$$

hence

$$p_0 = \inf_{x \in X} \sup_{(r,z) \in \mathbf{R}_+ \times Z} L(x, r, z), \tag{6}$$

$$d_0 = \sup_{(r,z) \in \mathbf{R}_+ \times Z} \inf_{x \in X} L(x, r, z). \tag{7}$$

Since the inf sup is never smaller than the sup inf, we obtain the weak duality

$$d_0 \leq p_0.$$

Thus, if $G(\bar{x}) \in M$ and $F(\bar{x}) = H(\bar{r}, \bar{z})$, then $p_0 = F(\bar{x}) = H(\bar{r}, \bar{z}) = d_0$.

We give first conditions for $d_0 = p_0$, starting from the assumption $d_0 > -\infty$ (and hence $p_0 > -\infty$).

In the next few results we shall employ the following requirement:

There exists $\bar{r} \geq 0$ such that $\varphi(u + z) - 2\varphi(z) \leq \bar{r}\varphi(u) \ \forall u \in Z, z \in Z.$
$$\tag{8}$$

Example. Condition (8) is satisfied if φ is convex and positively homogeneous of degree $1 \leq k \leq 2$, since then

$$\varphi(u + z) \leq \frac{1}{2}\varphi(2u) + \frac{1}{2}\varphi(2z)$$
$$\text{(from convexity)}$$

$$= 2^{k-1}\varphi(u) + 2^{k-1}\varphi(z)$$
$$\text{(from homogeneity)}$$

$$\leq 2^{k-1}\varphi(u) + 2\varphi(z).$$
$$\text{(from nonnegativity)}$$

In particular, (8) holds if Z is a normed space and $\varphi(\cdot) = \|\cdot\|$ or $\varphi(\cdot) = \|\cdot\|^2$.

We consider the following property:

there exist $\tilde{q} \in \mathbf{R}$ and $\tilde{r} \in \mathbf{R}_+$ such that $\tilde{q} \leq p(u) + \tilde{r}\varphi(u) \ \forall u \in Z.$
$$\tag{9}$$

We say that (P) satisfies the φ-growth condition iff (9) holds.

Lemma 2. *Assume that* (8) *holds. Then* $-\infty < d_0$ *if, and only if,* (9) *holds.*

Proof. Let (9) hold. Then, from (16),

$$\inf_{x \in X} L(x, \tilde{r}, 0) = \inf_{u \in Z} (p(u) + \tilde{r}\varphi(u)) \geq \tilde{q},$$

hence $d_0 \geq \tilde{q} > -\infty$.

Conversely, if $d_0 > -\infty$, then there exist $q \in \mathbf{R}, r \in \mathbf{R}_+, z \in Z$ such that

$$q \leq \inf_{x \in X} L(x, r, z).$$

By (16), this implies

$$q \le \inf_{u \in Z} \left(p(u) + r\varphi(u + z) - 2r\varphi(z) \right),$$

and then from (8) follows

$$q \le \inf_{u \in Z} \left(p(u) + r\bar{r}\varphi(u) \right).$$

Thus (9) holds with $\tilde{r} := r\bar{r}$ and $\tilde{q} := q$.

□

Theorem 1. *Assume that (8) holds, and let $d_0 > -\infty$. Then*

$$d_0 \ge \liminf_{u \to 0} p(u). \tag{10}$$

Equality holds, if, in addition, φ is upper semicontinuous at 0.

Proof. In order to prove (10) we may assume $\liminf_{u \to 0} p(u) > -\infty$. Observe that from $-\infty < d_0$ follows (9), by Lemma 2. Let $q \in \mathbb{R}$ be such that $q < \liminf_{u \to 0} p(u)$. Then there exists a zero neighborhood U such that $p(u) \ge q$ if $u \in U$, and from (1) there exists $\alpha > 0$ such that $\varphi(u) \ge \alpha$ if $u \notin U$. Let $\tilde{q}, \tilde{r}$ be as in (9). Choose $r \in \mathbb{R}$ satisfying $r - \tilde{r} \ge 0$ and $(r - \tilde{r})\alpha \ge q - \tilde{q}$. If $u \notin U$, then $(r - \tilde{r})\varphi(u) \ge (r - \tilde{r})\alpha \ge q - \tilde{q}$, hence

$$q - r\varphi(u) \le \tilde{q} - \tilde{r}\varphi(u) \le p(u). \tag{from (9)}$$

If $u \in U$, then $q \le p(u) \le p(u) + r\varphi(u)$. Hence

$$q \le \inf_{u \in Z} \left(p(u) + r\varphi(u) \right).$$

Since $q < \liminf_{u \to 0} p(u)$ was arbitrary, we obtain

$$\liminf_{u \to 0} p(u) \le \inf_{u \in Z} \left(p(u) + r\varphi(u) \right) = H(r, 0) \le d_0.$$
$$\text{(from (16))}$$

Now we show that $d_0 \le \liminf_{u \to 0} p(u)$ holds, if φ is upper semicontinuous at 0. From (8), we have

$$d_0 = \sup_{(r,z) \in \mathbb{R}_+ \times Z} \inf_{u \in Z} \left(p(u) + r\varphi(u + z) - 2r\varphi(z) \right)$$

$$\le \sup_{r \in \mathbb{R}_+} \inf_{u \in Z} \left(p(u) + r\bar{r}\varphi(u) \right).$$

Let $r \in \mathbb{R}_+$; we may assume $r\bar{r} > 0$. Let $\varepsilon > 0$. Since φ is upper semicontinuous at 0 and $\varphi(0) = 0$, there exists a zero neighborhood U such that $\varphi(u) \le \varepsilon/r\bar{r}$ for all $u \in U$. Then

$$\inf_{u \in Z} \left(p(u) + r\bar{r}\varphi(u) \right) \le \inf_{u \in U} \left(p(u) + r\bar{r}\varphi(u) \right) \le \inf_{u \in U} p(u) + \varepsilon \le \liminf_{u \to 0} p(u) + \varepsilon.$$

Hence $\sup\limits_{r \in \mathbb{R}_+} \inf\limits_{u \in Z} (p(u) + r\bar{r}\varphi(u)) \leq \liminf\limits_{u \to 0} p(u)$.

$\square$

Corollary 1. *Assume that* (8) *holds, and let $d_0 > -\infty$. If*

$$p_0 = \liminf_{u \to 0} p(u), \qquad (11)$$

then $p_0 = d_0$.

Proof. From Theorem 1 and (11) we obtain $d_0 \geq p_0$. Since $d_0 \leq p_0$ always holds, we have $d_0 = p_0$.

$\square$

Problem (P) is said to be inf-stable iff (11) holds, i.e., $p(u)$ is lower semi-continuous at $u = 0$. In order to proceed further we need a stronger requirement than inf-stability, namely:

$$p_0 > -\infty, \text{ and there exists } (\bar{r}, \bar{z}) \in \mathbb{R}_+ \times Z \text{ such that}$$
$$p_0 \leq p(u) + \bar{r}\varphi(u + \bar{z}) - 2\bar{r}\varphi(\bar{z}) \text{ for all } u \in Z. \qquad (12)$$

If (12) is fulfilled, we say that the augmented Kuhn-Tucker condition for (P) is satisfied. If φ is well-behaved, then (12) implies (9).

Theorem 2. *Let $p_0 > -\infty$ and $(\bar{r}, \bar{z}) \in \mathbb{R}_+ \times Z$. Then*

$$p_0 = d_0 = H(\bar{r}, \bar{z})$$

if, and only if, $(\bar{r}, \bar{z})$ fulfills (12).

Proof. Let $(\bar{r}, \bar{z})$ satisfy (12). Then, using (16), we obtain $p_0 \leq L(x, \bar{r}, \bar{z})$ for all $x \in X$, hence $p_0 \leq H(\bar{r}, \bar{z})$. Since trivially $p_0 \geq d_0 \geq H(\bar{r}, \bar{z})$, we obtain $p_0 = d_0 = H(\bar{r}, \bar{z})$.

Conversely, if $p_0 = H(\bar{r}, \bar{z})$, then $p_0 \leq L(x, \bar{r}, \bar{z})$ for all $x \in X$, and using (16) we obtain (12).

$\square$

Theorem 3. *Let $\bar{x} \in X$, $G(\bar{x}) \in M$, and $(\bar{r}, \bar{z}) \in \mathbb{R}_+ \times Z$. Then $(\bar{x}, (\bar{r}, \bar{z}))$ is a saddle point of L, i.e.,*

$$L(\bar{x}, r, z) \leq L(\bar{x}, \bar{r}, \bar{z}) \leq L(x, \bar{r}, \bar{z}) \quad \forall x \in X, \ (r, z) \in \mathbb{R}_+ \times Z,$$
$$(13)$$

if, and only if, $p_0 = F(\bar{x}) = H(\bar{r}, \bar{z}) = d_0$.

Proof. This is obvious.

$\square$

4   Regularization

Here we discuss regularization for the general augmented Lagrangians as presented in the Introduction. We follow the approach of [8] and [3]. Let

$$L(x, r, z) = F(x) + \varphi(x, r, z), \quad x \in X, r \in R, z \in Z.$$

We define primal and dual values as

$$p_0 := \inf_{x \in X} \sup_{(r,z) \in R \times Z} L(x, r, z),$$

$$d_0 := \sup_{(r,z) \in R \times Z} \inf_{x \in X} L(x, r, z).$$

For all $x \in X$, $r \in R$, let

$$\psi(x, r) := \sup_{z \in Z} \varphi(x, r, z).$$

We assume that the set R of multipliers is a cone, and that $\psi(x, r)$ is positively homogeneous in r. We consider regularized primal and dual values defined as

$$p_0^* := \sup_{r \in R} \inf\{F(x) \mid x \in X, \ \psi(x, r) \le 1\},$$

$$d_0^* := \sup_{r \in R} \inf_{x \in X} (F(x) + \psi(x, r)).$$

We study first the case $R = I\!R_+$.

Theorem 4. *Suppose that X is convex, F is convex, $R = I\!R_+$, and $\psi(\cdot, r)$ is convex. Let $p_0 < +\infty$ or $d_0 > -\infty$. Then $d_0 \le d_0^* = p_0^* \le p_0$. In particular, $p_0 = d_0$ if, and only if, $p_0 = p_0^*$ and $d_0 = d_0^*$.*

Proof. We show that $d_0 \le d_0^* = p_0^* \le p_0$.

(i) $p_0^* \le p_0$: Since $\psi(x, r)$ is positively homogeneous in r, there holds

$$\sup_{r \in R_+} \psi(x, r) < \infty \quad \Longleftrightarrow \quad \sup_{r \in R_+} \psi(x, r) = 0 \quad \Longleftrightarrow \quad \psi(x, r) \le 0 \ \forall r \in I\!R_+.$$

Therefore

$$
\begin{aligned}
p_0 &= \inf_{x \in X} (F(x) + \sup_{r \in R_+} \psi(x, r)) \\
&= \inf\{F(x) \mid x \in X, \ \psi(x, r) \le 0 \ \forall r \in I\!R_+\} \\
&\ge \sup_{r \in R_+} \inf\{F(x) \mid x \in X, \ \psi(x, r) \le 0\} \\
&\ge \sup_{r \in R_+} \inf\{F(x) \mid x \in X, \ \psi(x, r) \le 1\} = p_0^*.
\end{aligned}
$$

(ii) $d_0 \leq d_0^*$: Indeed,

$$
\begin{aligned}
d_0 &= \sup_{r \in \mathbb{R}_+} \sup_{z \in Z} \inf_{x \in X} L(x, r, z) \\
&\leq \sup_{r \in \mathbb{R}_+} \inf_{x \in X} \sup_{z \in Z} L(x, r, z) \\
&= \sup_{r \in \mathbb{R}_+} \inf_{x \in X} (F(x) + \psi(x, r)) = d_0^*.
\end{aligned}
$$

(iii) $d_0^* \leq p_0^*$: Let $p_0^* < \alpha < \beta$ and $r \in \mathbb{R}_+$. We choose $\lambda > 0$ such that $\lambda \leq \beta - \alpha$. From $p_0^* < \alpha$ there exists $x \in X$ with $F(x) < \alpha$ and $\psi(x, r/\lambda) \leq 1$. It follows that $\psi(x, r) \leq \lambda$, hence $F(x) + \psi(x, r) < \beta$. This shows that $d_0^* \leq \beta$.

(iv) $p_0^* \leq d_0^*$: We consider first the case that for every $r \in \mathbb{R}_+$ there exists $x \in X$ with $\psi(x, r) < 1$. This means that the convex program

$$
\inf\{F(x) \mid x \in X, \ \psi(x, r) \leq 1\}
$$

satisfies the Slater condition. Consequently, for every $r \in \mathbb{R}_+$ such that $\inf_{\psi(x,r) \leq 1} F(x) > -\infty$, the Kuhn-Tucker condition holds: There exists $\lambda \geq 0$ such that

$$
\begin{aligned}
\inf_{\psi(x,r) \leq 1} F(x) &\leq \inf_{x \in X} (F(x) + \lambda(\psi(x, r) - 1)) \\
&\leq \inf_{x \in X} (F(x) + \lambda \psi(x, r)) = \inf_{x \in X} (F(x) + \psi(x, \lambda r)) \leq d_0^*.
\end{aligned}
$$

Now suppose that there exists $r \in \mathbb{R}_+$ with $\psi(x, r) \geq 1$ for all x. Then $r > 0$ and $\psi(x, 2r) \geq 2$, hence $p_0 \geq p_0^* \geq \inf_{\psi(x,2r) \leq 1} F(x) = +\infty$. From the assumptions of the theorem, $d_0^* \geq d_0 > -\infty$, so there exists $s \in \mathbb{R}_+$ with $\inf_{x \in X} (F(x) + \psi(x, s)) > -\infty$. Hence $\inf_{x \in X} (F(x) + \psi(x, nr)) \to +\infty$ if $n \to +\infty$, thus $d_0^* = +\infty$.

$\square$

Let us turn to the case that R is an arbitrary cone. Replacing $\mathbb{R}_+$ by R, every step in the proof of Theorem 4 remains valid with the exception of the very last step, which from $p_0^* = +\infty$ infers $d_0^* = +\infty = p_0^*$. This step is not needed if we replace the hypothesis "$p_0 < +\infty$ or $d_0 > -\infty$" by "$p_0^* < +\infty$". Then Theorem 4 remains valid for R an arbitrary cone.

References

1. Bertsekas, D. P. (1976) Multiplier methods: A survey. Automatica **12**, 521–544
2. Boukari, D., Fiacco, A. V. (1995) Survey of penalty, exact-penalty and multiplier methods from 1968 to 1993. Optimization **32**, 301–334
3. Gwinner, J., Oettli, W. (1989) Duality and regularization for inf-sup problems. Results in Mathematics **15**, 227–237

4. Holmes, R. B. (1975) Geometric Functional Analysis and its Applications. Springer-Verlag, New York

5. Ito, K., Kunisch, K. (1990) The augmented Lagrangian method for equality and inequality constraints in Hilbert spaces. Mathematical Programming **46**, 341–360

6. Khanh, P. Q., Nuong, T. H., Théra, M. (1999) On duality in nonconvex vector optimization in Banach spaces using augmented Lagrangians. Positivity (to appear)

7. Oettli, W. (1982) Optimality conditions for programming problems involving multivalued mappings. In: Korte B. (Ed.) Modern Applied Mathematics –Optimization and Operations Research. North-Holland, Amsterdam, 195–226

8. Oettli, W. (1985) Régularisation et stabilité pour les problèmes "inf-sup". Mannheimer Berichte **26**, 9–10

9. Rockafellar, R. T. (1974) Augmented Lagrange multiplier functions and duality in nonconvex programming. SIAM Journal on Control **12**, 268–285

10. Rockafellar, R. T. (1976) Augmented Lagrangians and applications of the proximal point algorithm in convex programming. Mathematics of Operations Research **1**, 97–116

Using Analytic Center and Cutting Planes Methods for Nonsmooth Convex Programming[*]

Paulo Roberto Oliveira[1] and Marcos Augusto dos Santos[2]

[1] PESC/COPPE - Universidade Federal do Rio de Janeiro, Brazil
poliveir@cos.ufrj.br
[2] DCC/ICEX - Universidade Federal de Minas Gerais, Brazil
marcos@dcc.ufmg.br

Abstract. We propose a method for unconstrained minimization of nondifferentiable convex functions. At each iteration, the algorithm includes a reinitialization procedure, due to a knew cut, and some iterations of the analytic center method, those necessary to lower the upper bound of the best iterate function value. The convergence and the polynomial complexity by iteration are established, and numerical tests with typical problems from the literature are presented. The results are similar to those obtained by bundle methods and the algorithms that use analytic center.

Key words: nondifferentiable optimization, interior point method, analytic center, cutting plane

1 Introduction

Convex nonsmooth minimization problems have produced a lot of papers in the last decades. The interest on this well-established area follows both from problems that are naturally contained in that class, and, perhaps the most important, from the application of decomposition schemes in linear and nonlinear programming. It is known that subgradient and bundle methods solve a large spectrum of real problems, see [18] and [22]. Otherwise, the interior-point methodology and related algorithms, has shown their large possibilities in nonsmooth convex programming and feasibility problems where the cutting plane approach [19] is used. Many papers, some theoretical, other with implementable algorithms, demonstrate that such class is a rich field of research, see, e.g., [35], [13], [32], [11], [1], [2], [27], [21], [33], [8], [9], [12], [14], [15], [16], [17], [20], [23], [29], [30], [36], [38], [40],[41]. The reference [17] is the ACCPM library.

The cutting plane algorithm is the common basis for bundle and the majority of interior-point methods. Besides, in the last case, the using of the analytic center at each knew polyhedron is also a characteristic of most papers. Clearly, this procedure does not guarantee the descent property of the

---

[*] This research is partially supported by CNPq and FAPEMIG, Brazil

cost function or some upper-bound of it. Therefore some scheme is necessary to overcome this difficulty, the papers quoted above providing different approaches.

In this paper, we use the analytic center of the current polyhedron as the start point of the analytic center method, applied to the local cutting plane linear programming model. Only some steps are performed, those necessary to lower the upper bound of the best function value at the current (outer) iteration. As we will see, this procedure links the analytic center algorithm to the cutting plane methodology, in such a way that from the local descent property of the first and the global convergence of the second, the convergence to the optimum of the best function value is assured.

A crucial aspect of the methods that use analytic centers is the reinitialization procedure, where the (approximate) center must be recalculated after the adding of a knew constraint. Several authors have contributed with solutions for the build-up problem, in a direct form, or as a column generation technique, see, e. g. [26], [13], [39], [4], [37], [7], [6] and [10]. Here, we consider the build-up as a box where we can put any adequate algorithm. Nevertheless, for computational experiments, we were motivated by last paper.

Further, this paper presents the complexity analysis of a typical outer iteration - the lowering of the upper bound procedure. Our results are consequence of the theory developed for the analytic center algorithm applied to linear programming problems. Particularly, we use [6]. We obtain a polynomial complexity bound. As an outcome, we get an alternative proof of convergence.

Finally, we apply the algorithm to some tests of the nonsmooth convex programming literature. Our results are comparable with those that use bundle methods, and also, with [11].

This paper is divided in seven sections. In the following, we review some basic concepts related to the analytic center method. The algorithm is discussed in section 3, followed by the proof of its convergence in section 4. We establish the complexity for outer iterations and an alternative proof of convergence in section 5. Some numerical experiments are presented in section 6. Finally we have the conclusions and the bibliographical references.

2   The Analytic Center Method

Let f be a convex nonsmooth functional on the n-dimensional euclidean space and consider the problem:

$$\min f(x), \, x \in S \subset R^n \tag{1}$$

where S is some convex and compact set containing at least one solution of the unconstrained problem, supposed to have a finite solution in R^n.

As usually, we make the assumption that, given x, it is known the oracle $f(x)$ and some subgradient s of f at x.

Now consider the cutting plane algorithm [19] applied to problem (1). Let suppose that the iterates x^1, x^2, ..., x^k have been generated together with the corresponding function values $f(x^1)$, $f(x^2)$, ..., $f(x^k)$ and the subgradients $s^1 \in \partial f(x^1)$, $s^2 \in \partial f(x^2)$, ..., $s^k \in \partial f(x^k)$. The cutting plane approximation of f associated with this bundle of information is the piece-wise affine function:

$$x \in R^n \rightarrow f^k(x)$$
$$f^k(x) = \max\left\{f(x^i) + \langle s^i, x - x^i \rangle, \ i = 1, ..., k\right\}$$

and the next iterate x^{k+1} is obtained solving the problem:

$$\min f^k(x) \ x \in S. \tag{2}$$

The problem (2) is equivalent to:

$$\min r \tag{3}$$
$$r \geq f(x^i) + \langle s^i, x - x^i \rangle, \ i = 1, ..., k$$
$$x \in S, r \in R$$

At each step, the algorithm computes x^{k+1} and $s^{k+1} \in \partial f(x^{k+1})$.

Now suppose that the region S in R^n is given by

$$S = \{x \in R^n : \langle a_i, x \rangle \leq b_i, \ i = 1, ..., m\}$$

where $a_i \in R^n$, $b_i \in R$, $i = 1, ..., m$, and we assume that the interior of S, denoted by $int(S)$, is nonempty and bounded. Let denote

$$z = \left(x^T, r\right)^T \in R^{n+1}$$
$$w^i = \left(\left(s^i\right)^T, -1\right)^T, i = 1, ..., k$$
$$d^i = \langle s^i, x^i \rangle - f(x^i), i = 1, ..., k$$
$$S^k = \left\{\left(x^T, r\right)^T \in R^{n+1} : x \in S, \langle w^i, z \rangle \leq d^i, \ i = 1, ..., k\right\}$$

Then (3) can be written as

$$\min r \tag{4}$$
$$z \in S^k$$

The analytic center method [6] associates to (4) a family of nonlinear programs given by

$$\min \varphi_{\rho^k}(z) \qquad\qquad (5)$$
$$z \in int\left(S_{\rho^k}\right)$$

where the distance function is given by

$$\varphi_{\rho^k}(z) = -q \ln \left(\rho^k - r\right) - \sum_{j=1}^{m} \ln \left(b_j - \langle a_j, x\rangle\right) - \sum_{j=1}^{k} \ln \left(d_j - \langle w^j, z\rangle\right),$$

$\rho^k \in R$ is a given upper bound to the current r value (as to the optimum, which we denote by r_*^k), q is a positive integer, $q \geq m + k$, and

$$S_{\rho^k} = S^k \cap \left\{z \in R^{n+1} : r \leq \rho^k \text{ , replicated } q \text{ times }\right\}.$$

Problem (5) being approximately solved in a precise sense (see the quoted author for details), the points $z\left(\rho^k\right)$ form a smooth continuous differentiable curve known as the Central Path. The analytic center method closely follows this curve, converging to the optimal of (4).

Finally, we assume that are known ρ^0, an initial upper bound of the objective function value, and a point x^0 in the interior of the region S. We observe that the sets S^k, $k \geq 1$, can be made compact by including in their definitions the following bounds to r:

$$r \leq f\left(x^0\right)$$

and

$$r \geq \langle s^o, x - x^0\rangle + f\left(x^0\right), \ x \in S$$

where $s^0 \in \partial f\left(x^0\right)$. Indeed, we have, from last inequality,

$$r \geq -2LK + f\left(x^0\right),$$

L being the Lipschitz constant of f restricted to S, and K being the maximum Euclidean norm of the points of S.

3   The Algorithm

We propose a new cutting plane algorithm that partially follows the central path of each approximate polytope of the cutting plane model, for the unconstrainned nonsmooth convex programming problem.

Given $z^k = \left(\left(x^k \right)^T, r^k \right)^T$, close to the analytic center of S_{ρ^k}, an oracle call returns an arbitrary subgradient $s^{k+1} \in \partial f \left(x^k \right)$. Recalling the notation $w^i = \left(\left(s^i \right)^T, -1 \right)^T$, $i \geq 0$, it allows to compute the cut

$$\langle w^{k+1}, z \rangle \leq d^{k+1} = \langle s^{k+1}, x^k \rangle - f \left(x^k \right)$$

If the new set $S_{\rho^k} \cap \{z \in R^{n+1} : \langle w^{k+1}, z \rangle \leq d^{k+1}\}$ has a non empty interior, the restarting procedure will get a point, close to the central path, in the perturbed region. Call this point $z_c^k = \left(\left(x_c^k \right)^T, r_c^k \right)^T$.

Next, by denoting the best known value of the objective function by

$$f_m^k = \min \left\{ f \left(x^i \right), i \leq k \right\}$$

we propose the new upper bound according to

$$\rho^{k+1} = \rho^k - \theta \left(\rho^k - f_m^k \right)$$

for some $0 < \theta < 1$. Thus, the algorithm follows the central path of the current polytope until ρ^k is reduced to the value ρ^{k+1}. After this reduction, we call the new set $S_{\rho^{k+1}}$. That choice imitates the updating of the analytic center method. Notice also that it can happen that $\rho^{k+1} > r_c^k$, at the current central point. In that case only one centralization is necessary. Finally, note that if we entirely solve the problem (4) at each iteration, we will mime the Kelley's cutting plane method.

For the sake of simplicity, the algorithm does not depict the situation when $x^k = \arg \min f \left(x \right)$. If a cut is computed at this point, it will result in a perturbed region with an empty interior. This is detected in the reinitialization procedure: the auxiliary problem (4) is entirely solved.

Next, we present the algorithm through its principal steps. The formulation is rather conceptual.

algorithm NSACM
   Inicialization:
   Compute $\left(\left(x^0 \right)^T, r^0 \right)^T$ near the center of S_{ρ^0}
   Set $f_m^0 = f \left(x^0 \right)$. Choose $\rho^0 > f_m^0$, $\theta \in (0, 1)$, and $\varepsilon > 0$
   $k = 0$

   repeat
       if $\rho^k - r^k \leq \dfrac{72\varepsilon}{91}$
           then stop
       end if
       New cut: compute $s^{k+1} \in \partial f \left(x^k \right)$ and $d^{k+1} = \langle s^{k+1}, x^k \rangle - f \left(x^k \right)$

Reinitialization procedure:

Update the polytope:

$$S_{\rho^k} = S_{\rho^k} \cap \left\{ \left((x)^T, r \right)^T \in R^{n+1} : \langle s^{k+1}, x \rangle - r \leq d^{k+1} \right\}$$

Update the upper bound: $\rho^{k+1} = \rho^k - \theta \left(\rho^k - f_m^k \right)$

Apply ACM to S_{ρ^k}, until ρ^k is reduced to ρ^{k+1},

achieving $\left(\left(x^{k+1} \right)^T, r^{k+1} \right)$, close to the center of $S_{\rho^{k+1}}$

Compute $f \left(x^{k+1} \right)$

Compute $f_m^{k+1} = \min \left\{ f_m^{k+1}, f \left(x^{k+1} \right) \right\}$

$k = k + 1$

end repeat

end algorithm.

4   Convergence

In the following, we prove the convergence of the sequences $\{\rho^k\}_{k \geq 0}$ and $\{f_m^k\}_{k \geq 0}$ to the minimum value $f(x^*)$. Also, it is proved that the algorithm stops at some iteration K when an ε-solution is produced, meaning that

$$f_m^K \leq f(x^*) + \varepsilon$$

At iteration k, $\{s^{k+1} \in \partial f(x^k), f(x^k)\}$ is the oracle at x^k. Also, we suppose that the algorithm never finishes, that is $x^k \neq \arg\{\min f(x), x \in S\}$, for all $k \geq 0$.

Lemma 1. *The sequence $\{\rho^k\}_{k \geq 0}$ is strictly decreasing, and $\rho^k > f_m^k$. More precisely, we have*

$$\rho^k - f_m^k \geq (1 - \theta)^k \left(\rho^0 - f_m^0 \right)$$
$$\rho^k - \rho^{k+1} \geq (1 - \theta)^k \left(\rho^0 - f_m^0 \right).$$

Proof. From the updating formula, it holds

$$\rho^{k+1} - f_m^{k+1} = \rho^k - f_m^{k+1} - \theta \left(\rho^k - f_m^k \right) \qquad (6)$$
$$\geq \rho^k - f_m^k - \theta \left(\rho^k - f_m^k \right)$$

where the inequality derives from the definition of f_m^k (which makes $\{f_m^k\}_{k \geq 0}$ a non-decreasing sequence). Now (6) implies, recursively, the desired result. $\qquad \square$

Lemma 2. *The sequences $\{\rho^k\}_{k\geq 0}$ and $\{f_m^k\}_{k\geq 0}$ converge, and the limits are equal to ρ^*, that verifies*

$$\rho^* \geq f(x^*) = \min\{f(x), x \in S\}.$$

Proof. ¿From the preceding lemma, and the definition of f_m^k, we have, for all $k \geq 0$,

$$\rho^k \geq f_m^k \geq f(x^*).$$

Now, this lower bound, besides the monotonicity of the sequences $\{\rho^k\}$ and $\{f_m^k\}$ ensure, respectively, the existence of limits, call them ρ^* and f_m^*, such that

$$\rho^* \geq f_m^* \geq f(x^*),$$

showing the second part of the lemma.

By the other hand, taking the limits in both sides of the updating formula,

$$\rho^{k+1} = \rho^k - \theta\left(\rho^k - f_m^k\right),$$

gives

$$\rho^* = \rho^* - \theta\left(\rho^* - f_m^*\right),$$

and the conclusion follows, because $\theta \neq 0$. □

Next lemma is a result that comes from the cutting plane methodology. It simply assures the obvious: the model is better, as the iterations progress.

Lemma 3. *Given $\varepsilon > 0$, there exists $K(\varepsilon) \in N$ such that , for all $k \geq K(\varepsilon)$, it holds*

$$f\left(x^k\right) - r^k \leq \varepsilon.$$

Proof. Let us suppose, for contradiction, that there exists $\varepsilon > 0$ such that $f\left(x^k\right) - r^k > \varepsilon$, for all $k > 0$. By the other hand, the cutting plane model says that

$$r^k \geq f\left(x^i\right) + \left\langle s^{i+1}, x^k - x^i \right\rangle, i = 1, ..., k - 1$$

hence

$$f\left(x^k\right) - \varepsilon > f\left(x^i\right) + \left\langle s^{i+1}, x^k - x^i \right\rangle, i = 1, ..., k - 1.$$

Now, let L be the Lipschitz constant associated to f in the bounded set S. We have

$$2L \left\| x^k - x^i \right\| > \varepsilon \ \ for \ all \ i < k = 1, 2, ...$$

This is incompatible with the fact that $\{x^k\} \subset S$, S being a bounded set.

□

Lemma 4. *Suppose that we have the condition obtained in the preceding lemma. Given ε, let k sufficiently large to get $f\left(x^k\right) - r^k \leq \varepsilon/2$. Then, there exists $K\left(\varepsilon\right)$, such that for $k \geq K\left(\varepsilon\right)$,*

$$\rho^k - r^k \leq \varepsilon. \tag{7}$$

Proof. The hypothesis guarantees that for k sufficiently large,

$$\rho^k - r^k \leq \rho^k - f\left(x^k\right) + \frac{\varepsilon}{2} \leq \rho^k - f_m^k + \frac{\varepsilon}{2},$$

the second inequality stemming from the definition of f_m^k. We can write

$$\rho^k - r^k \leq \rho^k - \rho^* + \rho^* - f_m^k + \frac{\varepsilon}{2} \leq \rho^k - \rho^* + \frac{\varepsilon}{2}.$$

As ρ^k converges to ρ^*, there exists $K\left(\varepsilon\right)$ such that both the hypothesis and (7) be verified.

□

Before the convergence theorem, we need a result from the theory of analytic center methods. We just repeat the results of lemmas 3.2 and 3.6 of [6], adapted to our (local) linear programming problem. As seen above, when the iteration k is finished, the procedure of adding a new constraint leads to a point called $z_c^k = \left(\left(x_c^k \right)^T , r_c^k \right)^T$. Therefore, at iteration $k+1$, the problem to be considered by the ACM is

$$\min t \tag{8}$$
$$t \geq f\left(y^i\right) + \langle s^i, y - y^i \rangle, \ i = 1, ..., k+1$$
$$y \in S, t \in R$$

where the starting point is $\left(\left(y^0\right)^T, t^0 \right)^T = \left(\left(x_c^k\right)^T, r_c^k \right)^T$, and we suppose that $\rho^k > r_c^k$. For (8) we define the updating of the upper bound $u^j, j \geq 0$ as

$$u^j = u^{j-1} - \beta\left(u^{j-1} - t^{j-1}\right)$$

for some $0 < \beta < 1$. Now we can present the

Lemma 5. *Let consider the linear problem (8). Call its optimal solution $\left(y_*^{k+1}, r_*^{k+1}\right)$. Denote m^{k+1} the number of constraints of (8). Suppose that $q \geq 2\sqrt{m^{k+1}}$, and the proximity at all iterations is $1/2$. Then*

$$u^{j-1} - r_*^{k+1} \leq \left(1 + \frac{m^{k+1}}{q}\right)\left(1 + \frac{2\sqrt{m^{k+1}}}{q}\right)\left(u^{j-1} - t^{j-1}\right). \qquad (9)$$

Proof. See the quoted lemmas of [6]. $\qquad \square$

Clearly, from (9), it turns out, for a choice $q = 12m^{k+1}$, that

$$\rho^k - r_*^k \leq \frac{91}{72}\left(\rho^k - r^k\right) \qquad (10)$$

for all iterations $k \geq 1$.

Theorem 1. *The sequences $\{\rho^k\}_{k \geq 0}$ and $\{f_m^k\}_{k \geq 0}$ converge to $f(x^*)$, and the algorithm terminates with an ε-solution of (1), in the sense that, for some $\overline{k} \geq 0$,*

$$f_m^{\overline{k}} \leq f(x^*) + \varepsilon.$$

Proof. Due to lemma 4 the algorithm stops for sufficiently large k. Now, using (10) and the stopping criterion, we have

$$\rho^k - r_*^k \leq \frac{91}{72}\left(\rho^k - r^k\right) \leq \varepsilon.$$

By the other hand we know that $\lim_{k \to \infty} \rho^k = \rho^*$, and from the cutting plane theory, $\lim_{k \to \infty} r_*^k = f(x^*)$, thus, taking the limit in expression above, gives

$$\rho^* - f(x^*) \leq \varepsilon.$$

Because $\varepsilon > 0$ was arbitrary, we have actually just proved that $\rho^* \leq f(x^*)$. However, we know that the reverse inequality is also true (lemma 2), so the convergence is proved.

Now suppose that the algorithm stops at $k = \overline{k}$. Using (10), this means that

$$\frac{72}{91}\left(\rho^{\overline{k}} - r_*^{\overline{k}}\right) \leq \rho^{\overline{k}} - r^{\overline{k}} \leq \frac{72}{91}\varepsilon.$$

Lemma 1 and again the convergence of cutting plane method furnish, successively,

$$f_m^{\overline{k}} - r_*^{\overline{k}} < \rho^{\overline{k}} - r_*^{\overline{k}} \leq \varepsilon,$$

and

$$f_m^{\overline{k}} \leq r_*^{\overline{k}} + \varepsilon \leq f(x^*) + \varepsilon,$$

as desired. $\qquad \square$

The last result of this section shows that the points generated by the algorithm are different, unless optimality is achieved.

Proposition 1. *The points of the sequence $\left\{x^k\right\}_{k\geq 0}$ are different.*

Proof. Suppose that $x^j = x^k$ for some $j < k$. Then

$$r^k \geq f\left(x^j\right) + \left\langle s^{j+1}, x^k - x^j\right\rangle = f\left(x^j\right),$$

the inequality originating from the cutting plane model. This corresponds to lemma 3, with $\varepsilon = 0$.   □

5   Complexity Results

Our aim is to estimate the maximum number of iterations of ACM for each outer iteration, as the lowering procedure of upper bound ρ^k is concerned. The estimate will be obtained by using the theory developed for primal analytic center methods, as we find in [6]. The first background result is lemma 5 above. An interesting consequence of the developed theory is an alternative proof of convergence.

5.1   Complexity of the Lowering Procedure

The lemma shown below is a partial result arriving in the proof of the principal complexity theorem of the quoted book (theorem 3.1 of [6]).

Lemma 6. *Let consider the same notations and hypothesis of lemma 5. Then*

$$u^j \leq t^j + \left(1 - \frac{\beta}{2\left(1 + \frac{m^{k+1}}{q}\right)}\right)^j \left(\rho^k - r_*^{k+1}\right) \tag{11}$$

Proof. We have

$$\frac{u^j - r_*^{k+1}}{u^{j-1} - r_*^{k+1}} = \frac{u^{j-1} - \beta\left(u^{j-1} - t^{j-1}\right) - r_*^{k+1}}{u^{j-1} - r_*^{k+1}}$$
$$= 1 - \beta\frac{u^{j-1} - t^{j-1}}{u^{j-1} - r_*^{k+1}}$$

Now use (9) to get

$$\frac{u^j - r_*^{k+1}}{u^{j-1} - r_*^{k+1}} \leq 1 - \frac{\beta}{2\left(1 + \frac{m^{k+1}}{q}\right)} := 1 - \beta'.$$

Consequently, after j AC iterations we have:

$$u^j - t^j \leq u^j - r_*^{k+1}$$
$$\leq (1 - \beta') \left(u^{j-1} - r_*^{k+1} \right)$$
$$\leq (1 - \beta')^j \left(u^0 - r_*^{k+1} \right)$$
$$= (1 - \beta')^j \left(\rho^k - r_*^{k+1} \right)$$

as desired.   □

We repeat below the main theorem of complexity of [6], theorem 3.1.

Lemma 7. *After a maximum of*

$$\frac{2}{\beta} \left(1 + \frac{m^{k+1}}{q} \right) \ln \frac{4 \left(1 + \frac{m^{k+1}}{q} \right) \left(\rho^k - r_*^{k+1} \right)}{\varepsilon} \tag{12}$$

iterations of AC, the algorithm of center terminates with an ε-solution of problem (8).

Proof. See [6].   □

We arrive to the main result of this section.

Theorem 2. *Under the same hypothesis of lemma 5, after a maximum of*

$$\frac{2}{\beta} \left(1 + \frac{m^{k+1}}{q} \right) \ln \frac{8 \left(1 + \frac{m^{k+1}}{q} \right) \left(\rho^k - r_*^{k+1} \right)}{\rho^{k+1} - r_*^{k+1}} \tag{13}$$

AC iterations, the bound ρ^k decreases to the value ρ^{k+1}.

Proof. The aim is to calculate j such that

$$u^j \leq \rho^{k+1}.$$

According to (11) this certainly holds if

$$t^j + (1 - \beta')^j \left(\rho^k - r_*^{k+1} \right) \leq \rho^{k+1}. \tag{14}$$

The worst case analysis implies the supposition that $\rho^k \leq r_c^k$. By the other hand, observe that the left side of (14) converges to r_*^{k+1}, as $j \to \infty$, so we have to show that

$$\rho^{k+1} > r_*^{k+1}, \tag{15}$$

so that (14) has a solution, for sufficiently large j. However r_*^{k+1} is the optimal solution of the cutting plane model, therefore $r_*^{k+1} \leq f(x^*)$. Furthermore, we have shown (lemmas 1 and 2) that, for all i, $\rho^i > f_m^i \geq f(x^*)$, then

$$\rho^{k+1} > f(x^*) \geq r_*^{k+1}$$

which is (15).

Now we intend to solve (14), considering j that verifies simultaneously

$$t^j - r_*^{k+1} \leq \frac{1}{2}\left(\rho^{k+1} - r_*^{k+1}\right) \tag{16}$$

and

$$(1 - \beta')^j \left(\rho^k - r_*^{k+1}\right) \leq \frac{1}{2}\left(\rho^{k+1} - r_*^{k+1}\right). \tag{17}$$

Substitute $\varepsilon = \left(\rho^{k+1} - r_*^{k+1}\right)/2$ in lemma 7 to solve (16) for

$$j > \frac{2}{\beta}\left(1 + \frac{m^{k+1}}{q}\right)\ln\frac{8\left(1 + \frac{m^{k+1}}{q}\right)\left(\rho^k - r_*^{k+1}\right)}{\rho^{k+1} - r_*^{k+1}}. \tag{18}$$

Moreover (17) is verified for

$$j > \frac{2}{\beta}\left(1 + \frac{m^{k+1}}{q}\right)\ln\frac{2\left(\rho^k - r_*^{k+1}\right)}{\rho^{k+1} - r_*^{k+1}},$$

a bound which is smaller than (18), the desired result.   □

5.2   An Alternative Proof of Convergence

Our proof of convergence is consequence of part of the foregoing theory.

Let denote n^k the number of analytic center iterations necessary to lower ρ^k to ρ^{k+1}, at iteration $k \geq 0$. We know, from theorem 2, that, for $m^k/q = 12$, $k \geq 0$,

$$n^k \leq \frac{13}{6\beta}\ln\frac{\frac{26}{3}\left(\rho^k - r_*^{k+1}\right)}{\rho^{k+1} - r_*^{k+1}}.$$

Using the notation for the local upper bound, this means that $\rho^{k+1} \geq u^{n^k}$. In order to simplify the analysis, we suppose for now on that $\rho^{k+1} = u^{n^k}$.

Theorem 3. *Choose* $1 > \beta \geq 19/42$. *Then the sequences* $\{\rho^k\}_{k\geq 0}$ *and* $\{f_m^k\}_{k\geq 0}$ *converge to*

$$\rho^* = f(x^*). \tag{19}$$

Furthermore, the algorithm ends with an ε-solution, in the sense that, for some $\overline{k} \geq 0$,

$$f_m^{\overline{k}} \leq f(x^*) + \varepsilon$$

Proof. Apply (9), lemma 5, for $j - 1 = n^k$. We get

$$\rho^{k+1} - r_*^{k+1} \leq \frac{13}{12}\left(1 + \frac{1}{6\sqrt{m^{k+1}}}\right)\left(\rho^{k+1} - t^{j-1}\right). \tag{20}$$

Likeness, lemma 6 furnishes

$$\rho^{k+1} - t^{j-1} \leq \left(1 - \frac{6\beta}{13}\right)^{n^k}\left(\rho^k - r_*^{k+1}\right). \tag{21}$$

Now (20) and (21) immediately implies

$$\rho^{k+1} - r_*^{k+1} \leq \frac{91}{72}\left(1 - \frac{6\beta}{13}\right)^{n^k}\left(\rho^k - r_*^{k+1}\right),$$

which gives, after substituting the expression of ρ^{k+1}, and some simple manipulations:

$$\left(1 - \theta - \frac{91}{72}\left(1 - \frac{6\beta}{13}\right)^{n^k}\right)\rho^k + \theta f_m^k \leq \left(1 - \frac{91}{72}\left(1 - \frac{6\beta}{13}\right)^{n^k}\right)r_*^{k+1}. \tag{22}$$

In the left side of (22), we are interested in the sequence denoted by $\{\alpha^k\}_{k\geq 0}$,

$$\alpha^k = 1 - \frac{91}{72}\left(1 - \frac{6\beta}{13}\right)^{n^k}.$$

Due to the hypothesis on β this yields

$$\frac{23}{72} < \alpha^k < 1, \ k \geq 0.$$

Consequently, there exists some convergent subsequence $\{\alpha^{k_i}, \ k_i \in N, \ i \geq 0\}$. We can also consider, for the same indexes, the subsequences $\{\rho^{k_i}\}$, $\{f_m^{k_i}\}$ and $\{r_*^{k_i+1}\}$. Lemma 2 allows write

$$\lim_{i \to \infty} \rho^{k_i} = \lim_{i \to \infty} f_m^{k_i} = \rho^*. \tag{23}$$

Moreover, we know that the secant plane method converge, so

$$\lim_{i \to \infty} r_*^{k_i+1} = f(x^*). \tag{24}$$

Now, we take the limits in both sides of (22). Let $\alpha^* > 0$ be an accumulation point of $\{\alpha^{k_i}\}$. From (23) and (24), we have

$$(\alpha^* - \theta)\rho^* + \theta\rho^* \le \alpha^* f(x^*),$$

or

$$\rho^* \le f(x^*).$$

But the reverse inequality is true, again from lemma 2, thus producing the equality.

For the second part, see theorem 1.   □

6   Numerical experiments

The algorithm was implemented using [25] in an ordinary microcomputer with no special resource. To show the reliability of our method, some numerical experiments with classical nonsmooth convex test problems will be reported. The numerical tests that we considered are shown in Table 1.

It is important to observe that the number of oracle calls in bundle methods corresponds essentially to an equal number of calls of some quadratic linear solver, while in algorithms that use analytic centers, each oracle call, demands another kind of procedure - Newton steps for restoring and some centralization procedure. Therefore a strict comparison between bundle and analytic centers methods demands cpu times. Conversely, as ACCPM [17] is an analytic center code, we do make a more real comparison. Table 2 shows, for the tests above, the number of oracle calls and Newton steps for the main procedure for both methods. In our case, main signifies the lowering of the upper bound through the analytic center method, while in ACCPM, it means the application of the weighted projective algorithm for weighted analytic centers, applied to the dual of the original problem [11].

For the reinitialization procedure we used the results from [10]. We observed that, summing up the figures of the tests, the number of Newton steps represented approximately 3/5 of the total cost. This certainly could be improved with faster reinitialization procedures as in [11]. Also, we did not analyze the influence of the parameters in the behavior of the method. In particular, the updating parameters θ and β were fixed at the value $1/2$. Another fact is that we achieved the optimum predicted values through the main procedure. It appears that the using of some purification method could lower the number of Newton steps. Nevertheless, we get comparable results with ACCPM. In terms of oracle calls, our results are in general better than those obtained by bundle methods, as by ACCPM.

Table 1. Numerical tests

Test	Dimension	Reference
CB2, CB3	2	Charalambous, Conn [3]
Dem	2	Demyanov, Malozemov [5]
QL, LQ	2	Makela, Neittaanmaki [24]
Rosen	4	
Maxq, Max1	20	
Goffin	50	
Mifflin1, Mifflin2	2	Lemaréchal [22]
Maxquad1, Maxquad2	10	
Shor	5	Shor [34]
GO 0, TR 48	48	

7   Conclusions

We presented an algorithm that uses analytic centers to solve nonsmooth convex programming problems. From our limited computational experience it appears that the effectiveness of the algorithm depends strongly on the reinitialization procedure. Perhaps a dual approach, as in [11] could get better results.

The algorithm has, at least, one important feature. It produces a monotone strictly decreasing sequence to progress the iterations, following part of the central path associated to the minimization of the model function. As a consequence the convergence is assured for the unconstrained convex problem.

The main application of convex programming is in decomposition algorithms for large scale programming. A forthcoming paper [31] applies the ideas developed here.

Table 2. Newton steps and oracle calls

Test	NSACM Newton	Niter	ACCPM Newton	Niter
CB2	37	10		
CB3	43	11		
Dem	80	11		
QL	49	10		
LQ	26	8		
Mifflin 1	36	10		
Mifflin 2	37	14		
Rosen	38	19		
Maxq	321	170		
Max1	68	7		
Goffin	16	1		
Maxquad 2	128	106		
Maxquad 1	79	79	95 to 2661	87 to 289
Shor	70	39	39 to 331	29 to 52
GO 0	84	18	69 to 430	48 to 52
TR 48	911	186	266 to 8457	237 to 537
Partial sum	1144	322	469 to 11879	401 to 930
Sum	2023	699		

References

1. Atkinson, D. S., Vaidya, P.M. (1995) A cutting plane algorithm for convex programming that uses analytic centers. Mathematical Programming **69**, 1–43
2. Bahn, O., Merle O. du, Goffin, J. L., Vial, J. -P. (1995) A cutting plane method from analytic centers for stochastic programming. Mathematical Programming **69**, 45–73
3. Charalambous, C., Conn, A. R. (1978) An efficient method to solve the minimax problem directly. SIAM Journal on Numerical Analysis **15**, 162–187
4. Dantzig, G. B., Ye, Y. (1990) A build-up interior method for linear programming: affine scaling form. T. R. Sol 90-4, Systems Optimization Laboratory, Dept. of Operations Research, Stanford, USA
5. Demyanov, V. F., Malozemov V. N. (1986) Introduction to minimax. Wiley, New York
6. Den Hertog D. (1992) Interior point approach to linear, quadratic and convex programming: algorithms and complexity. Ph. D. Thesis, Delft University of Technology, Delft, The Netherlands

7. Den Hertog, D., Roos, C., Terlaky, T. (1992) A build-up variant of the path following method for LP. Operations Research Letters **12**, 181–186

8. du Merle, O., Goffin, J. -L., Vial, J. -Ph. (1998) On the improvements to the analytic center cutting plane method. Computational Optimization and Applications **11**, 37–52

9. Elzinga, J., Moore, T. G. (1975) A central cutting plane algorithm for the convex programming problem. Mathematical Programming **8**, 134–145

10. Feijoo B., Sanchez, A., Gonzaga C. C. (1997) Maintaining closedness to the analytic center of a polytope by perturbing added hyperplanes. Applied Mathematics and Optimization **35**, 2, 139–144

11. Goffin, J. -L., Haurie, A., Vial, J. -Ph. (1992) Decomposition and nondifferentiable optimization with the projective algorithm. Management Science **38**, 284–302

12. Goffin, J. -L., Luo, Z. -Q., Ye, Y. (1996) Complexity analysis of an interior cutting plane for convex feasibility problems. SIAM Journal on Optimization **6**, 638–652

13. Goffin, J. -L., Vial, J. -Ph. (1990) Cutting planes and column generation techniques with projective algorithm. Journal of Optimization Theory and Applications **65**, 409–429

14. Goffin, J. -L., Vial, J. -Ph. (1996) Shallow, deep and very deep cuts in the analytic center cutting plane method. Logilab Technical Report 96-3, Department of Management Studies, University of Geneva, Switzerland. Revised June 1997, to appear in Mathematical Programming

15. Goffin, J. -L., Vial, J. -Ph. (1997) A two-cut approach in the analytic center cutting plane method. Logilab Technical Report 97-20, Department of Management Studies, University of Geneva, Switzerland (to appear in Mathematical Methods in Operations Research **47**, 1999)

16. Goffin, J. -L., Vial, J. -Ph. (1998) Multiple cuts in the analytic center cutting plane method. Logilab HEC Working Paper 98.10, Department of Management Studies, University of Geneva, Switzerland

17. Gondzio, J., Merle, O. du, Sarkissian, R., Vial, J. -Ph. (1997) A library for convex optimization based on an analytic center cutting plane method. European Journal of Operational Research **94**, 206–211

18. Hiriart-Urruty, J. B., Lemaréchal, C. (1993) Convex Analysis and Algorithms II. Springer-Verlag Berlin Heidelberg

19. Kelley, J. E. (1960) The cutting plane method for solving convex programming. J. Soc. Indus. Appl. Math. VIII **4**, 703–712

20. Khachiyan, I., Todd, M. J. (1993) On the complexity of approximating the maximal inscribed ellipsoid for a polytope. Mathematical Programming **61**, 137–160

21. Kiwiel, K. C. (1996) Complexity of some cutting plane methods that use analytic centers. Mathematical Programming **74**, 47–54

22. Lemaréchal, C. (1982) Numerical experiments in nonsmooth optimization. In: Nurminski E.A. (Ed.) Progress in Nondifferentiable Optimization, International Institute for Applied Systems Analysis, Laxenburg, Austria, 61–84

23. Levin, A. (1965) An algorithm of minimization of convex functions. Soviet Mathematics Doklady **160**, 1244–1247

24. Makela, M.M., Neittaanmaki, P. (1992) Nonsmooth Optimization, World Scientific Publishing Co. Pte. Ltd., Singapore

25. Matlab (1994) High- performance numeric computation and visualization software. The MathWorks Inc., Massassuchets, USA
26. Mitchell, J. E. (1988) Karmarkar's algorithm and combinatorial optimization problems. PhD thesis, Cornell University
27. Nesterov, Y. (1995) Complexity estimates of some cutting plane methods based on analytic barrier. Mathematical Programming **69**, 149–176
28. Nesterov, Y., Nemirovsky, A. (1994) Interior point polynomial algorithms in convex programming: theory and applications. SIAM, Philadelphia
29. Nesterov, Y., Péton, O., Vial, J. -Ph. (1998) Homogeneous analytic center cutting plane methods with approximate centers. HEC Technical Report 98-3, Department of Management Studies, University of Geneva, Switzerland. Revised June 1998 (to appear in Computational Optimization and Applications)
30. Nesterov, Y., Vial, J. -Ph. (1997) Homogeneous analytic center cutting plane methods for convex problems and variational inequalities. Logilab Technical Report 1997.4, Department of Management Studies, University of Geneva, Switzerland (to appear in SIAM Journal of Optimization)
31. Oliveira, P. R., dos Santos, M. A. (1998) Interior point algorithms for decompositition of Dantzig-Wolfe. T. R., PESC/COPPE, UFRJ, Brazil
32. Sapatnekar, S. S., Rao, V. B., Vaidya, P. M. (1991) A convex optimization approach to transistor sizing for CMOS circuits. Proceedings of the IEEE International Conference on Computer Aided Design, IEEE Computer Society Press, Los Alamitos, CA, 482–485
33. Sharif-Mokhtarian, F. and Goffin, J. -L. (1998) A nonlinear analytic center cutting plane method for a class of convex programming problems. SIAM Journal of Optimization **8**, 4, 1108–1131
34. Shor, N. Z. (1985) Minimization Methods for non-differentiable functions. Springer Verlag, Berlin
35. Sonnevend, G. (1985) An analytical centre for polyhedrons and new classes of global algorithms for linear (smooth, convex) programming. Lecture Notes in Control and Information Sciences **84**, Springer Verlag, New York, 866–876
36. Tarasov, S., Khachiyan, L. G., Erlich, I. (1988) The method of inscribed ellipsoids. Soviet Mathematics Docklady **37**, 226–230
37. Tone, K. (1991) An active-set strategy in interior point methods for linear programming. Working paper, Graduate school, Saita University, Urawa, Japan
38. Vaidya, P. M. (1996) A new algorithm for minimizing convex functions over convex sets. Mathematical Programming **73**, 291–347
39. Ye, Y. (1992) A potential reduction algorithm allowing column generation. SIAM Journal of Optimization **2**, 7–29
40. Ye, Y. (1997) Complexity analysis of the analytic center cutting plane method that uses multiple cuts. Mathematical Programming **78**, 85–104
41. Ye, Y. (1997) Interior point algorithms: theory and analysis. John Wiley & sons, New York, USA

Recent Advances on Second-Order Optimality Conditions

Jean-Paul Penot

Laboratoire des Mathématiques Appliquées, Faculté des sciences, Université de Pau et des Pays de l' Adour, Avenue de l' Université, 64000 PAU, France.
E-mail: Jean-Paul.Penot@univ-pau.fr

Abstract. We compare some recent proposals dealing with necessary conditions and sufficient conditions for optimization problems with explicit or implicit constraints. We add a new condition which involves an additional term which bears on the constraint only, and for this reason is simpler than the known conditions. We also introduce two new types of second-order generalized derivatives and examine their uses for such a question.

Key words: derivative, epi-derivative, Lagrangian, mathematical programming, multiplier, necessary conditions, optimality conditions, projective tangent set, second-order conditions, second-order derivative, sufficient conditions

AMS subject classifications 49K27, 90C30, 46A20, 46N10, 52A05, 52A40

1 Introduction

The topic of optimality conditions is a lively subject, although its basic results have been given several years ago ([3], [20], [36], [52], [51], [64]...). Both theoretical concerns and algorithmic purposes justify the importance of the question. For instance, for many algorithms one can only ensure convergence to a point satisfying some necessary optimality conditions and such a point may not be a local minimizer (see [27] for a recent contribution and references). Error estimates (see [31]) and sensitivity analysis (see [6]-[12], [46], [66]-[68] for recent contributions) also require insights into optimality conditions. However, most textbooks do not present up-to-date conditions, although simple conditions exist, at least in the case of a finite number of scalar constraints (see Theorem 1 below, recalled from [3]). When the constraints are defined by a convex cone in a functional space, the situation is much more involved (see [3], [39], [38], [40], [43], [42], [44], [62], [65]...).

Therefore, many important developments are on the way: the needs of various problems justify such an interest: problems with non polyhedral constraints (in particular problems with functional constraints, continuous time problems, semi-definite programming, semi-infinite programming, generalized semi-infinite programming), nonsmooth problems (in particular multilevel problems), problems with equilibrium constraints, optimal control problems are exciting challenges. Therefore the literature on the subject is rich (see [5], [30], [34], [35], [36], [47], [50], [52], [53], [54], [77] as a sample). Structured problems such as mathematical programming problems, optimal control problems, continuous time problems, semi-infinite programming problems, require a particular attention because the constraints are not necessarily defined by a finite number of scalar functions. This lack of polyhedrality causes a gap between necessary conditions and sufficient conditions (see for instance [43]). Moreover, the conditions cannot be given the simple and aesthetic form of the cases in which the constraints are defined by a polyhedral cone, as in [4], [3], [14], for instance: additional terms have to be added to the second derivative of the Lagrangian in order to take into account what Kawasaki has called the envelop effect ([43]-[44]).

In [64] we endeavoured to reduce this gap to an acceptable extend: when the decision space X is finite dimensional, the sufficient condition differs from the necessary condition by the replacement of an inequality by a strict inequality. As is well-known, this difference is unavoidable, even in the unconstrained case. However, the second-order conditions of [64] are complex and so are the conditions of [36], [47], [53]. More handable conditions have been presented in [67] and [18]. Here we present a simplification of the approach of [67]: instead of dealing with a "compound tangent set" in a product space we just consider a compound tangent set in the space of constraints. This simplification allows to get closer to the properties of the cone defining the constraints.

Let us give a rough, general view of the topic; for the sake of simplicity, we do not deal with higher order conditions but the second-order ones. Also, we do not study the case the spaces are endowed with two topologies, although this situation is of crucial importance for problems involving functional spaces such as optimal control problems (see [5], [51] for instance). First of all, we consider it may be useful to link the subject of optimality conditions for smooth problems to nonsmooth analysis. Besides the fact that several problems such as optimal control problems and multi-level optimization problems lead to such a track, the conversion of a constrained optimization problem into a nonsmooth unconstrained problem by means of the indicator function of the feasible set is a strong incentive to follow this direction, at least from a theoretical point of view (for algorithmic purposes, other devices have to be used). In a nonsmooth framework, simple tools such as the *radial derivative* of a function $f : X \to \overline{\mathbb{R}}$ on a normed vector space X defined at each point x of $f^{-1}(\mathbb{R})$ by

$$f'_r(x, u) := d_r f(x, u) = \liminf_{t \to 0_+} \frac{1}{t} [f(x + tu) - f(x)]$$

or the *genuine parabolic derivative* of [3] given by

$$d_p^2 f(x, u, w) := \liminf_{t \to 0_+} \frac{2}{t^2} \left[f(x + tu + \frac{t^2}{2}w) - f(x) - f'_r(x, tu) \right]$$

are just too rough (at least in the non Lipschitzian case). Simple variants such as the *Hadamard lower derivatives*

$$f'(x, u) := df(x, u) = \liminf_{(t,v)\to(0_+,u)} \frac{1}{t}[f(x + tv) - f(x)], \tag{1}$$

$$d^2 f(x, u, w) := \liminf_{(t,z)\to(0_+,w)} \frac{2}{t^2}\left[f(x + tu + \frac{t^2}{2}z) - f(x) - f'(x, tu)\right] \tag{2}$$

are more adapted. Still more efficient is the *lower dual (epi-) derivative* introduced in the convex case by A. Seeger ([73], [74], [75]) where it can be given a dual form ([33]) and in the nonconvex nonsmooth case by Rockafellar ([70], [71]) by

$$f''(x, y, u) := \liminf_{(t,v)\to(0_+,u)} \frac{2}{t^2}[f(x + tv) - f(x) - \langle y, tv\rangle] \tag{3}$$

for $y \in X^*$, the dual space of X and $u \in X$. Calculus rules for such derivatives are of crucial importance for applications. It did not occur to the author when reading [37] and preparing [62] that a formula for the second derivative of a composite function is the key to optimality conditions; that viewpoint has been clearly put in light by A.D. Ioffe in [38] and still more in [39] (see also [65]). Let us note that besides the use of sequences of positive numbers which was a bit mysterious in [37], a crucial new element of Ioffe's approach has been an especial convergence of the second differential quotients; a geometrical approach and an explanation in terms of convexity have been brought to light in [62]. In simple terms, it can be said that this "compound convergence" takes into account the specific nature of the problem in which there are both a constraint space Z and a decision space X.

Both analytical and geometrical versions take into account the specificities of the situation: in particular, the first-order direction is allowed to move in a space which is not the space of the second-order direction. Let us note that in their "image space approach" A. Cambini, L. Martein and R. Cambini [16] (see also [15], [17], [18], [19], [29]) also make use of a convergence in a space which is not the image space in which sit their tangent cones; however the concepts are different (see [18] and [67] for some comparisons).

As our aim is not to give a complete picture of the question of second-order optimality conditions, we do not evoke the ideal case of quadratic problems with linear constraints for which necessary conditions hold ([13], [21]) nor cases for which the usual constraint qualifications are not satisfied ([26], [45]...) which are more involved.

In the following section we present two new second-order derivatives which are related to the preceding definitions and to a concept of projective tangent cone of order two introduced in [67]. The related optimality conditions are displayed in section 3 along with some comparisons with previous results.

Mathematical programming problems are considered in section 4; in particular, we prove a necessary condition and a sufficient optimality condition which are simpler than the conditions given in [64] (Theorems 3, 4).

In the talk we delivered, we also dealt with the use of second-order optimality conditions for sensitivity analysis (see [6]-[12], [46], [66], [68] and their references). This subject is not reported here, although the study of parametrized optimization problems is a strong incentive to get a clear analysis of optimality conditions and of the role of multipliers. Let us just note one of our conclusions in [68]: while second-order necessary conditions are important for the analysis of perturbed optimization problems, the role given to sufficient optimality conditions in the literature on sensitivity analysis seems to us to be too heavy and too technical. This opinion does not mean that the topic of optimality conditions should be neglected.

2   Projective tangent sets and projective derivatives

Throughout we denote by $\mathbb{P}$ (resp. $\mathbb{R}_+$) the set of positive (resp. non negative)real numbers. The closed ball with center x and radius r in a normed vector space (n.v.s.) X is denoted by $B(x, r)$. The closure of a subset F of X is denoted by clF.

The following definition has been introduced in [18], [67]. An apparently closely related notion is given in [45]; but in fact it differs significantly from the present concept (see the comments in [67]).

Definition 1. Given a subset F of a n.v.s. X, $x \in cl\ F$, $v \in X$, the second order projective tangent cone to F at (x, v) is the set $\widehat{T}^2(F, x, v)$ of pairs $(w, r) \in X \times \mathbb{R}_+$ such that there exist sequences (t_n), (r_n) in $\mathbb{P}$ with limits 0 and r respectively, $(w_n) \to w$ such that $(r_n^{-1} t_n) \to 0$ and

$$x_n := x + t_n v + \frac{t_n^2}{2r_n} w_n \in F \tag{4}$$

for each n.

Some variants may be of interest: one can take weak convergence instead of strong convergence, or weak* convergence if X is a dual space; one can also use nets or bounded nets. In [67] a related definition of second order projective incident cone is introduced in which relation (4) is required for any sequences (t_n), (r_n) in $\mathbb{P}$ with limits 0 and r respectively such that $(r_n^{-1} t_n) \to 0$ and for some sequence $(w_n) \to w$.

Clearly, if (w, r) belongs to $\widehat{T}^2(F, x, v)$ then, for each $s > 0$, the pair (sw, sr) belongs to $\widehat{T}^2(F, x, v)$. The preceding definition carries an abuse of terminology: in [67] it is the image $P\left(\widehat{T}^2(F, x, v)\right)$ by the projective projection p of $\widehat{T}^2(F, x, v)$ which is called the projective second-order tangent set of order two. Here we do not intend to insist on such aspects. Let us just recall that the *projective space* $P(X)$ associated with a vector space (or a cone) X is the set of equivalence classes of pairs $(v, r) \in X \times \mathbb{R}_+$ for the relation

$$(v, r) \sim (v', r') \text{ if } (v', r') = (tv, tr) \text{ for some } t > 0.$$

We denote by p the quotient mapping $p : X \times \mathbb{R}_+ \to P(X)$ and we call it the projective projection. Clearly, $P(X)$ can be decomposed as the union

$$P(X) = X_1 \cup X_0$$

where X_1 (resp. X_0) is the image of $X \times \{1\}$ (resp. $X \times \{0\}$).

Correspondingly, the set $\widehat{T}^2(F, x, v)$ can be split into two parts: this set is the union of $\mathbb{P}\left(F^2(x, v) \times \{1\}\right)$ and $\mathbb{P}\left(F_0^2(x, v) \times \{0\}\right)$ where

$$F^2(x, v) = \left\{ w \in X : \exists (t_n) \to 0_+, \exists (w_n) \to w, \ x + t_n v + \frac{1}{2} t_n^2 w_n \in F \ \forall n \right\}$$

is the familiar *second-order tangent set* to F at (x, v) and $F_0^2(x, v)$ is the set

$$\left\{ w \in X : \exists (t_n), (r_n), \left(\frac{t_n}{r_n}\right) \to 0_+, \exists (w_n) \to w, \forall n \ x + t_n v + \frac{t_n^2}{2r_n} w_n \in F \right\}$$

which can be called the *asymptotic second-order tangent cone* to F at (x, v). Similar decompositions hold for higher order projective tangent sets (see [67] for the definition in the higher order case).

Although the second-order tangent set to F may be empty, as shown by the example

$$F := \left\{ (r, s) \in \mathbb{R}^2 : r^2 = s^3 \right\}, \ x = (0, 0), \ v := (1, 0),$$

the following result asserts that, when X is finite dimensional (resp. is reflexive), the projective tangent set of order two is always nonempty when it is taken with respect to a tangent vector v (resp. and considered in the weak topology).

Proposition 1. *Let $v \in F'(x) := T(F, x) := \limsup_{t \searrow 0} t^{-1}(F - x)$, the (sequential weak) tangent cone to F at x, where F is an arbitrary subset of the finite dimensional space X and $x \in \mathrm{cl}\, F$. Then either $F^2(x, v)$ or $F_0^2(x, v)$ is nonempty, hence $\widehat{T}^2(F, x, v)$ is nonempty.*

EXAMPLE 1. Suppose F is the graph of a twice differentiable mapping $g : U \to V$ in $X = U \times V$, where U and V are n.v.s. Then, an easy calculation shows that for $(u_0, v_0) \in F$, $(u, v) \in F'(u_0, v_0)$, $(w, z) \in X$, $r > 0$ one has

$$((w, z), r) \in \widehat{T}^2 F((u_0, v_0), (u, v)) \Leftrightarrow z = g'(u_0) w + r g''(u_0) uu.$$

The following proposition shows the concept of projective tangent set is invariant under C^2-diffeomorphisms and thus can be extended to subsets of C^2-manifolds.

Proposition 2. *Let X_0 be an open subset X_0 of X, let $g : X \to Y$ be a twice differentiable mapping, let B be a subset of X, $x \in X_0 \cap \mathrm{cl}\, B$, and let C be a subset of Y such that $g(B) \subset C$. Then for each $v \in X$, $(w, r) \in \widehat{T}^2(B, x, v)$ one has*

$$(g'(x)w + rg''(x)vv, r) \in \widehat{T}^2(C, g(x), g'(x)v).$$

One can give a converse of the preceding property which is useful to compute the projective tangent cone of order two to the feasible set $F := g^{-1}(C)$ of a mathematical programming problem. It uses a condition of directional metric regularity introduced in [59] which is a weakening of the following more classical metric regularity condition:

(MR) there exist $\mu > 0$, $\delta > 0$ such that for each $x' \in B(x, \delta)$ one has

$$d\left(x', g^{-1}(C)\right) \leq \mu\, d(g(x'), C).$$

Here, for $z \in Z$, we set $d(z, C) = \inf\limits_{c \in C} \|z - c\|$ to denote the distance function to C and we adopt a similar notation for subsets of X.

Lemma 1. *Suppose the following directional metric regularity condition is satisfied for $x \in X$, $u \in X$:*

$(MR(u))$ *there exist μ, $\rho > 0$ such that for $(t, v) \in (0, \rho) \times B(u, \rho)$ one has*

$$d\left(x + tv,\, g^{-1}(C)\right) \leq \mu\, d(g(x + tv), C).$$

Then, for $F = g^{-1}(C)$, one has

$$(w, r) \in \widehat{T}^2(F, x, u) \Leftrightarrow (g'(x)w + rg''(x)uu, r) \in \widehat{T}^2(C, g(x), g'(x)u).$$

Condition (MR) (hence condition $(MR(u))$) is known to be a consequence of the classical Mangasarian-Fromovitz qualification [49], [32] and of its extension to the infinite dimensional case in [69], [55], [78] which can be written

$$(R^r) \qquad g'(x)(X) - \mathbb{R}_+(C - g(x)) = Z.$$

When C is a closed convex cone and the interior $\mathrm{int}\, C$ of C is nonempty, it has be shown in [55] that the radial tangent cone $T^r(C, x) := \mathbb{R}_+(C - g(x))$ in the preceding condition can be replaced by the usual tangent cone $T(C, g(x)) = \mathrm{cl}\,(T^r(C, g(x)))$:

$$(R) \qquad g'(x)(X) - T(C, g(x)) = Z.$$

However, in general, the transversality condition (R) is weaker than condition (R^r) and does not yield (MR).

The following property will be useful. For $r = 1$ it has some similarity with a property observed in [20] Proposition 3.1 and [57].

Proposition 3. *Let C be a convex subset of X and let $x \in C$, $v \in T(C, x)$. Then, for any $z \in T(T(C, x), v)$, $(w, r) \in \widehat{T}^2(C, x, v)$ one has*

$$(w + z, r) \in \widehat{T}^2(C, x, v).$$

The notion of projective tangent cone enables us to introduce a concept of projective derivative of a multimapping (or set-valued mapping).

Definition 2. *The second order projective derivative of a multimapping F : $X \rightrightarrows Y$ between two normed vector spaces at $z := (x, y)$ in the graph $G(F)$ of F, in the direction $(u, v) \in X \times Y$, is the multimapping $\widehat{D}^2 F\left((x, y), (u, v)\right)$ whose graph is the set $\widehat{T}^2\left(G(F), (x, y), (u, v)\right)$.*

Given a function $f : X \to \overline{\mathbb{R}}$ finite at $x \in X$, setting $y := f(x)$ and taking for F the multimapping whose graph is the epigraph

$$E(f) := \{(u, r) \in X \times \mathbb{R} : r \geq f(x)\}$$

of f, we see that for each $(w, z, r) \in \widehat{T}^2\left(E(f), (x, y), (u, v)\right)$ and each $p \in \mathbb{R}_+$ one has $(w, z + p, r) \in \widehat{T}^2\left(E(f), (x, y), (u, v)\right)$. This fact leads us to set

$$\widehat{d}^2 f(x, u, v, w, r) := \inf\left\{z \in \mathbb{R} : (w, z, r) \in \widehat{T}^2\left(E(f), (x, y), (u, v)\right)\right\}.$$

We see that

$$\widehat{d}^2 f(x, u, v, w, r) := \liminf_{\substack{(t, w', r') \to (0_+, w, r_+) \\ t/r' \to 0}} \frac{2r'}{t^2}\left[f(x + tu + \frac{t^2}{2r'}w') - f(x) - tv\right].$$

Note that for the sake of simplicity we write $(t, w', r') \to (0_+, w, r_+)$ instead of the two conditions $(t, w', r') \to (0_+, w, r)$, $r' > 0$ but that this replacement does not restrict the generality. When $v = f'(x, u)$, as defined in (1) we simply write $\widehat{d}^2 f(x, u, w, r)$.

Since $\widehat{d}^2 f(x, u, v, w, r) = r\widehat{d}^2 f(x, u, v, r^{-1}w, 1)$ for $r > 0$, it suffices to know $\widehat{d}^2 f(x, u, v, \cdot, 0)$ and $\widehat{d}^2 f(x, u, v, \cdot, 1)$ to determine $\widehat{d}^2 f(x, u, v, \cdot, r)$. For $r = 1$, we recover the second-order lower derivative of parabolic type

$$d^2 f(x, u, v, w) := \liminf_{(t, z) \to (0_+, w)} \frac{2}{t^2}\left[f(x + tu + \frac{t^2}{2}z) - f(x) - tv\right]$$

introduced in [58] as a variant of the Ben Tal-Zowe ([3]) second-order derivative (for which the direction w is not allowed to move).

Now let us introduce a projective version of the dual second-order epi-derivatives of [70]. Let us first recall that the (lower) *dual second-order epi-derivative* of f mentioned above is defined for $x \in \operatorname{dom} f$, $u \in X$, $y \in X^*$ by

$$f''(x, y, u) := \liminf_{(t, v) \to (0_+, u)} f_{x, y}^t(v)$$

where

$$f_{x, y}^t(v) := \frac{2}{t^2}\left[f(x + tv) - f(x) - \langle y, tv \rangle\right].$$

The changes we make for the projective version are in agreement with the modifications we introduced for the projective version of the parabolic derivative:

$$\widehat{f''}(x,y,u,w,r) := \liminf_{(t,r',t/r',w')\to(0_+,r_+,0,w)} r'f^t_{x,y}\left(u + \frac{t}{2r'}w'\right).$$

We note that for $r = 1$ we recover the *mixed second derivative* $f''(x,y,u,w)$ of [60]. When f is twice differentiable at x, we get, for $y = f'(x)$,

$$\widehat{f''}(x,y,u,w,r) = rf''(x)uu.$$

A connection with the second-order projective derivative of f is given in the following statement which is a projective version of [60] Proposition 3.2. Here we use the *Hadamard* (or Bouligand or contingent) *subdifferential* of f at x and its subset of elements exposed by $u \in X$ given respectively by

$$\partial f(x) := \{y \in X^* : y \le f'(x,\cdot)\},$$
$$\partial f(x)_u := \{y \in \partial f(x) : \langle y, u\rangle = f'(x,u)\}.$$

Proposition 4. *Given f, x, u as above one has $\widehat{f''}(x,y,u,w,r) = \infty$ whenever $r > 0$, $w \in X$, $y \in \partial f(x)\backslash\partial f(x)_u$. Moreover, for $y \in \partial f(x)_u$ one has for any $r \ge 0$, $w \in X$*

$$\widehat{f''}(x,y,u,w,r) := \widehat{d^2}f(x,u,w,r) - \langle y, w\rangle. \tag{5}$$

Proof. The first assertion is an easy consequence of the definition. Given f, x, y, u, w', $r' > 0$, we have for $f^t_{x,u}(z) := \frac{2}{t^2}\left[f(x + tu + \frac{t^2}{2}z) - f(x) - f'(x,tu)\right]$,

$$r'f^t_{x,y}\left(u + \frac{t}{2r'}w'\right) = r'f^t_{x,u}\left(\frac{1}{r'}w'\right) + \frac{2r'}{t^2}\left[f'(x,tu) - \langle y,tu\rangle\right] - \langle y, w'\rangle$$

so that the second assertion follows from a passage to the liminf. $\square$

The following calculus rules will be important. We leave their elementary proofs (which use Taylor expansions) to the reader.

Lemma 2. *Suppose $f = g + h$, where g is finite at x and h is twice differentiable at x. Then, for any $u, v, w \in X$, $y \in X^*$, $r \ge 0$ one has*

$$\widehat{d^2}f(x,u,w,r) = \widehat{d^2}g(x,u,w,r) + h'(x)(w) + rh''(x)(u,u),$$
$$\widehat{f''}(x,y,u,w,r) = \widehat{g''}(x,y - h'(x),u,w,r) + rh''(x)(u,u).$$

In particular, when f is twice differentiable at x, and $y = f'(x)$ one has

$$\widehat{d^2}f(x,u,w,r) = f'(x)(w) + rf''(x)(u,u),$$
$$\widehat{f''}(x,y,u,w,r) = rf''(x)(u,u).$$

3   Optimality conditions

The following necessary optimality conditions for an unconstrained problem do not assume any smoothness of the objective function. This generality is justified by the fact that the objective function may incorporate implicit constraints or may be the performance function of a first stage optimization problem depending on parameters. Here we do not consider two levels optimization problems (for which a vast literature exists) but we show how optimality conditions for constrained problems can be derived from this general condition for unconstrained problems. We refer the reader to [19], [18], [67], [72] for some views on the way the ideas about this topic emerged.

We first deal with derivatives of parabolic type and then consider dual epiderivatives.

Proposition 5. *Suppose* $f : X \to \overline{\mathbb{R}}$ *is finite at* $x \in F$ *and attains a (local) minimum on* X *at* x. *Then*

$$f'(x, u) \geq 0 \text{ for each } u \in X \tag{6}$$

and when $f'(x, u) = 0$

$$\widehat{d^2} f(x, u, w, r) \geq 0 \text{ for each } (w, r) \in X \times \mathbb{R}_+. \tag{7}$$

Proof. The assertion $f'(x, u) \geq 0$ for each $u \in X$ is immediate. The second one is also a direct consequence of the definition of $\widehat{d^2} f(x, u, w, r) := \widehat{d^2} f(x, u, v, w, r)$, taking $v := f'(x, u) = 0$.   □

Proposition 6. *Suppose* $f : X \to \overline{\mathbb{R}}$ *is finite at* $x \in F$ *and attains a (local) minimum on* X *at* x. *Then* $0 \in \partial f(x)$ *and, when* $f'(x, u) = 0$, *one has*

$$\widehat{f''}(x, 0, u, w, r) \geq 0 \text{ for each} (w, r) \in X \times \mathbb{R}_+. \tag{8}$$

Proof. The first assertion is obvious. The second one is a consequence of the preceding necessary relation and of relation (5): $\widehat{f''}(x, 0, u, w, r) = \widehat{d^2} f(x, u, w, r)$ when $f'(x, u) = 0$. In fact, the very definition shows that $\widehat{f''}(x, 0, u, w, r) \geq 0$ for any $(u, w, r) \in X^2 \times \mathbb{R}_+$.   □

As $\widehat{f''}(x, 0, u, w, r) \geq rf''(x, 0, u)$ for any $(u, w, r) \in X^2 \times]0, \infty[$, and $\widehat{f''}(x, 0, u, w, 0) \geq 0$ whenever $f''(x, 0, u) > -\infty$, the preceding condition is also a consequence of the necessary condition of [71]:

$$f''(x, 0, u) \geq 0. \tag{9}$$

Necessary conditions for optimization over a feasible set F follow easily from the preceding results and the calculus rule displayed above. This observation enlightens the condition given in [67].

Corollary 1. *Suppose $f : X \to \mathbb{R}$ is twice differentiable at $x \in F$ and attains a (local) minimum on $F \subset X$ at x. Then*

$$f'(x)\, u \geq 0 \text{ for each } u \in F'(x) := T(F, x) \tag{10}$$

and whenever $u \in F'(x) \cap \ker f'(x)$ one has

$$f'(x)\, w + r f''(x)\, uu \geq 0 \text{ for each} (w, r) \in \widehat{T}^2(F, x, u). \tag{11}$$

Proof. Let us apply the preceding result to the function f_F given by $f_F :=$ $f + \iota_F$, where ι_F is the *indicator function* of F defined by $\iota_F(z) = 0$ when $z \in F$, $\iota_F(z) = +\infty$ when $z \in X \backslash F$. We easily see that

$$f_F'(x, u) = f'(x)u + \iota_F'(x, u) = f'(x, u) + \iota_{T(F, x)}(u),$$

so that the first-order condition of the preceding result can be transcribed as in condition (10). Similarly, the sum rule of Lemma 2 yields for $u \in F'(x) \cap \ker f'(x)$, $(w, r) \in X \times \mathbb{R}_+$

$$\widehat{d}^2 f_F(x, u, w, r) = \widehat{d}^2 \iota_F(x, u, w, r) + f'(x)(w) + r f''(x)(u, u)$$
$$= \iota_{\widehat{T}^2(F, x, u)}(w, r) + f'(x)(w) + r f''(x)(u, u),$$

so that relation (11) follows immediately from the preceding result. □

Remark. Conversely, it can be shown that Proposition 5 is a consequence of Corollary 1. It suffices to observe that when x is an unconstrained minimizer of f, then $x_f := (x, f(x))$ is a minimizer of $p : (v, s) \mapsto s$ on $E := E(f) :=$ $\{(v, s) \in X \times \mathbb{R} : s \geq f(v)\}$. Thus, for each $u \in X$ we have

$$f'(x, u) = \inf \{s : (u, s) \in E'(x_f)\} \geq 0,$$

as $p'(x_f)(u, s) \geq 0$ for each $(u, s) \in E'(x_f)$ by relation (10). Moreover, when $f'(x, u) = 0$, i.e. when $p'(x_f)(u_f) = 0$ for $u_f := (u, f'(x, u))$, we can write relation (11) as

$$p'(x_f)\, w_f \geq 0 \text{ for each } (w_f, r) \in \widehat{T}^2(E, x_f, u_f)$$

since $p''(x_f) = 0$; setting $w_f := (w, z)$ we get $z \geq 0$ for each $(w, z, r) \in$ $\widehat{T}^2(E, x_f, u_f)$ or $\widehat{d}^2 f(x, u, w, r) \geq 0$ whenever there exists $z \in Z$ with (w, z, r) in $\widehat{T}^2(E, x_f, u_f)$. When there is no such z, we have $\widehat{d}^2 f(x, u, w, r) = \infty$, so that relation (7) holds. □

EXAMPLE 2. Let $F = \left\{(x_1, x_2) \in \mathbb{R}^2 : (x_1 - x_2^3)(x_2 - x_1^3) = 0\right\}$. Then $F'(0)$ $= \mathbb{R} \times \{0\} \cup \{0\} \times \mathbb{R}$ and, as for $x = 0$, $u \in F'(0)$, the set $\widehat{T}^2(F, x, u)$ contains $(w, 1)$, with $w = 0$, a necessary optimality condition for f on F at 0 is $f'(0) = 0$, $f''(0)\, uu \geq 0$ for each $u \in F'(0)$. □

It may be interesting to split the condition of Corollary 1 into two parts, using the decomposition of $\widehat{T}^2(F, x, v)$ we described above.

Corollary 2. *If* $f : X \to \mathbb{R}$ *is twice differentiable at* $x \in F$ *and attains a local minimum on* F *at* x, *then* $f'(x) v \geq 0$ *for each* $v \in F'(x)$ *and when* $v \in F'(x) \cap \ker f'(x)$ *one has*

$$f'(x) w + f''(x) vv \geq 0 \text{ for each } w \in F^2(x, v),$$

$$f'(x) w \geq 0 \text{ for each } w \in F_0^2(x, v).$$

The first condition is well known but the second one is new.

EXAMPLE 3 Let $F = \{(r, s) \in \mathbb{R}^2 : s = |r|^\alpha\}$ where $\alpha \in {]}1, 2{[}$. Then for $x = (0, 0)$, $v = (1, 0)$, the set $F^2(x, v)$ is empty but $F_0^2(x, v)$ contains $w = (0, 1)$. Thus a necessary condition for $(0, 0)$ to be a minimizer of f on F is $f'(x) = 0$ and $f''(x) uu \geq 0$.

EXAMPLE 4 Given a subset F of the space X, $x \in F$, $v \in X \backslash \{0\}$, given $0 < p < q$, let us denote by $T^{q/p}(F, x, v)$ the set of vectors w such that, for some sequences $(s_n) \to 0_+$, $(w_n) \to w$, one has $x + s_n^p v + s_n^q w_n \in F$ for each n. Then, if $q > 2p$, one has $0 \in F^2(x, v)$ whenever $T^{q/p}(F, x, v)$ is nonempty, while for $q = 2p$ one has $F^2(x, v) = T^{q/p}(C, x, v)$; for $q < 2p$ and for $w \in T^{q/p}(F, x, v)$ one has $(w, 0) \in \widehat{T}^2(F, x, v)$, as one can see by taking $(t_n) := (s_n^p)$, $(r_n) := (s_n^{2p-q})$. In the last case, a necessary condition for f to attain a local minimum on F at x is $f'(x) w \geq 0$ whenever $f'(x) v = 0$ and $w \in T^{q/p}(F, x, v)$. When p and q are integers, one may expect some relationship with higher-order optimality conditions as in [22], [23]. $\square$

It is shown in [67] that the necessary optimality condition of [64] :

$$\frac{1}{2} f''(x) uu + \liminf_{\substack{(t, v) \to (0, u) \\ t > 0, \ x + tv \in F}} f'(x) t^{-1}(v - u) \geq 0 \ \forall u \in F'(x) \cap \ker f'(x) \tag{12}$$

implies the necessary condition of Corollary 1. In turn, condition (12) is equivalent to condition (9) applied to the function $f_F := f + \iota_F$ in view of the following calculus rule when f is twice differentiable at x:

$$\begin{aligned} f_F''(x, 0, u) &= f''(x) uu + \iota_F''(x, -f'(x), u) \\ &= f''(x) uu + 2 \liminf_{\substack{(t, v) \to (0, u) \\ t > 0, x + tv \in F}} f'(x) t^{-1}(v - u). \end{aligned}$$

Example 3 shows that condition (12) is stronger than condition (11). In that example, for $x := (0, 0)$, $u := (1, 0)$ we have $\widehat{T}^2(F, x, u) = \mathbb{R}_+ w \times \{0\}$ where $w := (0, 1)$ and $f'(x) = 0$, so that condition (11) is satisfied. However, for f given by $f(x_1, x_2) = -x_1^2$, the origin is not a local minimizer. This fact is detected by condition (12) as $f''(x) uu < 0$ and $f'(x) = 0$.

However, one can still associate with Proposition 5 and Corollary 1 sufficient conditions of the same type (see also [18] and [19] Theorem 2 for a closely related result).

Proposition 7. *If X is finite dimensional and if the following conditions hold, then x is a local strict minimizer of f on F:*

   (a) $f'(x, u) \geq 0$ *for each* $u \in X$;

   (b) if $f'(x, u) = 0$, $u \neq 0$ *then* $\widehat{d^2} f(x, u, w, r) > 0$ *for each* $(w, r) \in X \times \mathbb{R}_+ \setminus \{(0,0)\}$.

Proof. Suppose on the contrary there exists a sequence (x_n) of $F \setminus \{x\}$ with limit x such that $f(x_n) \leq f(x)$ for each integer n. Let $t_n := \|x_n - x\|$, $u_n := t_n^{-1}(x_n - x)$. Taking a subsequence if necessary, we may suppose (u_n) has a limit u with norm 1. Then, by (a), we have $f'(x, u) = 0$. Let $s_n := \|u_n - u\|$, $w_n := s_n^{-1}(u_n - u)$ when $s_n > 0$, $w_n = 0$ when $s_n = 0$. When $s_n = 0$ for infinitely many n's, from the inequality $2t_n^{-2}(f(x + t_n u + \frac{1}{2}t_n^2 s_n w_n) - f(x)) \leq 0$ for infinitely many n's we get $\widehat{d^2} f(x, u, 0, 1) = 0$, a contradiction. Thus we may suppose $s_n > 0$ for each n and assume that the sequence (r_n) given by $r_n := (2s_n)^{-1} t_n$ has a limit r in $\mathbb{R}_+ \cup \{\infty\}$ and the sequence $(w_n) := (s_n^{-1}(u_n - u))$ has a limit w with norm 1. Then $(r_n^{-1} t_n) = (2s_n) \to 0$ and

$$x + t_n u + \frac{1}{2} r_n^{-1} t_n^2 w_n = x_n \quad \forall n.$$

When $r = \infty$, setting $w'_n := r_n^{-1} w_n$ we get again $\widehat{d^2} f(x, u, 0, 1) = 0$, a contradiction. When r is finite, we obtain $\widehat{d^2} f(x, u, w, r) \leq 0$, a contradiction, as $w \neq 0$. □

Observing that $\widehat{d^2} f(x, u, w, r) = \widehat{f''}(x, 0, u, w, r)$ whenever $f'(x, u) = 0$, we get the following statement which involves the dual epiderivative.

Proposition 8. *If X is finite dimensional, if the following conditions hold, then x is a local strict minimizer of f on F:*

   (a) $f'(x, u) \geq 0$ *for each* $u \in X$;

   (b) if $f'(x, u) = 0$, $u \neq 0$ *then* $\widehat{f''}(x, 0, u, w, r) > 0$ *for each* $(w, r) \in X \times \mathbb{R}_+ \setminus \{(0,0)\}$.

An argument similar to the one in the proof of Corollary 1 (or a direct argument as in [67]) yields the following sufficient optimality condition for a constrained minimization problem.

Corollary 3. *If X is finite dimensional, if f is twice differentiable at $x \in F$ and if the following conditions hold, then x is a local strict minimizer of f on F :*

   (a) $f'(x)u \geq 0$ *for each* $u \in F'(x)$;

   (b)   *if* $u \in F'(x) \cap \ker f'(x)$, $u \neq 0$ *then* $f'(x)w + rf''(x)uu > 0$ *for each* $(w, r) \in \widehat{T}^2(F, x, u) \setminus \{(0,0)\}$.

It is shown in [67] that the preceding sufficient condition is in fact a consequence of the sufficient condition of [64] Theorem 1.7: under the assumptions of the preceding corollary, for each $u \in F'(x) \cap \ker f'(x)$, $u \neq 0$ the condition

$$f'(x)w + rf''(x)uu > 0 \text{ for each } (w,r) \in \widehat{T}^2(F,x,u) \setminus \{(0,0)\} \tag{13}$$

implies the condition

$$\frac{1}{2}f''(x)uu + \liminf_{\substack{(t,v) \to (0,u) \\ t > 0, \; x+tv \in F}} f'(x)t^{-1}(v-u) > 0 \quad \forall u \in F'(0) \cap \ker f'(x). \tag{14}$$

However, this implication cannot be reversed, as the following example shows. Hence [64] Theorem 1.7 is a stronger result than Corollary 3.

EXAMPLE 5 Let $F := \{(x_1, x_2) \in \mathbb{R}^2 : x_1^6 = x_2^4\}$ and let f be a positive quadratic form on $X = \mathbb{R}^2$. Let $x := (0,0)$ $u := (1,0)$, $w := (0,1)$, so that $(w,0) \in \widehat{T}^2(F,x,u) \setminus \{(0,0)\}$. As $f'(x) = 0$ and as $\frac{1}{2}f''(x)uu = f(u) > 0$ condition (14) is satisfied. However condition (13) is not satisfied.

Genuine second dual derivatives not only can help to detect local minimizers, but also enable one to characterize local minimizers of second order, as the following result (which is just a variant of [71] Theorem 2.2) shows. Recall that x is said to be a *local minimizer of second order* of f if there exists $\alpha > 0$, $\beta > 0$ such that

$$f(u) \geq f(x) + \frac{1}{2}\alpha \|u - x\|^2 \text{ for any } u \in B(x, \beta). \tag{15}$$

Proposition 9. *Suppose x is a local minimizer of second order of f. Then, there exists $\alpha > 0$ such that for each $u \in X$ one has $f''(x, 0, u) \geq \alpha \|u\|^2$.*

Conversely, if X is finite dimensional and if the preceding property holds, then x is a local minimizer of second order.

Proof. The first assertion is an immediate consequence of the definition of a local minimizer of second order (and one can take α as in 15). Conversely, if for each $u \in X$ one has $f''(x, 0, u) \geq \alpha \|u\|^2$ and if for some $\alpha' < \alpha$ there exists a sequence $(x_n) \to x$ such that $f(x_n) < f(x) + \frac{1}{2}\alpha'\|x_n - x\|^2$, setting $t_n := \|x_n - x\|$ and taking a limit point u of $(u_n) := (t_n^{-1}(x_n - x))$, we get $f''(x, 0, u) \leq \alpha' \|u\|^2$, a contradiction, as $\|u\| = 1$. □

It would be interesting to know whether minimizers of second order can be detected by means of projective derivatives.

4   Application to mathematical programming

Let us apply what precedes to the mathematical programming problem

$$(\mathcal{M}) \quad \text{minimize} \quad f(x) : x \in F := g^{-1}(C),$$

where X, Z are Banach spaces, $f : X \to \mathbb{R}$, $g : X \to Z$ are twice differentiable mappings, C is a closed convex subset of the predual Z of a Banach space Y. Such a formulation encompasses problems in which equality and inequality constraints are present. It would be possible to add a basic constraint $x \in B$, where B is a closed subset of X; it would suffice to replace Z by $X \times Z$, C by $B \times C$, g by (I_X, g).

Since the conditions we present use multipliers, we need a duality result which is akin to a Farkas-Minkowski lemma. Here we denote by Q^0 the polar cone $\{y \in X^* : \forall x \in Q \ \langle y, x \rangle \leq 0\}$ of a cone Q of X.

Lemma 3. *Let P and Q be closed convex cones of the Banach spaces X and Z respectively and let $A : X \to Z$, $c : X \to \mathbb{R}$ be linear and continuous and such that $A(P) - Q = Z$, and for some $m \in \mathbb{R}, b \in Z$*

$$c(x) \geq m \text{ for each } x \in P \cap A^{-1}(b + Q).$$

Then, there exists $y \in Q^0$ such that for each $x \in P$

$$c(x) + \langle y, Ax - b \rangle \geq m.$$

Since P is a cone, the conclusion can be written $0 \in c + y \circ A + P^0$ and $\langle y, -b \rangle \geq m$. When $P = X$, we have $c + y \circ A = 0$. Taking $m = 0$, $b = 0$ we get a Farkas lemma:

$$-c \in (A^{-1}(Q) \cap P)^0 \Rightarrow \exists y \in Q^0 : -(c + y \circ A) \in P^0.$$

We also need a notion of critical vector (or critical direction). Now, such a notion is well accepted (but one or two decades ago it was not clear): we say that $u \in X$ is a *critical vector* at x if $f'(x)u = 0$, $g'(x)u \in T(C, g(x))$, and we write $u \in K(x)$. We will use a second-order qualification condition which generalizes the Ben-Tal qualification condition ([2]):

$$(TR(u)) \quad g'(x)(X) - T(T(C, g(x)), g'(x)u) = Z$$

in which u is a given vector of X; it is weaker than (R) since for $z := g(x)$ one has $T(C, z) \subset T(T(C, z), g'(x)u)$.

The simplest optimality conditions hold for problems with polyhedral constraints; they involve the Lagrangian ℓ defined on $X \times Y$ by $\ell(x, y) := f(x) + \langle y, g(x) \rangle$ and the set of *multipliers*

$$M(x) := \{y \in N(C, g(x)) : f'(x) + y \circ g'(x) = 0\}.$$

They can also be formulated with the help of the extended Lagrangian of F. John and of the associated multipliers.

Theorem 1. *([3]) Let x be a (local) solution to problem $(\mathcal{M})$. Suppose C is a polyhedral cone and conditions (MR) and (R) are satisfied at x. Then for each non null critical vector $u \in K(x)$, there exists some $y \in M(x)$ such that*

$$f''(x)uu + \langle y, g''(x)uu \rangle \geq 0.$$

Note that the multiplier y depends on the critical direction u; examples have been given showing that in general one cannot expect that the multiplier y does not depend on u (see [2], [3], [36]). The one we present below is adapted from [36].

EXAMPLE 6 Let $X := \mathbb{R}^3$, $Z := \mathbb{R}^3$, $C := -\mathbb{R}^3_+$, $f(x_1, x_2, x_3) = x_3$, $g(x_1, x_2, x_3) := (x_1^2 - x_2^2 - x_3, -x_1^2 + x_1 x_2 - x_3, -x_1^2 + x_1 x_2 - x_3)$.

When the cone C is non polyhedral, one cannot expect to get such a simple condition, as the following example inspired by [37] shows.

EXAMPLE 7 Let $X := \mathbb{R}^2$, $Z := \mathbb{R}^3$, $C := \left\{ (z_1, z_2, z_3) : z_3 \geq \sqrt{z_1^2 + z_2^2} \right\}$,

$f(x_1, x_2) = x_2$, $g(x_1, x_2) := (1 - x_1^2, 2x_1, x_2 + 1)$. Then $x := (0, 0)$ is a solution to problem $(\mathcal{M})$, as the feasible set is $F = \left\{ (x_1, x_2) : x_2 \geq x_1^2 \right\}$, and $u := (1, 0)$ is a critical vector. An element $y := (y_1, y_2, y_3)$ of $N(C, g(x))$ is a multiplier iff it satisfies $y_2 = 0$, $y_3 = -1$. Since $(y_1, 0, -1) \in N(C, g(x))$ entails $y_1 = 1$, the relation $\ell''(x, y)uu \geq 0$ cannot be satisfied, as $\ell''(x, y)uu = -2y_1 u_1 u_1$.

A natural way to get a necessary optimality condition consists in applying the unconstrained rule to the modified objective function f_F given by

$$f_F(x) := f(x) + \iota_F(x) = f(x) + \iota_C(g(x)).$$

The calculus rule of Lemma 2 yields for any vector $u \in T(F, x) \cap (f'(x))^{-1}(0)$

$$f_F''(x, 0, u) = f''(x)uu + (\iota_C \circ g)''(x, -f'(x), u) \geq 0.$$

When the cone C is polyhedral and condition (R) holds, one can use [70] Theorem 4.5 to get for any $u \in K(x)$

$$(\iota_C \circ g)''(x, -f'(x), u) = \max_{y \in M(x)} \langle y, g''(x)uu \rangle,$$

and we obtain the preceding theorem (see [70], [71] for details).

When C is not polyhedral, one can apply one of the two formulae devised by Ioffe ([38], [39]) for the second derivative of a composite function $i \circ g$, where $g : X \to Z$ is a mapping of class C^2 and $i : Z \to \overline{\mathbb{R}}$ is a closed proper convex function. The one in [38] involves the function $i_g'' : Z \times Y \times X \times Z \to \overline{\mathbb{R}}$ (called in [64], [65] the *compound derivative* of i) given by $i_g''(z, y, u, w) :=$

$$\liminf_{\substack{t \to 0_+ \\ (u', w') \to (u, w)}} \frac{2}{t^2} \left[i(z + tAu' + \frac{t^2}{2}w') - i(z) - \langle y, tAu' + \frac{t^2}{2}w' \rangle \right], \tag{16}$$

where $A := g'(x)$; it reads as follows

$$(i \circ g)''(x, -f'(x), u) = i_g''(g(x), y, u, g''(x)uu) + \langle y, g''(x)uu \rangle \tag{17}$$

provided $y \in \partial i(g(x))$ is such that $y \circ g'(x) = -f'(x)$ (i.e. $y \in M(x)$) and condition $(MR(u))$ is satisfied. In fact, in [38], [39], a more restrictive qualification condition is used; but one can show that the proof of [65] can be adapted to the assumption that $(MR(u))$ holds. Note that formula (17) above is similar to the usual chain rule formula for second derivatives: when i is twice differentiable at $z := g(x)$ and $y = i'(z)$ we have

$$i''_g(z, y, u, w) = i''(z) Au Au,$$

as easily checked. Then we get for $i := \iota_C$ and for any $u \in K(x)$, $y \in M(x)$

$$f''(x) uu + \langle y, g''(x) uu \rangle + i''_g(g(x), y, u, g''(x)uu) \geq 0.$$

Another method consists in applying the dual formula of [39].

None of these methods corresponds to the way we got the following necessary condition in [62] (reproduced in [64]). We rather used a Farkas lemma as in Lemma 3 above and the following geometrical concept.

Definition 3. Given n.v.s. X, W, a continuous linear map $A : X \to W$, a subset E of W, $e \in E$, $u \in X$, the compound tangent set to E at e with respect to u and A is the set

$$E''_A(e, u) := \limsup_{(t,u') \to (0_+, u)} \frac{2}{t^2} [E - e - A(tu')].$$

Thus, a vector $w \in W$ is in $E''_A(e, u)$ iff there exists a sequence $s := (s_n) \to 0_+$ such that w belongs to the set $E''_{A,s}(e, u)$ of limits of sequences (w_n) for which there exists a sequence (u_n) converging to u with $e + A(s_n u_n) + \frac{1}{2} s_n^2 w_n \in E$ for each n. Let us note that if E is convex the set $E''_{A,s}(e, u)$ is convex but in general the convexity of $E''_A(e, u)$ is not ensured: $E''_A(e, u)$ is just an union of closed convex sets.

Theorem 2. Let x be a (local) solution to problem $(\mathcal{M})$. Let $E := C \times (-\infty, f(x)] \subset Z \times \mathbb{R}$, $e := (g(x), f(x))$, $A := (g'(x), f'(x))$. Suppose conditions (MR) and (R) are satisfied at x. Then, for each critical vector $u \in K(x) \backslash \{0\}$ and for each convex subset Q of $E''_A(g(x), u)$ there exists some $y \in M(x)$ such that

$$f''(x) uu + \langle y, g''(x) uu \rangle \geq \sup \{\langle y, z \rangle + r : (z, r) \in Q\}.$$

Among the possible choices for Q one can take a singleton, or the set $E''_{A,s}(g(x), u)$ for some sequence $s = (s_n) \to 0_+$ or some convex subset of $T^2(E, e, A(u))$. In particular, one can take the second-order incident set

$$T^{ii}(E, e, A(u)) := \liminf_{t \to 0_+} \frac{2}{t^2} (E - e - tA(u)).$$

Let us give a simplified version of the preceding result (or rather of its consequence when one takes for Q a singleton). Its advantage is that it involves $W := Z$, $E := C$, $e := g(x)$, $A := g'(x)$ in Definition 3 above, rather than products $Z \times \mathbb{R}$, $C \times (-\infty, f(x)]$, $e := (g(x), f(x))$, $A := (g'(x), f'(x))$. Thus, a knowledge of the geometrical properties of the cone C is directly applicable to our new process. Given sequences $s = (s_n) \to 0_+$, $(u_n) \to u$ and $w \in C_A''(g(x), u)$ it will be convenient to set

$$q_s((u_n)) := \liminf_n \frac{2}{s_n} f'(x) u_n,$$

$$q(w) := \inf \left\{ q_s((u_n)) : s = (s_n) \to 0_+, \ (u_n) \to u, \ (w_n) \in C_{A,s}''((u_n)) \right\},$$

where $C_{A,s}''((u_n))$ is the set of sequences (w_n) with limit w such that

$$g(x) + s_n g'(x) u_n + \frac{1}{2} s_n^2 w_n \in C \qquad (18)$$

for each n. Thus, for $i := \iota_C$, $A := g'(x)$, $y \in M(x)$ we have

$$q(w) = i_A''(g(x), y, u, w) + \langle y, w \rangle. \qquad (19)$$

The following lemma is similar to [64] Lemma 3.2. We give the proof for the sake of completeness since some changes are in order.

Lemma 4. *Let x be a solution to $(\mathcal{M})$, let $u \in K(x)$ be such that condition $(MR(u))$ holds, and let $w \in C_A''(g(x), u)$. Then, for each $v \in X$ such that $z := g'(x)v + g''(x)uu - w \in T(C, g(x))$ one has $f'(x)v + f''(x)uu + q(w) \geq 0$.*

Proof. We may suppose $q(w) < \infty$. Then, for any $r > q(w)$ we can find sequences $(s_n) \to 0_+$, $(u_n) \to u$, $(w_n) \in C_A''((s_n), (u_n))$ such that $r > 2f'(x)s_n^{-1}u_n$ and (18) holds for each n. Moreover, we can find a sequence $(z_n) \to z$ such that $g(x) + \frac{1}{2}s_n z_n \in C$ for each n. Let $w_n' := g'(x)v + g''(x)u_n u_n - z_n$, so that $(w_n') \to w$. Let us set $u_n' := (1 - s_n)u_n + \frac{1}{2} s_n v$,

$$c_n := g(x) + s_n g'(x) u_n' + \frac{1}{2} s_n^2 [g''(x) u_n u_n + (1 - s_n) w_n - w_n']$$

$$= (1 - s_n) \left[g(x) + s_n g'(x) u_n + \frac{1}{2} s_n^2 w_n \right] + s_n \left[g(x) + \frac{1}{2} s_n z_n \right],$$

so that $c_n \in C$ for each n and $(u_n') \to u$. It follows from condition $(MR(u))$ that for some $c > 0$ and some sequence $(\alpha_n) \to 0$ one has

$$d(x + s_n u_n', F) \leq c \| g(x + s_n u_n') - c_n \|$$

$$\leq \frac{1}{2} c s_n^2 \| g''(x) u_n u_n + (1 - s_n) w_n - w_n' - g''(x) u_n' u_n' \| + \alpha_n s_n^2.$$

Thus, there exists some $(\varepsilon_n) \to 0$ and some (u_n'') such that $x + s_n u_n'' \in F$ and $\|u_n'' - u_n'\| \le \varepsilon_n s_n^2$.

Relation (12) implies that

$$
\begin{aligned}
0 &\le f''(x)uu + 2 \liminf_n f'(x)s_n^{-1}(u_n'' - u) \\
&\le f''(x)uu + 2 \liminf_n f'(x)s_n^{-1}u_n' \\
&\le f''(x)uu + 2 \liminf_n f'(x)s_n^{-1}(1 - s_n)u_n + f'(x)v \\
&\le f''(x)uu + r + f'(x)v.
\end{aligned}
$$

Since r is arbitrarily close to $q(w)$, the result is proved. $\square$

We are ready to prove the following simplified necessary condition. It would be interesting to know whether one can use condition $(TR(u))$ instead of condition (R) in that statement.

Theorem 3. *Let x be a (local) solution to problem $(\mathcal{M})$. Then, for each critical vector $u \in K(x)$, $u \neq 0$ for which conditions $(MR(u))$ and (R) are satisfied and for each $w \in C_A''(g(x), u)$ there exists some $y \in M(x)$ such that*

$$ f''(x)\,uu + \langle y, g''(x)\,uu \rangle - \langle y, w \rangle + q(w) \ge 0. $$

Note that by (19) the additional term $q(w) - \langle y, w \rangle$ is nothing but the second compound derivative $(\iota_C)_A''(g(x), y, u, w)$.

Proof. Given $u \in K(x)$, $u \neq 0$, $w \in C_A''(g(x), u)$ we apply Lemma 3 with $A := g'(x)$, $b := w - g''(x)uu$, $m := -f''(x)uu - q(w)$ so that the conclusion follows from the inequality $\langle y, -b \rangle - m \ge 0$ for some $y \in M(x)$. $\square$

The accuracy of the preceding criterion can be appreciated when compared with the following sufficient condition.

Theorem 4. *When X is finite dimensional, the following conditions ensure that an element x of F is a strict local minimizer:*

(a) the set $M(x) = \{y \in N(C, g(x)) : f'(x) + y \circ g'(x) = 0\}$ of multipliers at x is nonempty;

(b) for each $u \in F'(x) \setminus \{0\}$ with $f'(x)u = 0$ and for each $w \in C_A''(g(x), u)$ there exists some $y \in M(x)$ such that

$$ f''(x)\,uu + \langle y, g''(x)\,uu \rangle - \langle y, w \rangle + q(w) > 0. $$

Proof. Suppose, on the contrary, that there exists a sequence (x_n) with limit x such that $f(x_n) \le f(x)$ and $s_n := \|x_n - x\| > 0$. Without loss of generality, we may suppose the sequence (u_n) given by $u_n := s_n^{-1}(x_n - x)$ has a limit $u \neq 0$. Let us define a sequence (w_n) with limit $w := g''(x)uu$ by

$$ g(x + s_n u_n) = g(x) + s_n g'(x)u_n + \frac{1}{2}s_n^2 w_n. $$

Then, with the notation of (18) we have $(w_n) \in C''_{A,s}((u_n))$, hence

$$\liminf_n \frac{2}{s_n} f'(x) u_n \geq q(w) > \langle y, w \rangle - f''(x) uu - \langle y, g''(x) uu \rangle.$$

As $w := g''(x) uu$ we get

$$\liminf_n \frac{2}{s_n} f'(x) u_n + f''(x) uu > 0,$$

a contradiction with the Taylor expansion of $f(x + s_n u_n)$ which yields

$$\liminf_n \frac{2}{s_n} f'(x) u_n + f''(x) uu = \liminf_n \frac{2}{s_n^2} [f(x + s_n u_n) - f(x)] \leq 0.$$

□

Now, let us present optimality conditions involving the projective tangent cones; for the proofs we refer to [67].

Theorem 5. *([67]) Let x be a (local) solution to problem $(\mathcal{M})$. Then, for each critical vector $u \in K(x)$, $u \neq 0$ for which conditions $(MR(u))$ and $(TR(u))$ are satisfied and for each $(z,r) \in \widehat{T}^2 (C, g(x), g'(x) u)$ there exists some $y \in M(x)$ such that*

$$r(f''(x) uu + \langle y, g''(x) uu \rangle) \geq \langle y, z \rangle.$$

Note that the first order condition implies that $y \in N(T(C, g(x)), g'(x)u)$, a stronger statement than $y \in N(C, g(x))$; this follows from the fact that $\langle y, g'(x) u \rangle = -f'(x)u = 0$ for any multiplier y.

Let us present a variant of the preceding necessary condition.

Theorem 6. *Let x be a (local) solution to problem $(\mathcal{P})$. Suppose conditions $(MR(u))$ and $(TR(u))$ are satisfied at x. Then for each non null critical vector $u \in K(x)$ and each nonempty closed convex subcone $\widehat{Q}$ of $\widehat{T}^2 (C, g(x), g'(x) u)$ not contained in $Z \times \{0\}$ there exists some $y \in M(x)$ such that*

$$\inf_{(z,r) \in \widehat{Q}} [r(f''(x) uu + \langle y, g''(x) uu \rangle) - \langle y, z \rangle] \geq 0.$$

It is shown in [67] that the preceding optimality condition is a consequence of the necessary condition of [64] reported in Theorem 2.

The corresponding sufficient condition is as follows.

Theorem 7. *When X is finite dimensional, the following conditions ensure that an element x of F is a strict local minimizer:*

(a)   the set $M(x)$ of multipliers at x is nonempty;

(b)   for each $u \in F'(x) \setminus \{0\}$ with $f'(x) u = 0$ and each $(w, r) \in X \times \mathbb{R}_+ \setminus \{(0,0)\}$ such that $(z, r) := (g'(x) w + r g''(x) uu, r) \in \widehat{T}^2 (C, g(x), g'(x)u)$ there exists $y \in M(x)$ such that

$$r(f''(x) uu + \langle y, g''(x) uu \rangle) > \langle y, z \rangle.$$

Proof. The existence of a multiplier y ensures condition (a) of Corollary 3 since for any $u \in F'(x)$ one has $g'(x)u \in T(C, g(x))$ and $y \in N(C, g(x))$, hence $\langle y, g'(x)u \rangle \le 0$ and $f'(x)u \ge 0$.

In order to check condition (b) of Corollary 3, let us consider $u \in F'(x) \cap \ker f'(x)$ with $u \ne 0$ and $(w, r) \in \widehat{T}^2(F, x, u)$ with $(w, r) \ne (0, 0)$. Then Proposition 2 ensures that $(z, r) \in \widehat{T}^2(C, g(x), g'(x)u)$ for $z = g'(x)w + rg''(x)uu$. Taking $y \in M(x)$ as in assumption (b) we get

$$f'(x)w + rf''(x)uu > -\langle y, g'(x)w \rangle + \langle y, z \rangle - r\langle y, g''(x)uu \rangle = 0$$

and condition (b) of Corollary 3 is satisfied. $\square$

References

1. Auslender, A. (1984) Stability in mathematical programming with nondifferentiable data. SIAM J. Control and Optim. **22**, 239–254
2. Ben-Tal, A. (1980) Second order and related extremality conditions in nonlinear programming. J. Optim. Th. Appl. **31**, 143–165
3. Ben Tal, A., Zowe, J. (1982) A unified theory of first and second-order conditions for extremum problems in topological vector spaces. Math. Programming Study **19**, 39–76
4. Ben Tal, A., Zowe, J. (1982) Necessary and sufficient optimality conditions for a class of nonsmooth minimization problems. Math. Programming **24**, 70–91
5. Bergounioux, M. (1998) Optimal control of problems governed by abstract elliptic variational inequalities with state constraints. SIAM J. Control and Opt. **36**, 1, 273–289
6. Bonnans, J.F., Cominetti, R. (1996) Perturbed optimization in Banach spaces I: a general theory based on a weak directional constraint qualification. SIAM J. Control Opt. **34**, 4, 1151–1171
7. Bonnans, J.F., Cominetti, R. (1996) Perturbed optimization in Banach spaces II: a theory based on a strong directional constraint constraint qualification. SIAM J. Control Opt. **34**, 4, 1172–1189
8. Bonnans, J.F., Cominetti, R. (1996) Perturbed optimization in Banach spaces III: semi-infinite optimization. SIAM J. Control Opt. **34** 5, 1555–1567
9. Bonnans, J.F., Cominetti, R., Shapiro, A. (1998) Sensitivity analysis of optimization problems under abstract constraints. Math. Oper. Res. **23**, 806–831
10. Bonnans, J.F., Cominetti, R., Shapiro, A. (1999) Second order optimality conditions based on parabolic second order tangent sets. SIAM J. Optim. **9**, 466–492
11. Bonnans, J.F., Shapiro, A. (1998) Optimization problems with perturbations: a guided tour. SIAM Review **40**, 228–264
12. Bonnans, J.F., Shapiro, A. Perturbation Analysis of Optimization Problems, book to appear
13. Borwein, J. (1982) Necessary and sufficient conditions for quadratic minimality. Num. Funct. Anal. Opt. **5**, 2, 137–140

14. Burke, J. (1987) Second order necessary and sufficient conditions for convex composite NDO. Math. Programming **38**, 287–302

15. Cambini, A., Martein, L. (1995) Second order necessary optimality conditions in the image space: preliminary results. In: Castagnoli E., Giorgi G. (Eds.) Scalar and vector optimization in economic and financial problems, Univ. Bocconi, Milano, 27–38

16. Cambini, A., Martein, L., Cambini, R. (1995) A new approach to second order optimality conditions in vector optimization. Technical report 103, Department of Statistics and Applied Math., Univ. of Pisa

17. Cambini, R. (1996) Second order optimality conditions in the image space. Technical report 99, Department of Statistics and Applied Math., Univ. of Pisa

18. Cambini, A., Komlosi, S., Martein, L. (1996) Recent developments in second order necessary optimality conditions. In: Crouzeix J.-P., J.-E. Martinez-Legaz and M. Volle (Eds.) Generalized Convexity, Generalized Monotonicity 1996, Kluwer, Dordrecht, 347-356

19. Cambini, A., Martein, L., Vlach, M. (1997) Second order tangent sets and optimality conditions. Preprint, Japan Advanced Study of Science and Technology, Hokuriku, Japan, June 1997

20. Cominetti, R. (1990) Metric regularity, tangent sets and second-order optimality conditions. Applied Math. and Optim. **21**, 265–287

21. Contesse, L. (1980) Une caractérisation complète des minima locaux en programmation quadratique. Numer. Math. **34**, 315–332

22. Dedieu, J.-P. (1995) Third and fourth-order optimality conditions in optimization. Optimization **33**, 2, 97–105

23. Dedieu, J.-P., Janin, R. (1995) A propos des conditions d'optimalité d'ordre trois et quatre pour une fonction de plusieurs variables. Preprint Univ. Poitiers and Toulouse

24. Dmitruk, A., Milyutin, A., Osmolovski, N. (1980) Lyusternik's theorem and the theory of extrema. Russian Math. Surveys **35**, 11–51

25. Do, C.N. (1992) Generalized second-order derivatives in reflexive Banach spaces. Trans. Amer. Math. Soc. **334**, 281–301

26. El Boukhari, A. (1997) Necessary optimality conditions for an optimization problem with equality constraint. Optim. **39**, 253–274

27. Facchinei, F., Lucidi, S. (1998) Convergence to second order stationary points in inequality constrained optimization. Math. Oper. Res. **23**, 3, 743–766

28. Fletcher, R., Watson, G.A. (1980) First and second order conditions for a class of nondifferentiable optimization problems. Math. Prog. **18**, 291–307

29. Giannessi, F. (1984) Theorems of the alternative and optimality conditions. J. Optim. Th. Appl. **42**, 331–365

30. Gollan, B. (1981) Higher order necessary conditions for an abstract optimization problem. Math. Programming Study 14, 69–76

31. Hager, W.W., Gowda, M.S. (1999) Stability in the presence of degeneracy and error estimation. Math. Prog. A **85**, 1, 181–192

32. Han, S.P., Mangasarian, O.L. (1979) Exact penalty functions in nonlinear programming. Math. Programming **17**, 251–269

33. Hiriart-Urruty, J.-B. (1982) Approximating a second order directional derivative for nonsmooth convex functions. SIAM J. Control and Optim. **10**, 783–807

34. Hiriart-Urruty, J.-B., Strodiot, J.J., Nguyen, V.H. (1984) Generalized Hessian matrix and second-order optimality conditions for problems with $C^{1,1}$ data. Applied Math. Optim. **11**, 43–56
35. Hoffmann, K.H., Kornstaedt, H.J. (1978) Higher order necessary conditions in abstract mathematical programming. J. Optim. Th. Appl. **26**, 533–569
36. Ioffe, A.D. (1979) Necessary and sufficient conditions for a local minimum 3: second-order conditions and augmented Lagrangians. SIAM J. Control and Opt. **17**, 2, 266–288
37. Ioffe, A.D. (1989) On some recent developments in the theory of second-order optimality conditions. In: Dolecki S. (Ed.) Optimization. Proc. Conf. Varetz, 1988, Lecture Notes in Maths. **1405**, Springer Verlag, Berlin, 55-68
38. Ioffe, A.D. (1991) Variational analysis of a composite function: a formula for the lower second-order epi-derivative. J. Math. Anal. Appl. **160**, 2, 379–405
39. Ioffe, A.D. (1992) Dual representation of the lower second-order epi-derivative of a composite function. In: Ioffe A., Marcus M., Reich S. (Eds.) Optimization and Nonlinear Analysis. Pitman Res. Notes in Maths **244**, Longman, Harlow, 145–154
40. Ioffe, A.D. (1994) On sensitivity analysis of nonlinear programs in Banach spaces. SIAM J. Optimization **4**, 1–44
41. Janin, R. (1995) Conditions d'optimalité d'ordre supérieur en programmation mathématique. Preprint, Univ. of Poitiers, France
42. Kawasaki, H. (1988) Second-order necessary conditions of the Kuhn-Tucker type under new constraint qualification. J. Optim. Th. Appl. **57**, 2, 253–264
43. Kawasaki, H. (1988) An envelop-like effect of infinitely many inequality constraints on second-ordernecessary conditions for minimization problems. Math. Programming **41**, 73–96
44. Kawasaki, H. (1992) Second-order necessary and sufficient optimality conditions for minimizing a sup type function. Applied Math. Optim. **26**, 195–220
45. Ledzewicz, U., Schaettler, H. (1995) Second-order conditions for extremum problems with nonregular equality constraints. J. Optim. Th. Appl. **86**, 1, 113–144
46. Levitin, E.S. (1994) Perturbation theory in mathematical programming and its applications. John Wiley, Chichester
47. Levitin, E.S., Milyutin, A.A., Osmolovskii, N.P. (1978) Higher order conditions for a local minimum in problems with constraints. Uspehi Math. Nauk **33**, 85–148
48. Levy, A.B. (1993) Second-order epi-derivatives of integral functionals. Set-Valued Anal. **1**, 4, 379–392
49. Mangasarian, O.L., Fromovitz, S. (1967) The Fritz-John necessary opptimality condition in the presence of equality and inequality constraints. J. Math. Anal. Appl. **7**, 37–47
50. Maruyama, Y. (1990) Second-order necessary conditions for nonlinear optimization problems in Banach spaces and their application to an optimal control problem. Math. Oper. Res. **15**, 3, 467–482
51. Maurer, H. (1981) First and second-order sufficient optimality conditions in mathematical programming and optimal control. Math. Programming Study **14**, 163–177
52. Maurer, H., Zowe, J. (1979) First and second-order necessary and sufficient optimality conditions for infinite-dimensional programming problems. Math. Programming **16**, 98–110

53. Pales, Z., Zeidan, V.M. (1994) Nonsmooth optimum problems with constraints. SIAM J. Control and Opt. **32**, 5, 1476–1502
54. Pales, Z., Zeidan, V. (1994) First- and second-order conditions for control problems with constraints. Trans. Amer. Math. Soc. **346**, 2, 421–453
55. Penot, J.-P. (1982) On regularity conditions in mathematical programming. Math. Programming Study **19**, 167–199
56. Penot, J.-P. (1984) A view of second-order extremality conditions. In: Lemaréchal C. (Ed.) Third Franco-German Conference in Optimization, July 1984, INRIA, Rocquencourt, France, 34–39
57. Penot, J.-P. (1984) A geometric approach to higher order necessary conditions. Manuscript, Univ. of Pau, July 1984
58. Penot, J.-P. (1984) Generalized higher order derivatives and higher order optimality conditions. Preprint, Univ. Santiago, Nov. 1984
59. Penot, J.-P. (1984) Differentiability of relations and differential stability of perturbed optimization problems. SIAM Journal Control and Optimization **22**, 4, 529–551
60. Penot, J.-P. (1992) Second-order generalized derivatives: comparisons of two types of epi-derivatives. In: Oettli W., Pallaschke D. (Eds.) Advances in Optimization, Proceedings, Lambrecht FRG, 1991, Lecture Notes in Econ. and Math. Systems **382**, Springer-Verlag, Berlin, 52–76
61. Penot, J.-P. (1993) Second-order generalized derivatives: relationships with convergence notions. In: Giannessi F. (Ed.) Nonsmooth Analysis, Methods and Applications, Gordon and Breach, London, 303–322
62. Penot, J.-P. (1990) Optimality conditions in mathematical programming. Preprint, Univ. of Pau, France
63. Penot, J.-P. (1990) Optimality conditions for composite functions. Preprint, Univ. of Pau, France
64. Penot, J.-P. (1994) Optimality conditions in mathematical programming and composite optimization. Math. Prog. **67**, 225–245
65. Penot, J.-P. (1995) Sequential derivatives and composite optimization. Revue Roumaine Math. Pures Appl. **40**, 501–519
66. Penot, J.-P. (1997) Central and peripheral results in the study of marginal and performance functions. In: Fiacco A. (Ed.) Mathematical programming with data perturbations, Marcel Dekker, New York, 305–337
67. Penot, J.-P. (1998) Second-order conditions for optimization problems with constraints. SIAM J. Control and Optim. **37**, 1, 303–318
68. Penot, J.-P. (1999) Points de vue sur l'analyse de sensibilité en programmation mathématique. In Decarreau A., Janin R., Philippe R., Pietrus A., (Eds.) Actes des sixièmes journées du groupe Mode, Atlantique, Poitiers, 176–203
69. Robinson, S.M. (1976) Stability theory for systems of inequalities Part 2: differentiable nonlinear systems. SIAM J. Numer. Anal. **13**, 497–513
70. Rockafellar, R.T. (1988) First and second-order epi-differentiability in nonlinear programming. Trans. Amer. Math. Soc. **307**, 1, 75–108
71. Rockafellar, R.T. (1989) Second-order optimality conditions in nonlinear programming obtained by way of epi-derivatives. Math. Oper. Research **14**, 3, 462–484
72. Rockafellar, R.T., Wets, R. J.B. (1998) Variational analysis, Springer, New York
73. Seeger, A. (1986) Analyse du second ordre de problèmes non différentiables. Thesis, Univ Toulouse I, France

74. Seeger, A. (1992) Limiting behavior of the approximate second-order subdifferential of a convex function. J. Optim. Th. Appl. **74**, 527–544

75. Seeger, A. (1992) Second derivatives of a convex function and and of its Legendre-Fenchel transformate. SIAM J. on Optim. **2**, 405–424

76. Stuniarski, M. (1991) Second-order necessary conditions for optimality in nonsmooth nonlinear programming. J. Math. Anal. Appl. **154**, 303–317

77. Zeidan, V. (1984) First and second order sufficient conditions for optimal control and the calculus of variations. Appl. Math. Optim. **11**, 209–226

78. Zowe, J., Kurcyusz, S. (1979) Regularity and stability for the mathematical programming problem in Banach space. Applied Math. and Optim. **5**, 49–62

On Some Relations Between Generalized Partial Derivatives and Convex Functions

Peter Recht

University of Dortmund, Operations Research and Wirtschaftsinformatik,
Vogelpothsweg 87,
D 44 224 Dortmund, Germany

Abstract. In [8] and [9] a generalization of the concept of the ordinary gradient was presented, which exists for a broad class of nondifferentiable functions. This concept differs from traditional "set-valued" generalizations by an expansion of the directional derivative into a special orthogonal series.

By investigating the coefficients of this series and interpreting them as a kind of *partial derivatives* formal analogies to properties of classical partial derivatives can be drawn. Moreover such an approach reveals relations between questions appearing within the theory of optimization and results coming from other branches of mathematics, especially from the theory of harmonic and subharmonic functions.

Applying the concept of generalized partial derivatives to the case that the functions under considerations are convex, a characterization of convexity, differentiablilty properties and a necessary optimality condition in terms of generalized partial derivatives can be given.

Key words: analysis of nondifferentiable functions, generalized gradients, convex functions, subharmonic functions

1 Introduction

In classical calculus the *derivative* and the *gradient* (as a vector of special *partial derivatives*) of a function f are essentially related to each other. In the case that f is *Frechet*-differentiable at some point x_0, the representation of the derivative df_{x_0} as a *linear* function "reduces" to the computation of a couple of one-dimensional derivatives. The "play" with these figures (of partial derivatives) can deliver a variety of information concerning the local behaviour of f. They provide therefore a main tool within the analysis of functions.

It is well known that a convex function f does not need to be *Frechet*-differentiable over the interior of its domain. But convexity implies continuity, local *Lipschitz*-continuity and directional differentiability. These properties provide that f can be approximated in a first order way by its *sublinear* directional derivatives. The correspondence between sublinear functions and convex, compact sets then immediately delivers the concept of subdifferentials. This concept can be regarded as the essential tool within the framework of convex analysis. In a powerful way it generalizes the concept of gradients for ordinary differentiation to the case of convex functions.

Nevertheless, to exploit information on f carried by the subdifferentials, in general leads to perform operations on closed convex sets. Apart from some special situations, these kind of operations can, of course, cause a lot of difficulties in their calculation. It seems that just the set-valued character of a derivative makes it impossible to "simply play with figures", as we are used to do in classical differentiation when calculating partial derivatives.

In [8] and [9] a concept of *generalized partial derivatives* was introduced, which for locally *Lipschitz*-continuous, directionally differentiable functions f allows to characterize the local behaviour of f by "a couple of real numbers". For this class of functions it can be regarded as the "nondifferentiable counterpart" of the ordinary partial derivatives. These numbers can then be used as well for deriving optimality conditions for f (see [10]) as for approaching to a nondifferentiable version of the *Lemma of Poincare* (see [11]).

The aim of this contribution is to use the concept of generalized partial derivatives for a characterization of convex functions: it will look for conditions - in terms of generalized partial derivatives - that a given function f, locally *Lipschitz*-continuous and directionally differentiable, is a convex function.

The paper is organized as follows:

in section 2 we deal with the general construction scheme for the directional derivative df of a locally *Lipschitz*-continuous, directionally differentiable function f. The continuity of df then allows a *Fourier*-series-like representation of the directional derivative.

It, moreover, leads to an extension of df in terms of "elementary", positively homogeneous functions $P_k^{(i)} \in C^\infty(\mathbb{R}^n \setminus 0)$. Additionally, this extension reveals an analogy to *Frechet*-differentiable functions f, since in this case df can be respresented by a special subset of these "elementary" functions $P_k^{(i)}$, namely just the linear ones. Therefore the multiples of the $P_k^{(i)}$'s give reason to interpret them as *generalized partial derivatives of* f. The investigation of these numbers seems reasonable, as they carry information about the function f of order one and so can be used as a tool for the local characterization of f.

Especially questions, concerning the way in which *convexity information* can be gained from this generalized gradient seem to be of importance, since convexity often plays a crucial part as well from the theoretical point of view as from the view of application in practical problems.

In section 3 we will collect some properties of the class of harmonic and subharmonic functions. In connection with generalized partial derivatives it turns out that it is helpful to deal with these types of functions for two reasons:

first, they can be regarded as "natural" generalizations of affine and sublinear functions, respectively, which in classical, real (or convex) analysis appear as the most simple classes of functions that carry first order information.

Secondly, the positively homogeneous functions $P_k^{(i)}$ themselves are originally derived from the class of harmonic polynomials. Therefore features of those harmonic functions might suggest properties for the $P_k^{(i)}$'s and, consequently, for df.

Later the problem of *Dirichlet* will be crucial for our proofs. This problem asks for the exsistence of an harmonic function in a domain with preassigned boundary values. An essential part in this connection will also play *Green's* function as well as the *Riesz* representation theorem of subharmonic functions. In our case, this theorem uniquely relates sublinear functions to *Borel* measures.

Equipped with these tools, the essential purpose of this paper is then elaborated in section 4.

Here the concept of generalized partial derivatives, as worked out in chapter 1, and the properties of harmonic and subharmonic functions, as cited in the previous chapter, are connected.

For convex functions f it will turn out that the first generalized partial derivative $\partial f_{x_0}/\partial P_0^{(1)}$ is of special meaning. This single number carries as well a necessary condition for the convexity of f, as it determines *Frechet*-differentiability of f. Together with a "condition of monotonicity" it, moreover, allows a complete characterization of convexity. With its help also a necessary optimality condition for (nondifferentiable) convex functions can be deduced.

2   Constructing generalized partial derivatives

In the sequel let $U \subset \mathbb{R}^n$ be an open set, let $x_0 \in U$ be an arbitrary point and $f : U \longrightarrow \mathbb{R}$ a locally *Lipschitz*, directionally differentiable function. I.e. we will consider functions f having the properties

- for all $x_0 \in U$ there exists a neighbourhood $V(x_0)$ of x_0 and $L > 0$ such that for all $x \in V(x_0)$ the inequality $|f(x_0) - f(x)| \le L \cdot \|x_0 - x\|$ holds,
- for all $x_0 \in U$ and all $h \in \mathbb{R}^n$ the limit $df_{x_0}(h) = \lim_{t \searrow 0} \frac{1}{t} \cdot [f(x_0 + t \cdot h) - f(x_0)]$ exists.

Locally *Lipschitz* continuity of f implies *Lipschitz* continuity of df_{x_0} (with respect to the direction h). Hence, the restriction $df_{x_0}|_{S^{n-1}}$ to the unit-sphere $S^{n-1} := \{h \mid \|h\| = 1\}$ is an element of $L^2\left(S^{n-1}, < \cdot, \cdot >_{S^{n-1}}\right)$.

Here $L^2\left(S^{n-1}, < \cdot, \cdot >_{S^{n-1}}\right)$ denotes the *Hilbert*-space of all square-integrable functions on S^{n-1}, for which we assume, that the inner product $< \cdot, \cdot >_{S^{n-1}}$ on S^{n-1} is canonically given by

$$< f, g >_{S^{n-1}} = c_n^{-1} \cdot \text{int }_{S^{n-1}} f(\zeta) \cdot g(\zeta) d\sigma,$$

where $c_n := 2 \cdot \pi^{\frac{n}{2}}/\Gamma(\frac{n}{2})$ and $d\sigma$ denotes the "surface element"[1] on S^{n-1}.

We will use this fact to look for a helpful development of df_{x_0} in terms of special elements of this *Hilbert*-space.

We construct this kind of representation as follows:

Let $H_k : \mathbb{R}^n \longrightarrow \mathbb{R}$ be an harmonic[2] polynomial of degree equal k, in the variables $x = (x_1, x_2, x_3, ...x_n)$, i.e. $H_k(\lambda x) = \lambda^k H_k(x)$ for $\lambda \in \mathbb{R}$. The function S_k will then be defined to be the restriction of H_k to the sphere S^{n-1} and will be called a *spheric harmonic polynomial of degree k*.

For spheric harmonic polynomials the following result is known:

Lemma 1. *Let $k \in \mathbb{N}_o$ be fixed then*
(i.) the space $Im_k := \{S_k | S_k$ is a spheric harmonic polynomial of degree $k\}$ is a linear space.

(ii.) $dim(Im_k) := V(n,k) := \binom{k+n-1}{n-1} - \binom{k+n-3}{n-1}$, where $\binom{r}{s} := 0$ if $r < s$.

(iii.) For $k := 0, 1, 2, ...$ there are $V(n,k)$ orthonormalized [3] spheric harmonic polynomials $S_k^{(i)}; 1 \leq i \leq V(n,k)$, such that the set $\left(S_k^{(i)}\right)_{\substack{i=1,2,...,V(n,k) \\ k=0,1,2,...;}}$ is a complete, orthonormalized system in $L^2\left(S^{n-1}, < \cdot, \cdot >_{S^{n-1}}\right)$.

Proof. The proofs can be found in [1],or [6], respectively.   □

Now, for $k = 0, 1, 2, ...$; and $1 \leq i \leq V(n,k)$ we will define by $P_k^{(i)}$ the *positively homogeneous extension* of each of $S_k^{(i)}$ to the space $\mathbb{R}^n$, i.e. the function $P_k^{(i)} : \mathbb{R}^n \longrightarrow \mathbb{R}$,

$$P_k^{(i)}(h) := \begin{cases} \|h\|_2 \cdot S_k^{(i)}(\frac{h}{\|h\|_2}), \text{if } h \neq 0 \\ 0, \text{ if } h = 0 \end{cases}$$

These functions will provide the desired representation of df_{x_0}:

Proposition 1. *Let df_{x_0} be the directional derivative of f at x_0.*

---

[1] induced by a local parametrization of a suitable submanifold of S^{n-1} and the $n-1$-dimenisional *Lebesgue* measure

[2] for an open, connected subset $D \subset \mathbb{R}^n$ a function $h \in C^2(D)$, solving the homogeneous *Laplace* equation $\Delta h := \sum_{i=1}^{n} \frac{\partial^2 h}{\partial x_i^2} \equiv 0$ in D is called harmonic (in D)

[3] orthonormalized with respect to $< \cdot, \cdot >_{S^{n-1}}$ on $L^2\left(S^{n-1}, < \cdot, \cdot >_{S^{n-1}}\right)$

Then there is a unique sequence $\nabla f_{x_0} = \left(\dfrac{\partial f_{x_0}}{\partial P_k^{(i)}} \right)_{\substack{k=0,1\dots \\ i=1,2,\dots,V(n,k)}}$ $\in l_2$, *such*

that df_{x_0} *can be represented by:*

$$df_{x_0} = \sum_{k=0}^{\infty} \sum_{i=1}^{V(n,k)} \frac{\partial f_{x_0}}{\partial P_k^{(i)}} \cdot P_k^{(i)}.$$

Proof. For the proof see [8]. □

Remark 1. For $k = 1$, the positively homogeneous extensions $P_1^{(i)}, i = 1, \dots,$ $V(n,1) = n$ obviously remain harmonic polynomials of degree 1, consequently, they are just linear functions. Hence, in the special case that the directional derivative df_{x_0} is linear [4], the above representation coincides with the "classical" one

$$df_{x_0} = \sum_{i=1}^{n} \frac{\partial f_{x_0}}{\partial P_1^{(i)}} \cdot P_1^{(i)}.$$

This motivates to speak of the sequence $\nabla f|_{x_0} = \left(\dfrac{\partial f_{x_0}}{\partial P_k^{(i)}} \right)_{\substack{k=0,1\dots \\ i=1,2,\dots,V(n,k)}}$

to be the *generalized gradient* of f at x_0 and of the coefficients $\dfrac{\partial f_{x_0}}{\partial P_k^{(i)}}$ to be

generalized partial derivatives of f at x_0.

Remark 2. For $n = 1$, the only harmonic polynomials are the affine functions of the real line (i.e. $k \in \{0,1\}$). Therefore, following the construction scheme above, for each point $x_0 \in U \subset \mathbb{R}$ there are uniquely determined numbers $\alpha_o(x_0)$ and $\alpha_1(x_0)$, such that for the directional derivative $df|_{x_0}(h) :=$ $\alpha_o(x_0) \cdot |h| + \alpha_1(x) \cdot h$ holds. In such a way, the generalized gradient ∇f_{x_0} is isomorphic to an element of $\mathbb{R}^2$.

Remark 3. An analogous approach can also be made for other types of directional derivatives, provided that it is positively homogeneous and an element of $L^2\left(S^{n-1}, < \cdot, \cdot >_{S^{n-1}}\right)$ (these properties are, e.g. satisfied for the *Clarke* directional derivative $d_{Cl}f_{x_0}(h) = \lim_{\substack{t \searrow 0 \\ x \to x_0}} \frac{1}{t} \cdot [f(x + t \cdot h) - f(x)])$.

3   Harmonic and subharmonic functions

The motivation for this section is reasoned by the fact that there is a close relation between *convex* and *subharmonic* functions. Our idea is to exploit

---

[4] e.g., if f is *Gateaux*-differentiable or *Frechet*-differntiable

some of these relations and use them to characterize properties of a convex function f by means of its generalized partial derivatives.

Starting from a geometric point of view, we are usually characterizing convex functions as follows:

Let $U \subset \mathbb{R}^n$ be an open, convex set and $f : U \longrightarrow \mathbb{R}$. Then f is convex if and only if for every open, convex, relatively compact subset $I \subset U$ with boundary ∂I and every *affine* function a the following relation holds:

$$f \leq a \text{ on } \partial I \Longrightarrow f \leq a \text{ on } I$$

Conversely, affine functions can be characterized as *maximal* convex functions of the space, i.e. a is affine on $\mathbb{R}^n$ if and only if for every open, convex subset $U \subset \mathbb{R}^n$ and each convex function $f : \mathbb{R}^n \longrightarrow \mathbb{R}$ the above inclusion holds. Now, if we admit a broader class of functions as upper bounds than the class of affine functions, we arrive at the definition of subharmonicity. Similar as with convex functions, there exists a number of equivalent definitions of subharmonic functions. We choose the following:

Definition 1. Let $D \subset \mathbb{R}^n$ be a domain, i.e. an open, connected set and let $s : U \longrightarrow \mathbb{R}$ be an upper semi-continuous function. Then s is called *subharmonic in D* if and only if for every open, connected, relatively compact subset $I \subset D$ with boundary ∂I and every *harmonic* function h in I the following relation holds:

$$s \leq h \text{ on } \partial I \Longrightarrow s \leq h \text{ in } I$$

The properties of subharmonic functions are extensively studied in the fields of function theory and potential theory, respectively.

In the following we now collect some basic properties that are needed for our further considerations. This collection reveals not only the formal similarity between subharmonic and convex functions, it also cites that the latter class of functions is contained in the first one. Just from this point of view it seems reasonable to use tools that are available for the investigation of subharmonicity also for the research of convex functions. The properties, collected within the following lemma are chosen in such a way, that they are particulary used in the reminder of the paper.

Lemma 2. *Let $D \subset \mathbb{R}^n$ be a domain $h, s : D \longrightarrow \mathbb{R}$ functions defined on D. Then*

i. *If h is harmonic in D, then $h \in C^\infty(D)$.*
ii. *h is harmonic in D if and only if h and $-h$ are subharmonic in D.*
iii. *If s is subharmonic and h is harmonic, respectively, then their sum $s + h$ is also subharmonic in D.*
iv. *If $D = \mathbb{R}^n$ and h is an affine function, then h is harmonic.*
v. *If D is a convex set and s is a convex function, then s is subharmonic.*

vii. If $s \in C^2(D)$ then s is subharmonic if and only if $\Delta s \geq 0$ in D.

vii. If s is subharmonic and D is regular[5] such that its boundary ∂D has a zero n-dimensional Lebesgue measure, then $h^* := \inf\{h | s \leq h, h$ harmonic on $D\}$ exists and is a harmonic function on D (least harmonic majorant of s).

viii. If s is subharmonic, then it cannot have local maxima in some point $x_0 \in D$, except s reduces to the constant $s(x_0)$ in a neighbourhood of x_0.

ix. If s is subharmonic in D, then it cannot have a global maximum in some point $x_0 \in D$, except s reduces to a constant in D^6.

Proof. The proofs are very basic within the theory of harmonic and subharmonic functions. Here, we refer to references [4], [5], or [7] .   □

Apart from elementary properties in connection with subharmonicity there are also deep and fundamental results concerning this class of functions. One of these results is due to *Riesz* and concerns a special representation via a *Borel*-measure. Since we will also need this representation later for an alternate representation of df_{x_0} in the unit ball, we will quote the representation theorem within this section. But first, as a preparation, we have to recall the definition of *Greens'* function and its meaning for the problem of *Dirichlet*.

Definition 2. Let $B := \{z | \; \|z\| \leq 1\}$ be the unit ball in $\mathbb{R}^n$. A function $G : B \times int(B) \longrightarrow \mathbb{R}$ is called *Green's function for* $int(B)$ if for every (fixed) $\zeta \in int(B)$ the following three properties are satisfied:

i. $G(x, \zeta)$ is a harmonic function of x in the domain $int(B)$, except at the points $x = \zeta$.

ii. $G(x, \zeta)$ is a continuous function of x in B, except at the points $x = \zeta$, and $G(x, \zeta) = 0$ for $x \in S^{n-1}$.

iii. for $n = 2$ the function $G(x, \zeta) + log\,(\|x - \zeta\|)$ remains harmonic at the points $x = \zeta$.

for ≥ 3 the function $G(x, \zeta) - \|x - \zeta\|^{2-n}$ remains harmonic at the points $x = \zeta$.

For the special case we will need later, namely that D is the interior $int(B)$ of the unit ball $B := \{z | \|z\| \leq 1\}$, it can be shown, that *Green's* function exists for this domain and is uniquely defined. Furthermore, $G(x, \zeta) > 0$ in $int(B)$ ([4]).

Now, let $D \subset \mathbb{R}^n$ be a bounded domain with boundary ∂D and let $r : \partial D \longrightarrow \mathbb{R}$ be a continuous function.

The *Dirichlet*-problem (for D with boundary values r) then consists in finding a continuous function $h : D \cup \partial D \longrightarrow \mathbb{R}$ such that $h|_D$ is harmonic and $h|_{\partial D} \equiv r$.

---

[5] for the definition of a *regular* domain see [4]

[6] This property is usually called *the maximum principle*

It is a well known fact that if a solution of the *Dirichlet*-problem exists then it is unique.

A general question is, for which kind of domains D the *Dirichlet*-problem has a solution, given an arbitrary continuous boundary function r. If D is a bounded domain with a boundary that has n-dimensional *Lebesgue* measure zero, then a sufficient condition for solving the *Dirichlet*-problem for D is the existence of *Green's* function for this domain ([13]). Hence, if $D = int(B)$, the solution h exists and, moreover, in this special case it can be computed by

$$h(x) := \frac{\Gamma(\frac{n}{2})}{2\pi^{\frac{n}{2}}} \mathrm{int} \ _{\xi \in S^{n-1}} r(\xi) \cdot \frac{1 - \|x\|^2}{\|x - \xi\|^n} d\sigma(\xi).$$

One of the most important and deep results on subharmonic functions is due to *Riesz* and allows a special local representation of a subharmonic function s regardless, whether s is smooth or not ([4]). This representation theorem is crucial for a characterization of convexity or differentiability properties, respectively, for functions f in terms of their generalized partial derivatives. We will need the following

Lemma 3. *Let* $B := \{z | \|z\| \leq 1\}$ *be the closed unit ball in* $\mathbb{R}^n$ *and let* $s : \mathbb{R}^n \longrightarrow [-\infty, \infty[$ *be a subharmonic function which is continuous on the boundary* S^{n-1} *of* B. *Then*

  i. *there exists a uniquely determined Borel-measure* μ *on* $\mathbb{R}^n$, *such that* s *is given in* $int(B)$ *by* $s(x) = h^*(x) - int \ _{int(B)}G(x,\zeta)d\mu(\zeta)$, *where* h^* *is the least harmonic majorant of* s *in* $int(B)$ *and* G *is the* Green's *function for this domain*[7].
 ii. *The function* h^* *in this representation is just the solution of the* Dirichlet-*Problem for* $int(B)$ *with boundary values* $s|_{S^{n-1}}$.

Proof. i. Since $D = int(B)$ is a bounded, regular domain in $\mathbb{R}^n$ and its boundary S^{n-1} has a zero n-dimensional *Lebesgue* measure, we can apply Theorems 3.14 and 3.15, respectively in [4], obtainig the desired result.
 ii. This result is proved in [4].   □

4   Analyzing convex functions by generalized partial derivatives

In this section we now want to combine our considerations of the previous chapters. We will focus on the characterization of proper convex functions f, defined on a convex set $U \subset \mathbb{R}^n$ in terms of their generalized partial derivatives. Convexity obviously implies *Lipschitz*-continuity and directional differentiability at every point $x_0 \in U$ with a sublinear directional derivative df_{x_0}. Consequently df_{x_0} is a subharmonic function in the domain $\mathbb{R}^n$. Using this property we can then prove:

---

[7] this representation is sometimes called the Poisson-Jensen formula

Proposition 2. *Let $U \subset \mathbb{R}^n$ be an open convex set and $f : U \to \mathbb{R}$ be a proper convex function. Let $\nabla f|_{x_0} = \left(\dfrac{\partial f_{x_0}}{\partial P_k^{(i)}} \right)_{\substack{k=0,1\ldots \\ i=1,2,\ldots,V(n,k)}}$ be the generalized gradient of f at x_0.*

Then

(i) for all $x_0 \in U$ the estimation $\left(\dfrac{\partial f_{x_0}}{\partial P_0^{(1)}} \right) \geq 0$ holds.

(ii) f is Frechet-differentiable if and only if $\dfrac{\partial f_{x_0}}{\partial P_0^{(1)}} = 0$.

Proof.   i. Since df_{x_0} is subharmonic there exists a uniquely determined Borel-measure μ, such that for $g \in int(B)$: $df_{x_0}(g) = h^*(g) - \int_B G(g, \zeta) d\mu(e_\zeta)$ holds. Here h^* is the least harmonic majorant of df_{x_0} in $int(B)$ and G is the *Green's* function for this domain. We now represent $g \in int(B)$ by $g = r \cdot \xi$, with suitable $r \in [0, 1[$ and $\xi \in S^{n-1}$. Since the least harmonic majorant $h^*(g)$ of df_{x_0} is the solution of the *Dirichlet*-problem for $int(B)$, given the continuous function $df_{S^{n-1}}$ on the boundary S^{n-1}, this solution then has the representation: $h^*(g) = h^*(r \cdot \xi) = \sum_{k=0}^{\infty} r^k \sum_{i=1}^{V(n,k)} \dfrac{\partial f_{x_0}}{\partial P_k^{(i)}} \cdot P_k^{(i)}(\xi)$ ([6]). Observe now, that $-\int_B G(g, \zeta) d\mu(e_\zeta)$ is a subharmonic function, since df_{x_0} has this property and h^* is harmonic. Moreover it is non-positive, since *Green's* function is positive in $int(B)$ and μ is a *Borel* measure. We evaluate df_{x_0} at $g = 0$ and obtain: $0 = df_{x_0}(0) = \dfrac{\partial f_{x_0}}{\partial P_0^{(1)}} - \int_B G(g, \zeta) d\mu(e_\zeta)$, from which $\dfrac{\partial f_{x_0}}{\partial P_0^{(1)}} \geq 0$ immediately follows.

ii. If f is *Frechet*-differentiability at x_0 then, in the convex case, this is equivalent to the linearity of df_{x_0}, i.e. df_{x_0} is a harmonic polynomial of degree equal 1, hence $\dfrac{\partial f_{x_0}}{\partial P_0^{(1)}} = 0$. If, on the other hand, $\dfrac{\partial f_{x_0}}{\partial P_0^{(1)}} = 0$ holds, then $0 = df_{x_0}(0) = -\int_B G(0, \zeta) d\mu(e_\zeta)$.

I.e. the non-positive, subharmonic function $-\int_B G(g, \zeta) d\mu(e_\zeta)$ attains its maximal value 0 in the interior of B. The *maximum principle* then implies that it must vanish identically in $int(B)$. Therefore df_{x_0} coincides with its least harmonic majorant h^* on $int(B)$.

To show that h^* is a polynomial of degree equal 1 we use positive homogeneity of df: for all $\lambda \in [0, 1]$ and for all $g = r \cdot \xi \in int(B)$ we have:

$df_{x_0}(\lambda \cdot g) = h^*(\lambda \cdot g) = h^*(\lambda \cdot r \cdot \xi) = \sum_{k=1}^{\infty} \lambda^k \cdot r^k \sum_{i=1}^{V(n,k)} \dfrac{\partial f_{x_0}}{\partial P_k^{(i)}} \cdot P_k^{(i)}(\xi)$. The last expression coincides with $\lambda \cdot df_{x_0}(g)$ if and only if it coincides with

$$\lambda \cdot \sum_{k=1}^{\infty} r^k \sum_{i=1}^{V(n,k)} \frac{\partial f_{x_0}}{\partial P_k^{(i)}} \cdot P_k^{(i)}(\xi).$$ But this is possible if and only if $\dfrac{\partial f_{x_0}}{\partial P_k^{(i)}} = 0$

for all $k \geq 2$ and all $1 \leq i \leq V(n, k)$. Therefore df_{x_0} is a harmonic polynomial of degree 1 in $int(B)$. Since df_{x_0} coincides on the space $\mathbb{R}^n$ with the positively homogeneous extension of the restriction $df_{x_0}|_{int(B)}$ in the interior of B, it is a harmonic polynomial of degree 1 in the whole space, i.e. df_{x_0} is linear. $\qquad\square$

The last proposition reveals a special meaning of the first generalized partial derivative $\partial f_{x_0}/\partial P_0^{(1)}$ for f: its nonnegativity is necessary for convexity of f and it exactly determines the property of the function being *Frechet*-differentiable at x_0 or not. We also deduce immediately, that the set $\Omega := \{x | \partial f_{x_0}/\partial P_0^{(1)} \neq 0\}$ is a set of *Lebesgue* measure zero.

These properties are, of course, not sufficient to imply convexity since they do not take into account any behaviour of the remaining generalized partial derivatives of f. Nevertheless they turn out to be very substantial for a complete caracterization of convex functions by means of generalized gradients.

To show this, we will first consider the one-dimensional case:

Proposition 3. *Assume that $f : [a, b] \longrightarrow \mathbb{R}$ is a locally* Lipschitz *directionally differentiable function on $[a, b]$ For $x \in \,]a, b[$ let $\nabla f_x = ((\alpha_0(x), \alpha_1(x))$ denote its generalized gradient. Then*
f is a convex function if and only if the following two conditions are satisfied:

i. $\alpha_0(x) \geq 0$ *on* $]a, b[$.
ii. $\alpha_1(x)$ *is a nondecreasing function on* $[a, b]$.

Proof. Assume that condition *i.* and *ii.* hold. Obviously $\alpha_1(x) - \alpha_0(x) \leq \alpha_1(x) \leq \alpha_1(x) + \alpha_0(x)$ is satisfied for all $x \in [a, b]$, where equalities appear for all, but at most a set of points of *Lebesgue* measure zero.

Since f is locally *Lipschitz* on a compact interval it is absolutely continuous. Therefore, $f(x)$ can be evaluated by $f(x) := \text{int }_a^x \alpha_1(\zeta) d\zeta + c, \quad c \in \mathbb{R}$. But, by (ii.), then f must be convex.

If, on the other hand, f is assumed to be convex, then for all $x \in [a, b]$ the relations $-df_x(-1) = \alpha_1(x) - \alpha_0(x) \leq \alpha_1(x) \leq \alpha_1(x) + \alpha_0(x) = df_x(1)$ hold. Hence $\alpha_0(x) \geq 0$ on $]a, b[$. For convex functions the set of points, where $-df_x(-1) \neq df_x(1)$ is a set of measure zero. This set are just the points where $\alpha_0(x) > 0$. Moreover, $df_x(1)$ and $-df_x(-1)$ are both nondecreasing functions on $]a, b[$. Therefore the function $\alpha_1(x) = \dfrac{1}{2} \cdot df_x(1) - df_x(-1)$ has the same property. $\qquad\square$

Hence, in the one-dimensional case convexity of a function is characterized by nonnegativity of the generalized partial derivative of degree *zero* and monotonicity of the generalized partial derivative of degree *one*.

For higher dimensions we can prove a very similar convexity condition using nonnegativity and "monotonicity" in terms of generalized partial derivatives. For this, note that the functions $P_k^{(i)}$, appearing in the representation of df_x have the property, that $P_k^{(i)}(-h) = (-1)^k \cdot P_k^{(i)}(h)$. Therefore, analogously to the one-dimensional case, df_x allows a representation $df_x = A_x^0 + A_x^1$. Here

$$A_x^0 := \sum_{k=0,2,4,\dots} \sum_{i=1}^{V(n,k)} \frac{\partial f_x}{\partial P_1^{(i)}} \cdot P_1^{(i)}$$

refers to the partial derivatives *even* degree and

$$A_x^1 := \sum_{k=1,3,5,7,\dots} \sum_{i=1}^{V(n,k)} \frac{\partial f_x}{\partial P_1^{(i)}} \cdot P_1^{(i)}$$

collects the terms of the partial derivatives of *odd* degree.

Using this, we get:

Proposition 4. *Let $f : U \longrightarrow \mathbb{R}$ be a locally-Lipschitz directionally differentiable function, where $U \subset \mathbb{R}^n$ is an open, convex set.*
Then
f is convex if and only if the following properties hold:

i. $A_x^0 \geq 0$ *for all* $x \in U$.
ii. $\left[A_{x_1}^1 (x_2 - x_1) + A_{x_2}^1 (x_1 - x_2) \right] \leq 0$
holds for each pair of points $x_1, x_2 \in U$.

Proof. To prove the necessity of the conditions we first observe, that convex functions are continuously differentiable almost everywhere in U. Hence, by proposition 2 of this chapter, the set Ω is contained in a set of *Lebesgue* measure zero. To show that $A_x^0 \geq 0$ we make use of the fact that f is convex on U if and only if it is convex on every line segment passing U. Now, let $x \in U$ and $\zeta \in S^{n-1}$ be arbitrarily fixed, such that $x + t \cdot \zeta \in U$ for $t \in [a,b]$, $a < 0 < b$.
The function $f^\zeta : [a,b] \longrightarrow \mathbb{R}$, defined by $f^\zeta(t) := f(x + t \cdot \zeta)$, is a convex function in t and its directional derivative can be represented by $df_x^\zeta(s) := \alpha_0(x) \cdot |s| + \alpha_1(x) \cdot s$. On the other hand $df_x^\zeta(s) = df_x(s \cdot \zeta) = A_x^0(s \cdot \zeta) + A_x^1(s \cdot \zeta)$. This yields $df_x^\zeta(1) = \alpha_0(x) + \alpha_1(x) = A_x^0(\zeta) + A_x^1(\zeta)$ and $df_x^\zeta(-1) = \alpha_0(x) - \alpha_1(x) = A_x^0(\zeta) - A_x^1(\zeta)$, from which $\alpha_0(x) = A_x^0(\zeta)$ follows. Since, by the last theorem, $\alpha_0(x) \geq 0$, condition (i.) is proved.
Convexity of f, moreover, implies

$$f(x_1) \geq f(x_2) + \sum_{k=0}^{\infty} \sum_{i=1}^{V(n,k)} \frac{\partial f_{x_1}}{\partial P_k^{(i)}} \cdot P_k^i(x_2 - x_1) \geq f(x_2) + A_{x_1}^1(x_2 - x_1)$$

and

$$f(x_2) \geq f(x_1) + \sum_{k=0}^{\infty} \sum_{i=1}^{V(n,k)} \frac{\partial f_{x_2}}{\partial P_k^{(i)}} \cdot P_k^i(x_1 - x_2) \geq f(x_1) + A_{x_2}^1(x_1 - x_2),$$

respectively, for arbitrary points x_1 and $x_2 \in U$.
Combining these inequalities, we get

$$\left[A_{x_1}^1(x_2 - x_1) + A_{x_2}^1(x_1 - x_2) \right] \leq 0,$$

which implies condition (ii.).
For proving sufficiency, we again consider the function $f^\varsigma : [a, b] \longrightarrow \mathbb{R}$ on the line segment passing U. The function f^ς is directional differentiable on $]a, b[$. For its directional derivative $df_x^\varsigma(s) := \alpha_o(x) \cdot |s| + \alpha_1(x) \cdot s$ we get, using condition (i.) and the last considerations: $\alpha_0(x) = A_x^0(\varsigma) \geq 0$.

Now, let x_1 and x_2 be two points on the line segment such that $x_1 = x + t_1 \cdot \varsigma$ and $x_2 = x + t_2 \cdot \varsigma$. W.l.o.g. let $t_1 < t_2$, then, using (ii.), $\alpha(x_2) - \alpha(x_1) = A_{x_2}^1(\varsigma) - A_{x_1}^1(\varsigma) = \frac{1}{t_2 - t_1} \cdot A_{x_1}^1(x_2 - x_1) + \frac{1}{t_2 - t_1} \cdot A_{x_2}^1(x_1 - x_2) = \frac{1}{t_2 - t_1} \cdot \left[A_{x_1}^1(x_2 - x_1) + A_{x_2}^1(x_1 - x_2) \right] \leq 0$, i.e. $\alpha(x)$ is a nondecreasing function on $]a, b[$. Using the same argumentation as in the previous proposition we will obtain f^ς to be a convex function on $[a, b]$. Hence, f is convex on U.   □

In the following section we will prove a result on necessary optimality conditions for convex functions. Similar to previous proofs, it also gives an example, how the criterion can be deduced from the behaviour of the solution of the related *Dirichlet*-problem.

We will need a Lemma first.

Lemma 4. *Let s be a subharmonic function in the domain $\mathbb{R}^n$. Then for all $0 < \rho < R$ the following inequalities hold:*

$$\frac{\rho^{1-n}}{c_n} \cdot int_{\xi \in S^{n-1}(0,\rho)} s^+(\xi) d\sigma(\xi) \quad \leq \quad sup_{\|x\| = \rho} s^+(x)$$

$$\leq \frac{(R + \rho)}{c_n \cdot R \cdot (R - \rho)^{n-1}} \cdot int_{\xi \in S^{n-1}(0,R)} s^+(\xi) d\sigma(\xi)$$

where $s^+(x) := max\{0, s(x)\}$ and $d\sigma$ describes the "surface element" on $S^{n-1}(x_0, r) := \{\varsigma \| \|\varsigma - x_0\| = r\}$ induced by the $n - 1$ dimensional Lebesque measure.

Proof. The proof can be found in [4].   □

Proposition 5. *Let $U \subset \mathbb{R}^n$ be an open, convex set and $f : U \to \mathbb{R}$ be a convex function. If x_0 is a minimizer of f, then for the generalized gradient the estimation*

$$\frac{\partial f_{x_0}}{\partial P_0^{(1)}} \geq \left(\left[\frac{(n+1)^{n-1} \cdot (n+2)}{n^{n-1}} \right]^2 - 1 \right)^{-\frac{1}{2}} \cdot \left(\sum_{k=1}^{\infty} \sum_{i=1}^{V(n,k)} \left[\frac{\partial f_{x_0}}{\partial P_k^{(i)}} \right]^2 \right)^{\frac{1}{2}}$$

holds.

Proof. The minimizer x_0 implies, that $df\,x_0$ is nonnegative, i.e. $df\,x_0 = df^+|x_0$. Using the inequalities of the last Lemma and taking into account that $df\,x_0$ is positively homogeneous, we arrive at the following estimation. For $0 < \rho < R$ we obtain:

$$\rho^{1-n} \cdot c_n^{-1} \cdot \text{int }_{\xi \in S^{n-1}(0,\rho)} df^+|_{x_0}(\xi) d\sigma(\xi)$$

$$= c_n^{-1} \cdot \rho \text{int }_{\xi \in S^{n-1}} df_{x_0}(\xi) d\sigma(\xi)$$

$$\leq c_n^{-1} \cdot \rho \text{int }_{\xi \in S^{n-1}} \sup\{df_{x_0}(\xi) \mid \|\xi\| = 1\} d\sigma(\xi)$$

$$= \sup\{df_{x_0}(\xi) \mid \|\xi\| = \rho\} \cdot c_n^{-1} \cdot \text{int }_{\xi \in S^{n-1}} 1 \cdot d\sigma(\xi)$$

$$= \sup\{df^+|_{x_0}(\xi) \mid \|\xi\| = \rho\} \cdot 1$$

$$\leq \frac{(R+\rho)}{R \cdot (R-\rho)^{n-1}} \cdot c_n^{-1} \cdot \text{int }_{\xi \in S^{n-1}(0,R)} df^+|_{x_0} d\sigma(\xi)$$

$$\leq \frac{R^{n-2}(R+\rho)}{(R-\rho)^{n-1}} \cdot c_n^{-1} \cdot R \cdot \text{int }_{\xi \in S^{n-1}} df_{x_0}(\xi) d\sigma(\xi).$$

Setting special values $\rho := 1$, $R := n + 1$ we get

$$\sup\{df_{x_0}(\xi) \mid \|\xi\| = 1\} \leq$$

$$\leq \frac{(n+1)^{n-2} \cdot (n+2) \cdot (n+1)}{n^{n-1}} \cdot c_n^{-1} \cdot \text{int }_{\xi \in S^{n-1}} df_{x_0}(\xi) d\sigma(\xi) \leq$$

$$\leq \frac{(n+1)^{n-1} \cdot (n+2)}{n^{n-1}} \cdot c_n^{-1} \cdot \text{int }_{\xi \in S^{n-1}} df_{x_0}(\xi) d\sigma(\xi).$$

Now, as before, let F be the solution of the Dirichlet-problem in $\text{int}(B(0, 1))$ induced by the boundary condition $df_{x_0}|_{S^{n-1}}$. The mean value theorem for harmonic functions allows to compute $F(0)$ by:

$$F(0) := c_n^{-1} \cdot \text{int }_{\xi \in S^{n-1}} df_{x_0}(\xi) d\sigma(\xi).$$

Since, on the other hand, $F(0) = \dfrac{\partial f_{x_0}}{\partial P_0^{(1)}} \cdot P_0^{(1)}(\xi) = \dfrac{\partial f_{x_0}}{\partial P_0^{(1)}} \cdot 1$, we have for

$\xi \in S^{n-1}$:

$$\frac{\partial f_{x_0}}{\partial P_0^{(1)}} \cdot \|\xi\| \geq \left[\frac{(n+1)^{n-1} \cdot (n+2)}{n^{n-1}} \right]^{-1} \cdot \sup\{df_{x_0}(\xi) \mid \|\xi\| = 1\}$$

$$\geq \left[\frac{(n+1)^{n-1} \cdot (n+2)}{n^{n-1}} \right]^{-1} \cdot df_{x_0}(\xi)$$

The last inequality, together with the nonnegativity of $df_{x_0}(\xi)$ then implies:

$$
\left[\frac{\partial f_{x_0}}{\partial P_0^{(1)}}\right]^2 = \left\langle \frac{\partial f_{x_0}}{\partial P_0^{(1)}} \cdot P_0^{(1)}, \frac{\partial f_{x_0}}{\partial P_0^{(1)}} \cdot P_0^{(1)} \right\rangle_{L^2(S^{n-1})}
$$

$$
= \left\langle \frac{\partial f_{x_0}}{\partial P_0^{(1)}} \cdot \|\cdot\|, \frac{\partial f_{x_0}}{\partial P_0^{(1)}} \cdot \|\cdot\| \right\rangle_{L^2(S^{n-1})}
$$

$$
\geq \left[\frac{(n+1)^{n-1} \cdot (n+2)}{n^{n-1}}\right]^{-2} \cdot \langle df_{x_0}, df_{x_0}\rangle_{L^2(S^{n-1})}
$$

$$
= \left[\frac{(n+1)^{n-1} \cdot (n+2)}{n^{n-1}}\right]^{-2} \cdot \left(\sum_{k=0}^{\infty} \sum_{i=1}^{V(n,k)} \left[\frac{\partial f_{x_0}}{\partial P_k^{(i)}}\right]^2\right),
$$

i.e.

$$
\left[\frac{\partial f_{x_0}}{\partial P_0^{(1)}}\right]^2 \left(1 - \left[\frac{(n+1)^{n-1} \cdot (n+2)}{n^{n-1}}\right]^{-2}\right)
$$

$$
\geq \left[\frac{(n+1)^{n-1} \cdot (n+2)}{n^{n-1}}\right]^{-2} \cdot \left(\sum_{k=1}^{\infty} \sum_{i=1}^{V(n,k)} \left[\frac{\partial f_{x_0}}{\partial P_k^{(i)}}\right]^2\right).
$$

Obviously $0 < \left[\dfrac{n^{n-1}}{(n+1)^{n-1} \cdot (n+2)}\right] < 1$. Hence,

$$
\frac{\partial f_{x_0}}{\partial P_0^{(1)}} \geq \left(\left[\frac{(n+1)^{n-1} \cdot (n+2)}{n^{n-1}}\right]^2 - 1\right)^{-\frac{1}{2}} \cdot \left(\sum_{k=1}^{\infty} \sum_{i=1}^{V(n,k)} \left[\frac{\partial f_{x_0}}{\partial P_k^{(i)}}\right]^2\right)^{\frac{1}{2}}.
$$

$\square$

Remark 4. Note, that in this proof sublinearity and nonnegativity of the directional derivative df_{x_0} is exploited. In such a way a similar proposition can be obtained also for *locally Lipschitz* functions (not necessarily directionally differentiable) if we replace the directional derivative df_{x_0} by the *Clarke* directional derivative $df_{Cl}|_{x_0}$.

Then an inequality

$$
\frac{\partial_{Cl} f_{x_0}}{\partial P_0^{(1)}} \geq \left(\left[\frac{(n+1)^{n-1} \cdot (n+2)}{n^{n-1}}\right]^2 - 1\right)^{-\frac{1}{2}} \cdot \left(\sum_{k=1}^{\infty} \sum_{i=1}^{V(n,k)} \left[\frac{\partial_{Cl} f_{x_0}}{\partial P_k^{(i)}}\right]^2\right)^{\frac{1}{2}}
$$

holds for each of the cases, that x_0 minimizes or maximizes the locally *Lipschitz* function f, since in both situations $d_{Cl}f_{x_0} \geq 0$.

References

1. Axler, S. et al. (1992) Harmonic Function Theory. Springer
2. Clarke, F.H. (1983) Optimization and nonsmooth analysis. Wiley, New York
3. Demyanov, V.F., Rubinov,A.M. (1980) On quasidifferentiable functionals. Dokl. of USSR Acad.of Sci. **250**, No.1, 21–25
4. Hayman, W.K., Kennedy, P.B. (1976) Subharmonic functions. Acad.Press, New York
5. Klimek, M. (1991) Pluripotential Theory. Claredon Press
6. Müller, C. (1966) Spherical Harmonics. Lecture Notes in Math. **17**
7. Rado, T. (1937) Subharmonic functions. Berlin
8. Recht, P. (1992) On generalized Gradients. ZOR **36**, 201–210
9. Recht, P. (1993) Generalized Derivatives : an approach to new gradient in nonsmooth optimization. Math. Systems in Economics **136**
10. Recht, P. Generalized partial derivatives and optimality conditions in nonsmooth optimization, submitted to Optimization
11. Recht, P. (1997) On the reconstruction problem for nondifferentiable functions. In: Gritzmann P., Horst R., Sachs E., Tichatschke R. (Eds.) Recent Advances in Optimization, Lect. Notes in Econ. and Math. Systems **452**, 261–281
12. Rockafellar, R.T. (1970) Convex Analysis. Princeton University Press
13. Wermer, J. (1974) Potential Theory. Springer

A Perturbed and Inexact Version of the Auxiliary Problem Method for Solving General Variational Inequalities with a Multivalued Operator

Geneviève Salmon, Van Hien Nguyen, and Jean-Jacques Strodiot

Facultés Universitaires N.-D. de la Paix, Département de Mathématique,
5000 Namur, Belgium

Abstract. We consider general variational inequalities with a multivalued maximal monotone operator in a Hilbert space. For solving these problems, Cohen developed several years ago the auxiliary problem method. Perturbed versions of this method have been already studied in the literature for the single-valued case. They allow to consider for example, barrier functions and interior approximations of the feasible domain. In this paper, we present a relaxation of these perturbation methods by using the concept of ε-enlargement of a maximal monotone operator. We prove that, under classical assumptions, the sequence generated by this scheme is bounded and weakly convergent to a solution of the problem. Strong convergence is also obtained under additional conditions.
In the particular case of nondifferentiable convex optimization, the ε-subdifferential will take place of the ε-enlargement and some assumptions for convergence will be weakened. In the nonperturbed situation, our scheme reduces to the projected inexact subgradient procedure.

Key words: variational inequalities, multivalued maximal monotone operator, perturbed auxiliary problem principle, ε-enlargement of a maximal monotone operator, nondifferentiable convex optimization, ε-subdifferential

1 Introduction

Let H be a Hilbert space identified to its topological dual H^* and let F be a maximal monotone multivalued operator defined on H. Let, in addition, $\varphi : H \to I\!R \cup \{+\infty\}$ be a lower semicontinuous proper convex function. We consider in this paper the following general variational inequality problem:

$$(P) \begin{cases} \text{find } x^* \in H \text{ and } r(x^*) \in F(x^*) \text{ such that, for all } x \in H, \\ \langle r(x^*), x - x^* \rangle + \varphi(x) - \varphi(x^*) \geq 0. \end{cases}$$

The resolution of such problems is a research topic in several fields of applied mathematics (see, for example, [14], [15], [16], [17], [20], [23], [30]).

Concerning sufficient conditions for the existence of a solution of (P), we refer the reader, for example, to [5], [15].

When F is the subdifferential of a finite-valued convex continuous function f defined on H, problem (P) reduces to the nondifferentiable convex optimization problem:

$$(OP) \quad \min_{x \in H} \{f(x) + \varphi(x)\}.$$

In the particular case where φ is the indicator function of a nonempty closed convex subset C of H, problem (OP) amounts to minimize f onto C.

For solving (P), Cohen developed in [11] a method based on the so-called auxiliary problem principle. In [28] and [33], this method is combined with a perturbation of φ in order to get a basic family of perturbation methods. More precisely, we consider an auxiliary function $K : H \to \mathbb{R}$ supposed to be continuously differentiable and strongly convex, positive numbers $\{\lambda_k\}_{k \in \mathbb{N}}$ and a sequence $\{\varphi^k\}_{k \in \mathbb{N}}$ of lower semicontinuous proper convex functions from H into $\mathbb{R} \cup \{+\infty\}$ that approximate φ. So, the problem considered at iteration k can be expressed as:

$$(PP^k) \begin{cases} \text{find } x^{k+1} \in H \text{ such that, for all } x \in H, \\[2mm] \langle r(x^k) + \lambda_k^{-1}(\nabla K(x^{k+1}) - \nabla K(x^k)), x - x^{k+1} \rangle \\[2mm] \qquad\qquad + \varphi^k(x) - \varphi^k(x^{k+1}) \geq 0, \\[2mm] \text{with } r(x^k) \in F(x^k). \end{cases}$$

This problem can also be equivalently written under the following minimization form:

$$\begin{cases} x^{k+1} \in \operatorname{argmin}_{x \in H} \{ \lambda_k^{-1} K(x) + \varphi^k(x) + \langle r(x^k) - \lambda_k^{-1} \nabla K(x^k), x - x^k \rangle \}, \\[2mm] \text{with } r(x^k) \in F(x^k). \end{cases}$$

To our knowledge, the influence of a variational perturbation of φ on the convergence of the auxiliary problem method has only been studied in the case of a single-valued operator F. For optimization problems, we refer to [24], while for variational inequalities, we refer to [28] and [37] if F is strongly monotone and to [33] if F has the Dunn property. In these works, the sequence $\{\varphi^k\}_{k \in \mathbb{N}}$ is assumed to epiconverge to φ, what is denoted by $\varphi^k \overset{epi}{\to} \varphi$. This means that the following conditions hold for all $w \in H$:

(i) for every sequence $\{w^k\}_{k \in \mathbb{N}}$ weakly converging to w, one has

$$\underline{\lim}_{k \to \infty} \varphi^k(w^k) \geq \varphi(w);$$

(ii) there exists a sequence $\{w^k\}_{k \in I\!\!N}$ strongly converging to w such that

$$\overline{\lim}_{k \to \infty} \varphi^k(w^k) \leq \varphi(w).$$

For more details about the theory of epiconvergence, see, for example, [3]. In [33], an additional condition on the speed of convergence of the sequence $\{\varphi^k\}_{k \in I\!\!N}$ is also needed. More precisely, it is assumed besides that for each solution x^* of problem (P), there exist a constant $\nu > 1$ and a sequence $\{w^k\}_{k \in I\!\!N}$ strongly converging to x^* such that

$$\lim_{k \to \infty} k^\nu \|w^k - x^*\| = 0, \quad \text{and}$$

$$\lim_{k \to \infty} k^\nu \|\varphi^k(w^k) - \varphi(x^*)\| = 0. \tag{1}$$

We focus also more specially on interior approximation of the function φ. A typical example is the sequence of barrier functions associated with a feasible set described by inequality constraints. For this example, the epiconvergence is satisfied (see [3]) and condition (1) holds provided that the barrier parameters converge sufficiently fast to infinity (see [33]).

Our purpose in this paper is to present convergence results for the perturbed auxiliary problem scheme when F is multivalued. Without perturbation on φ, there exist some convergence results in the litterature.
In [11], it is shown that the sequence generated by the nonperturbed method strongly converges to the unique solution of problem (P) provided that F is strongly monotone and satisfies

$$\exists\, a, b > 0 \; : \; \|r(x)\| \leq a\|x\| + b, \; \forall x \in H, \; \forall r(x) \in F(x), \tag{2}$$

and the sequence $\{\lambda_k\}_{k \in I\!\!N}$ is such that

$$\lambda_k > 0 \; \forall k \in I\!\!N, \; \sum_{k=0}^{+\infty} \lambda_k = +\infty \quad \text{and} \quad \sum_{k=0}^{+\infty} \lambda_k^2 < +\infty. \tag{3}$$

As we can see in the proof, the selection rule (3) ensures that the stepsizes are small enough to guarantee boundedness of the sequence but not too small to ensure convergence to a solution of the problem. This rule is also considered in the literature for nonsmooth minimization problems (see [2], [4], [12], [13], [31],...).
When F is the subdifferential of a convex function f, in [12], Cohen and Zhu prove under assumptions (2) and (3), that $\lim_{k \to \infty} (f + \varphi)(x^k) = \inf_{x \in H} \{f(x) + \varphi(x)\}$, the sequence $\{x^k\}_{k \in I\!\!N}$ is bounded and each of its weak limit point is a solution of problem (OP).
Moreover, when φ is the indicator function of a closed convex subset C of H and the auxiliary function K is defined, for each $x \in H$, by $K(x) = (1/2)\, x^T x$,

then the auxiliary problem method reduces to the classical projected subgradient procedure:

$$\begin{cases} x^{k+1} = Proj_C \, (x^k - \lambda_k r(x^k)), \\ \\ \text{with } r(x^k) \in \partial f(x^k). \end{cases}$$

We refer to [2] and the references cited therein for details on convergence of this method. In that paper, the projected subgradient method is relaxed by allowing inexact computation of the subgradients in the sense that $r(x^k)$ can be chosen in the ε^k-subdifferential of f at x^k. Remind that for $\varepsilon \geq 0$, the ε-subdifferential of f at $x \in H$ is the set $\partial_\varepsilon f(x)$ defined by:

$$\partial_\varepsilon f(x) = \{ u \in H \; : \; f(y) \geq f(x) + <u, y - x> -\varepsilon, \; \forall y \in H \, \}.$$

The introduction of the parameter ε produces an enlargement of $\partial f(x)$ with good continuity properties. This generally preserves the convergence properties of the method while giving more latitude and more robustness with respect to numerical errors. This concept is studied, for example, in [6], [7], [18], [19], [29]. It is applied to develop methods of ε-descent in [19], bundle methods in [25], [26], [34], [38], [39] or also to devise an inexact proximal point method with generalized Bregman distances in [22].

In [2], the sequence generated by the projected inexact subgradient method is proved to be weakly convergent to a minimizer if the problem has at least one solution and unbounded otherwise. This result is obtained under the conditions that the sequence $\{\lambda_k\}_{k \in \mathbb{N}}$ satisfies (3), the sequence $\{\varepsilon^k\}_{k \in \mathbb{N}}$ is such that $\varepsilon^k \leq \varepsilon \lambda_k \leq \bar{\varepsilon}$, for some $\varepsilon, \bar{\varepsilon} > 0$, for all k, and $\partial_{\bar{\varepsilon}} f$ is bounded on bounded sets.

Following the same idea as in [13], we can see that the projected inexact subgradient algorithm reduces to a projected proximal scheme like that studied in [27] by choosing adequately the sequence $\{\varepsilon^k\}_{k \in \mathbb{N}}$. Note also that inexact computation of the subgradients in the framework of the projected subgradient method is considered in a different way in [1], [36]. In these papers, the iteration is of the form $x^{k+1} = Proj_C \, (x^k - \lambda_k [r(x^k) + v^k])$ with $r(x^k) \in \partial f(x^k)$ and $\{v^k\}_{k \in \mathbb{N}}$ such that $lim_{k \to \infty} v^k = 0$ in [1] or $||v^k|| \leq \tau$ where τ denotes a maximal error of magnitude in [36].

Similarly, for the general variational inequality, we allow an inexact computation of an element of $F(x^k)$ in subproblems (PP^k). This is done by choosing $r(x^k)$ in the ε^k-enlargement of F at x^k introduced in [8]. For $\varepsilon \geq 0$, the ε-enlargement of the maximal monotone operator F at $x \in H$ is defined by:

$$F^\varepsilon(x) = \{ u \in H \; : \; < r(y) - u, y - x > \geq -\varepsilon, \; \forall y \in H, \; \forall r(y) \in F(y) \, \}.$$

It is easy to verify that $F = F^0 \subset F^\varepsilon$ for all $\varepsilon \geq 0$. We refer to [8], [9] for other relevant properties of F^ε. Note that this enlargement is applied in [10] to propose a bundle method to find a zero of a maximal monotone operator. It is also used to devise an inexact proximal point method with Bregman distances for variational inequalities in [8] and to construct a hybrid approximate extragradient-proximal point algorithm in [35]. These works show that when F is maximal monotone, F^ε inherits most properties from the ε-subdifferential and plays the role of this one in nonsmooth optimization. However, observe that when $F = \partial f$ for some convex function f, we have that $\partial_\varepsilon f \subset F^\varepsilon$ for all $\varepsilon \geq 0$ but we do not get in general that $\partial_\varepsilon f = F^\varepsilon$. Examples with a strict inclusion are presented in [8].

So, in this paper, we study the convergence of an inexact version of the perturbed auxiliary problem method for solving general variational inequality problems with a multivalued operator. We propose successively conditions that ensure boundedness of the sequence generated by the algorithm, weak and strong convergence of this sequence to a solution of problem (P). One of the key conditions for ensuring weak convergence is that F be strongly monotone with modulus $\alpha > 0$ over dom φ. Under this condition, we introduce a new enlargement of F contained in F^ε, for $\varepsilon \geq 0$. It will be called the α-ε-enlargement of F, denoted by F_α^ε and defined at $x \in H$ by:

$$F_\alpha^\varepsilon(x) = \{\, u \in H : <r(y) - u, y - x> \geq \alpha\|y - x\|^2 - \varepsilon,$$

$$\forall y \in H, \ \forall r(y) \in F(y) \,\}.$$

As we will see, in the general multivalued case, we restrict ourselves to use elements in the α-ε-enlargement F_α^ε of F. However, when F is single valued and Lipschitz continuous, we can work with elements in the ε-enlargement F^ε of F.

When F is the subdifferential of a convex function f, we take elements in the ε-subdifferential of f. In this last case, weak convergence of the method is obtained without requiring strong convexity of f. Note that this condition is however needed if we want to ensure strong convergence.

The paper is organized as follows. We present in Section 2 the convergence properties of the perturbed inexact auxiliary problem method for solving general variational inequalities with a multivalued operator. We also discuss how these results generalize or improve previously known results. In Section 3, we study the optimization problem (OP). For this particular case, we show that some assumptions of Section 2 can be weakened for convergence and we explain how previously known results as those obtained for the projected inexact subgradient method can be extended.

Throughout this paper, $\|\cdot\|$ denotes the l_2–norm for vectors and $\langle \cdot, \cdot \rangle$ denotes the Euclidean scalar product of H.

Let T be a multivalued operator defined on H. The operator T is said to be monotone on H if

$$\langle r(x) - r(y), x - y \rangle \geq 0, \ \forall x, y \in H, \ \forall (r(x), r(y)) \in T(x) \times T(y).$$

T is maximal monotone if, in addition, its graph is not contained in the graph of any other monotone operator. We say that T is strongly monotone on H if there exists a positive constant α such that

$$\langle r(x) - r(y), x - y \rangle \geq \alpha \|x - y\|^2, \ \forall x, y \in H, \ \forall (r(x), r(y)) \in T(x) \times T(y).$$

The mapping T has the Dunn property over H with modulus $\gamma > 0$ if

$$\langle r(x) - r(y), x - y \rangle \geq \gamma \|r(x) - r(y)\|^2, \ \forall x, y \in H, \ \forall (r(x), r(y)) \in T(x) \times T(y).$$

T is Lipschitz continuous with Lipschitz constant A over H if

$$\|r(x) - r(y)\| \leq A \|x - y\|, \ \forall x, y \in H, \ \forall (r(x), r(y)) \in T(x) \times T(y).$$

Note that if T has the Dunn property or is Lipschitz continuous over H, then it is single valued.
We denote by $\Gamma_0(H)$ the set of proper, convex, lower semicontinuous functions from H into $\mathbb{R} \cup \{+\infty\}$. Let $f \in \Gamma_0(H)$, we say that f is strongly convex with modulus α if

$$f(y) \geq f(x) + \ <r(x), y - x > +(\alpha/2)\|y - x\|^2, \ \forall x, y \in H, \ \forall r(x) \in \partial f(x).$$

It is well known that ∂f is strongly monotone with modulus α if and only if f is strongly convex with modulus α.
Finally, let C be a closed convex subset of H, we denote by $Proj_C$ the function defined for all $x \in H$ by:

$$Proj_C(x) = \inf_{y \in C} \|y - x\|.$$

Any other undefined term or usage should be taken as in the Ekeland and Teman book [15] and the Rockafellar book [32].

2   Convergence of the perturbed inexact auxiliary problem method for solving (P)

In this section, we analyze the convergence of a perturbed and inexact version of the auxiliary problem scheme for solving the general variational inequality problem (P) with a maximal monotone and multivalued operator F.

Let us introduce a sequence of positive numbers $\{\varepsilon^k\}_{k \in \mathbb{N}}$ converging to zero such that, for some $\bar{\varepsilon} > 0$, $\varepsilon^k \leq \bar{\varepsilon}$, for all $k \in \mathbb{N}$, and a sequence of multivalued operators $\{G^k\}_{k \in \mathbb{N}}$ defined onto H such that $G^k \subset F^{\varepsilon^k}$, for all

$k \in I\!N$. Then, for a given iterate x^k, the problem considered at iteration k will be the following:

$$(IPP^k) \begin{cases} \text{find } x^{k+1} \in H \text{ such that, for all } x \in H, \\[2mm] \langle \eta_k^{-1} r^k(x^k) + \lambda_k^{-1}(\nabla K(x^{k+1}) - \nabla K(x^k)), x - x^{k+1} \rangle \\[2mm] \qquad\qquad + \eta_k^{-1}(\varphi^k(x) - \varphi^k(x^{k+1})) \geq 0, \\[2mm] \text{with } r^k(x^k) \in G^k(x^k) \subset F^{\varepsilon^k}(x^k), \\[2mm] \text{and } \eta_k = \begin{cases} max\{1, \|r^0(x^0)\|\} & \text{if } k = 0 \\[2mm] max\{\eta_{k-1}, \|r^k(x^k)\|\} & \text{if } k \geq 1. \end{cases} \end{cases}$$

First, we prove that under suitable assumptions, the sequence $\{x^k\}_{k \in I\!N}$ is bounded and converges weakly to a solution of problem (P) and secondly, that under additional assumptions, it converges strongly. The selection rule for the stepsizes $\{\lambda_k\}_{k \in I\!N}$ will be like in (3). For ensuring the weak convergence of $\{x^k\}_{k \in I\!N}$, we impose that

(A) F is strongly monotone with modulus $\alpha > 0$ over dom φ;

(B) $F^{\bar{\varepsilon}}$ is bounded on bounded subsets of H.

Observe that if F is maximal monotone, then $F^{\bar{\varepsilon}}$ is locally bounded (see [9]). In finite dimension, if in addition dom F is closed, then condition (B) is satisfied (see [8]).

When condition (A) holds and F is multivalued, we will use the α-ε-enlargement of F instead of its ε-enlargement in the subproblems (IPP^k), $k \in I\!N$. In other words, we will take $r^k(x^k)$ in $G^k(x^k)$ with the operator G^k restricted to satisfy $G^k \subset F_\alpha^{\varepsilon^k}$.

To better situate this new enlargement, let us compare F_α^ε with $\partial_\varepsilon f$ in the case where $F = \partial f$ with f some strongly convex function.

Proposition 1. *If $F = \partial f$, where f is a strongly convex function with constant $\alpha > 0$, then*

$$\partial_\varepsilon f \subset F_{\alpha/2}^\varepsilon, \quad \forall \varepsilon \geq 0.$$

Proof. Let $\varepsilon \geq 0$, $x \in H$. Take $r^\varepsilon(x) \in \partial_\varepsilon f(x)$. Consider any $y \in H$, $r(y) \in F(y)$. We have to show that

$$< r(y) - r^\varepsilon(x), y - x > \geq (\alpha/2)\|y - x\|^2 - \varepsilon$$

This follows immediately from the facts that

- $r^\varepsilon(x) \in \partial_\varepsilon f(x)$ such that
  $< r^\varepsilon(x), x - y > \geq f(x) - f(y) - \varepsilon;$
- f is strongly convex with constant α such that
  $< r(y), y - x > \geq f(y) - f(x) + (\alpha/2)\|y - x\|^2.$ □

However, we do not have in general that $\partial_\varepsilon f = F^\varepsilon_{\alpha/2}$ as it is shown in the following example.

Example 1. In $\mathbb{R}^n$, let us consider the quadratic function

$$f(x) = (1/2)x^T A x + b^T x + c, \quad \forall x \in \mathbb{R}^n,$$

with $A \in \mathbb{R}^{n \times n}$ a symmetric and positive definite matrix, $b \in \mathbb{R}^n$, $c \in \mathbb{R}$.

We know that f is strongly convex with constant $\lambda min(A) > 0$, where $\lambda min(A)$ denotes the minimum eigenvalue of A. Then, explicit computation of $\partial_\varepsilon f(x)$, $F^\varepsilon(x)$, $F^\varepsilon_{\lambda min(A)/2}(x)$, for any $x \in \mathbb{R}^n$, gives:

$$\partial_\varepsilon f(x) = \{\, Ax + b + w : \ w^T A^{-1} w \leq 2\varepsilon \,\};$$

$$F^\varepsilon(x) = \{\, Ax + b + w : \ w^T A^{-1} w \leq 4\varepsilon \,\};$$

$$F^\varepsilon_{\lambda min(A)/2}(x) = \{\, Ax + b + w : \ w^T [2A - \lambda min(A)I]^{-1} w \leq 2\varepsilon \,\}.$$

We see that

$$\partial_\varepsilon f \subset F^\varepsilon_{\lambda min(A)/2} \subset F^\varepsilon = \partial_{2\varepsilon} f.$$

The first inclusion is an equality only in the case where $A = aI$, $a \in \mathbb{R}$.

Theorem 1. *Assume that F is a maximal monotone multivalued mapping defined on H, that the solution set of problem (P) is nonempty and that the following conditions are satisfied:*

(i) *$K : H \to \mathbb{R}$ is continuously differentiable and strongly convex with modulus $\beta > 0$ over dom φ;*

(ii) *∇K is a Lipschitz continuous mapping with Lipschitz constant Λ over dom φ;*

(iii) *$\{\lambda_k\}_{k \in \mathbb{N}}$ is a sequence of positive numbers such that*

$$\lambda_{k+1} \leq \lambda_k, \ \forall k \in \mathbb{N} \ \text{and} \ \lim_{k \to \infty} \lambda_k = 0;$$

(iv) *the sequence $\{\varepsilon^k\}_{k \in \mathbb{N}}$ is such that*

$$\exists \bar\varepsilon > 0 : \ 0 < \varepsilon^k \leq \bar\varepsilon, \ \forall k \in \mathbb{N} \ \text{and} \ \lim_{k \to \infty} \varepsilon^k = 0;$$

(v) $\{\varphi^k\}_{k\in\mathbb{N}}, \varphi \in \Gamma_0(H)$ are such that $\varphi^k \overset{epi}{\to} \varphi$ and $\varphi \leq \varphi^k$ for all k. Moreover, for some solution x^* of problem (P), there exist $\nu > 1$ and a sequence $\{w^k\}_{k\in\mathbb{N}}$ converging to x^* such that (1) holds.

If $\sum_{k=0}^{+\infty}\lambda_k^2 < +\infty$ and there exists $\varepsilon > 0$ such that $\varepsilon^k \leq \lambda_k\varepsilon$ for all $k \in \mathbb{N}$, then, provided that $x^0 \in \text{dom } \varphi$, the sequence $\{x^k\}_{k\in\mathbb{N}}$ is bounded.

If, in addition, $\sum_{k=0}^{+\infty}\lambda_k = +\infty$, F satisfies conditions (A) and (B) and

either $G^k \subset F_\alpha^{\varepsilon^k}$ for all k,

or F is Lipschitz continuous on H with Lipschitz constant A,

then, the sequence $\{x^k\}_{k\in\mathbb{N}}$ weakly converges to x^*, the unique solution of problem (P).

Proof. Let x^* be a solution of problem (P) used in assumption (v). The necessary and sufficient optimality conditions satisfied by x^* and x^{k+1} are

$$\exists \, r(x^*) \in F(x^*) \, : \, \langle r(x^*), x - x^* \rangle + \varphi(x) - \varphi(x^*) \geq 0, \, \forall x \in H, \qquad (4)$$

and

$$\langle \eta_k^{-1}r^k(x^k) + \lambda_k^{-1}(\nabla K(x^{k+1}) - \nabla K(x^k)), x - x^{k+1}\rangle$$
$$+\eta_k^{-1}(\varphi^k(x) - \varphi^k(x^{k+1})) \geq 0, \forall x \in H, \qquad (5)$$

with $r^k(x^k) \in G^k(x^k) \subset F^{\varepsilon^k}(x^k)$.

We consider the sequence of Lyapunov functions $\{\Gamma^k(x^*, \cdot)\}_{k\in\mathbb{N}}$ defined on H by

$$\Gamma^k(x^*, x) = K(x^*) - K(x) - \langle \nabla K(x), x^* - x \rangle$$
$$+(\lambda_k/\eta_k)[\langle r(x^*), x - x^* \rangle + \varphi(x) - \varphi(x^*)]. \qquad (6)$$

From the strong convexity of K and inequality (4), we obtain that, for all $x \in \text{dom } \varphi$ and all $k \in \mathbb{N}$,

$$\Gamma^k(x^*, x) \geq (\beta/2)\|x - x^*\|^2. \qquad (7)$$

Using the definition of the Lyapunov function and the facts that $\lambda_{k+1} \leq \lambda_k$ and $\eta_{k+1} \geq \eta_k$ for all $k \in \mathbb{N}$, we can write

$$\Gamma^{k+1}(x^*, x^{k+1}) - \Gamma^k(x^*, x^k) \leq \Gamma^k(x^*, x^{k+1}) - \Gamma^k(x^*, x^k) = s_1 + s_2 + s_3, \qquad (8)$$

with

$$s_1 = K(x^k) - K(x^{k+1}) + \langle \nabla K(x^k), x^{k+1} - x^k \rangle,$$
$$s_2 = \langle \nabla K(x^k) - \nabla K(x^{k+1}), x^* - x^{k+1} \rangle,$$
$$s_3 = (\lambda_k/\eta_k)[\langle r(x^*), x^{k+1} - x^k \rangle + \varphi(x^{k+1}) - \varphi(x^k)].$$

For s_1, we derive easily from the strong convexity of K that

$$s_1 \leq -(\beta/2)\|x^{k+1} - x^k\|^2. \tag{9}$$

Now, using the sequence $\{w^k\}_{k \in \mathbb{N}}$ given in assumption (v), we can write s_2 as the sum of the two following terms:

$$s_{21} = \langle \nabla K(x^k) - \nabla K(x^{k+1}), x^* - w^k \rangle,$$
$$s_{22} = \langle \nabla K(x^k) - \nabla K(x^{k+1}), w^k - x^{k+1} \rangle.$$

From the Lipschitz continuity of K, we deduce that

$$s_{21} \leq \Lambda \|x^{k+1} - x^k\| \, \|x^* - w^k\|$$
$$\leq (\tau/2)\|x^{k+1} - x^k\|^2 + (\Lambda^2/(2\tau))\|x^* - w^k\|^2, \tag{10}$$

where the second inequality holds for any $\tau > 0$.
Using (5) with $x = w^k$, we obtain

$$s_{22} \leq (\lambda_k/\eta_k)[\langle r^k(x^k), w^k - x^{k+1} \rangle + \varphi^k(w^k) - \varphi^k(x^{k+1})]$$
$$= (\lambda_k/\eta_k)[\langle r^k(x^k), w^k - x^* \rangle + \langle r^k(x^k), x^* - x^k \rangle$$
$$+ \langle r^k(x^k), x^k - x^{k+1} \rangle + \varphi^k(w^k) - \varphi^k(x^{k+1})]. \tag{11}$$

From the definition of the sequence $\{\eta_k\}_{k \in \mathbb{N}}$ and the fact that $\lambda_k \leq \lambda_0$ for all k, we have

$$(\lambda_k/\eta_k)\langle r^k(x^k), w^k - x^* \rangle \leq \lambda_0 \|w^k - x^*\|, \tag{12}$$

$$(\lambda_k/\eta_k)\langle r^k(x^k), x^k - x^{k+1} \rangle \leq \lambda_k \|x^{k+1} - x^k\|$$
$$\leq \lambda_k^2/(2\gamma) + (\gamma/2)\|x^{k+1} - x^k\|^2, \tag{13}$$

where the last inequality holds for any $\gamma > 0$. And we also have, for the first term in s_3, that

$$(\lambda_k/\eta_k)\langle r(x^*), x^{k+1} - x^k \rangle \leq \lambda_k \|r(x^*)\| \, \|x^{k+1} - x^k\|$$
$$\leq (\lambda_k^2/(2\mu))\|r(x^*)\|^2 + (\mu/2)\|x^{k+1} - x^k\|^2, \tag{14}$$

with μ any positive number.
Gathering the fact that $\varphi \leq \varphi^k$ for all k with inequalities (8)–(9) and rearranging the terms, we deduce that

$$\Gamma^{k+1}(x^*, x^{k+1}) - \Gamma^k(x^*, x^k) \leq -c\|x^{k+1} - x^k\|^2 + T^k + U^k$$
$$+ (\lambda_k/\eta_k)[\langle r^k(x^k), x^* - x^k \rangle + \varphi(x^*) - \varphi(x^k)], \tag{15}$$

with $c = (1/2)(\beta - \tau - \gamma - \mu)$,
$T^k = \lambda_0 \|w^k - x^*\| + \lambda_0 \|\varphi^k(w^k) - \varphi(x^*)\| + (\Lambda^2/(2\tau))\|w^k - x^*\|^2$,
$U^k = \lambda_k^2 [(1/(2\gamma)) + (1/(2\mu))\|r(x^*)\|^2]$,
$\tau,\ \gamma,\ \mu > 0$ such that $\tau + \gamma + \mu < \beta$.

- We will use this inequality to prove that the sequence $\{x^k\}_{k \in \mathbb{N}}$ is bounded. Since $r^k(x^k) \in F^{\varepsilon^k}(x^k)$, we have

$$\langle r^k(x^k), x^* - x^k \rangle \le \langle r(x^*), x^* - x^k \rangle + \varepsilon^k.$$

Hence, from (4) and the facts that $\varepsilon^k \le \varepsilon \lambda_k$ and $\eta_k \ge 1$ for all k, we deduce that

$$(\lambda_k/\eta_k)[\langle r^k(x^k), x^* - x^k \rangle + \varphi(x^*) - \varphi(x^k)] \le \varepsilon \lambda_k^2.$$

With this inequality, we derive from (15) that

$$\Gamma^{k+1}(x^*, x^{k+1}) - \Gamma^k(x^*, x^k) \le T^k + U^k + \varepsilon \lambda_k^2. \tag{16}$$

From assumption (v) and the convergence of the series $\sum\limits_{k=0}^{+\infty} \lambda_k^2$, it follows that the series $\sum\limits_{k=0}^{+\infty} T^k$ and $\sum\limits_{k=0}^{+\infty} (U^k + \varepsilon \lambda_k^2)$ are convergent. Inequality (16) then ensures that the sequence $\{\Gamma^k(x^*, x^k)\}_{k \in \mathbb{N}}$ is a Cauchy sequence, which implies that it is convergent in H. Using inequality (7), we conclude that the sequence $\{x^k\}_{k \in \mathbb{N}}$ is bounded.

- Now, we will show that the following inequality holds for all $k \in \mathbb{N}$:

$$\Gamma^{k+1}(x^*, x^{k+1}) - \Gamma^k(x^*, x^k) \le -\kappa \lambda_k \|x^k - x^*\|^2 + T^k + \lambda_k V^k, \tag{17}$$

with $\kappa > 0$ and $\{V^k\}_{k \in \mathbb{N}}$ such that $\lim\limits_{k \to \infty} V^k = 0$.

First, we study the case where $G^k \subset F_\alpha^{\varepsilon^k}$ for all k. Since $r^k(x^k) \in F_\alpha^{\varepsilon^k}(x^k)$, we have that

$$\langle r^k(x^k), x^* - x^k \rangle \le \langle r(x^*), x^* - x^k \rangle - \alpha \|x^k - x^*\|^2 + \varepsilon^k.$$

Using (4) and the fact that $\eta_k \ge 1$ for all k, we deduce that

$$\begin{aligned}(\lambda_k/\eta_k)[\langle r^k(x^k), x^* - x^k \rangle + \varphi(x^*) - \varphi(x^k)] \\ \le -\alpha(\lambda_k/\eta_k)\|x^k - x^*\|^2 + \varepsilon^k \lambda_k.\end{aligned} \tag{18}$$

Moreover, since the sequence $\{x^k\}_{k \in \mathbb{N}}$ is bounded and $F^{\bar{\varepsilon}}$ is bounded on bounded sets, the sequence $\{r^k(x^k)\}_{k \in \mathbb{N}}$ is bounded and there exists a constant $\bar{\eta} > 1$ such that $\|r^k(x^k)\| \le \bar{\eta}$ for all k. Therefore, for all k,

$$\eta_k \le max(1, max_{i=1,\dots,k}\|r^i(x^i)\|) \le \bar{\eta}.$$

With this observation and inequality (18), we derive from (15) that

$$\Gamma^{k+1}(x^*, x^{k+1}) - \Gamma^k(x^*, x^k) \leq -(\alpha/\bar{\eta})\lambda_k \|x^k - x^*\|^2 + T^k + \lambda_k V^k,$$
(19)

with $V^k = \lambda_k[(1/(2\gamma)) + (1/(2\mu))\|r(x^*)\|^2] + \varepsilon^k$.
Remark that $\lim\limits_{k \to \infty} V^k = 0$.

Secondly, let us turn to the case where F is Lipschitz continuous on H. Since F is strongly monotone over dom φ with constant α, we have that, for any $r(x^k) \in F(x^k)$:

$$\langle r(x^k), x^* - x^k \rangle \leq \langle r(x^*), x^* - x^k \rangle - \alpha \|x^k - x^*\|^2.$$

Therefore,

$$\begin{aligned}
\langle r^k(x^k), x^* - x^k \rangle \leq {} & \langle r(x^*), x^* - x^k \rangle - \alpha \|x^k - x^*\|^2 \\
& + \langle r^k(x^k) - r(x^k), x^* - x^k \rangle.
\end{aligned}$$
(20)

Since $r^k(x^k) \in F^{\varepsilon^k}(x^k)$ and F is maximal monotone, an extension of the Brønsted and Rockafellar's Theorem (see Theorem 2.1 of [9]) ensures that, for any k:

$$\begin{aligned}
& \exists y^k \in H \text{ and } \exists r(y^k) \in F(y^k) \text{ such that} \\
& \|x^k - y^k\| \leq \sqrt{\varepsilon^k} \text{ and } \|r^k(x^k) - r(y^k)\| \leq \sqrt{\varepsilon^k}.
\end{aligned}$$
(21)

Hence,

$$\begin{aligned}
\langle r^k(x^k) - r(x^k), x^* - x^k \rangle & \\
= \langle r^k(x^k) - r(y^k), x^* - x^k \rangle & + \langle r(y^k) - r(x^k), x^* - x^k \rangle \\
\leq \sqrt{\varepsilon^k}(1 + A) \|x^k - x^*\|, &
\end{aligned}$$
(22)

where the inequality comes from (21) and the Lipschitz continuity of F. Using (20), (4), (22) and the fact that $\eta_k \geq 1$, we deduce that

$$\begin{aligned}
(\lambda_k/\eta_k)[\langle r^k(x^k), x^* - x^k \rangle + \varphi(x^*) - \varphi(x^k)] \leq {} & -\alpha(\lambda_k/\eta_k)\|x^k - x^*\|^2 \\
& + \lambda_k \sqrt{\varepsilon^k}(1 + A)\|x^k - x^*\|.
\end{aligned}$$
(23)

Moreover, since the sequence $\{x^k\}_{k \in \mathbb{N}}$ is bounded, there exists $e > 0$ such that $\|x^k - x^*\| \leq e$ for all k. And by the same reasoning as before, we also have that there exists $\bar{\eta} > 1$ such that $\eta_k \leq \bar{\eta}$ for all k. With these observations and inequality (23), we derive from (15) that

$$\Gamma^{k+1}(x^*, x^{k+1}) - \Gamma^k(x^*, x^k) \leq -(\alpha/\bar{\eta})\lambda_k\|x^k - x^*\|^2 + T^k + \lambda_k V^k,$$
(24)

with $V^k = \lambda_k[(1/(2\gamma)) + (1/(2\mu))\|r(x^*)\|^2] + \sqrt{\varepsilon^k}(1 + A)e$.
Note that $\lim\limits_{k\to\infty} V^k = 0$.

So, in the two cases, we obtain an inequality as announced in (17).

• It remains to prove that the sequence $\{x^k\}_{k\in I\!N}$ weakly converges to x^*.

First, let us show that $\underline{\lim}_{k\to\infty}\|x^k - x^*\| = 0$. Obviously, we have that $\underline{\lim}_{k\to\infty}\|x^k - x^*\| \geq 0$. Assume for contradiction that there are $\delta > 0$ and $k_0 \in I\!N$ such that

$$\|x^k - x^*\| > \delta, \ \forall k \geq k_0.$$

Since $\lim\limits_{k\to\infty} V^k = 0$, there also exists $k_1 \in I\!N$ such that

$$V^k < \kappa\delta^2/2, \ \forall k \geq k_1.$$

So, from (17), it follows that

$$\Gamma^{k+1}(x^*, x^{k+1}) - \Gamma^k(x^*, x^k) \leq T^k - \lambda_k(\kappa\delta^2/2), \ \forall k \geq max(k_0, k_1) \equiv \bar{k}.$$

Summing up, this gives for all $N \geq \bar{k}$:

$$0 \leq \Gamma^{N+1}(x^*, x^{N+1}) \leq \Gamma^{\bar{k}}(x^*, x^{\bar{k}}) + \sum_{k=\bar{k}}^{N} T^k - (\kappa\delta^2/2)\sum_{k=\bar{k}}^{N} \lambda_k.$$

So, if we take the limit on N, we obtain a contradiction with the fact that the series $\sum\limits_{k=0}^{+\infty} \lambda_k$ is divergent.

Now, let $\bar{x}$ be a weak limit point of $\{x^k\}_{k\in I\!N}$. Since the norm is weakly lower semicontinuous, we have

$$\underline{\lim}_{k\to\infty}\|x^k - x^*\| \geq \|\bar{x} - x^*\|.$$

It follows that $\bar{x} = x^*$ and therefore, all the sequence $\{x^k\}_{k\in I\!N}$ weakly converges to x^*. □

Remark 1. As we can see in the proof, when the sequence $\{x^k\}_{k\in I\!N}$ is known to be bounded beforehand (for example when dom φ is bounded), then the weak convergence of the sequence is ensured without the conditions $\sum\limits_{k=0}^{+\infty} \lambda_k^2 < +\infty$ and $\varepsilon^k \leq \lambda_k\varepsilon$ for all $k \in I\!N$.

To prove the strong convergence of $\{x^k\}_{k\in I\!N}$, we have to impose the following condition on sequence $\{\varphi^k\}_{k\in I\!N}$:

(C) $\forall k \in I\!N, \varphi \leq \varphi^{k+1} \leq \varphi^k$ and, $\forall x, y \in$ dom φ^k,

$$\exists c_1, c_2 > 0 : \|\varphi^k(x) - \varphi^k(y)\| \leq [c_1 max(\|x\|, \|y\|) + c_2] \|x - y\|.$$

When φ is the indicator function of a closed convex subset C of H and φ^k is the indicator function of a closed convex subset C^k of C such that $C^k \subset C^{k+1} \subset C$ for all $k \in I\!N$, then condition (C) is satisfied.

Theorem 2. *Assume that F is a maximal monotone multivalued operator, that the solution set of (P) is nonempty and that assumptions (i)–(v) of Theorem 1 are satisfied.*

If $\displaystyle\sum_{k=0}^{+\infty} \lambda_k^2 < +\infty$ and $\displaystyle\sum_{k=0}^{+\infty} \lambda_k = +\infty$, conditions (A), (B) and (C) hold and

$\qquad$ *either $G^k \subset F_\alpha^{\varepsilon^k}$, $\forall k \in I\!N$ and $\exists \varepsilon > 0: \varepsilon^k \leq \lambda_k \varepsilon$, $\forall k \in I\!N$,*

$\qquad$ *or F is Lipschitz continuous on H with Lipschitz constant A and*
$\qquad\quad \exists \varepsilon > 0: \varepsilon^k \leq \lambda_k^2 \varepsilon$, $\forall k \in I\!N$,

then, the sequence $\{x^k\}_{k \in I\!N}$ strongly converges to the unique solution of problem (P).

Proof. Since the assumptions of Theorem 1 are satisfied, all the conclusions of this theorem hold and we can use any intermediary result appearing in its proof. So, from (5) with $x = x^k$, we can write

$$\langle \nabla K(x^{k+1}) - \nabla K(x^k), x^{k+1} - x^k \rangle$$
$$\leq (\lambda_k/\eta_k)[\langle r^k(x^k), x^k - x^{k+1}\rangle + \varphi^k(x^k) - \varphi^k(x^{k+1})].$$

By using the strong convexity of K, the definition of η_k, condition (C) and boundedness of the sequence $\{x^k\}_{k \in I\!N}$, we deduce that there exists some constant $L > 0$ such that

$$\beta \|x^{k+1} - x^k\|^2 \leq \lambda_k(\|x^{k+1} - x^k\| + L\|x^{k+1} - x^k\|).$$

Hence,

$$\|x^{k+1} - x^k\| \leq \lambda_k(1+L)/\beta. \qquad (25)$$

On the other hand, from (19) with $\varepsilon^k \leq \varepsilon\lambda_k$ or (24) with $\varepsilon^k \leq \varepsilon\lambda_k^2$, we have that

$$(\alpha/\bar\eta)\lambda_k\|x^k - x^*\|^2 \leq \Gamma^k(x^*, x^k) - \Gamma^{k+1}(x^*, x^{k+1}) + T^k + \lambda_k^2 v,$$

with

$$v = \begin{cases} (1/(2\gamma)) + (1/(2\mu))\|r(x^*)\|^2 + \varepsilon & \text{in the case of (19)} \\ (1/(2\gamma)) + (1/(2\mu))\|r(x^*)\|^2 + \sqrt{\varepsilon}(1+A)e & \text{in the case of (24).} \end{cases}$$

Summing up, this gives for all $N \geq 0$:

$$(\alpha/\bar{\eta}) \sum_{k=0}^{N} \lambda_k \|x^k - x^*\|^2 \leq \Gamma^0(x^*, x^0) + \sum_{k=0}^{N} T^k + v \sum_{k=0}^{N} \lambda_k^2.$$

And if we take the limit on N, it follows that

$$\sum_{k=0}^{+\infty} \lambda_k \|x^k - x^*\|^2 < +\infty. \tag{26}$$

Since the function $h(x) \equiv \|x - x^*\|^2$ is Lipschitz continuous on a bounded convex hull of the sequence $\{x^k\}_{k \in \mathbb{N}}$, since $\sum_{k=0}^{+\infty} \lambda_k = +\infty$ and we have (25) and (26), we can apply Lemma 4 of [12] to conclude that $\lim_{k \to \infty} \|x^k - x^*\| = 0$.

$\square$

Remark 2. If we do not use the sequence $\{\eta_k\}_{k \in \mathbb{N}}$ in subproblems (IPP^k), $k \in \mathbb{N}$, the same conclusions as in Theorems 1 and 2 can be obtained provided that one of the two additional conditions is satisfied:

(D) $\exists\, a, b > 0 \, : \, \|\bar{r}(x)\| \leq a\|x\| + b, \; \forall x \in H, \; \forall \bar{r}(x) \in F^{\bar{\varepsilon}}(x),$

(E) F is Lipschitz continuous on H and $\exists \varepsilon > 0 : \; \varepsilon^k \leq \lambda_k^2 \varepsilon, \; \forall k \in \mathbb{N}.$

Observe that condition (D) on $F^{\bar{\varepsilon}}$ implies condition (B). And when condition (E) holds, the proof can be established without condition (B).
If $\varepsilon^k = 0$ for all k, condition (D) reduces to (2). Under this assumption, in the nonperturbed case, Theorem 2 reduces to the convergence result of [11] mentioned in the Introduction.

Remark 3. When F is strongly monotone and Lipschitz continuous, we can also prove the convergence of the method for a sequence of stepsizes $\{\lambda_k\}_{k \in \mathbb{N}}$ that does not converge to zero but that is adequately bounded. In this case, some assumptions can be weakened as it is shown in the following Theorem. Note that the setting of this Theorem suppose that we do not use the sequence $\{\eta_k\}_{k \in \mathbb{N}}$.

Theorem 3. *Assume that F is a maximal monotone operator, that the solution set of (P) is nonempty and that assumptions (i), (ii) and (iv) of Theorem 1 are satisfied. In addition, suppose that F satisfies condition (A) and is Lipschitz continuous on H with Lipschitz constant $A > 0$.*
If the sequence $\{\lambda_k\}_{k \in \mathbb{N}}$ is such that

$$\exists\, \underline{\lambda}, \bar{\lambda} > 0 : \; 0 < \underline{\lambda} \leq \lambda_k \leq \bar{\lambda} < 2\alpha\beta/A^2, \; \forall k \in \mathbb{N},$$

and $\{\varphi^k\}_{k\in\mathbb{N}}, \varphi \in \Gamma_0(H)$ are such that $\varphi^k \overset{epi}{\to} \varphi$ and $\varphi \leq \varphi^k$ for all k, then, the sequence $\{x^k\}_{k\in\mathbb{N}}$ generated by the problems (IPP^k), $k \in \mathbb{N}$, strongly converges to the unique solution of problem (P).

This result generalizes Theorem 4.1 of [28] by allowing an inexact calculation of $F(x^k)$ by means of the ε-enlargement.

3   A perturbed version of the projected inexact subgradient method

In this section, we consider the minimization problem (OP). For solving it, we use the scheme defined by problems (IPP^k), $k \in \mathbb{N}$, but with the operator $G^k \subset \partial_{\varepsilon^k} f$ instead of $(\partial f)^{\varepsilon^k}$. So, the problem to be solved at iteration k is the following:

$$(OIPP^k) \begin{cases} \text{find } x^{k+1} \in H \text{ such that, for all } x \in H, \\[2mm] \langle \eta_k^{-1} r^k(x^k) + \lambda_k^{-1}(\nabla K(x^{k+1}) - \nabla K(x^k)), x - x^{k+1} \rangle \\[2mm] \qquad\qquad + \eta_k^{-1}(\varphi^k(x) - \varphi^k(x^{k+1})) \geq 0, \\[2mm] \text{with } r^k(x^k) \in G^k(x^k) \subset \partial_{\varepsilon^k} f(x^k), \\[2mm] \text{and } \eta_k = \begin{cases} max\{1, \|r^0(x^0)\|\} \text{ if } k = 0 \\[2mm] max\{\eta_{k-1}, \|r^k(x^k)\|\} \text{ if } k \geq 1. \end{cases} \end{cases}$$

Note that since f is convex, continuous and its effective domain is H, $\partial_\varepsilon f(x)$ is nonempty for all $\varepsilon \geq 0$ and all $x \in H$ (see, for example, [15]). Moreover, as it is mentioned in [2], $\partial_\varepsilon f$ is always locally bounded. In finite dimension, this implies that $\partial_\varepsilon f$ is bounded on bounded sets while it is not generally the case in a general Hilbert space. However, a sufficient condition for this to hold is that $|f|$ is bounded on bounded sets.

Under the same type of assumptions as in Theorem 1 except that the strong convexity of f is not required, we prove that the sequence $\{x^k\}_{k\in\mathbb{N}}$ is bounded, minimizing (in the sense that $\underline{\lim}_{k\to\infty}(f + \varphi)(x^k) = \inf_{x\in H}\{f(x) + \varphi(x)\}$), and weakly convergent to a solution of (OP). If f is strongly convex and assumptions like those of Theorem 2 are considered, then we also obtain the strong convergence of $\{x^k\}_{k\in\mathbb{N}}$.

Theorem 4. *Assume that f is a convex, continuous and finite-valued function defined on H, that problem (OP) admits at least one solution and that assumptions (i)–(v) of Theorem 1 are satisfied.*

If $\sum_{k=0}^{+\infty} \lambda_k^2 < +\infty$ *and there exists* $\varepsilon > 0$ *such that* $\varepsilon^k \leq \lambda_k \varepsilon$ *for all* $k \in \mathbb{N}$,
then, provided that $x^0 \in \operatorname{dom} \varphi$, *the sequence* $\{x^k\}_{k \in \mathbb{N}}$ *is bounded.*

If, in addition, $\sum_{k=0}^{+\infty} \lambda_k = +\infty$ *and* $\partial_{\bar{\varepsilon}} f$ *is bounded on bounded subsets of* H,
then, $\lim_{k \to \infty} (f + \varphi)(x^k) = \inf_{x \in H} \{f(x) + \varphi(x)\}$. *Moreover, if* $f + \varphi$ *is weakly lower semicontinuous on* $\operatorname{dom} \varphi$, *then each weak limit point of the sequence* $\{x^k\}_{k \in \mathbb{N}}$ *is a solution of problem (OP) and if* ∇K *is weakly continuous on* $\operatorname{dom} \varphi$, *the sequence* $\{x^k\}_{k \in \mathbb{N}}$ *weakly converges to some solution of problem (OP).*

Proof. • In order to show that the sequence $\{x^k\}_{k \in \mathbb{N}}$ is bounded, the proof is the same as in Theorem 1 with $F = \partial f$. (Recall that $\partial_{\varepsilon^k} f \subset (\partial f)^{\varepsilon^k} = F^{\varepsilon^k}$ for all k.)

• Further, let us treat inequality (15) of Theorem 1 by taking advantage of the particular case of optimization. Since $r^k(x^k) \in \partial_{\varepsilon^k} f(x^k)$, we have

$$\langle r^k(x^k), x^* - x^k \rangle \leq f(x^*) - f(x^k) + \varepsilon^k. \tag{27}$$

Moreover, since the sequence $\{x^k\}_{k \in \mathbb{N}}$ is bounded and $\partial_{\bar{\varepsilon}} f$ is bounded on bounded sets, the sequence $\{r^k(x^k)\}_{k \in \mathbb{N}}$ is bounded. Therefore, there exists a constant $\bar{\eta} > 1$ such that $\eta_k \leq \bar{\eta}$ for all k. With this observation, the fact that $\eta_k \geq 1$ and inequality (27), we deduce from (15) that

$$\begin{aligned} \Gamma^{k+1}(x^*, x^{k+1}) &- \Gamma^k(x^*, x^k) \\ &\leq -(1/\bar{\eta})\lambda_k[(f + \varphi)(x^k) - (f + \varphi)(x^*)] + T^k + \lambda_k V^k, \end{aligned} \tag{28}$$

where $V^k = \lambda_k[(1/(2\gamma)) + (1/(2\mu))\|r(x^*)\|^2] + \varepsilon^k$.

• Now, we show that $\lim_{k \to \infty} (f + \varphi)(x^k) = \inf_{x \in H} \{f(x) + \varphi(x)\}$. Obviously, we have that $\lim_{k \to \infty} (f + \varphi)(x^k) \geq \inf_{x \in H} \{f(x) + \varphi(x)\}$. Assume for contradiction that there are $\delta > 0$ and $k_0 \in \mathbb{N}$ such that, for all $k \geq k_0$:

$$(f + \varphi)(x^k) > (f + \varphi)(x^*) + \delta.$$

Since V^k in (28) converges to zero, there also exists $k_1 \in \mathbb{N}$ such that, for all $k \geq k_1$:

$$V^k < \delta/(2\bar{\eta}).$$

Hence, from (28), we deduce that, for all $k \geq max(k_0, k_1) \equiv \bar{k}$:

$$\Gamma^{k+1}(x^*, x^{k+1}) - \Gamma^k(x^*, x^k) \leq T^k - \lambda_k \delta/(2\bar{\eta}).$$

Summing up, this gives for all $N \geq \bar{k}$:

$$0 \leq \Gamma^{N+1}(x^*, x^{N+1}) \leq \Gamma^{\bar{k}}(x^*, x^{\bar{k}}) + \sum_{k=\bar{k}}^{N} T^k - (\delta/(2\bar{\eta})) \sum_{k=\bar{k}}^{N} \lambda_k.$$

If we take the limit on N in this last inequality, we obtain a contradiction with the fact that the series $\sum_{k=0}^{+\infty} \lambda_k$ is divergent.

• It is then easy to prove that any weak limit point of $\{x^k\}_{k \in \mathbb{N}}$ is a solution of problem (OP). Indeed, let $\bar{x}$ be a weak limit point of $\{x^k\}_{k \in \mathbb{N}}$, since $f + \varphi$ is weakly lower semicontinuous, we have that

$$\underline{\lim}_{k \to \infty} (f + \varphi)(x^k) \geq (f + \varphi)(\bar{x}).$$

It follows that $(f + \varphi)(\bar{x}) = \inf_{x \in H} \{f(x) + \varphi(x)\}$ which means that $\bar{x}$ solves (OP).

• To finish the proof, it remains to verify that the sequence $\{x^k\}_{k \in \mathbb{N}}$ weakly converges to some solution of problem (P). For contradiction, let us suppose that the sequence $\{x^k\}_{k \in \mathbb{N}}$ has two different weak limit points x^1 and x^2 such that $\{x^{l(k)}\}_{k \in \mathbb{N}}$ is a subsequence of $\{x^k\}_{k \in \mathbb{N}}$ weakly converging to x^1 and $\{x^{n(k)}\}_{k \in \mathbb{N}}$ is a subsequence weakly converging to x^2. We have just shown that x^1 and x^2 are solutions of problem (OP) so that, by the first part of the proof, there exist $l_1, l_2 \in \mathbb{R}$ such that

$$\lim_{k \to \infty} \Gamma^k(x^1, x^k) = l_1 \text{ and } \lim_{k \to \infty} \Gamma^k(x^2, x^k) = l_2.$$

By the definition of the Lyapunov function,

$$\Gamma^{n(k)}(x^1, x^{n(k)}) - \Gamma^{n(k)}(x^2, x^{n(k)}) = K(x^1) - K(x^2) - \langle \nabla K(x^{n(k)}), x^1 - x^2 \rangle$$
$$+ (\lambda_{n(k)}/\eta_{n(k)})[\langle r(x^1), x^{n(k)} - x^1 \rangle - \langle r(x^2), x^{n(k)} - x^2 \rangle + \varphi(x^2) - \varphi(x^1)].$$

∇K being weakly continuous, if we take the limit on k in the last inequality, we obtain

$$l_1 - l_2 = K(x^1) - K(x^2) - \langle \nabla K(x^2), x^1 - x^2 \rangle. \tag{29}$$

Since the role of x^1 and x^2 is symmetric, we also have that

$$l_1 - l_2 = K(x^1) - K(x^2) - \langle \nabla K(x^1), x^1 - x^2 \rangle. \tag{30}$$

Comparing (29) and (30), we obtain

$$\langle \nabla K(x^1) - \nabla K(x^2), x^1 - x^2 \rangle = 0.$$

Since ∇K is strongly monotone, this inequality implies that $x^1 = x^2$ and the proof is complete. □

Theorem 5. *Assume that f is a convex, continuous and finite-valued function defined on H, that the solution set of (OP) is nonempty and that assumptions (i)–(v) of Theorem 1 hold.*
If

$$\sum_{k=0}^{+\infty} \lambda_k^2 < +\infty \ \text{ and } \ \sum_{k=0}^{+\infty} \lambda_k = +\infty;$$
$$\exists \varepsilon > 0 : \ \varepsilon^k \leq \lambda_k \varepsilon, \ \forall k \in I\!N;$$

$\partial_{\bar{\varepsilon}} f$ is bounded on bounded subsets of H;

$f + \varphi$ is Lipschitz continuous on bounded subsets of dom φ;

$\{\varphi^k\}_{k \in I\!N}, \varphi$ satisfy condition (C),

then, $\lim_{k \to \infty} (f + \varphi)(x^k) = \inf_{x \in H} \{f(x) + \varphi(x)\}$. If, in addition, f is strongly convex on dom φ, then the sequence $\{x^k\}_{k \in I\!N}$ strongly converges to the unique solution of problem (OP).

Proof. From the definition of the sequence $\{x^k\}_{k \in I\!N}$ with $x = x^k$, we can write

$$\langle \nabla K(x^{k+1}) - \nabla K(x^k), x^{k+1} - x^k \rangle$$
$$\leq (\lambda_k/\eta_k)[\langle r^k(x^k), x^k - x^{k+1} \rangle + \varphi^k(x^k) - \varphi^k(x^{k+1})].$$

By using the strong convexity of K, the definition of η_k, condition (C) and boundedness of $\{x^k\}_{k \in I\!N}$, we deduce that there exists some constant $L > 0$ such that

$$\beta \|x^{k+1} - x^k\|^2 \leq \lambda_k (\|x^{k+1} - x^k\| + L\|x^{k+1} - x^k\|).$$

Therefore,

$$\|x^{k+1} - x^k\| \leq \lambda_k (1 + L)/\beta. \tag{31}$$

On the other hand, the assumptions allow us to obtain relation (28) by the same arguments as in Theorem 4. So, from (28) with the fact that $\varepsilon^k \leq \varepsilon \lambda_k$ for all k, we obtain that

$$(1/\bar{\eta})\lambda_k[(f+\varphi)(x^k) - (f+\varphi)(x^*)] \leq \Gamma^k(x^*, x^k) - \Gamma^{k+1}(x^*, x^{k+1}) + T^k + \lambda_k^2 v,$$

with $v = (1/(2\gamma)) + (1/(2\mu))\|r(x^*)\|^2 + \varepsilon$.

Summing up, this gives for all $N \geq 0$:

$$(1/\bar{\eta}) \sum_{k=0}^{N} \lambda_k [(f+\varphi)(x^k) - (f+\varphi)(x^*)] \leq \Gamma^0(x^*, x^0) + \sum_{k=0}^{N} T^k + v \sum_{k=0}^{N} \lambda_k^2.$$

And if we take the limit on N, we obtain that

$$\sum_{k=0}^{+\infty} \lambda_k [(f + \varphi)(x^k) - (f + \varphi)(x^*)] < +\infty. \tag{32}$$

Since the function $f + \varphi$ is Lipschitz continuous on a bounded convex hull of the sequence $\{x^k\}_{k \in I\!\!N}$, since $\sum_{k=0}^{+\infty} \lambda_k = +\infty$ and we have (31) and (32), we can apply Lemma 4 of [12] to conclude that

$$\lim_{k \to \infty} (f + \varphi)(x^k) = (f + \varphi)(x^*) = \inf_{x \in H} \{f(x) + \varphi(x)\}.$$

If in addition, f is strongly convex with constant α on dom φ, then problem (OP) admits a unique solution and

$$f(x^k) - f(x^*) \geq \langle r(x^*), x^k - x^* \rangle + (\alpha/2)\|x^k - x^*\|^2.$$

By this last inequality together with (4), we obtain

$$(f + \varphi)(x^k) - (f + \varphi)(x^*) \geq (\alpha/2)\|x^k - x^*\|^2.$$

Since $\lim_{k \to \infty} (f + \varphi)(x^k) = (f + \varphi)(x^*)$, it implies that $\{x^k\}_{k \in I\!\!N}$ strongly converges to x^* and this completes the proof. $\square$

Remark 4. If, for all $x \in H$, $K(x) = (1/2)x^T x$, and for all $k \in I\!\!N$, $\varphi^k = \varphi = \Psi_C$, where Ψ_C denotes the indicator function of a closed convex subset C of H, then the method defined by problems $(OIPP^k)$, $k \in I\!\!N$, reduces to the projected inexact subgradient process:

$$\begin{cases} x^{k+1} = Proj_C \left[x^k - (\lambda_k/\eta_k)r^k(x^k)\right], \\[2mm] \text{with } r^k(x^k) \in \partial_{\varepsilon^k} f(x^k), \\[2mm] \text{and } \eta_k = \begin{cases} max\{1, \|r^0(x^0)\|\} \text{ if } k = 0 \\[2mm] max\{\eta_{k-1}, \|r^k(x^k)\|\} \text{ if } k \geq 1. \end{cases} \end{cases}$$

If problem (OP) is solvable, Theorem 4 reduces, for this scheme, to the convergence result of [2]. The case where the problem could have no solution is discussed in the next Remark.

Remark 5. When f is Lipschitz continuous on dom φ and the assumptions of Theorem 4 hold, we can omit to suppose that the problem has some solution and however obtain the following results:

- $\underline{\lim}_{k \to \infty} (f + \varphi)(x^k) = \inf_{x \in H} \{f(x) + \varphi(x)\};$

- If problem (OP) admits at least one solution, then the sequence $\{x^k\}_{k \in \mathbb{N}}$ weakly converges to some solution;
- If the solution set of problem (OP) is empty, then the sequence $\{x^k\}_{k \in \mathbb{N}}$ is unbounded.

Remark 6. We can formulate for the optimization case the same kind of comment as in Remark 2. If the sequence $\{\eta_k\}_{k \in \mathbb{N}}$ does not appear in subproblems $(OIPP^k)$, $k \in \mathbb{N}$, the conclusions of Theorems 4 and 5 still hold if we impose in addition that:

(F) $\exists\, a, b > 0 \;:\; \|\bar{r}(x)\| \leq a\|x\| + b, \; \forall x \in H, \; \forall \bar{r}(x) \in \partial_{\bar{\varepsilon}} f(x).$

Condition (F) implies that $\partial_{\bar{\varepsilon}} f$ is bounded on bounded sets of H.
In this context, if $\varphi^k = \varphi$ and $\varepsilon^k = 0$, for all k, Theorem 4 completes the convergence result of [12] where the weak convergence of the sequence $\{x^k\}_{k \in \mathbb{N}}$ is not proved. Note also that Theorem 5 reduces to their strong convergence result.

References

1. Alber, Ya.I. (1983) Recurrence Relations and Variational Inequalities. Soviet Mathematics Doklady **27**, 511–517
2. Alber, Ya.I., Iusem, A.N., Solodov, M.V. (1998) On the Projected Subgradient Method for Nonsmooth Convex Optimization in a Hilbert Space. Math. Program. **81**, 23–35
3. Attouch, H. (1984) Variational Convergence for Functions and Operators. Pitman, London
4. Auslender, A. (1976) Optimisation. Méthodes Numériques. Masson, Paris, France
5. Aubin, J.P., Ekeland, I. (1984) Applied Nonlinear Analysis. Wiley, New York, New York
6. Bertsekas, D.P., Mitter, S.K. (1973) A Descent Numerical Method for Optimization Problems with Nondifferentiable Cost Functionals. SIAM J. Cont. **11**, 637–652
7. Brønsted, A., Rockafellar, R.T. (1965) On the Subdifferentiability of Convex Functions. Proc. Am. Math. Soc. **16**, 605–611
8. Burachik, R.S., Iusem, A.N., Svaiter, B.F. (1997) Enlargement of Maximal Monotone Operators with Application to Variational Inequalities. Set-Valued Anal. **5**, 159–180
9. Burachik, R.S., Sagastizábal, C.A., Svaiter, B.F. (1998) ε-Enlargement of a Maximal Monotone Operator: Theory and Applications. In: Fukushima M., Qi L. (Eds.) Reformulation - Nonsmooth, Piecewise Smooth, Semismooth and Smoothing Methods. Kluwer, Academic Publishers, 25–44
10. Burachik, R.S., Sagastizábal, C.A., Svaiter, B.F. Bundle Methods for Maximal Monotone Operators. Submitted to Math. Prog.
11. Cohen, G. (1988) Auxiliary Problem Principle Extended to Variational Inequalities. J. Optim. Theory Appl. **59**, 325–333

12. Cohen, G., Zhu, D.L. (1984) Decomposition Coordination Methods in Large Scale Optimization Problems: The Nondifferentiable case and the Use of Augmented Lagrangians. In: Cruz J.B. (Ed.) Advances in Large Scale Systems Theory and Applications 1. JAI Press, Greenwich, Connecticut, USA, 203–266

13. Correa, R., Lemaréchal, C. (1993) Convergence of Some Algorithms for Convex Minimization. Math. Program. **62**, 261–275

14. Cottle, R.W., Giannessi, F., Lions, J.L. (1980) Variational Inequalities and Complementarity Problems: Theory and Applications. Wiley, New York, New York

15. Ekeland, I., Temam, R. (1976) Convex Analysis and Variational Inequalities. North–Holland, Amsterdam

16. Glowinski, R., Lions, J.L., Tremolieres, R. (1981) Numerical Analysis of Variational Inequalities. North–Holland, Amsterdam

17. Harker, P.T., Pang, J.S. (1990) Finite–Dimensional Variational Inequality and Nonlinear Complementarity Problems. A Survey of Theory, Algorithms and Applications. Math. Program. **48**, 161–220

18. Hiriart–Urruty, J-B. (1982) ε-Subdifferential Calculus. Convex Analysis and Optimization. Res. Notes in Math. **57**, 1–44

19. Hiriart–Urruty, J.-B., Lemaréchal, C. (1993) Convex Analysis and Minimization Algorithms. Springer, Berlin

20. Kinderlehrer, D., Stampacchia, G. (1980) An Introduction to Variational Inequalities and their Application. Academic Press, New York, New York

21. Kiwiel, K.C. (1990) Proximity Control in Bundle Methods for Convex Nondifferentiable Minimization. Math. Program. **46**, 105–122

22. Kiwiel, K.C. (1997) Proximal Minimization Methods with Generalized Bregman Functions. SIAM J. Cont. Optim. **35**, 1142–1168

23. Lions, P.-L., Mercier, B. (1979) Splitting Algorithms for the Sum of Two Nonlinear Operators. SIAM J. Num. Anal. **16**, 964–979

24. Lemaire, B. (1988) Coupling Optimization Methods and Variational Convergence. In: Hoffmann K.H., Hiriart–Urruty J.B., Lemaréchal C., Zowe J. (Eds.) Trends in Mathematical Optimization International Series of Numerical Mathematics **84** Birkhäuser Verlag, Basel, 163–179

25. Lemaréchal, C., Nemirovskii, A., Nesterov, Yu. (1995) New Variants of Bundle Methods. Math. Program. **69**, 111–148

26. Lemaréchal, C., Strodiot, J-J., Bihain, A. (1981) On a Bundle Method for Nonsmooth Optimization. In: Mangasarian O.L., Meyer R.R., Robinson S.M. (Eds.) Nonlinear Programming 4, Academic Press, 245–282

27. Mahey, P., Nguyen, V.H., Pham, D.T. (1992) Proximal Techniques for the Decomposition of Convex Programs. Rapport de Recherches Laboratoire ARTEMIS RR **877**-M, 21–37

28. Makler–Scheimberg, S., Nguyen, V.H., Strodiot, J.J. (1996) Family of Perturbation Methods for Variational Inequalities. J. Optim. Theory Appl. **89**, 423–452

29. Nurminski, E.A. (1986) ε-Subgradient Mapping and the Problem of Convex Optimization. Cybernetics **21(6)**, 796–800

30. Pang, J.S. (1985) Asymmetric Variational Inequality Problems over Product Sets: Applications and Iterative methods. Math. Program. **31**, 206–219

31. Polyak, B.T. (1987) Introduction to Optimization. Optimization Software, New York, New York

32. Rockafellar, R.T. (1970) Convex Analysis. Princeton University Press, Princeton, New Jersey

33. Salmon, G., Nguyen, V.H., Strodiot, J.J. (1999) Coupling the Auxiliary Problem Principle and the Epiconvergence Theory for Solving General Variational Inequalities. To appear in J. of Optim. Theory and Appl.
34. Schramm, H., Zowe, J. (1992) A Version of the Bundle Idea for Minimizing a Nonsmooth Function: Conceptual Idea, Convergence Analysis, Numerical Results. SIAM J. Optim. **2**, 121–152
35. Solodov, M.V., Svaiter, B.F. A Hybrid Approximate Extragradient-Proximal Point Algorithm Using the Enlargement of a Maximal Monotone Operator. Submitted to Set-Valued Anal.
36. Solodov, M.V., Zavriev, S.K. (1998) Error-Stability Properties of Generalized Gradient-type Algorithms. J. Optim. Theory Appl. **98**, 663–680
37. Sonntag, Y. (1982) Convergence au Sens de Mosco: Théorie et Applications à l'Approximation des Solutions d'Inéquations. PhD Thesis, Université de Provence, France
38. Strodiot, J.J., Nguyen, V.H. (1988) On the Numerical Treatment of the Inclusion $0 \in \partial f(x)$. In: Moreau J.J, Panagiotopoulos P.D., Strang G. (Eds.) Topics in Nonsmooth Mechanics. Birkhäuser Verlag, 267–294
39. Strodiot, J.J., Nguyen, V.H., Heukemes, N. (1983) ε–Optimal Solutions in Nondifferentiable Convex Programming and some Related Questions. Math. Program. **25**, 307–328

Contributions to a Multiplier Rule for Multiobjective Optimization Problems

Matthias Sekatzek

Martin-Luther-Universität Halle–Wittenberg,
Fachbereich Mathematik und Informatik,
Institut für Optimierung und Stochastik,
D–06099 Halle/Saale, Germany

Abstract. In this paper, we will consider a necessary optimality condition in form of a multiplier rule for so–called K-derived sets, given by BRECKNER in 1994. For this multiplier rule, we present an alternative method of proof, without making use of the theorem of HOANG TUY that was needed in BRECKNER's original paper. Furthermore, for this rule it will be derived an additional complementarity relation as well as regularity conditions for the multipliers, the latter according to an idea of HESTENES for the scalar case.

Key words: multiobjective optimization, necessary optimality condition, derived set, multiplier rule, complementarity condition, regularity condition

1 Introduction

For to solve a nonlinear optimization problem succesfully, one needs in general such classical assumptions as differentiability or convexity. In recent time, much work has been done in weakening those assumptions for special problems in an appropriate way. The optimality condition considered in the present paper is a multiplier rule given by BRECKNER in [1]. Some applications of this rule can be found in [2]. In the latter paper it is also shown that both the differentiable case and the convex case are covered by BRECKNER's theorem.

BRECKNER's multiplier rule deals with so–called derived sets. The term of a derived set was introduced by HESTENES in [4]. There are some optimality conditions related with the theory of derived sets given in the papers [3], [5], [6], and [8]. But all these theorems aim at optimization problems with only one objective function. In [1], HESTENES' definition has been extended onto problems with a multidimensional objective function. For distinction, BRECKNER called the new objects K-derived sets, where K is the ordering cone of the image space. Going this way, BRECKNER received a multiplier rule for a multiobjective optimization problem.

Note, that in [1] the author originally deals with a vector–maximizing problem. For compatibility reasons, in the given paper the terminology is transformed for a minimization problem. Correctly spoken, all K-derived sets considered here are $(-K)$-derived sets in the sense of [1].

Consider the optimization problem

$$f_1(x) \longrightarrow K_1 - Min! \tag{1}$$

subject to

$$S := \{x \in X; \ f_2(x) \in -K_2, f_3(x) \in -K_3\}, \tag{2}$$

where X is a nonempty subset of a topological space, $f_1 : X \to R^{m_1}$, $f_2 : X \to R^{m_2}$, $f_3 : X \to R^{m_3}$ are some functions, and $K_1 \subseteq R^{m_1}$, $K_2 \subseteq R^{m_2}$, $K_3 \subseteq R^{m_3}$ are convex cones, of which K_1 and K_2 have a nonempty interior, while K_2 and K_3 are closed. With K_1^*, K_2^*, and K_3^* we describe the respective dual cones. Furthermore, let

$$L := \{y \in K_2^*; \ \langle y, f_2(x_0) \rangle = 0\}. \tag{3}$$

For short, we agree upon this:

$$m := m_1 + m_2 + m_3,$$

$$f(x) := \begin{pmatrix} f_1(x) \\ f_2(x) \\ f_3(x) \end{pmatrix} : X \to R^m,$$

$$K := K_1 \times K_2 \times K_3 \subseteq R^m.$$

According to this convention, every vector $y \in R^m$ is understood as the triple $y = (y_1, y_2, y_3) \in R^{m_1} \times R^{m_2} \times R^{m_3}$.

Definition 1. A point $x_0 \in X$ is called a local weakly K_1-minimal point for the function f_1 over S if there is a neighbourhood V of x_0 such that

$$(f_1(x_0) - int \ K_1) \cap f_1[S \cap V] = \emptyset.$$

Definition 2. An convex cone $D \subseteq R^m$ is called a K-derived convex cone for f at x_0 if for every $(m+1)$-tuple $\{d_1, \ldots, d_{m+1}\}$ of points in D there exist a number $r > 0$, a function $\omega : B_+^{m+1}(r) \to X$ $(B_+^{m+1}(r) := \{x \in R_+^{m+1}; \ \|x\| \leq r\})$, and a function $\varrho = (\varrho_1, \varrho_2, \varrho_3) : B_+^{m+1}(r) \to R^{m_1} \times R^{m_2} \times R^{m_3}$ such that the following conditions are satisfied:

(A) we have

$$f(\omega(t)) - f(x_0) - (t^1 d_1 + \cdots + t^{m+1} d_{m+1}) - \|t\| \varrho(t) \in -K$$
$$\forall t = (t^1, \ldots, t^{m+1}) \in B_+^{m+1}(r),$$

(B) $\omega(0) = x_0$ and ω is continuous at 0,

(C1) there exists a point $y_1^0 \in K_1$ so that for each number $\varepsilon > 0$ there is a number $r_\varepsilon \in (0, r]$ such that

$$\varrho_1(t) \in \varepsilon y_1^0 - K_1 \qquad \forall t \in B_+^{m+1}(r_\varepsilon),$$

(C2) there exists a point $y_2^0 \in K_2$ so that for each number $\varepsilon > 0$ there is a number $r_\varepsilon \in (0, r]$ such that

$$\varrho_2(t) \in \varepsilon y_2^0 - K_2 \qquad \forall t \in B_+^{m+1}(r_\varepsilon),$$

(C3) $\varrho_3(0) = 0$ and ϱ_3 is continuous.

Now we are able to state the theorem to be considered.

Theorem 1. *Let $x_0 \in X$ be a local weakly K_1-minimal point for the function f_1 over S and let $D \in R^m$ be an K-derived convex cone for f at x_0. Then there exists a multiplier vector $\lambda = (\lambda_1, \lambda_2, \lambda_3) \in K_1^* \times K_2^* \times K_3^*$, $\lambda \neq 0$, such that*

$$\langle \lambda, d \rangle \geq 0 \qquad \forall d \in D, \tag{4}$$
$$\langle \lambda_2, f_2(x_0) \rangle = 0. \tag{5}$$

In section 2, we will give some remarks about the proof of BRECKNER's theorem. On the one hand, we show that the multiplier rule can be proved without making use of a rather complicated linearization theorem of TUY. Such a theorem as well BRECKNER in [1] as NIEUWENHUIS in [8] needed. On the other hand, we give a proving idea, which seems to be more evident than the method BRECKNER used. In section 3, for this rule it will be derived the additional complementarity condition.

$$\langle \lambda_3, f_3(x_0) \rangle = 0, \tag{6}$$

missing in Theorem 1. Finally, in section 4, a regularity condition ensuring $\lambda_1 \neq 0$ in Theorem 1 is given.

2   Alternative proving methods

2.1   Proof without using the theory of nonlinear inequalities

Proving the multiplier rule, BRECKNER states the following assertion.

Proposition 1. *Let t_0 be a vector in the space R^m, let K be a closed convex cone in the space R^p, and let $F : R^m \to R^p$ be a continuous function, differentiable at the origin, satisfying the following conditions:*

(i) $F(0) \in -K$,

(ii) $F'(0)(t_0) \in -K$,

(iii) $F'(0)[R^m] - K = R^p$.

Then there exist a sequence $\{a_n\}$ $(n = 1, 2, \ldots)$ of positive numbers with $\lim_{n \to \infty} a_n = 0$, and a sequence $\{t_n\}$ $(n = 1, 2, \ldots)$ of vectors in the space R^m with $\lim_{n \to \infty} t_n = t_0$, such that

$$F(a_n t_n) \in -K \qquad \forall n.$$

For the proof of this proposition, BRECKNER [1] refers to a result given by TUY in [9]. In the latter paper, the author considers systems of nonlinear inequalities. Indeed, under the assumptions of proposition 1, the system

$$\{x \in R^m; \; F(0) + F'(0)(x) \in -K\}$$

is a local approximation at 0 of the system

$$\{x \in R^m; \; F(x) \in -K\}$$

in the terminology introduced in [9].

To see this, one must study the theory appearing in TUY's paper. But the result is only a very special case of Theorem 1 in [9]. Furthermore, proposition 1 can be formulated without any connection to inequality systems. So we asked whether it is possible to prove only the needed special case without using the theory in [9].

In the following, we will give such a proof. Before, we have to prove a lemma.

Lemma 1. *Let $U \subset R^m$ be a nonempty, convex, and compact set, let $A : U \to R^p$ be a continuous function, $B : R^m \to R^p$ be a linear function, and let K be a convex closed cone of the space R^p.*
Then the set–valued mapping $\Gamma : U \to 2^U$ given by

$$\Gamma(u) := \{\tilde{u} \in U; \; A(u) \in B(\tilde{u}) - K\}$$

possesses a fixed point, if for every $u \in U$ the set $\Gamma(u)$ is nonempty.

Proof. We will show that the assumptions of KAKUTANI's fixed point theorem are fulfilled. So, we have to prove the mapping Γ to be closed, and the images $\Gamma(u)$ to be convex and closed for every $u \in U$.
a) *Closedness of Γ.* Consider two converging sequences $\{u_n\} \subset U$ and $\{\tilde{u}_n\} \subset U$ with $\tilde{u}_n \in \Gamma(u_n) \; \forall n$, or equivalently

$$A(u_n) \in B(\tilde{u}_n) - K \qquad \forall n.$$

Let be $u := \lim_{n \to \infty} u_n$ and $\tilde{u} := \lim_{n \to \infty} \tilde{u}_n$. Then $u \in U$ and $\tilde{u} \in U$. By continuity of A and B and the closedness of K, it follows

$$A(u) \in B(\tilde{u}) - K,$$

what means $\tilde{u} \in \Gamma(u)$.
b) *Convexity of $\Gamma(u) \; \forall u \in U$.* Let $u \in U$ be fixed. Consider two arbitrary points $\tilde{u}_1$ and $\tilde{u}_2$ in $\Gamma(u)$. Since U is convex, we have $\lambda \tilde{u}_1 + (1 - \lambda)\tilde{u}_2 \in U$ $\forall \lambda \in [0, 1]$. Furthermore it is

$$
\begin{aligned}
A(u) &= \lambda A(u) + (1 - \lambda)A(u) \\
&\in \lambda B(\tilde{u}_1) + (1 - \lambda)B(\tilde{u}_2) - K \\
&= B(\lambda \tilde{u}_1 + (1 - \lambda)\tilde{u}_2) - K \qquad \forall \lambda \in [0, 1],
\end{aligned}
$$

hence $\lambda \tilde{u}_1 + (1 - \lambda)\tilde{u}_2 \in \Gamma(u)$ for all $\lambda \in [0, 1]$.

c) *Closedness of $\Gamma(u)$ $\forall u \in U$.* Let $u \in U$ be fixed. Consider a converging sequence $\{\tilde{u}_n\} \subset \Gamma(u)$ with $\tilde{u} := \lim\limits_{n \to \infty} \tilde{u}_n$. Since $\tilde{u}_n \in U$ $\forall n$ and the closedness of U, we have $\tilde{u} \in U$. From

$$A(u) \in B(\tilde{u}_n) - K \qquad \forall n$$

it follows by the continuity of B and the closedness of K

$$A(u) \in B(\tilde{u}) - K,$$

what means $\tilde{u} \in \Gamma(u)$. $\qquad\qquad\qquad\qquad\qquad\qquad\qquad\qquad\qquad$ $\square$

Remark 1. Lemma 1 is still valid, even if the linearity of function B is replaced by K-concavity.

Proof. (of Proposition 1) If $t_0 = 0$, the proposition is fulfilled trivially choosing $\{t_n\} = \{0, 0, \dots\}$ and an arbitrary sequence $\{a_n\}$ of positive numbers converging to zero. So, in the following we assume $t_0 \neq 0$.

For every natural number $n = 1, 2, \dots$ let be $U_n := \{u \in R^m; \|u\| \leq \frac{1}{2n}\}$. It is easily seen, that all sets U_n are convex, closed, and bounded, hence compact. Due to the linearity of $F'(t_0)(.)$ and condition (iii), the set

$$V_n := -(F'(t_0)[U_n] + K)$$

is a neighbourhood of the origin of R^p. Taking into account

$$\lim_{x \to 0} \frac{1}{\|x\|} [F(x) - F(0) - F'(0)(x)] = 0,$$

we find for every n a constant $\varepsilon_n > 0$, such that

$$F(x) - F(0) - F'(0)(x) \in \|x\| V_n \qquad \forall x : \|x\| < \varepsilon_n.$$

Hence we have

$$F(x) \in F(0) + F'(0)(x) - \|x\| \left(F'(t_0)[U_n] - K\right)$$
$$= F(0) + F'(0)(x) - \|x\| F'(t_0)[U_n] - K \qquad \forall x : \|x\| < \varepsilon_n \ \forall n.$$

By (i), this implies

$$F(x) \in F'(0)(x) - \|x\| F'(t_0)[U_n] - K \qquad \forall x : \|x\| < \varepsilon_n \ \forall n. \qquad (7)$$

Let $\{\alpha_n\}$ be a sequence of positive numbers converging to zero with

$$0 < \alpha_n \leq \varepsilon_n \qquad \forall n.$$

From

$$\left\|\frac{t_0}{2\|t_0\|} + u\right\| \le \frac{\|t_0\|}{2\|t_0\|} + \|u\| \le \frac{1}{2} + \frac{1}{2n} < 1 \qquad \forall u \in U_n \quad \forall n \qquad (8)$$

therefore we get

$$\left\|\alpha_n\left(\frac{t_0}{2\|t_0\|} + u\right)\right\| < \varepsilon_n \qquad \forall u \in U_n \quad \forall n.$$

So from (7) yields

$$F\left(\alpha_n\left(\frac{t_0}{2\|t_0\|} + u\right)\right)$$

$$\in F'(0)\left(\alpha_n\left(\frac{t_0}{2\|t_0\|} + u\right)\right) - \alpha_n\left\|\frac{t_0}{2\|t_0\|} + u\right\|F'(t_0)[U_n] - K$$

$$= \frac{\alpha_n}{2\|t_0\|}F'(0)(t_0) + \alpha_n F'(0)(u) - \alpha_n F'(t_0)\left[\left\|\frac{t_0}{2\|t_0\|} + u\right\|U_n\right] - K$$

$$\forall u \in U_n \quad \forall n.$$

By (ii), we have

$$\frac{\alpha_n}{2\|t_0\|}F'(0)(t_0) \in -K \qquad \forall n,$$

and by (8)

$$\left\|\frac{t_0}{2\|t_0\|} + u\right\|U_n \subseteq U_n \qquad \forall u \in U_n \quad \forall n.$$

Hence, it holds

$$F\left(\alpha_n\left(\frac{t_0}{2\|t_0\|} + u\right)\right) \in \alpha_n F'(0)(u) - \alpha_n F'(t_0)[U_n] - K \quad \forall u \in U_n \quad \forall n. \qquad (9)$$

For every natural number n let $\Gamma_n : U_n \to 2^{U_n}$ be the set–valued mapping defined by

$$\Gamma_n(u) := \left\{\tilde{u} \in U_n; \ F\left(\alpha_n\left(\frac{t_0}{2\|t_0\|} + u\right)\right) \in \alpha_n F'(0)(u - \tilde{u}) - K\right\}.$$

Due to (9), for every n and each $u \in U_n$, the set $\Gamma_n(u)$ is nonempty. Furthermore, the mapping

$$F\left(\alpha_n\left(\frac{t_0}{2\|t_0\|} + \cdot\right)\right) - \alpha_n F'(0)(.)$$

is continuous, and the mapping

$$-\alpha_n F'(0)(.)$$

is linear. Applying Lemma 1, for every n the mapping Γ_n has a fixed point $u_n \in U_n$ fulfilling $u_n \in \Gamma_n(u_n)$, what means

$$F\left(\alpha_n\left(\frac{t_0}{2\|t_0\|} + u_n\right)\right) \in -K \qquad \forall n. \qquad (10)$$

So, the sequences $\{a_n\}$ and $\{t_n\}$ defined by

$$a_n := \frac{\alpha_n}{2\|t_0\|} \qquad \forall n$$
$$t_n := t_0 + 2\|t_0\|u_n \qquad \forall n$$

have the properties claimed in the proposition. $\qquad\qquad\qquad\qquad$ $\square$

2.2 A direct approach by using a separation theorem

In this section we will show that derived cones constructed at an optimal point never can contain a vector belonging to the interior of the ordering cone. ¿From that property, BRECKNER's multiplier rule can be deduced using a simple separation theorem. Also in the original proof, there is performed a separation of two sets. But these sets are constructed rather artificially. So, our method of proof seems to be more evident.

For simplicity of the proof, we assume $K_3 = \{0^{m_3}\}$. As we will see in section 3, any problem that meets the requirements initially made can be reformulated equivalently fulfilling this assumption. By this, in the proof a result like proposition 1 can be replaced by a classical implicit function theorem.

All we have to do is to show the following proposition being true.

Proposition 2. *Let $x_0 \in X$ be a local weakly K_1-mimimal point for the function f_1 over S and let $D \in R^m$ be an K-derived convex cone for f at x_0. Then there exists a sequence $\{b_k\} \subset R^m$ $(k = 1, 2, \dots)$ converging to zero, such that*

$$D \cap (-b_k - (K_1 \times L^* \times \{0^{m_3}\})) = \emptyset \qquad \forall k. \qquad (11)$$

Remark 2. Property (11) implies the origin of R^m being an extremal point of the system $\{D, (-K_1) \times (-L^*) \times \{0^{m_3}\}\}$ in the sense of KRUGER and MORDUKHOVICH[1] [7].

---

[1] A point x_0 is called to be an extremal point of the system $\{C_1, \dots, C_n\}$, if $x_0 \in \bigcap_{i=1}^{n} C_i$ and there are sequences $\{a_{ik}\}$ $(k = 1, 2, \dots)$ converging to zero such that

$$\bigcap_{i=1}^{n}(C_i - a_{ik}) = \emptyset \qquad (k = 1, 2, \dots).$$

By (11), we can separate[2] the convex sets D and $-b_k + (-K_1) \times (-L^*) \times \{0^{m_3}\}$. Hence, there exists for each $k \in \{1, 2, \dots\}$ a multiplier vector $\lambda_k = (\lambda_{k1}, \lambda_{k2}, \lambda_{k3}) \in R^{m_1} \times R^{m_2} \times R^{m_3}$, $\|\lambda_k\| = 1$, fulfilling

$$\langle \lambda_k, d \rangle \geq 0 \qquad \forall d \in D,$$
$$\langle \lambda_{k1}, y \rangle \geq \langle \lambda_{k1}, b_{k1} \rangle \qquad \forall y \in K_1,$$
$$\langle \lambda_{k2}, y \rangle \geq \langle \lambda_{k2}, b_{k2} \rangle \qquad \forall y \in L^*.$$

Because of the compactness of the finite–dimensional unit ball, the sequence $\{\lambda_k\}$ converges to a certain vector $\lambda = (\lambda_1, \lambda_2, \lambda_3) \in R^{m_1} \times R^{m_2} \times R^{m_3}$, $\lambda \neq 0$, with

$$\langle \lambda, d \rangle \geq 0 \qquad \forall d \in D,$$
$$\langle \lambda_1, y \rangle \geq 0 \qquad \forall y \in K_1,$$
$$\langle \lambda_2, y \rangle \geq 0 \qquad \forall y \in L^*,$$

hence $\lambda_1 \in K_1^*$, $\lambda_2 \in L \subseteq K_2^*$, and

$$\langle \lambda_2, f_2(x_0) \rangle = 0.$$

For to prove proposition 2, we need the following lemma.

Lemma 2. *Let $D \in R^m$ be an K-derived convex cone for a function $f : X \to R^m$ at $x_0 \in S$. If there are some $d_1 \in -int\ K_1$, and $d_2 \in -int\ L^*$ with*

$$(d_1, d_2, d_3) \in D \qquad \forall d_3 \in R^{m_3},$$

so there exists a sequence $\{x_n\} \in X$ ($n = 1, 2, \dots$) converging to x_0 such that

$$f_1(x_n) - f_1(x_0) \in -int\ K_1 \qquad \forall n, \tag{12a}$$
$$f_2(x_n) \in -K_2 \qquad \forall n, \tag{12b}$$
$$f_3(x_n) = 0 \qquad \forall n. \tag{12c}$$

Proof. Let

$$a_i := \begin{pmatrix} d_1 \\ d_2 \\ 0^{m_3} \end{pmatrix} + \begin{pmatrix} 0^{m_1} \\ 0^{m_2} \\ e_i \end{pmatrix} \qquad (i = 1, \dots m_3)$$

and

$$a_0 := \begin{pmatrix} d_1 \\ d_2 \\ 0^{m_3} \end{pmatrix} - \sum_{i=1}^{m_3} \begin{pmatrix} 0^{m_1} \\ 0^{m_2} \\ e_i \end{pmatrix}.$$

---

[2] Relation (11) was received searching for fundamental properties of K-derived sets at optimal points. The autor wishes to express his thanks to Prof. W. W. Breckner for the idea of applying a separation theorem on this in order to prove Theorem 1.

(Here, e_i denotes the i-th unit vector of the space R^{m_3}.) By assumptions, all a_i $(i = 0, 1, \ldots, m_3)$ belong in D. According to Definition 2, for the $(m+1)$-tuple $\{a_0, a_1, \ldots, a_{m_3}, 0^{m_1}, 0^{m_2}\}$ we find some r, ω, ϱ with the properties (A), (B), (C1), (C2), (C3). For short, we set

$$\bar{\omega}(t) := \omega(t, 0^{m_1}, 0^{m_2}) : \ B_+^{m_3+1}(r) \to X,$$
$$\bar{\varrho}(t) := \varrho(t, 0^{m_1}, 0^{m_2}) : \ B_+^{m_3+1}(r) \to R^m.$$

Then, (A) implies especially for $(t, 0^{m_1}, 0^{m_2})$ with $t \in B_+^{m_3+1}(r)$

$$f_1(\bar{\omega}(t)) - f_1(x_0) - \left(\sum_{i=0}^{m_3} t^i\right) d_1 - \|t\|\bar{\varrho}_1(t) \in -K_1 \ \forall t \in B_+^{m_3+1}(r), \tag{13a}$$

$$f_2(\bar{\omega}(t)) - f_2(x_0) - \left(\sum_{i=0}^{m_3} t^i\right) d_2 - \|t\|\bar{\varrho}_2(t) \in -K_2 \ \forall t \in B_+^{m_3+1}(r), \tag{13b}$$

$$f_3(\bar{\omega}(t)) - \sum_{i=0}^{m_3} t^i a_{i3} - \|t\|\bar{\varrho}_3(t) = 0 \ \forall t \in B_+^{m_3+1}(r). \tag{13c}$$

Now, we can prove the following three assertions.

Assertion 1. *There is a constant $r_1 \in (0, r]$, such that*

$$f_1(\bar{\omega}(t)) - f_1(x_0) \in -int \ K_1 \qquad \forall t \in B_+^{m_3+1}(r_1) \setminus \{0\}. \tag{14}$$

By (C1), there exists a point $y_1^0 \in K_1$ so that $\forall \varepsilon > 0 \ \exists r_\varepsilon \in (0, r]$ with

$$\bar{\varrho}_1(t) \in \varepsilon y_1^0 - K_1 \qquad \forall t \in B_+^{m_3+1}(r_\varepsilon).$$

Due to $d_1 \in -int \ K_1$, one can find $\varepsilon_0 > 0$ sufficiently small such that

$$\frac{\sum_{i=0}^{m_3} t^i}{\|t\|} d_1 + \varepsilon_0 y_1^0 \in -int \ K_1 \quad \forall t = (t^0, t^1, \ldots, t^{m_3}) \in R_+^{m_3+1} \setminus \{0\}.$$

With $r_1 := r_{\varepsilon_0}$ it follows from (13a)

$$f_1(\bar{\omega}(t)) - f_1(x_0) \in \left(\sum_{i=0}^{m_3} t^i\right) d_1 + \|t\|\bar{\varrho}_1(t) - K_1$$

$$\subseteq \left(\sum_{i=0}^{m_3} t^i\right) d_1 + \varepsilon_0\|t\|y_1^0 - K_1$$

$$\subseteq -int \ K_1 \qquad \forall t \in B_+^{m_3+1}(r_1) \setminus \{0\}.$$

Assertion 2. *There is a constant $r_2 \in (0, r]$, such that*

$$f_2(\bar{\omega}(t)) \in -K_2 \qquad \forall t \in B_+^{m_3+1}(r_2). \tag{15}$$

Regarding (C2) and $d_2 \in -int\, L^*$, analogously to the proof of Assertion 1 one can find a constant $r' \in (0, r]$ such that

$$\left(\sum_{i=0}^{m_3} t^i \right) d_2 + \|t\| \bar{\varrho}_2(t) \in -int\, L^* \qquad \forall t \in B_+^{m_3+1}(r') \setminus \{0\}.$$

Given a multiplier $\mu \in L$, it follows

$$\left\langle \mu, f_2(x_0) + \left(\sum_{i=0}^{m_3} t^i \right) d_2 + \|t\| \bar{\varrho}_2(t) \right\rangle \leq 0 \qquad \forall t \in B_+^{m_3+1}(r').$$

For every multiplier $\mu \in K_2^* \setminus L$, we also find because of $\langle \mu, f_2(x_0) \rangle < 0$ a sufficient small constant $r_\mu > 0$ such that

$$\left\langle \mu, f_2(x_0) + \left(\sum_{i=0}^{m_3} t^i \right) d_2 + \|t\| \bar{\varrho}_2(t) \right\rangle \leq 0 \qquad \forall t \in B_+^{m_3+1}(r_\mu).$$

Thus, there exists to every $\mu \in K_2^*$ a $r_\mu \in (0, r]$, such that the above inequality holds. Since the cone K_2^* is generated by a compact base, we can fix a $r_2 \in (0, r]$ independent from μ such that

$$\left\langle \mu, f_2(x_0) + \left(\sum_{i=0}^{m_3} t^i \right) d_2 + \|t\| \bar{\varrho}_2(t) \right\rangle \leq 0 \quad \forall t \in B_+^{m_3+1}(r_2) \quad \forall \mu \in K_2^*.$$

It follows

$$f_2(x_0) + \left(\sum_{i=0}^{m_3} t^i \right) d_2 + \|t\| \bar{\varrho}_2(t) \in -K_2 \qquad \forall t \in B_+^{m_3+1}(r_2),$$

and by (13b)
$$f_2(\bar{\omega}(t)) \in -K_2 \qquad \forall t \in B_+^{m_3+1}(r_2).$$

Assertion 3. There is a sequence $\{t_n\} \subset B_+^{m_3+1}(r) \setminus \{0\}$ convering to zero, such that

$$f_3(\bar{\omega}(t_n)) = 0 \qquad \forall n. \tag{16}$$

This step requires an implicit function theorem. We will apply Theorem 8.2. in [6]. Set $h = (1, \ldots, 1) \in R^{m_3+1}$. Choose a $\delta > 0$, such that

$$\|\tau(h + \sigma)\| \leq r \qquad \forall \tau \in [0, \delta] \quad \forall \sigma \in B^{m_3+1}(1).$$

Because of $h + \sigma \in R_+^{m_3+1}$ for all $\sigma \in B^{m_3+1}(1)$, we also have

$$\tau(h + \sigma) \in B_+^{m_3+1}(r) \qquad \forall \tau \in [0, \delta] \quad \forall \sigma \in B^{m_3+1}(1).$$

Consider the vector–valued function $\Phi : B^{m_3+1}(1) \times [0, \delta] \to R^{m_3}$, defined componentwise by

$$\Phi^k(\sigma, \tau) := \begin{cases} \dfrac{1}{\tau} f_3^k(\bar{\omega}(\tau(h + \sigma))), & \text{if } \tau > 0 \\[3mm] \displaystyle\sum_{i=0}^{m_3} a_{i3}^k \sigma^i, & \text{if } \tau = 0 \end{cases} \qquad (k = 1, \ldots, m_3).$$

Obviously Φ is continuous everywhere, even in $\tau = 0$ cause it is regarding (13c)

$$\begin{aligned} \lim_{\tau \to +0} \Phi^k(\sigma, \tau) &= \lim_{\tau \to +0} \frac{f_3^k(\bar{\omega}(\tau(h + \sigma)))}{\tau} \\ &= \lim_{\tau \to +0} \left(\frac{\sum_{i=0}^{m_3} \tau(h^i + \sigma^i) a_{i3}^k}{\tau} + \frac{\|\tau(h + \sigma)\|}{\tau} \bar{\varrho}_3^k(\tau(h + \sigma)) \right) \\ &= \sum_{i=0}^{m_3} (h^i + \sigma^i) a_{i3}^k + \|(h + \sigma)\| \lim_{\tau \to +0} \bar{\varrho}_3^k(\tau(h + \sigma)) \\ &= \sum_{i=0}^{m_3} \sigma^i a_{i3}^k \qquad (k = 1, \ldots, m_3) \end{aligned}$$

uniformly for all $\sigma \in B^{m_3+1}(1)$. Moreover it holds

$$\Phi(\sigma, 0) = A\sigma$$

with

$$A = \begin{pmatrix} -1 & 1 & \cdots & 0 \\ \vdots & \vdots & \ddots & \vdots \\ -1 & 0 & \cdots & 1 \end{pmatrix} \in R^{m_3 \times (m_3+1)},$$

where the matrix A has rank m_3. Applying the implicit function theorem mentioned above on Φ, there exist a constant $\delta' \in (0, \delta]$, and a function $\sigma : [0, \delta'] \to B^{m_3+1}(1)$ with

$$\Phi(\sigma(\tau), \tau) = 0 \qquad \forall \tau \in [0, \delta'] \tag{17}$$

and

$$\lim_{\tau \to +0} \sigma(\tau) = \sigma(0) = 0. \tag{18}$$

Let $\{\tau_n\} \subset (0, \delta']$ $(n = 1, 2, \ldots)$ be a sequence of positive numbers converging to zero, and let $\{t_n\} \subset R^{m_3+1}$ be the sequence defined by

$$t_n = \tau_n(h + \sigma(\tau_n)) \qquad (n = 1, 2, \ldots).$$

This sequence fulfills $\{t_n\} \subset B_+^{m_3+1}(r)$, and due to $\tau_n > 0 \ \forall n$ and (18) moreover

$$t_n \in B_+^{m_3+1}(r) \setminus \{0\} \qquad \forall n.$$

Furthermore we have $\lim\limits_{n \to \infty} t_n = 0$, and by (17) finally

$$f_3(\bar{\omega}(t_n)) = f_3(\bar{\omega}(\tau_n(h + \sigma(\tau_n))))$$
$$= \tau_n \Phi(\sigma(\tau_n), \tau_n)$$
$$= 0 \qquad \forall n.$$

Since $\{t_n\}$ converges to zero, there is a number n_0 such that for all $n > n_0$ it holds $t_n \in B_+^{m_3+1}(\min\{r_1, r_2\})$. So, the sequence $\{x_n\} \subset X$ defined by

$$x_n := \bar{\omega}(t_{n+n_0}) \qquad (n = 1, 2, \dots).$$

converges by (B) to x_0 and fulfills by (14), (15), and (16) the relations stated in the lemma. $\qquad \square$

Proof. (of Proposition 2) Take a vector $b_1 \in int \, K_1$, and a vector $b_2 \in int \, L^*$. To b_1 and b_2, always one can find a vector $b_3 \in R^{m_3}$ such that

$$- \begin{pmatrix} b_1 + K_1 \\ b_2 + L^* \\ b_3 \end{pmatrix} \cap D = \emptyset. \tag{19}$$

Else, we would have

$$- \begin{pmatrix} b_1 + K_1 \\ b_2 + L^* \\ b_3 \end{pmatrix} \cap D \neq \emptyset \qquad \forall b_3 \in R^{m_3}.$$

Thus, there would be $d_1 \in -b_1 - K_1 \subseteq -int \, K_1$, and $d_2 \in -b_2 - L^* \subseteq -int \, L^*$ with

$$\begin{pmatrix} d_1 \\ d_2 \\ -b_3 \end{pmatrix} \in D \qquad \forall b_3 \in R^{m_3}.$$

But by Lemma 2, then there must exist a sequence $\{x_n\} \in X$ converging to x_0 such that

$$f_1(x_n) - f_1(x_0) \in -int \, K_1 \qquad \forall n,$$
$$f_2(x_n) \in -K_2 \qquad \forall n,$$
$$f_3(x_n) = 0 \qquad \forall n.$$

This contradicts the local K_1-minimality of x_0. Hence (19) must be true. The sequence $\{b_k\} \subset R^m$ defined by

$$b_k := \frac{1}{k} \begin{pmatrix} b_1 \\ b_2 \\ b_3 \end{pmatrix} \qquad (k = 1, 2, \dots)$$

converges to zero and fulfills by (19) property (11). $\qquad \square$

3   An additional complementarity condition

In BRECKNER's Multiplier rule, to the constraint $f_2(x) \in -K_2$, there corresponds a complementary slackness condition $\langle \lambda_2, f_2(x_0) \rangle = 0$. We asked if there is also such a property (6) valid for $f_3(x) \in -K_3$.

The only difference between the two constraints is the demand on K_2 to have a nonempty interior. If the cone K_3 has a nonempty interior, too, equation (6) must hold by reasons of analogy. If, on the other hand, we have $K_3 = \{0\}$, equation (6) holds trivially. So, the problem is to show the desired complementarity condition for cones K_3 with $int\, K_3 = \emptyset$, but $K_3 \neq \{0\}$. We show how that case can be reduced on the two easy cases.

Let $M := span(K_3)$ be the subspace of R^{m_3} spanned up by the elements of K_3. Set $k := dim(M)$. Obviously $0 \leq k \leq m_3$. (Here, if $int\, K_3 \neq \emptyset$, it is $k = m_3$, whereas if $K_3 = \{0\}$, it is $m_3 = 0$.) Let N be the orthogonal complement of M with respect to R^{m_3}. Choose an orthonormal basis $\{\eta_1, \ldots, \eta_k\}$ of M as well as orthonormal basis $\{\eta_{k+1}, \ldots, \eta_{m_3}\}$ of N. Then $\{\eta_1, \ldots, \eta_{m_3}\}$ is an orthonormal basis of R^{m_3}.

Consider now the orthonormal transformation[3] $A : R^{m_3} \to R^{m_3}$ defined by

$$A(\eta_i) = e_i \qquad (i = 1, \ldots, m_3),$$

where e_i is the i-th unit vector of the space R^{m_3}. Let be $\tilde{K}_3 \in R^k$ the set generated by the first k elements of the image of K_3 transformed by A;

$$\tilde{K}_3 := \left\{ \begin{pmatrix} A^1(y) \\ \vdots \\ A^k(y) \end{pmatrix} ; y \in K_3 \right\}.$$

Of course, $\tilde{K}_3$ is a closed convex cone, too. By construction, it is $span(\tilde{K}_3) = R^k$, hence $int\, \tilde{K}_3 \neq \emptyset$ relatively to the space R^k. Since $K_3 \subseteq M \perp N$, we have $A^i(y) = 0$ $(i = k+1, \ldots, m_3)$ for every $y \in K_3$. That means

$$A[K_3] = \tilde{K}_3 \times \{0^{m_3-k}\}. \tag{20}$$

We define a function $\tilde{f}_2 : X \to R^{m_2+k}$ by

$$\tilde{f}_2(x) := \begin{pmatrix} f_2(x) \\ A^1(f_3(x)) \\ \vdots \\ A^k(f_3(x)) \end{pmatrix} \qquad \forall x \in X,$$

---

[3] The author is indebted to an anonymous referee for a hint on formulating the transformation mapping in such a compact way.

as well as a function $\tilde{f}_3 : X \to R^{m_3-k}$ by

$$\tilde{f}_3(x) := \begin{pmatrix} A^{k+1}(f_3(x)) \\ \vdots \\ A^{m_3}(f_3(x)) \end{pmatrix} \qquad \forall x \in X,$$

and a cone $\tilde{K}_2 \subset R^{m_2+k}$ with $int\, \tilde{K}_2 \neq \emptyset$ by

$$\tilde{K}_2 := K_2 \times \tilde{K}_3.$$

Thus, the set

$$S = \{x \in X;\ f_2(x) \in -K_2, f_3(x) \in -K_3\} \tag{21}$$

can be described equivalently by

$$S = \{x \in X;\ \tilde{f}_2(x) \in -\tilde{K}_2, \tilde{f}_3(x) = 0^{m_3-k}\}, \tag{22}$$

where the cone $\tilde{K}_2$ has a nonempty interior relatively to R^{m_2+k}.

As shown above, every problem fulfilling the requirements initially made can be transformed into a problem with the feasible set in form (22). So, the property $K_3 = \{0\}$ can be assumed for each such problem without loss of generality. Then the question about a relation like (6) becomes uninterestingly. But, from an other point of view, we can construct by this transformation also such an additional complementarity relation for problems with a cone K_3 of general nature.

Given a problem with a feasible set in form (21), let $x_0 \in X$ be a local weakly K_1-mimimal point for the function f_1 over S, and let $D \in R^m$ be an K-derived convex cone for f at x_0. It is easily seen, that

$$\tilde{D} := D_1 \times \begin{pmatrix} D_2 \\ A^1[D_3] \\ \vdots \\ A^k[D_3] \end{pmatrix} \times \begin{pmatrix} A^{k+1}[D_3] \\ \vdots \\ A^{m_3}[D_3] \end{pmatrix}$$

is a $(K_1 \times \tilde{K}_2 \times \{0^{m_3-k}\})$-derived cone for the function $(f_1, \tilde{f}_2, \tilde{f}_3)$ at x_0.

Applying Theorem 1 on the transformed problem, there exists a multiplier vector $\tilde{\lambda} = (\tilde{\lambda}_1, \tilde{\lambda}_2, \tilde{\lambda}_3) \in K_1^* \times \tilde{K}_2^* \times R^{m_3-k}$, $\tilde{\lambda} \neq 0$, such that

$$\langle \tilde{\lambda}, d \rangle \geq 0 \qquad \forall d \in \tilde{D},$$
$$\langle \tilde{\lambda}_2, \tilde{f}_2(x_0) \rangle = 0.$$

Setting

$$\lambda_1 := \tilde{\lambda}_1,$$

$$\lambda_2 := \begin{pmatrix} \tilde{\lambda}_2^1 \\ \vdots \\ \tilde{\lambda}_2^{m_2} \end{pmatrix},$$

$$\lambda_3 := A^\top \begin{pmatrix} \tilde{\lambda}_2^{m_2+1} \\ \vdots \\ \tilde{\lambda}_2^{m_2+k} \\ \tilde{\lambda}_3 \end{pmatrix},$$

we get $\lambda_1 \in K_1^*$, $\lambda_2 \in K_2^*$, $\lambda = (\lambda_1, \lambda_2, \lambda_3) \neq 0$, and

$$\langle \lambda, d \rangle \geq 0 \qquad \forall d \in D,$$
$$\langle \lambda_2, f_2(x_0) \rangle = 0.$$

But furthermore, from (20) it follows

$$\lambda_3 \in A^\top [\tilde{K}_3^* \times R^{m_3-k}] \subseteq K_3^*,$$

and finally

$$\langle \lambda_3, f_3(x_0) \rangle = \langle A(\lambda_3), A(f_3(x_0)) \rangle$$
$$= \left\langle \begin{pmatrix} \tilde{\lambda}_2^{m_2+1} \\ \vdots \\ \tilde{\lambda}_2^{m_2+k} \end{pmatrix}, \begin{pmatrix} \tilde{f}_2^{m_2+1} \\ \vdots \\ \tilde{f}_2^{m_2+k} \end{pmatrix} \right\rangle + \langle \tilde{\lambda}_3, \tilde{f}_3(x_0) \rangle$$
$$= 0.$$

By this deduction we are able to propose:

Theorem 2. *Let $x_0 \in X$ be a local weakly K_1-minimal point for the function f_1 over S and let $D \in R^m$ be an K-derived convex cone for f at x_0. Then there exists a multiplier vector $\lambda = (\lambda_1, \lambda_2, \lambda_3) \in K_1^* \times K_2^* \times K_3^*$, $\lambda \neq 0$, such that*

$$\langle \lambda, d \rangle \geq 0 \qquad \forall d \in D,$$
$$\langle \lambda_2, f_2(x_0) \rangle = 0,$$
$$\langle \lambda_3, f_3(x_0) \rangle = 0.$$

4 Regularity conditions

Usually a multiplier rule is equipped with a so–called regularity condition. A regularity condition ensures that the multiplier corresponding with the objective function does not vanish. For Theorem 1 we were missing such a condition. The idea for the following result was taken from a regularity condition by HESTENES [6] (Corollary on page 377).

Theorem 3. *Let the assumptions of Theorem 1 be fulfilled.*

(a) If there is a vector $e \in int\ K_1$ with

$$(-e, 0, 0) \in D + K_1 \times L^* \times K_3,$$

so for the multiplier λ in Theorem 1 it holds $\lambda_1 = 0$.
(b) If there is an arbitrary vector $e \in R^{m_1}$ with

$$(-e, 0, 0) \notin D + K_1 \times L^* \times K_3,$$

so in Theorem 1 a multiplier λ can be chosen with $\lambda_1 \neq 0$.

Proof. *(a)* Since $(-e, 0, 0) \in D + K_1 \times L^* \times K_3$, there is a vector $(x_1, x_2, x_3) \in K_1 \times L^* \times K_3$, such that

$$(-e, 0, 0) - (x_1, x_2, x_3) \in D.$$

Let be λ the multiplier existing along Theorem 1. Then it is

$$\langle \lambda, (-e, 0, 0) - (x_1, x_2, x_3) \rangle \geq 0.$$

Because of $\lambda_1 \in K_1^*$, $\lambda_2 \in K_2^*$, and $\lambda_3 \in K_3^*$, it follows

$$
\begin{aligned}
\langle \lambda_1, e \rangle &= -\langle \lambda, (-e, 0, 0) \rangle \\
&\leq -\langle \lambda, (x_1, x_2, x_3) \rangle \\
&= -\langle \lambda_1, x_1 \rangle - \langle \lambda_2, x_2 \rangle - \langle \lambda_3, x_3 \rangle \\
&\leq 0.
\end{aligned}
$$

But we have $\langle \lambda_1, x \rangle \geq 0\ \forall x \in K_1$, hence $\langle \lambda_1, e \rangle = 0$. By $e \in int\ K_1$ it follows $\lambda_1 = 0$.
(b) Every such vector can be written as

$$(-e, 0, 0) = x - \lambda$$

with

$$x \in D + K_1 \times L^* \times K_3,$$
$$-\lambda \in -[D + K_1 \times L^* \times K_3]^*,$$

and $\langle \lambda, x \rangle = 0$. It follows $\lambda \neq 0$, else $(-e, 0, 0) = x \in D + K_1 \times L^* \times K_3$. Due to $D \subseteq D + K_1 \times L^* \times K_3$ or $[D + K_1 \times L^* \times K_3]^* \subseteq D^*$, it follows $\lambda \in D^*$, hence

$$\langle \lambda, d \rangle \geq 0 \qquad \forall d \in D.$$

Due to $K_1 \times L^* \times K_3 \subseteq D + K_1 \times L^* \times K_3$ or $[D + K_1 \times L^* \times K_3]^* \subseteq K_1^* \times L \times K_3^*$, we have

$$\lambda \in K_1^* \times L \times K_3^*.$$

Moreover it is

$$\langle \lambda_1, e \rangle = -\langle \lambda, (-e, 0, 0) \rangle = -\langle \lambda, x \rangle + \langle \lambda, \lambda \rangle = \|\lambda\|^2 > 0.$$

Assuming $\lambda_1 = 0$, we would get a contradiction.   □

Remark 3. Indeed, the conditions in (a) und (b) are contradictory. If there is a vector $e' \in int K_1$ with $(-e', 0, 0) \in D + K_1 \times L^* \times K_3$, so there cannot be a vector $e'' \in R^{m_1}$ with $(-e'', 0, 0) \notin D + K_1 \times L^* \times K_3$, and vice versa.

Proof. Let be given $e' \in int K_1$ with $(-e', 0, 0) \in D + K_1 \times L^* \times K_3$. That means $-e' \in D_1 + K_1$. Additionally we have $e' \in D_1 + int K_1 \subseteq int(D_1 + K_1)$. So, it is

$$0 = e' - e' \in int(D_1 + K_1) + (D_1 + K_1) \subseteq int(D_1 + K_1),$$

hence $D_1 + K_1 = R^{m_1}$. Let us assume there would be $e'' \in R^{m_1}$ with $(-e'', 0, 0) \notin D + K_1 \times L^* \times K_3$. That would mean $-e'' \notin D_1 + K_1$, contradicting to $D_1 + K_1 = R^{m_1}$. So, it is shown

$$\exists e' \in int K_1 : (-e', 0, 0) \in D + K_1 \times L^* \times K_3$$
$$\implies \not\exists e'' \in R^{m_1} : (-e'', 0, 0) \notin D + K_1 \times L^* \times K_3,$$

and by negation

$$\exists e'' \in R^{m_1} : (-e'', 0, 0) \notin D + K_1 \times L^* \times K_3$$
$$\implies \not\exists e' \in int K_1 : (-e', 0, 0) \in D + K_1 \times L^* \times K_3.$$

□

References

1. Breckner, W. W. (1994) Derived sets in multiobjective optimization. Z. Anal. Anw. **13**, 725–738
2. Breckner, W. W., Göpfert, A. (1996) Multiplier rules for weak pareto optimization problems. Optimization **38** No. 1, 23–37
3. Gittleman, W. (1971) A general multiplier rule. JOTA **7**, 29–38
4. Hestenes, M. R. (1965) On variational theory and optimal control theory. SIAM J. Control **3**, 23–48
5. Hestenes, M. R. (1966) Calculus of variations and optimal control theory. John Wiley & Sons, New York London Sydney Toronto

6. Hestenes, M. R. (1975) Optimization theory: The finite dimensional case. John Wiley & Sons, New York London Sydney Toronto
7. Kruger, A. Ya., Mordukhovich, B. Sh. (1978) Minimization of nonsmooth functionals in optimal control problems. Engineerg. Cybern. **16**, 126–133
8. Nieuwenhuis, J. W. (1980) A general multiplier rule. JOTA **31**, 167–176
9. Tuy, H. (1974) On the convex approximation of nonlinear inequalities. Math. Operationsforsch. Statist. **5**, Heft 6, 451–466

The Maximum Flow in a Time-Varying Network[*]

D. Sha[1], X. Cai[1], and C. K. Wong[2][**]

[1] Department of System Engineering and Engineering Management, The Chinese University of Hong Kong, Shatin, N.T., Hong Kong
[2] Department of Computer Science and Engineering, The Chinese University of Hong Kong, Shatin, N.T., Hong Kong

Abstract. The classical maximum flow problem is to send flow from the source to the sink as much as possible. The problem is static, in that arc capacity is a constant. Max-flow min-cut theorem of Fold and Fulkerson [1] is a fundamental result which reveals the relationship between the maximum flow and the minimum cut in a static network. It has played an important role not only in investigating the classical maximum flow problem, but also in developing graph theory, since many results could be regarded as its corollaries.

In practical situations, there are, however, numerous problems where the attributes of the network under consideration are time-varying. In such a network, a flow must take a certain time to traverse an arc. The transit time on an arc and the capacity of an arc are all time-varying parameters. To depart at the best time, a flow can wait at the beginning vertex of an arc, which is however constrained by a time-varying vertex capacity. For instance, consider a network in which several cargo-transportation services are available between a number of cities. These may include air transportation, sea transportation, and road transportation, which are available at different times and have different capacities. Waiting at a city is permitted, but it is limited by the space of the warehouse, and dependent upon the time. A question often asked is: what is the maximum flow that can be sent between two particular cities within a certain duration T?

We show that this time-varying maximum flow problem is NP-complete. We then establish the max-flow min-cut theorem in the time-varying version. With this result, we develop an approach, which can solve the time-varying maximum flow problem with arbitrary waiting time at vertices in a pseudopolynomial time.

Key words: maximum flow, time-varying network

1   Introduction

The classical maximum network flow problem is to send as much flow from one particular vertex (the source) to another (the sink) as possible. The problem

---

[*] This research was partially supported by the Research Grants Council of Hong Kong under Earmarked Grant No. CUHK 278/94E and Earmarked Grant No. CUHK 4135/97E.
[**] On leave from IBM T.J. Watson Research Center, Yorktown Heights NY 10598, U.S.A.

is static, in that both arc capacity and vertex capacity are fixed. In practical situations, there are, however, numerous problems where the attributes of the network under consideration are time-varying. For instance, consider a network in which several transportation services are available between a pair of cities. These may include air transportation, sea transportation, and road transportation, which are available at different times and have different capacities. A question often asked is: what is the maximum flow that can be sent between two particular cities within a certain duration T? This is a time-varying maximum flow problem. Its solution is essential to the planning of the network.

A nontrivial extension of the classical problem is the maximal dynamic flow model formulated and solved by Ford and Fulkerson [1], where the transit time to traverse an arc is taken into consideration. Nevertheless, their model still assumes that all attributes in the problem, including arc capacities and transit times, are static parameters which do not change over time. For this problem, Ford and Fulkerson have developed an efficient procedure to find the optimal solution, which first finds the static flow from the source to the sink, and then develops a set of temporally repeated flows, with the optimal flow decomposed into a set of chain flows. A further extension is addressed by Halpern [2], where arc capacities vary with time and storages at intermediate vertices may be prohibited at some times. When the time is considered as a variable taking discrete values, both these problems can be solved by constructing an equivalent, static time-expanded network (see, e.g., [4]).

In this paper, we will study such a maximal flow model, where transit times, arc capacities and vertex capacities are all time-varying. This is a new model, for which we have to develop new algorithms to find solutions. As we have mentioned above, both in Ford and Fulkerson's and Halpern's models, the transit times are not time-varying. The approaches they derived for these models cannot be applied to the time-varying model we want to address. In addition, although in our model we shall consider discrete times, the well-known approach of creating an equivalent static time-expanded network to model the dynamic network is also not applicable to solve our problem. The following example explains this argument. Consider the case where a waiting time is allowed at each vertex x. Thus, an optimal solution should specify how long a flow f should wait at x, when it arrives at time t. The waiting time is a decision variable, which depends on both f and the arriving time t. Such a decision variable can not be directly incorporated into the static time-expanded network as described in [1].

The organization of the remainder of this article is as follows. In Section 2, some basic concepts and the problem formulation will be introduced. The problem is shown to be NP-complete in Section 3. Section 4 will be devoted to develop label-setting algorithms which can find optimal solutions for time-varying maximum flow problem with or without vertex capacity limit, respectively. Finally, some concluding remarks will be given in Section 5.

2 Problem Formulation

Let $N = (V, A, b, c, l)$ be a network without parallel arcs and loops, where V is the set of vertices, A is the set of arcs, $b(x, y, t)$ is the transit time sending one unit of flow through arc (x, y), $c(x, y, t)$ is the capacity of arc (x, y), and $l(x, t)$ is the capacity of vertex x. Both $b(x, y, t)$ and $c(x, y, t)$ are functions of the departure time t when the flow starts to traverse the arc (x, y), where $t = 0, 1, ..., T$, and $T > 0$ is a given number. The vertex capacity $l(x, t)$ is a function of the time t when the flow arrives at the vertex x. We assume that c and l are nonnegative and the transit time b is positive. We further assume that there are two particular vertices s and ρ, being the source vertex and the sink vertex, respectively. Throughout the article, we let $n = |V|$ and $m = |A|$.

Without ambiguity, we let $f(x, y, t)$ be the value of the flow departing at time t to traverse the arc (x, y), and $f(\lambda, T)$ the total flow value under the solution λ, which specifies when and how to send flows from the source s to the sink ρ within the time limit T. Clearly,

$$f(\lambda, T) = \sum_{(x,\rho) \in A, t+b(x,\rho,t) \leq T} f(x, \rho, t)$$

is the value of flows sent from s to ρ no later than the time T.

The main problem of interest in this paper is to find the optimal λ^* so as to maximize $f(\lambda^*, T)$. We say the problem is *time-varying maximum flow problem* (TMFP).

Recall that the basic idea of the maximum flow algorithm (Ford and Fulkerson, 1956) is to find an f-augmenting path from the source node to the sink node in the residual network and then send as much flow along the path as possible. We will, in this paper, adopt a similar idea to tackle our problem. Nevertheless, as *time* plays an essential role in our model, many concepts will have to be generalized to incorporate those time-varying factors. Specifically, there are two major questions we have to answer:

(1) How to define and find *a feasible dynamic f-augmenting path*, where such a path is said to be feasible if the flow arriving times and departing times at all internal vertices are matched?

(2) How to define and generate *a dynamic residual network* after a feasible dymanic f-augmenting path is found?

We need the following definitions.

Definition 1. *Let $P(s, x) = (s = x_1, ..., x_r = x)$ be a directed path from s to x and assume that a flow traverses the path. Let $w(x_i), i = 1, ..., r$, be waiting* *times of the flow at vertices* $x_1, ..., x_r$, *respectively. Let* $\tau(x_1) = w(x_1)$ *and define recursively*

$$\tau(x_i) = w(x_i) + \tau(x_{i-1}) + b(x_{i-1}, x_i, \tau(x_{i-1})) \qquad \text{for } i = 2, ..., r.$$

The departure time of the flow at a vertex x_i *on* P *is defined as* $\tau(x_i)$ *if* $1 \leq i < r$. *Accordingly, the arrival time of the flow at a vertex* x_i *on* P *is defined as*

$$\alpha(x_1) = 0; \quad \alpha(x_i) = \tau(x_{i-1}) + b(x_{i-1}, x_i, \tau(x_{i-1})) \qquad \text{for } i = 2, ..., r.$$

P *with all waiting times* $w(x_i)$ *is said a dynamic path from* s *to* x.

Definition 2. *Let* $P(s, x)$ *be a dynamic path from* s *to* x. *The time of* P *is defined as* $\alpha(x)$. *A path is said to have time at most* t, *if the time of the path is no greater than* t.

Definition 3. *Let* $P(s, x) = (s = x_1, x_2, ..., x_r = x)$ *be a directed path from* s *to* x. *Given* $\tau(x_i)$, $\alpha(x_i)$ *and* $w(x_i)$, $i = 1, ..., r$, $P(s, x)$ *is said to be a dynamic f-augmenting path from* s *to* x *if, for* $i = 2, ..., r$, *it satisfies*
   (i) $\tau(x_{i-1}) + b(x_{i-1}, x_i, \tau(x_{i-1})) = \alpha(x_i)$;
   (ii) $\alpha(x_i) + w(x_i) = \tau(x_i)$;
   (iii) $c(x_{i-1}, x_i, \tau(x_{i-1})) > 0$;
   (iv) $\displaystyle\prod_{t=0}^{w(x_{i-1})-1} l(x_{i-1}, \alpha(x_{i-1}) + t) > 0$,
where $0 \leq \alpha(x_i) \leq T$ *and* $0 \leq \tau(x_i) \leq T$ *for* $i = 1, ..., r$.

Next we generalize the concept of *residual network*. Let us first create a new network as follows.

For every arc $(x, y) \in A$, create an artificial arc, denoted by $[y, x]$ (note that there may exist a $(y, x) \in A$ already, so we use a different symbol $[y, x]$ here). Associated with it are the transit time $b[y, x, u]$ and capacity $c[y, x, u]$. For arc $[y, x]$ and $t = 0, 1, ..., T$, let $c[y, x, t] = 0$ initially and define transit time $b[y, x, t]$ as:

$$b[y, x, t] = \begin{cases} -b(x, y, u) & 0 \leq t = u + b(x, y, u) \leq T, u = 0, 1, ..., T \\ +\infty & otherwise \end{cases}$$

Note that $b[y, x, t]$ would be a multiple valued function since for some t, there may exist more than one u satisfying $u + b(x, y, u) = t$. However, this will not affect both the correctness of our solution and the application of algorithms proposed in the rest of this paper.

For every vertex $x \in V$, we also define an artificial vertex capacity $l[x, t]$, to represent the capacity under which a flow can be "waiting" at x from time t to $t - 1$. This definition means that a flow may have a negative waiting time at a vertex x. In fact, similar to the definition of a negative transit time

$b[x, y, t]$ that provides a chance to retract a flow on an arc, $l[x, t]$ provides a chance to retract a waiting time of a flow at vertex x. Initially, let $l[x, t] = 0$ for each x and $t = 1, 2, ..., T$.

Obviously, the new network as creates above with the initial settings is equivalent to the original one, thus we still denote it by N.

The concept of a dynamic f-augmenting path (see Definition 3) is further generalized as follows:

Definition 4. Let $P(s, x) = (s = x_1, x_2, ..., x_r = x)$ be a directed path from s to x. Given $\tau(x_i)$, $\alpha(x_i)$ and $w(x_i)$ where $i = 1, ..., r$ and $w(x_i)$ can be negative, $P(s, x)$ is said to be a dynamic f-augmenting path from s to x if for $i = 2, ..., r$, it satisfies

(i) $\tau(x_{i-1}) + b(x_{i-1}, x_i, \tau(x_{i-1})) = \alpha(x_i)$ if (x_{i-1}, x_i) is not an artificial arc; otherwise, $\tau(x_{i-1}) + b[x_{i-1}, x_i, \tau(x_{i-1})] = \alpha(x_i)$;

(ii) $\alpha(x_i) + w(x_i) = \tau(x_i)$;

(iii) $c(x_{i-1}, x_i, \tau(x_{i-1})) > 0$, if (x_{i-1}, x_i) is not an artificial arc; or $c[x_{i-1}, x_i, \tau(x_{i-1})] > 0$, if (x_{i-1}, x_i) is an artificial arc; and

(iv) $\displaystyle\prod_{t=0}^{w(x_{i-1})-1} l(x_{i-1}, \alpha(x_{i-1}) + t) > 0$, if $w(x_{i-1}) > 0$; or

$\displaystyle\prod_{t=0}^{|w(x_{i-1})|-1} l[x_{i-1}, \tau(x_{i-1}) - t] > 0$, if $w(x_{i-1}) < 0$,

where $0 \le \alpha(x_i) \le T$, $0 \le \tau(x_i) \le T$ for $i = 1, ..., r$.

After a dynamic f-augmenting path is found, we can send an augmenting flow along it, and then construct a residual network by the following procedure:

Network Updating Procedure-UPNET

Let $N = (V, A, b, c, l)$ be the network considered. Let $P(s, \rho) = (s = x_1, x_2, ..., x_r = \rho)$ be a dynamic f-augmenting path (as defined in Definition 4) from s to ρ with $\tau(x_i)$, $w(x_i)$ and $\alpha(x_i)$, and $f_p > 0$ the flow value sent along $P(s, \rho)$. For $i = 1, ..., r - 1$, do:

Update arc capacity

Case I: (x_i, x_{i+1}) is not an artificial arc. Let

$$c(x_i, x_{i+1}, \tau(x_i)) := c(x_i, x_{i+1}, \tau(x_i)) - f_p$$
$$c[x_{i+1}, x_i, \alpha(x_{i+1})] := c[x_{i+1}, x_i, \alpha(x_{i+1})] + f_p$$

Case II: (x_i, x_{i+1}) is an artificial arc. Let

$$c(x_{i+1}, x_i, \alpha(x_{i+1})) := c(x_{i+1}, x_i, \alpha(x_{i+1})) + f_p$$
$$c[x_i, x_{i+1}, \tau(x_i)] := c[x_i, x_{i+1}, \tau(x_i)] - f_p$$

For $i = 2, 3, ..., T - 1$ do:

Update vertex capacity

Case I: $w(x_i) > 0$. Let

$$l(x_i, t) := l(x_i, t) - f_p, \quad t = \tau(x_i), \tau(x_i) + 1, ..., \alpha(x_{i+1}) - 1$$
$$l[x_i, t] := l[x_i, t] + f_p, \quad t = \alpha(x_{i+1}), \alpha(x_{i+1}) - 1, ..., \tau(x_i) + 1$$

Case II: $w(x_i) < 0$. Let

$$l[x_i, t] := l[x_i, t] - f_p, \quad t = \tau(x_i), \tau(x_i) - 1, ..., \alpha(x_{i+1}) + 1$$
$$l(x_i, t) := l(x_i, t) + f_p, \quad t = \alpha(x_{i+1}), \alpha(x_{i+1}) + 1, ..., \tau(x_i) - 1.$$

Definition 5. *The updated network generated by the procedure above is said to be a dynamic residual network.*

The problem in the original network and the problem in the dynamic residual network are equivalent in the sense that there is a one-to-one correspondence between their feasible solutions. Notice that, in the original network, we assume that all transit time $b > 0$. Thus, the first dynamic f-augmenting path will only contain arcs with positive transit time b and positive waiting time w. But in a dynamic residual network, the transit time associated with an artificial arc is a negative number, and a flow can be stored at a vertex with a negative waiting time. Thus a dynamic f-augmenting path found in the dynamic residual network may contain negative transit time b and negative waiting time w. An illustrative example is given in Appendix (see Example 1) to help understand the concepts and the procedure.

Next we generalize the concept of *minimum cut* to the time-varying situation we are considering. This is also an important concept in the studies of maximum network flow problems. We will use the generalized concept in the analysis of the efficiency of our algorithms.

Definition 6. *A generalized cut K separating vertex s and ρ is a set-valued function of time defined as*

$$K = \{K(t)|K(t) \subseteq V, s \in K(t), \rho \notin K(t), t = 0, 1, ..., T\}$$

Further, define the capacity of K as

$$capK = \sum_{t=0}^{T-1} \sum_{x \in K(t), x \notin K(t+1)} l(x, t) + \sum_{t=1}^{T} \sum_{x \in K(t), x \notin K(t-1)} l[x, t]$$
$$+ \sum_{t=0}^{T} \sum_{x \in K(t), y \notin K(t')} c(x, y, t) + \sum_{t=0}^{T} \sum_{x \in K(t), y \notin K(t'')} c[x, y, t]$$

if arbitrary waiting time are allowable at vertices, where $t' = t + b(x, y, t)$, $t'' = t + b[x, y, t]$.

The generalized cut can be better understood if one bears in mind the time-expanded network [4]. Similar to the well-known *Max-flow Min-cut Theorem*, a fundmental theorem for the time-varying maximum dynamic flow problem can be obtained as follows:

Theorem 1. Let v be the value of any dynamic flow f on N with waiting times at vertices, and $capK$ the value of any generalized cut K in N. Then, $v \leq capK$.

Proof. Since any generalized cut K separates s and ρ for any dynamic flow f, we have
$$v = f^+(K) - f^-(K)$$
where $f^+(K)$ and $f^-(K)$ are the flow values that flow out of and flow in K respectively. Because f is a dynamic flow on N, it must satisfy all capacity constraints. Let $f[x,t]$ denote the flow value stored at x from the time t to $t-1$, and $f[x,y,t]$ denote the flow value transitted on arc $[x,y]$ starting from time t. Then we have

$$
\begin{aligned}
&0 \leq f(x,t) \leq l(x,t), \\
&0 \leq f[x,t] \leq l[x,t], && \forall x \in V, \forall t \\
&0 \leq f(x,y,t) \leq c(x,y,t), && \forall (x,y) \in A, \forall t \\
&0 \leq f[x,y,t] \leq c[x,y,t], && \forall [x,y] \in A, \forall t.
\end{aligned}
$$

Thus, when arbitrary waiting time is permitted at each internal vertex, we have

$$
f^+(K) = \sum_{t=0}^{T-1} \sum_{x \in K(t), x \notin K(t+1)} f(x,t) + \sum_{t=0}^{T} \sum_{x \in K(t), x \notin K(t+b(x,y,t))} f(x,y,t)
$$

$$
+ \sum_{t=1}^{T} \sum_{x \in K(t), x \notin K(t-1)} f[x,t] + \sum_{t=0}^{T} \sum_{x \in K(t), x \notin K(t+b[x,y,t])} f[x,y,t]
$$

$$
\leq \sum\sum l(x,t) + \sum\sum l[x,t] + \sum\sum c(x,y,t) + \sum\sum c[x,y,t]
$$

$$
= capK.
$$

Noting that $f^-(K) \geq 0$, we have
$$v = f^+(K) - f^-(K) \leq capK.$$

This proves the theorem.   □

3   Computational Complexity

We will discuss the time complexity of TMFP in this section. It is shown that TMFP is NP-complete.

Theorem 2. The time-varying maximum flow problem is NP-complete.

Proof. Obviously, TMFP is in the class of NP. We show that 3-Dimensional Matching problem (3DM) polynomially reduces to TMFP.

3DM is defined as: A set $M \subseteq W \times X \times Y$, where W, X and Y are disjoint sets having the same number q of elements. Does M contain a matching, i.e., a subset $M' \subseteq M$, such that $|M'| = q$ and no two elements of M' agree in any coordinate?

The decision version of TMFP is: Given a time-varying network N, a time limit T, and an integer K, does there exist a dynamic flow f from s to ρ within time T such that $v(f) \geq K$?

For an instance of 3DM, we construct the instance of TMFP as follows: For each element in W, X and Y, we create vertices w_i, x_i and y_i, $1 \leq i \leq q$. All these vertices, added with additional vertices s, ρ, a and l, compose the vertex set V of N. Create arcs (s, a), (l, ρ), (a, w_i) and (y_i, l), $1 \leq i \leq q$, and create arcs (w_i, x_j) and (x_j, y_k) if $(w_i, x_j, y_k) \in M$. All of these arcs compose the arc set A of N. The structure of the network N constructed is shown as in Figure 1. The transit time b and the capacity c are set as follows:

$$b(u, v, t) = 1, \qquad \forall (u, v) \in A, \forall t,$$

$$c(l, \rho, t) = \begin{cases} q & t = 5 \\ 0 & otherwise, \end{cases}$$

$$c(s, a, t) = \begin{cases} q & t = 0 \\ 0 & otherwise, \end{cases}$$

$$c(u, v, t) = 1, \qquad \forall (u, v) \in A \backslash \{(l, \rho), (s, a)\}, \forall t.$$

Finally, let $T = 6$ and $K = q$. In what follows, we will prove that the answer to 3DM and to TMFP are the same under the reduction described above.

Suppose that $M' = \{m_1, m_2, ..., m_q\}$ is a matching of M. For each element $m_i = (w^i, x^i, y^i) \in M'$, we create a dynamic path $P^i = (s, a, w^i, x^i, y^i, l, \rho)$ with departure time 0 from s and arrival time 6 at ρ. Then we can send a subflow, f^i, along P^i with $v(f^i) = 1$, $1 \leq i \leq q$. Union these q subflows to obtain a dynamic flow f. It is a maximum flow in N, since $v(f) = q$ which is the maximum possible flow value in N.

Given a maximum dynamic flow f in N with $v(f) \geq K = q$, then must have $v(f) = q$, since q is the maximum possible flow in N. By the structure of N, there must have q disjoint paths, $P_i = (w^i, x^i, y^i)$ $(1 \leq i \leq q)$, from a to l. Let $m_i = (w^i, x^i, y^i)$. By the reduction, we know $m_i \in M$ and no two of them agree in any coordinate. Let $M' = \{m_1, m_2, ..., m_q\}$. It is a matching of M with $|M'| = q$.

In summary, we have completed the proof.   □

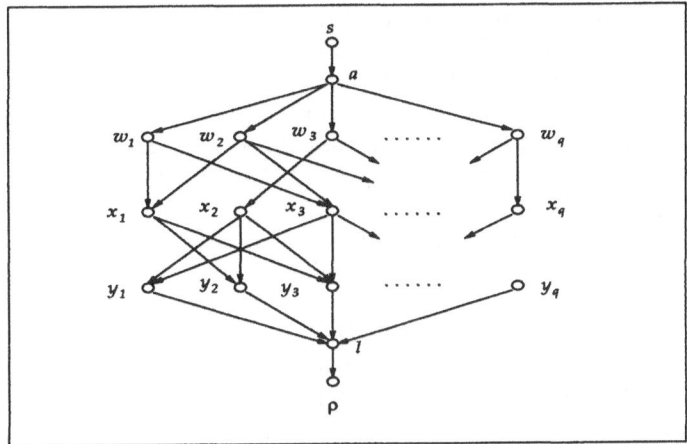

Fig. 1.

4 Algorithms

We consider, in this section, the problem where a flow is allowed to wait at any vertex without any restriction. The problem can be restated as: *Given a network $N(V, A, b, c, l)$ and a time limit T, find the maximum flow from the source vertex s to the sink vertex ρ within time T where waiting at any vertices is allowed without any constraints.*

We examine the case with no capacity limit at each vertex first, and then deal with the case where a capacity limit is imposed at each vertex.

4.1 The Time-Varying Maximum Flow with No Vertex Capacity Limit

The basic idea behind of the algorithm is:

(i) Maintain a set of labels $d(y, 0)$, $d(y, 1)$,..., $d(y, T)$ for each vertex y, where $d(y, t)$ is set to be the predecessor of y in a feasible dynamic f-augmenting path $P(s, y)$ of time at most t if this path exists (for source vertex s, define its predecessor to be 0); otherwise, if such a path does not exist, set $d(y, t) = null$. A vertex y is said to be "labeled" at time t if $d(y, t) \neq null$.

(ii) Initially, set all $d(s, t)$ $(t = 0, 1, ..., T)$ to be 0, and construct a queue, Q, to contain them. Then, while Q is not empty, perform a *labeling opration*: Choose the first label $d(x, t)$ from Q, and for any $(x, y) \in A$ (or $[x, y] \in A$), set $d(y, t') = \{x\}$ if the following conditions hold:

(1) $0 \leq t' = t + b(x, y, t) \leq T$ (or $0 \leq t' = t + b[x, y, t] \leq T$);
(2) $d(y, t') = null$;
(3) $c(x, y, t) > 0$ (or $c[x, y, t] > 0$).

Delete $d(x, t)$ from Q and append $d(y, t')$, if $d(y, t') = \{x\}$ (that is, the conditions above hold), to Q.

(iii) Since waiting time is permitted, a flow can be stored at an internal vertex for a period of time. Suppose that there is a feasible dynamic f-augmenting path $P(s, y)$ of time at most t in N. Then it must be a feasible dynamic f-augmenting path of time at most $t + 1$ (notice that here $l(y, t) = +\infty$ for any y). This means that if $d(y, t)$ is labeled, then for any t' such that $t \leq t' \leq T$, $d(y, t')$ can also be labeled. On the other hand, if $d(y, t)$ is labeled and $l[y, t] > 0$, $d(y, t - 1)$ can also be labeled since there must exist a previous flow waiting at x from the time t to $t - 1$. And then we can "push" it back using the dynamic residual network (see Example 1).

(iv) Check all labels $d(\rho, t)$ $(t = 1, 2, ..., T)$. If there is a label $d(\rho, t) \neq null$ $(t = 0, 1, ..., T)$, then there is a feasible dynamic f-augmenting path from s to ρ of time at most t. Choose one with the earliest arriving time t if there are multiple labels which are not null. Then send the maximum possible flow along this path, construct the dynamic residual network accordingly, reset all labels $d(y, t)$ to null and try to find another path; otherwise, stop the algorithm, and the sum of the flows found during the procedure gives the final solution.

We now describe the detailed algorithm we propose to solve the problem with waiting times, where the notation $g(t_{min})$ denotes the augmenting flow value along the feasible dynamic f-augmenting path $P(s, \rho)$ of time exactly t_{min} found in one iteration, and $maxf$ denotes the maximum flow value.

Algorithm 4.1

Begin
$maxf := 0$; $t_{min} := 0$;
While $t_{min} \leq T$ **do**
  Begin
    Initialization: $d(s, t) := 0, t = 0, 1, ..., T$; $Q := \{d(s, 0), d(s, 1), ..., d(s, T)\}$;
    While $Q \neq \emptyset$ **do**
      Begin
        Select the first label in Q, denoted by $d(x, t)$;
        For any y such that (x, y) or $[x, y] \in A$ **do:**
          Begin
            Let $t' := t + b(x, y, t)$ (or $t' := t + b[x, y, t]$);
            If $t < t' \leq T$ **then** $c' := c(x, y, t)$;
            Else $c' := c[x, y, t]$
            Endif;
            labeling:
            If $(0 \leq t' \leq T)$And$(d(y, t') = null)$And$(c' > 0)$ **then**
              Begin
                Let $\tau := t'$;
                While $(\tau \leq T)$And$(d(y, \tau) = null)$ **do**
                  $d(y, \tau) := x$; $Q := Q \cup \{d(y, \tau)\}$; $\tau := \tau + 1$;
                Let $\tau := t' - 1$;

While $(\tau \geq 0)$And$(d(y, \tau) = null)$And$(l[y, \tau + 1] > 0)$ **do**
$\qquad d(y, \tau) := x;\ Q := Q \cup \{d(y, \tau)\};\ \tau := \tau - 1;$
$\qquad$ **End;**
$\qquad$ **End;**
$\quad$ **Let** $Q := Q \backslash \{d(x, t)\}$
$\quad$ **End;**
Let t_{min} be the minimun t such that $d(\rho, t) \neq null$. If $d(\rho, t) = null$ for all t, then let $t_{min} = T + 1$;
$\quad$ **If** $t_{min} < T + 1$ **then**
$\qquad$ **Begin**
$\qquad$ Use the predecessor indices to identify the feasible dynamic f-augmenting path $P(s, \rho)$ of time at most t_{min};
$\qquad$ **Let** $g(t_{min}) :=$ the minimum capacity of arcs in $P(s, \rho)$;
$\qquad$ Augment $g(t_{min})$ units of flow along $P(s, \rho)$;
$\qquad$ Call procedure UPNET to update the arc and vertex capacities;
$\qquad$ **Let** $maxf := maxf + g(t_{min})$
$\qquad$ **End**
$\quad$ **End**
End.

Now let us examine the optimality and the time complexity of Algorithm 4.1.

Theorem 3. Algorithm 4.1 can optimally solve the time-varying maximum flow problem with arbitrary waiting time at each vertex.

Proof. We only need to prove that when the algorithm stops, it generates a generlized cut K such that $capK = 0$ in the last dynamic residual network.

For any t, let $K(t) = \{y | d(y, t) \neq null, y \in V\}$ and $K = \{K(t), t = 0, 1, ..., T\}$. Since $s \in K(t)$ and $\rho \notin K(t)$ for any t, K is a generalized cut by Definition 6. Notice that $l(x, t) = +\infty$ for any $x \in V$ and if $x \in K(t)$, then one must have $x \in K(t + 1)$. Therefore

$$capK = \sum_{t=0}^{T-1} \sum_{x \in K(t), x \notin K(t+1)} l(x, t) + \sum_{t=1}^{T} \sum_{x \in K(t), x \notin K(t-1)} l[x, t]$$

$$+ \sum_{t=0}^{T} \sum_{x \in K(t), y \notin K(t')} c(x, y, t) + \sum_{t=0}^{T} \sum_{x \in K(t), y \notin K(t'')} c[x, y, t]$$

$$= \sum_{t=1}^{T} \sum_{x \in K(t), x \notin K(t-1)} l[x, t] + \sum_{t=0}^{T} \sum_{x \in K(t), y \notin K(t')} c(x, y, t)$$

$$+ \sum_{t=0}^{T} \sum_{x \in K(t), y \notin K(t'')} c[x, y, t].$$

Clearly, for any $(x, y) \in A$ such that $x \in K(t)$ and $y \notin K(t')$, we must have $c(x, y, t) = 0$; otherwise, if $c(x, y, t) > 0$, by the labeling operation, $d(y, t')$ must be labeled, so $y \in K(t')$. This is a contradiction. Similarly, we can show $c[x, y, t] = 0$ for any $x \in K(t)$ and $y \notin K(t'')$. On the other hand, when $x \in K(t)$ and $x \notin K(t-1)$, if $l[x, t] > 0$, then $d(x, t-1)$ must be labeled by the labeling operation, and therefore $x \in K(t-1)$. This contradicts $x \notin K(t-1)$. Consequently, we must have $capK = 0$. This proves the theorem. □

Complexity Analysis

Consider one iteration. In the initialization block, the running time is $O(T)$. For the labeling operation, we may need to examine all arcs at all time t (in the worst case), so the running time is $O(Tm)$. The algorithm applied the procedure UPNET to update capacities, which needs a running time $O(Tm)$. Hence the total running time in one iteration is bounded by $O(Tm)$. Suppose that U is the maximum capacity of arcs. Then, the capacity of the generalized cut $K = \{K(t)|t = 0, 1, ..., T, s \in K(t)\}$ is at most nTU. Therefore, the maximum dynamic flow value is bounded by nTU. Each iteration at least augments one unit of flow, so the algorithm will terminate within nTU iterations. Thus, the total running time of Algorithm 4.1 is bounded by $O(UnmT^2)$.

4.2   The Time-Varying Maximum Flow with Vertex Capacity Limit

Now, we consider the case in which there is a capacity limit $l(y, t)$ at each vertex y, which is also a function of time t. Unlike the previous case, here a feasible dynamic f-augmenting path of time at most t, $P(y, t)$, may not necessarily imply a feasible path $P(y, t + 1)$ due to the vertex capacity $l(y, t)$ at y. Thus we have to check whether there is still capacity available at y, namely, whether $l(y, t) > 0$, when we are to label y with $d(y, t + 1)$ after it is labeled with $d(y, t)$. In fact, if y is labeled with $d(y, t)$, then, for any $t' > t$, y can be labeled with $d(y, t')$ if $\prod_{\tau=t}^{t'-1} l(y, \tau) > 0$. On the other hand, if y is labeled with $d(y, t)$, then, for any $t'' < t$, y can be labeled with $d(y, t'')$ if $\prod_{\tau=t}^{t''+1} l[y, \tau] > 0$.

In view of the above, we modify Algorithm 4.1 as follows:

Algorithm 4.2

All steps are the same as those of Algorithm 4.1 except the following labeling operation:

labeling:

If $(0 \le t' \le T)$And$(d(y, t') = null)$And$(c' > 0)$ then

Begin
$\quad\tau := t'; \; d(y,\tau) := x; \; Q := Q \cup \{d(y,\tau)\}; \; \tau := \tau + 1;$
$\quad$**While** $(\tau \leq T)\text{And}(d(y,\tau) = null)\text{And}(l(y,\tau-1) > 0)$ **do**
$\quad\quad d(y,\tau) := x; \; Q := Q \cup \{d(y,\tau)\}; \; \tau := \tau + 1;$
$\quad$**Let** $\tau := t' - 1;$
$\quad$**While** $(\tau \geq 0)\text{And}(d(y,\tau) = null)\text{And}(l[y,\tau+1] > 0)$ **do**
$\quad\quad d(y,\tau) := x; \; Q := Q \cup \{d(y,\tau)\}; \; \tau := \tau - 1;$
End;

Results similar to those for Algorithm 4.1 can be obtained similarly on the optimality and time complexity of Algorithm 4.2. The details are omitted here.

5 Conclusion

In this paper, we have studied a generalization of the classical maximum flow problem, to include the time-varying features in many practical situations. In our model, the transit times, the capacities of arcs, and the capacities of vertices are all time-varying. We show that this problem is NP-complete in ordinary sense. Two algorithms are proposed, which can obtain the optimal solutions with pseudopolynomial time complexity for the problem with or without vertex capacity limit, respectively.

Future reseach of interest includes problems where the transit time on each arc can be shortened within a certain range, at a cost, with the total cost being less than a given amount. These are more practical features in many real-world problems where we often desire to control the speed of flows at different vertices and arcs. Investigation of time-varying network to find the minimum cost solution, when there is a cost to send a unit of flow on each vertex and arc, should also be an interesting work.

Appendix

Example 1
$\quad$Consider a network as shown in Figure 2, where s is the source vertex and ρ is the sink vertex. The three numbers inside each pair of brackets associated with an arc are t, $b(x,y,t)$ and $c(x,y,t)$ respectively. For instance, $(0,1,2)$ near the arc (s,d) means at time 0, transit time $b(s,d,0)$ is 1 and capacity limit $c(s,d,0)$ is 2, and at other times, $b(s,d,t) = +\infty$ and $c(s,d,t) = 0$. We assume, in this example, that no waiting time at any internal vertex is allowed, so let $l(x,y,t) = 0$ for all x and t.

$\quad$Notice that $P_1(s,\rho) = (s,d,h,\rho)$ is a feasible dynamic f-augmenting path from s to ρ. Starting from time 0, at least two units of flow can be sent through arc (s,d) since the capacity of (s,d) at time 0 is 2 and the flow reaches the vertex d at time 1. In the same way, at least two units of flow

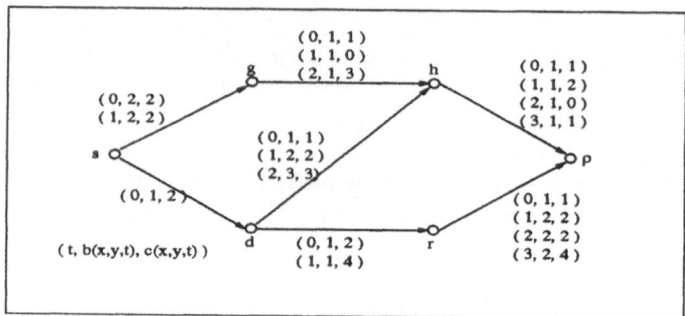

Fig. 2.

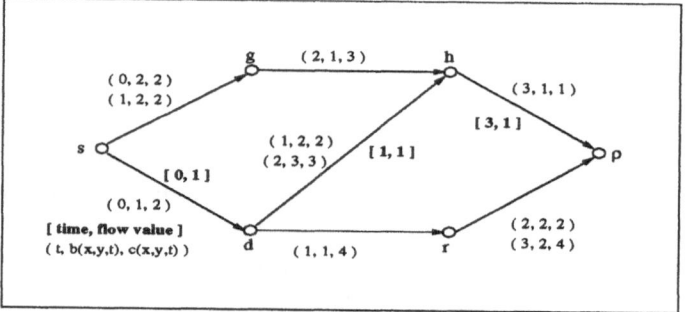

Fig. 3.

can be sent beginning at time 1 through arc (d, h) since $c(d, h, 1) = 2$, and the flow reaches the vertex h at time 3. Beginning at time 3, one unit of flow can be sent through arc (h, ρ) and it arrives at vertex ρ at time 6. Thus, we can send one unit of flow from s to ρ along $P_1(s, \rho)$ within time 4. In order to avoid confusion, we label each arc on P_1 with a pair of numbers $[t, v(f_1)]$ to denote that the flow value $v(f_1)$ is sent by this arc at departing time t. The new network is shown in Figure 3.

Then, we update the original network by the Network Updating Procedure. The dynamic residual network is shown as Figure 4.

Next, we find another feasible dynamic f-augmenting path $P_2(s, \rho) = (s, g, h, d, r, \rho)$ (notice that $[h, d]$ is an artificial arc with negative transit time). Let us check the time parameters and see whether they can be matched at each vertex. Starting from time 0, at least two units of flow can be sent by (s, g) since $c(s, g, 0) = 2$, and the flow arrives at g at time 2. Departing at time 2, two units of flow can be sent by (g, h) and the flow arrives at h at time 3. $[h, g]$ is an artificial arc and it means that there is one unit of flow value to be sent in previous path P_1 during time period $[1, 3]$; so we can "push" this one unit of flow value back to vertex d while reducing the time from 3 to 1. And then, since $c(d, r, 1) = 4 > 0$, we can send this one unit of flow through (d, r) beginning at time 1 and it reaches r at time 2. In the same way, we

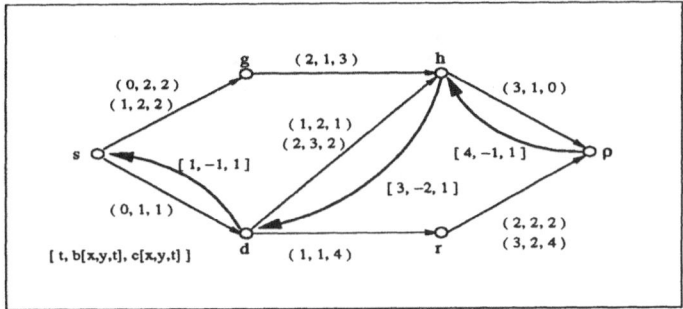

Fig. 4.

can send one unit of flow through (r, ρ) beginning at time 2 to arrive at ρ at time 4. As a conclusion, we! ! can "send" one unit of flow "along" P_2.

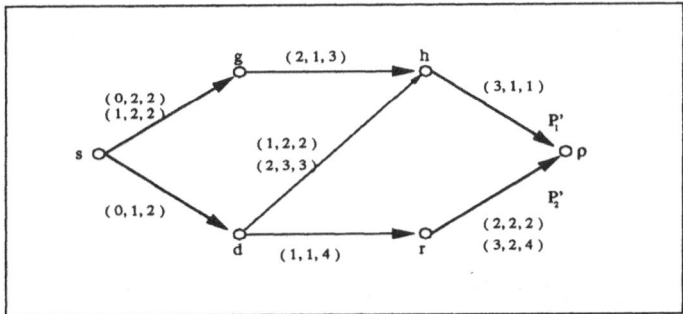

Fig. 5.

Obviously, P_2 is not a real path since it contains an artificial arc, but it can help us find the real path. In fact, it should be understood that we send the first flow along $P_1' = (s, d, r, \rho)$ and the second flow along $P_2' = (s, g, h, \rho)$. Total flow value we sent is $2 + 1 = 3$ (see Figure 5; pay attention to the time parameters).

Example 2

In Figure 6, the solid lines stand for the original arcs of network N, and the dotted lines stand for the artificial arcs. Initially, for all artificial arcs $[x, y]$ and all time t, let $c[x, y, t] = 0$ and $b[x, y, t] = -b(y, x, u)$ where $u = t + b(x, y, t)$. For all non-artificial arcs (x, y), we list their values of $c(x, y, t)$ and $b(x, y, t)$ in Table 1 and Table 2 respectively (given $T = 8$). Since we do not consider the vertex capacity limit here, let $l(y, t) = +\infty$ and $l[y, t] = 0$ for all $y \in V \backslash \{s\}$ and for all t.

First, we set $d(s, t) = 0$ $(t = 0, 1, ..., T)$ and all other labels $d(y, t) = null$. $Q = \{d(s, 0), d(s, 1), ..., d(s, T)\}$. Consider $d(s, 0)$ first.

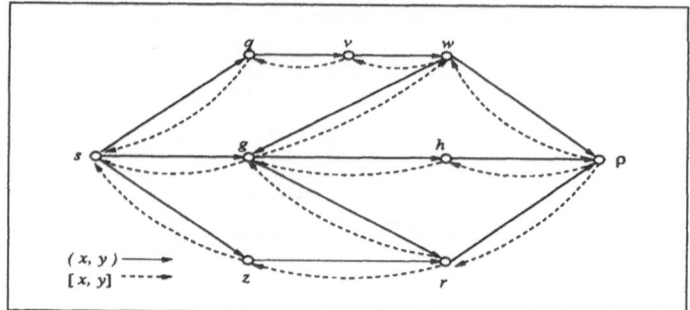

Fig. 6.

Table 1. $c(x, y, t)$

t	(s, q)	(s, g)	(s, z)	(q, v)	(v, w)	(g, h)	(g, r)	(z, r)	(w, g)	(w, ρ)	(h, ρ)	(r, ρ)
0	1	3	2	1	3	3	2	3	4	3	3	2
1	0	0	0	2	3	2	2	2	2	3	4	2
2	0	0	0	1	2	0	0	0	2	2	0	2
3	0	0	0	3	3	0	0	0	1	0	4	1
4	0	0	0	2	2	2	0	0	3	0	0	0
5	0	0	0	1	2	0	0	0	3	0	0	2
6	0	0	0	2	1	0	0	3	3	0	1	2
7	0	0	0	1	3	0	0	0	4	0	0	1
8	-	-	-	-	-	-	-	-	-	-	-	-

Table 2. $b(x, y, t)$

t	(s, q)	(s, g)	(s, z)	(q, v)	(v, w)	(g, h)	(g, r)	(z, r)	(w, g)	(w, ρ)	(h, ρ)	(r, ρ)
0	1	1	1	1	1	1	1	1	1	1	1	1
1	1	3	1	1	1	2	1	1	1	3	1	1
2	1	3	2	1	1	2	2	1	1	3	1	1
3	3	1	1	2	2	1	2	2	1	2	1	5
4	2	1	3	1	1	2	2	2	3	1	1	5
5	1	2	4	3	2	1	1	1	3	3	1	4
6	1	1	1	3	1	2	4	3	4	4	1	4
7	2	2	1	2	3	1	1	3	5	3	2	3
8	-	-	-	-	-	-	-	-	-	-	-	-

Since $b(s, q, 0) = 1$, $c(s, q, 0) = 1 > 0$, $d(q, 1)$ can be labeled, which we set as $\{s\}$. Moreover, since arbitrary waiting time at vertex is permitted, this flow can be stored at q (note that $l(q, t) = +\infty$ for any t). Thus, $d(q, 2)$, ..., $d(q, 8)$ can also be labeled. Similarly, $d(g, 1)$,...,$d(g, 8)$ and $d(z, 1)$,...,$d(z, 8)$ can be labeled as $\{s\}$ too. Delete $d(s, 0)$ from Q and append $d(q, 1)$,...,$d(q, 8)$, $d(g, 1)$,...,$d(g, 8)$ and $d(z, 1)$,...,$d(z, 8)$ in Q.

Consider $d(s, 1)$ (now it becomes the first elememt in Q). Noting $b(s, q, 1) = 1$, but $d(q, 2)$ is labeled already. Similarly, $d(g, 2)$ and $d(z, 2)$ are labeled already too. Therefore we just delete $d(s, 1)$ from Q only.

Following this process, we can obtain other labels. When Q becomes empty, this iteration is completed. The result is shown in Figure 7 (all other labels $d(y, t)$ which do not appear in the figure are nulls except $d(s, t)$).

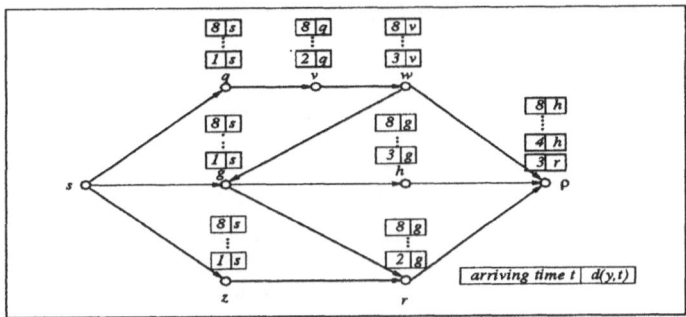

Fig. 7.

Notice that $d(\rho, 3) = \{s\}$, $d(\rho, 4) = \{h\} = \ldots = d(\rho, 7) = \{h\}$. The earliest arriving time for vertex ρ is 3. By a backward searching, we can find a feasible dynamic f-augmenting path $P_1 = \{s, g, r, \rho\}$ with $\alpha(P_1) = 3$. Let f_1 denote the dynamic flow along P_1; then we have

$$v(f_1) = \min\{c(s, g, 0), c(g, r, 1), c(r, \rho, 2)\} = \min\{3, 2, 2\} = 2$$

$$maxf = v(f_1) = 2$$

And then, update N. Let

$$c(s, g, 0) := c(s, g, 0) - v(f_1) = 1 \quad c[g, s, 1] := c[g, s, 1] + v(f_1) = 2$$

$$c(g, r, 1) := c(g, r, 1) - v(f_1) = 0 \quad c[r, g, 2] := c[r, g, 2] + v(f_1) = 2$$

$$c(r, \rho, 2) := c(r, \rho, 2) - v(f_1) = 0 \quad c[\rho, r, 3] := c[\rho, r, 3] + v(f_1) = 2$$

Other b and c remain unchanged.

From Figure 8 to Figure 9, we obtain P_2 and P_3. To highlight the previous feasible dynamic f-augmenting paths, we use dotted lines to represent them in each figure. To save space, we omit the network updating process for each iteration.

$$P_2 = \{s, g, h, \rho\} \qquad \alpha(P_2) = 4$$

$$v(f_2) = \min\{c(s, g, 0), c(g, h, 1), c(h, \rho, 3)\} = \min\{1, 2, 4\} = 1$$

$$maxf := maxf + v(f_2) = 2 + 1 = 3$$

$$P_3 = \{s, z, r, g, h, \rho\}, v(f_3) = 1$$

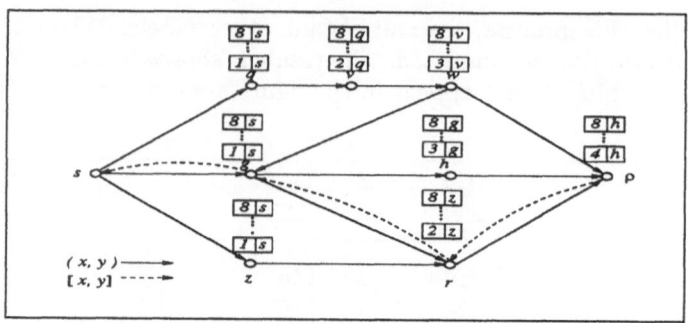

Fig. 8.

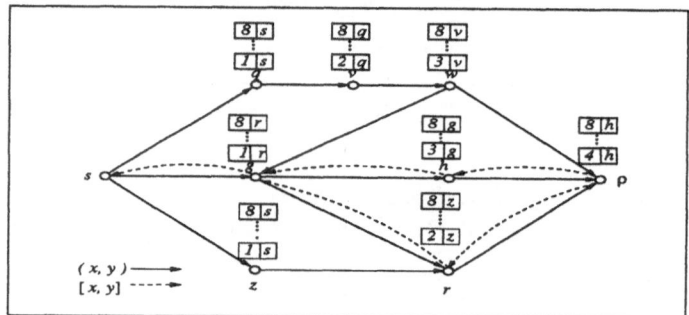

Fig. 9.

Table 3. $c(x,y,t)$

t	(s,q)	(s,g)	(s,z)	(q,v)	(v,w)	(g,h)	(g,r)	(z,r)	(w,g)	(w,ρ)	(h,ρ)	(r,ρ)
0	1	0	1	1	3	3	2	3	4	3	3	2
1	0	0	0	2	3	0	1	1	2	3	4	2
2	0	0	0	1	2	0	0	0	2	2	0	0
3	0	0	0	3	3	0	0	0	1	0	2	1
4	0	0	0	2	2	2	0	0	3	0	0	0
5	0	0	0	1	2	0	0	0	3	0	0	2
6	0	0	0	2	1	0	0	3	3	0	1	2
7	0	0	0	1	3	0	0	0	4	0	0	1
8	-	-	-	-	-	-	-	-	-	-	-	-

Table 4. $c[x,y,t]$

t	[q,s]	[g,s]	[z,s]	[v,q]	[w,v]	[h,g]	[r,g]	[r,z]	[g,w]	[ρ,w]	[ρ,h]	[ρ,r]
0	0	0	0	0	0	0	0	0	0	0	0	0
1	0	3	1	0	0	0	0	0	0	0	0	0
2	0	0	0	0	0	0	1	1	0	0	0	0
3	0	0	0	0	0	2	0	0	0	0	0	2
4	0	0	0	0	0	0	0	0	0	0	2	0
5	0	0	0	0	0	0	0	0	0	0	0	0
6	0	0	0	0	0	0	0	0	0	0	0	0
7	0	0	0	0	0	0	0	0	0	0	0	0
8	-	-	-	-	-	-	-	-	-	-	-	-

Up till now, all updated $c(x, y, t)$ and $c[x, y, t]$ are listed in Table 3 and Table 4. Now consider vertex s.

Since $b(s, q, 0) = 1$ and $c(s, q, 0) = 1 > 0$, it is clear that $d(q, 1)$ can be labeled as $\{s\}$. Thus, $d(q, 2), ..., d(q, 8)$ can also be labeled (see Figure 10). Nevertheless, $d(g, 1)$ can not be labeled as $b(s, g, 0) = 1$, but $c(s, g, 0) = 0$. Since $b(s, z, 0) = 1$ and $c(s, z, 0) > 0$, $d(z, 1)$ can be labeled, and $d(z, 3), ..., d(z, 8)$ can also be labeled.

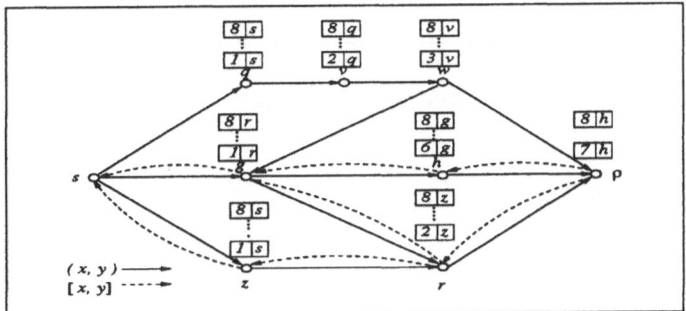

Fig. 10.

Similarly, we can label $d(v, 2), ..., d(v, 8), d(r, 2), ..., d(r, 8)$, and $d(w, 3), ..., d(w, 8)$. Now consider vertex r. Since $b[r, g, 2] = -1$ and $c[r, g, 2] = 2 > 0$, $d(g, 1)$ can be labeled. Also, $d(g, 2), ..., d(g, 8)$ can be labeled. When this iteration is finished, we will obtain all labels $d(y, t)$ as shown in Figure 10. As $d(\rho, 7)$ is labeled, we have found a feasible dynamic f-augmenting path $P_4 = \{s, z, r, g, h, \rho\}$, with $v(f_4) = 1$. Since f_4 arrives at g at time 1 and departs at time 4, the waiting time $w(g)$ is 3 units. Thus, we need updating $l[g, 4]$, $l[g, 3]$ and $l[g, 2]$ from 0 to 1.

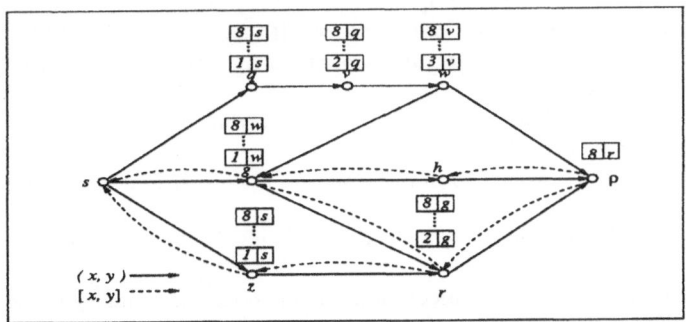

Fig. 11.

Figure 11 shows all labels after the 5th iteration is finished, where we can find a feasible dynamic f-augmenting path $P_5 = \{s, q, v, w, g, r, \rho\}$, with $v(f_5) = 1$. We only explain how $d(g, 1), ..., d(g, 8)$ are labeled. Since $d(w, 3)$ is labeled, $b(w, g, 3) = 1$ and $c(w, g, 3) > 0$, we know that $d(g, 4)$ can be labeled. Consequently, $d(g, 5), ..., d(g, 8)$ can be labeled accordingly. On the other hand, since $l[g, 4] = 1$, a flow can be stored during the time period [3,4], and thus $d(g, 3)$ can be labeled. Similarly, since $l[g, 3] = l[g, 2] = 1$, $d(g, 2)$ and $d(g, 1)$ can also be labeled. After that, $d(r, 2), ..., d(r, 8)$ can be labeled accordingly.

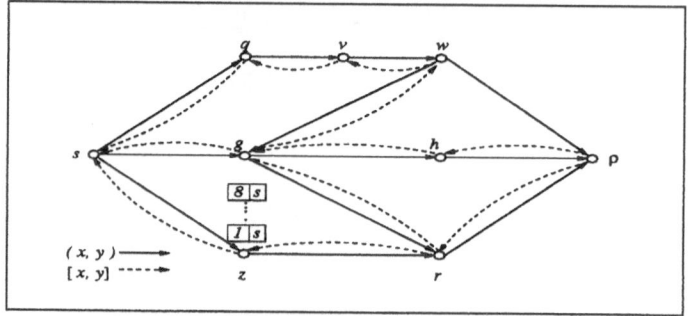

Fig. 12.

In the 6th iteration (see Figure 12), since $d(\rho, t)$ can not be labeled, the algorithm stops, which gives a maximum dynamic flow $f = \{f_1, f_2, ..., f_5\}$ with $v(f) = 6$. This completes this example.

References

1. Ford, L.R. and Fulkereon, D.R. (1962) Flows in networks. Princeton University Press, Princeton, New Jersey
2. Halpern, J. (1979) Generalized dynamic flows. Networks **9**, 133–167
3. Anderson, E.J., Nash, P., Philpott, A.B. (1982) A class of continuous network flow problems. Mathematics of Operations Research. **7**, 4, 501–514
4. Minieka, E. (1978) Optimization algorithms for networks and graphs. Marcel Dekker. INC., New York
5. Cai, X., Kloks, T., Wong, C.K. (1997) Time-varying shortest path problems with constraints. Networks. **29**, 3, 141–149
6. Gale, D. (1959) Transient flows in networks. Michigan Mathematical Journal **6**, 59–63
7. Minieka, E. (1973) Maximal, lexicographic and dynamic network flows. Oper. Res. **2**, 517–527
8. Wilkinson, W.L. (1971) An algorithm for universal maximal dynamic flows in a network. Oper. Res. **7**, 1602–1612

Transmission Network Planning Under Uncertainty with Benders Decomposition

Panagiota Tsamasphyrou[1], Arnaud Renaud[2], and Pierre Carpentier[3]

[1] École des Mines, Centre d'Automatique et Systèmes
   35, rue Saint-Honoré, 77305 Fontainebleau, France
   penny@cas.ensmp.fr
[2] Électricité de France, DER, MOS
   1, Avenue du Général de Gaulle, 92141 Clamart Cedex, France
   Arnaud.Renaud@edf.fr
[3] École des Mines, Centre d'Automatique et Systèmes
   35, rue Saint-Honoré, 77305 Fontainebleau, France
   carpentier@cas.ensmp.fr

Abstract. This paper treats the problem of the long term transmission expansion planning of an electrical network under uncertainty. A static form of this problem is considered; transmission constraints are represented in the DC approximation (linearized power flow). Uncertainty is taken into account to evaluate the operating cost (demand, availability); the investment variables are binary. The problem is therefore a large stochastic mixed integer one, requiring a considerable computational effort.

This article proposes a new formulation of the problem, which is well suited to the application of the Benders Decomposition. Several ways of dealing with the cuts which are obtained at each iteration are described and compared. Numerical tests show the practical interest of the algorithm finally obtained.

Key words: transmission expansion planning, electrical network, Benders decomposition, linearized power flow (DC), scenarios

1 Introduction

The objective of the long term transmission expansion problem is to determine when and where to build the new facilities (new lines) required while minimizing the sum of the capital cost of investments and the operating cost associated with the system expansion over the planning horizon.

The form of this problem most often encountered in the literature is a mixed integer (the investment variables are discrete) static, deterministic model, using the linearized power flow (DC) model for the transmission constraints [5,7,9–12]. However, neglecting the stochastic nature of the problem leads to significant errors in the evaluation of the operating cost (and thus in the evaluation of the optimal investment plan). The level of uncertainty on the future availability of generation units and transmission lines is very high: a deterministic model cannot be realistic. For this reason, Electricité de France as well as numerous other companies throughout the world use not optimization but simulation software [3]. In these models, investment plans are proposed by the user; operating cost is then evaluated in a stochastic framework. The work presented below aims mainly at developing and testing algorithms which allow the optimization of investment plans with a realistic evaluation of operating costs, in a stochastic framework.

In order to represent the stochastic side of the problem, the model we consider takes into account (see Sect. 2) several availability and demand scenarios. This way, the problem is expressed as a deterministic equivalent one [1]: the expected value of the operating cost is expressed as a sum of mappings, each of these mapping being related to one scenario.

To solve this optimization problem, Benders Decomposition seems to be a fairly natural approach: the master problem deals with the 0/1 investment variables; at the low level, the operating cost is computed for a given invest-ment plan in order to obtain new cuts. Nevertheless, the application of the Benders algorithm is not straightforward: a new formulation is necessary. A transformation of the problem is proposed in this paper. Comparisons with another equivalent formulation which have been proposed in the literature have shown the superiority of this transformation (see [17]).

Because of the additive structure of the criterion, several ways of imple-menting the Benders algorithm may be considered. The decomposition can be applied directly to the sum or it can be applied to each term of the sum separately, each implementation having its own advantages and disadvan-tages. A compromise between these two main implementations is proposed here. Numerical tests show the significance of this approach.

The presentation is structured as follows. First of all, the model is in-troduced (see Sect. 2). A formulation of the problems which allows the ap-plication of the Benders algorithm is then given (see Sect. 3). Several ways of dealing with the cuts which are obtained at each iteration are proposed (Sect. 4). Numerical tests and comparisons, over several networks, are pre-sented (Sect. 5), to show the utility of the proposed approach; several trans-formations and heuristics are finally introduced in order to reduce the com-putation time.

2   Model

The model of the transmission planning problem we consider is the following:

$$\left. \begin{aligned} & \min_{x,\,y} a^\top y + \sum_{s=1}^{n_s} \pi_s w(y, \xi_s) \\ & \text{s.t. } y \in \mathbf{Y} \subseteq \{0,1\}^{n_l}, \end{aligned} \right\} P$$

where:

- n_l is the total number of buses (constructed or not);
- n_s is the number of scenarios;
- π_s is the probability of scenario s;
- ξ_s denotes one realization of random variables (availability of lines and generation units, demand at each node of the network);
- y is a vector of binary variables related to investment decisions: $y_l = 1$, if the line l is built; $y_l = 0$, if not. We will assume that, if $y_l = 1 \; \forall l \in \{1, \ldots, n_l\}$, the network is connected;
- $\mathbf{Y}$ is the set of possible investments decisions;
- $a \in \mathbb{R}^{n_l}$ is the vector of the construction costs of the buses;
- $w(y, \xi_s)$ is the operating cost of the system for a specific realization of random variables ξ_s.

For every ξ, the operating cost mapping w is defined over $\{0,1\}^{n_l}$ as follows :

$$w(y, \xi) \stackrel{\text{def}}{=} \min c^\top p + h^\top q$$

s.t.

$$p + q - St = d(\xi) \tag{1}$$

$$v = S^\top \theta \tag{2}$$

$$\forall l \leq n_l \, , v_l = \begin{cases} r_l t_l & \text{if } y_l = 1, \\ v_l < \infty & \text{if } y_l = 0 \end{cases} \tag{3}$$

$$|t| \leq y\bar{t}(\xi) \tag{4}$$

$$0 \leq p \leq \bar{p}(\xi) \tag{5}$$

$$0 \leq q \leq d(\xi) \, , \tag{6}$$

where:

- S denotes the node-branch incidence matrix;
- $\mathcal{N}$ is the set of the nodes of the network, of cardinal n_n;
- $p \in \mathbb{R}^{n_n}$, with n_n the number of nodes of the network, is the production vector;
- $q \in \mathbb{R}^{n_n}$ is the shortage vector;
- $c \in \mathbb{R}^{n_n}$ and $h \in \mathbb{R}^{n_n}$ denote respectively the production and shortage cost;
- t and $v \in \mathbb{R}^{n_l}$ denote the transit and tension vector;
- $\theta \in \mathbb{R}^{n_n}$ is the vector of node voltage angle;
- r_l is the reactance of line l;
- $d(\xi) \in \mathbb{R}^{n_n}$ is the vector of demands at the nodes, when the random event ξ occurs;
- $\bar{t}(\xi)$ and $\bar{p}(\xi)$ are the production and transit maximal capacity vector in the situation ξ.

It is therefore a static — over one time period —, stochastic model in the DC approximation that we are dealing with.

Moreover, when the investment decisions y are fixed, problem P becomes a stochastic operating cost evaluation program, like the one in [3], used to simulate the impact of an investment plan on the performances of the network.

3   Application of Benders Decomposition

The w mappings are only defined on a discrete set (for $y \in \{0,1\}^{n_l}$). To apply the Benders Decomposition algorithm, it is necessary to extend every function w to a mapping w^δ which has to be convex, decomposable and defined over a convex set[1] containing $\{0,1\}^{n_l}$. Clearly enough, several extension of this type may be considered. The one we propose and that will be used in this paper is the following:

$$
\left.
\begin{array}{l}
\min_{x,\,y} a^\top y + \sum_{s=1}^{n_s} \pi_s w^\delta(y, \xi_s) \\
\text{s.t. } y \in \mathbf{Y} \subseteq \{0,1\}^{n_l} ,
\end{array}
\right\} P_1
$$

where:

$$
\begin{aligned}
w^\delta(y, \xi) &\stackrel{\text{def}}{=} \min c^\top p + h^\top q \\
&\text{s.t.} \\
&p + q - St = d(\xi) \\
&v = S^\top \theta \\
&v_l = r_l t_l + u_l , \quad \forall l \le n_l \\
&|u_l| \le \delta(1 - y_l) , \quad \forall l \le n_l \\
&|t_l| \le y_l \bar{t}_l(\xi) , \quad \forall l \le n_l \\
&0 \le p \le \bar{p}(\xi) \\
&0 \le q \le d(\xi) ,
\end{aligned}
$$

and where $u \in \mathbb{R}^{n_l}$ is a slack variable homogeneous to a tension.

Provided that δ is large enough, w and w^δ are equal over $\{0,1\}^{n_l}$. More precisely, it has been proved [17] that

$$
\forall \delta \ge \sum_{l=1}^{n_l} r_l \bar{t}_l, \; \forall y \in \{0,1\}^{n_l} , \forall \xi , \; w^\delta(y,\xi) = w(y,\xi) . \tag{7}
$$

---

[1] We remark here that the disjonctive constraint (3) make the extension to a convex set impossible. On the other hand, if we transform this constraint into the form $y_l(v_l - r_l t_l)$, we see that the problem is now defined $\forall y \in [0,1]$, but the constraint set is not linearly decomposable into two parts, one concerning the operation variables and one concerning the investment ones; moreover, the problem is not convex with respect to all the variables.

Another transformation of this problem to put it in a convex form may be found in the literature [5], where the additional variables are homogeneous to a transit, but the one proposed here yields much better computational results (see Appendix).

This problem is a linear programming one, separable in two kind of variables: the first-stage variables, y, and the second-stage (recourse), x. Furthermore, $w^\delta(.,.)$ is feasible $\forall y$ and $\forall \xi$ (the problem has *complete recourse*) and the operating problem obtained when the first-stage variables are fixed is bounded from below by 0. So, the standard Benders Decomposition [4] can be directly applied; moreover, we need only to form optimality cuts.

The Benders algorithm (iteration k) when w is convex takes the following form:

- *master program* : it is a problem in MIP variables; let us define $\tilde{w} : y \mapsto \sum_{s=1}^{n_s} \pi_s w(y, \xi_s)$, a cutting plane approximation of the total operating cost. This level may be written:

$$\min a^\top y + \tilde{w}^k(y)$$
$$\text{s.t.} \qquad y \in \mathbf{Y} ,$$

  with $\tilde{w}^k(y)$ the approximation obtained after k iterations. The proof that $\tilde{w}(y)$ is convex is well-known; see [18].
  We will see in the Appendix how these cuts can be formed in the case of an electrical network.

- *slave subproblem* : the investments are fixed and for each scenario the operating cost of the system is computed (evaluation of $w(y, \xi^s)$, with y and ξ^s fixed). For a given y, the problem P_w which computes the $w^\delta(y, \xi)$ may be written as follows:

$$\left. \begin{array}{l} \min c^\top p + h^\top q \\ \text{s.t. } p + q - St = d(\xi) \\ v = S^\top \theta \\ v_l = r_l t_l , \quad \forall l \in \mathbf{Y}_l \\ |t_l| \leq \bar{t}_l(\xi) , \quad \forall l \in \mathbf{Y}_l \\ 0 \leq p \leq \bar{p}(\xi) \\ 0 \leq q \leq d(\xi) , \end{array} \right\} P_w$$

  where $\mathbf{Y}_l \subseteq \mathbf{Y}$ is the set of lines for which $y_l = 1$.
  This optimization may be done with standard programs of cost evaluation, like [3]; this is one of the main interests for the application of the Benders Decomposition. The final cost and the dual variables associated to the demand constraint (1) and to the constraint concerning the bounds of transit variables (4) which are obtained at this level are returned to the master program via the Benders cuts.

4   Dealing with the cuts

In a stochastic framework, the expected value of the operating cost of the system is written as a sum of functions, $w(.,\xi_s)$, representing the operating cost for a given scenario. At each iteration and for each scenario, a cut (the value of the function and a subgradient) is obtained by solving the low-level subproblems (operating cost problems).

Several ways of dealing with these cuts are considered. To present them, let us define the following partition $\mathcal{P}_\mathcal{L} = \{L_1, \ldots, L_{n_g}\}$ of the set of scenarios $\mathbf{S} = \{1, 2, \ldots, n_s\}$, such that $L_g \neq \emptyset, \{L_g \subseteq S \mid \forall g \leq n_g\}$ and

$$\forall g_1, g_2 \leq n_g, \; g_1 \neq g_2 \;\; L_{g_1} \cap L_{g_2} = \emptyset, \; \bigcup_{g \leq n_g} L_g = \mathbf{S},$$

where:

- l_g is the cardinal of L_g, such that $l_g \leq n_s$;
- n_g is the number of sets L_g.

In general, a Benders cut may be formed for each set of scenarios L_g and at each Benders iteration n_g cuts are finally formed. The Benders master program becomes (we suppose that all the scenarios have the same probability of occurrence, $\pi_s = 1/n_s$):

$$\min a^\top y + \sum_{g=1}^{n_g} z_{0_g} \tag{8}$$

s.t.:

$$z_{0_g} \geq \pi_g \sum_{s \in L_g} w(y^i, \xi_s) + \sigma^\top (y - y^i), \; \forall g \leq n_g, \forall i < k \tag{9}$$

$$y \in \mathbf{Y},$$

where k is the number of master iterations so far and $\pi_g \stackrel{\text{def}}{=} \sum_{s \in L_g} \pi_s = l_g/n_s$.

To any partition $\mathcal{P}_\mathcal{L}$ of the set $\mathbf{S}$, an implementation of the Benders algorithm is associated. The most usual implementations ([6,?]) are:

- the *single-cut method* : only one cut (average cut) is formed per iteration. For this method, $n_g = 1$, $L_1 = \mathbf{S}$ and $l_1 = n_s$. The advantages are that, since all the information is gathered in one cut per Benders iteration, the master program remains easy to solve. On the other hand, a significant number of iterations may be necessary to converge (see [14]).
- the *multicut method* : one cut is formed per realization ξ^s; thus, we have n_s cuts at each Benders iteration. Therefore, $n_g = n_s$ and $l_g = 1, \forall g = \{1, \ldots, n_s\}$. The advantage is that the operating cost function is better

approximated; the method converges in less iterations. However, since too many cuts have to be taken into account in the master program, the total CPU time is not always reduced.

However, other cutting plane approximations of w may be used: by choosing $n_g \in]1, n_s[$, compromises between the single-cut and the multicut method can be obtained. This type of partitioning and the cutting plane approximation it imposes will be called hereafter "aggregated multicut algorithm" (AMA).

5 Implementation and numerical tests

5.1 Data sets

The tests were carried on on a network derived from a IEEE network: a network with (in total)

- 40 lines (35 connections),
- 24 nodes,
- 52 production units.

We consider 20 investment variables; in this case, $y_l = 1$ for the other lines. So, the set of investment decisions is $\mathbf{Y} = \{\mathbf{Y}_1 \cup \mathbf{Y}_2\}$, with $\mathbf{Y}_1$ such that $\forall y_l \in \mathbf{Y}_1, y_l = 1$ and where $\mathbf{Y}_2$ has a cardinal of 20.

We use the theoretical value of δ (see (7)), $\delta \approx 4500000$.

5.2 Implementation of AMA

Comparison of partition strategies To build the sets L_g we proceed as follows: we take the first $\lfloor n_s/n_g \rfloor$ scenarios to form the first set, L_1 and so forth; the last set is composed by the remaining scenarios.

The results obtained are shown in Table 1. Note that when treating larger networks, the reduction of the number of Benders iterations is even more significant. By now, the CPU time required for the operating cost evaluation was insignificant compared to the master program. However, for larger networks, the operating cost evaluation requires more CPU time; as for each Benders iteration the CPU time is the sum of the CPU time required for the operating cost evaluation plus the time required for the master program and since the total CPU time is the CPU time for each iteration multiplied by the number of Benders iterations, the significance of the reduction of the number of Benders iterations is clear.

Choice of sets L_g in AMA To improve the AMA method, a criterion based on the operating cost of each scenario is used to form, off-line, the sets L_g. We form one set L_g per scenario ($l_g = 1$) for the most "difficult" scenarios,

Table 1. Comparison of partition strategies

n_s	Partition	Objective	n_g	Time(sec)	Iter
1000	AMA	40208.26	11	2571	70
	Single-cut	40208.26	1	2650	106
	Multicut	40208.26	1000	45199	39
2000	AMA	40389.32	13	6727	81
	Single-cut	40389.32	1	18922	349
	Multicut	40389.32	2000	193557	39

whereas for the other ones the sets L_g formed put together realizations that have similar effect on the problem. The number of those sets, m, which have $l_g > 1$ is predefined by the user.

To determine the classes of similar costs, we proceed off-line: we run the operating cost program with no investments ($y = 0$). An operating cost is found for each realization ξ. We take the difference between the lowest and the highest value of operating cost and we divide it by $m + 1$; $m + 1$ intervals are defined. We then partition the realizations according to these intervals: we form one set L_g, $g \leq m$, gathering all scenarios providing cost belonging to one of them; finally, one set L_g is formed per scenario leading to a cost belonging to the last interval. We see then that the choice of m affects indirectly the number of sets with $l_g = 1$ as well.

The AMA algorithm with this choice of sets is denoted AMA-1. The AMA-1 algorithm yields better results (see Table 2) than the AMA. The results are obtained by taking $m = 2$ for the proposed choice of sets. The n_g is the same for both partitions; Iter is the number of Benders (master) iterations.

Table 2. Choosing the sets L_g

n_s	Partition	n_g	Time(sec)	Iter
1000	AMA-1	11	1617	46
	AMA	11	2571	70
2000	AMA-1	13	5010	66
	AMA	13	6727	81

5.3   Neglecting cuts

We have a partition with m sets L_g with cardinal $l_g \neq 1$ and $n_g - m$ sets of cardinal 1. Still, we know that the first Benders iterations contribute essentially to the definition of an investment plan without shortage (small operating cost) and are almost inefficient afterwards, as for all cutting plane techniques (see [15]). We propose here a technique partially inspired by the "critical scenarios" approach [13–16], that consists in neglecting the cuts formed by the sets L_g with $l_g > 1$ until a solution without shortage is met. So, for the first iterations, only cuts generated by the sets-singletons (formed by the scenarios corresponding to the highest operating costs) are taken into consideration. For the tests we took $m = 2$, the methods compared are the method using the AMA-1 algorithm and the one with the modifications proposed above, named AMA-2.

Table 3. Neglecting cuts

n_s	Method	Time(sec)	Iter
1000	AMA-1	1617	46
	AMA-2	1491	47
2000	AMA-1	5010	66
	AMA-2	4942	66

Table 3 shows that the strategy of taking only the cuts for the sets-singletons initially is somewhat better. From now on, all tests will be conducted using this strategy.

Choice of the value of m   To find the best value for the parameter m for this problem, we implement the AMA-2 method using different number of groups.

We conclude from the results of Tables 4 and 5 that the best value of m for this problem is $m = 2$ and this shall be the value used hereafter.

5.4   Initialization: induced constraints

The experience taught us that the first iterations of the problems aim essentially at finding a feasible solution; the cuts associated have little influence on the final iterations. It is particularly helpful to add *a priori* constraints, thus limiting the possible investment decisions to the most useful ones (the ones giving a solution potentially feasible) [2,8].

Table 4. Different values of m for $n_s = 2000$

m	n_g	Time(sec)	Iter
2	13	4942	66
9	13	8120	114
11	15	7684	109
14	18	7813	112

Table 5. Different values of m for $n_s = 1000$

m	n_g	Time(sec)	Iter
2	11	1491	47
9	11	3052	100
11	13	3123	99

To form these constraints, we begin with the equation $p - d = St$ and we bound it:

$$d_i - \bar{p}_i \leq \sum_{j=1}^{I} \bar{t}_{ij} y_{ij} \tag{10}$$

Since the formulation of the problem (P) includes potential shortage (q), the optimal solution may include shortage for some situations of availability. However, for realistic networks, the shortage cannot be different than zero for all situations, since that would mean that some regions (some nodes) are cut-off (have no electricity) for all situations; this is an unacceptable situation. The constraints (10) correspond to the most favorable scenario and state that, at least for this scenario, the maximum production at one node plus the production brought to this node by the lines must be at least equal to the demand at this node. One equation per node is generated off-line, excluding solutions clearly unacceptable (cutting off solutions that surely lead to situations producing shortage for all scenarios).

To strengthen the above constraints (10), we can use the stochastic information: instead of using the maximum production at each node to form the constraints, we use the maximum production taking into consideration the availabilities for all the scenarios. The constraint is as in (10), but $\bar{p}_i$

represents here the maximum bound of production for all scenarios at the node i ($\max \bar{p}(i, \xi)$).

These induced constraints can be considered as the feasibility cuts we should have added if the shortage variables were not taken into account. These variables render the problem always feasible, but the shortage may be zero for some scenarios at the optimal solution.

The resolution time is reduced substantially, as seen from the Tables 6 (for $n_s = 1000$) and 7 (for $n_s = 2000$); these tables present a comparison between the solution without and with initial constraints.

Table 6. Induced constraints–$n_s = 1000$

Method	Time(sec)	Iter
Without	1491	47
With	600	24

Table 7. Induced constraints–$n_s = 2000$

Method	Time(sec)	Iter
Without	4942	66
With	2391	44

The above results have been obtained using the CPLEX software to solve the master problem (a mixed integer problem), using a simple Branch&Bound. The tolerance for the integer solution is 10^{-9} (the variables must be strictly integer); for the Branch&Bound, we assume that a solution has been found if the relative error is 0.005 for the iterations with shortage and 0.00005 for the iterations without.

6   Conclusions

An exact method to solve problems of transmission planning under uncertainty is presented here. A way of aggregating the Benders cuts is also proposed. Numerical tests show that using this method divides the CPU time by a factor varying from 1.8 (for medium-sized problems, $n_s = 1000$) to 3.8

(for large problems, $n_s = 2000$). Moreover, by adding induced constraints, the CPU time is divided by an additional factor which ranges from 2. to 2.5.

The numerical results obtained throughout this paper are therefore very encouraging. Improvements to the efficiency of the algorithm, such as addition of initial constraints involving regions (and not only nodes) or choice of a suitable value for δ are also being tested. Researches based on this approach taking into account the security constraints are currently undertaken.

Appendix

Computation of the Benders cuts

The slave subproblem is in this case the operating cost problem. To resolve it, we fix the investment plan and minimize the following problem with respect to the operating (continuous) variables, for one scenario.

$$w(\tilde{y}, \xi^j) = \min \sum_{i=1}^{n_n} c_i p_i \tag{11}$$

$$\text{s.t.:} \tag{12}$$

$$p_i - d_i = S_i t \quad \lambda_i^1 \tag{13}$$

$$v_l = \theta_i - \theta_j \quad \lambda_l^2 \tag{14}$$

$$v_l = r_l t_l + u_l \quad \lambda_l^3 \tag{15}$$

$$u_l \leq \delta(1 - \tilde{y}_l \xi^s) \quad \mu_l^1 \tag{16}$$

$$-u_l \leq \delta(1 - \tilde{y}_l \xi^s) \quad \mu_l^2 \tag{17}$$

$$t_l \leq \bar{t}_l \tilde{y}_l \xi^s \quad \mu_l^3 \tag{18}$$

$$-t_l \leq \bar{t}_l \tilde{y}_l \xi^s \quad \mu_l^4 \tag{19}$$

$$p_i \leq \bar{p}_i \xi^j \quad \nu_i^1 \xi^s \tag{20}$$

$$-p_i \leq 0 \quad \nu_i^2. \tag{21}$$

Besides each constraint, the corresponding dual variable is shown. The dual program then becomes:

$$\max - \sum_{l=1}^{n_l} \left((\mu_l^1 + \mu_l^2)\delta(1 - \tilde{y}_l \xi^j) + (\mu_l^3 + \mu_l^4)\bar{t}_l \xi^s \tilde{y}_l \right)$$

$$- \sum_{i=1}^{n_n} \left(\lambda_i^1 d_i + \nu_i^1 \bar{p}_i \xi^s \right) \tag{22}$$

$$\text{s.t.:} \quad c_i + \lambda_i^1 + \nu_i^1 - \nu_i^2 = 0 \qquad \forall i \tag{23}$$

$$\lambda_j^1 - \lambda_i^1 - \lambda_l^3 r_l + \mu_l^3 - \mu_l^4 = 0 \; \forall l \tag{24}$$

$$\lambda_l^2 + \lambda_l^3 = 0 \qquad \qquad \forall l \tag{25}$$

$$\mu_l^1 - \mu_l^2 - \lambda_l^3 = 0 \qquad \qquad \forall l \tag{26}$$

$$\sum_{l=1}^{n_l} \lambda_l^2 S_i^l = 0 \qquad \qquad \forall i \tag{27}$$

$$\mu_l^1, \; \mu_l^2, \; \mu_l^3, \; \mu_l^4, \; \nu_i^1, \; \nu_i^2 \geq 0, \; \forall l, i \, . \tag{28}$$

So, the operating cost is $w(\tilde{y}, \xi^s)$ and the sensitivity of the $w(y, \xi)$ with respect to the variables y is $\sum_{l=1}^{n_l} \left((\tilde{\mu}_l^1 + \tilde{\mu}_l^2)\delta - (\tilde{\mu}_l^3 + \tilde{\mu}_l^4)\bar{t}_l \right)$. Consequently, the Benders cuts can be written as:

$$z_0 \geq w(\tilde{y}, \xi) + \sum_{l=1}^{n_l} \left((\tilde{\mu}_l^1 + \tilde{\mu}_l^2)\delta\xi^s - (\tilde{\mu}_l^3 + \tilde{\mu}_l^4)\bar{t}_l\xi^j \right) (y_l - \tilde{y}_l) \, . \tag{29}$$

We introduce in the constraints (16), (17), (18), (19) the variables z_l^1, z_l^2, z_l^3, $z_l^4 \geq 0$ respectively, so as to have:

$$u_l + z_l^1 = \delta(1 - y_l\xi^j) \tag{30}$$

$$-u_l + z_l^2 = \delta(1 - y_l\xi^j) \tag{31}$$

$$t_l + z_l^3 = \bar{t}_l y_l \xi^j \tag{32}$$

$$-t_l + z_l^4 = \bar{t}_l y_l \xi^j \, . \tag{33}$$

To have an optimal basis to the problem, at least one of the variables z_l^1 and z_l^2 (z_l^3 and z_l^4 respectively) must be of base; otherwise, two lines of the base matrix would be proportional.

So, we have, $\forall y \in \{0, 1\}^{n_l}$, $\mu_l^1 \mu_l^2 = 0$ and $\mu_l^3 \mu_l^4 = 0$.

- If $\tilde{y}_l = 0$, we have $|u_l| \leq \delta$. As we chose δ to be large enough for the constraints to be inactive, $\mu_l^1 = \mu_l^2 = 0$ and by the constraint (26) $\lambda_l^3 = 0$. By the constraint (24), we have $\mu_l^3 - \mu_l^4 = \lambda_i^1 - \lambda_j^1$. As $\mu_l^3, \mu_l^4 \geq 0$, if $\lambda_i^1 - \lambda_j^1 \geq 0$, $\mu_l^3 = \lambda_i^1 - \lambda_j^1$, $\mu_l^4 = 0$ and if $\lambda_i^1 - \lambda_j^1 < 0$, $\mu_l^4 = -(\lambda_i^1 - \lambda_j^1)$, $\mu_l^3 = 0$. So,

$$\mu_l^3 + \mu_l^4 = |\lambda_i^1 - \lambda_j^1| \, .$$

- If $\tilde{y}_l = 1$, we dispose of μ_l^3, μ_l^4. From the constraint (24), we have $\mu_l^1 - \mu_l^2 = \frac{1}{r_l}(\lambda_j^1 - \lambda_i^1 + \mu_l^3 - \mu_l^4)$. Following the same reasoning, we conclude that:

$$\mu_l^1 + \mu_l^2 = \frac{1}{r_l}|\lambda_j^1 - \lambda_i^1 + \mu_l^3 - \mu_l^4| \, .$$

- If $\xi^j = 0$ for the line l, then we have the same case as in $\tilde{y}_l = 0$ ($|u_l| \leq \delta$ and $|t_l| \leq 0$). If $\tilde{y}_l = 0$, the subdifferential is $|\lambda_i^1 - \lambda_j^1|\bar{t}_l\xi^j = 0$ and if $\tilde{y}_l = 1$, the subdifferential is equally $|\lambda_i^1 - \lambda_j^1|\bar{t}_l\xi^j = 0$. So, if $\xi^j = 0$ for the line l, the subdifferential for this line is zero, or $|\lambda_i^1 - \lambda_j^1|\bar{t}_l\xi^j$.

So, the cuts take the form:

$$z_0 \geq w(\tilde{y},\xi) + \sum_{l=1}^{n_l} \sigma_l(\xi)(y_l - \tilde{y}_l) ,$$

where

$$t_l(\xi^j) = 0.0 \; : \sigma_l(\xi^j) = -|\tilde{\lambda}_i^1 - \tilde{\lambda}_j^1| \; \bar{t}_l(\xi^j) = 0.0$$

$$\tilde{y}_l = 0 \; : \sigma_l(\xi^j) = -|\tilde{\lambda}_i^1 - \tilde{\lambda}_j^1| \; \bar{t}_l$$

$$\tilde{y}_l = 1 \; : \sigma_l(\xi^j) = \frac{\delta}{r_l}|\tilde{\lambda}_j^1 - \tilde{\lambda}_i^1 + \tilde{\mu}_l^3 - \tilde{\mu}_l^4|$$
$$-(\tilde{\mu}_l^3 + \tilde{\mu}_l^4)\bar{t}_l .$$

Comparison with other transformations

In Sect. Sect. 3 a transformation of the problem P_1 to obtain a convex and continuous form was presented. Another one has been proposed in the literature [5], yielding a form with the same characteristics. This transformation P_2 is seen hereafter:

$$\left.\begin{array}{l} \min_{x,\,y} a^\top y + \sum_{s=1}^{n_s} \pi_s w^t(y,\xi_s) \\ \text{s.t. } y \in \mathbf{Y} \subseteq \{0,1\}^{n_l} , \end{array}\right\} P_2$$

where

$$w^t(y,\xi) \stackrel{\text{def}}{=} \min c^\top p + h^\top q$$
$$\text{s.t. } p_i - d_i = S_i^0 t_i^0 + S_i(t - t^1) \; \forall i$$
$$v_l = \theta_i - \theta_j$$
$$t_l^0 - (1/r_l)v_l = 0$$
$$t_l - (1/r_l)v_l = 0$$
$$|t_l - t_l^1| \leq \delta y_l$$
$$|t_l^1| \leq \delta(1 - \tilde{y}_l)$$
$$|t_l| \leq \bar{t}_l + \delta(1 - \tilde{y}_l)$$
$$|t_l^0| \leq \bar{t}_l$$
$$0 \leq p_i \leq \bar{p}_i .$$

When $n_s = 1$ (deterministic case), these two transformations, applied to the IEEE test network for $n_l = 20$, yield the following results:

Table 8. Comparison of the two transformations

Transformation	Time (sec)	Iter
P_1	29.384	54
P_2	994.0453	281

References

1. Artstein, Z., Wets, R.J.B. (1995) Consistency of Minimizers and the SLLN for Stochastic Programs. Journal of Convex Analysis **2**, no.1/2, 1–17
2. Boyer, F. (1997) Conception et routage des réseaux de télécommunications, PhD Thesis, Université Blaise Pascal, France
3. Dodu, J.C. (1979) Recent Improvements of the Mexico Model for Probabilistic Planning Studies. Int. J. Electrical Power & Energy Systems **1**, 46–56
4. Geoffrion, A.M. (1972) Generalized Benders Decomposition. J. Opt. Tech. Alg. **10**, no.4, 237–260
5. Granville, S., Pereira, M.V.F. (1988) Mathematical Decomposition Techniques for Power System Expansion Planning, Vol.2: Analysis of the Linearized Power Flow Model Using the Benders Decomposition. EPRI EL-5299, Project 2473-6, Stanford University
6. Kall, P., Wallace, S.W. (1994) Stochastic Programming. Wiley-Interscience Series in Systems and Optimization
7. Latorre-Bayona, G., Pérez-Arriaga, I. (1988) CHOPIN, A Heuristic Model for Long-term Transmission Expansion Planning. IEEE Transactions on Power systems **9**, no. 4, 1886–1894
8. Mahey, P., Benchakroun, A., Boyer, F. (1997) Capacity and Flow Assignment of Data Networks by Generalized Benders Decomposition. Rapport de recherche LIMOS, Universite Blaise Pascal, submitted to Networks
9. Oliveira, G.C., Costa, A.P.C., Binato, S. (1995) Large-scale Transmission Network Planning Using Optimization and Heuristic Techniques. IEEE Transactions on Power systems **10**, no.4, 1828–1834
10. Pereira, M.V.F., Pinto, L.M.V.G., Cunha, S.H.F., Oliveira, G.C. (1985) A Decomposition Approach to Automated Generation/Transmission Expansion Planning. IEEE Transactions on Power apparatus and systems **PAS-104**, no. 11, 3074–3083
11. Pereira, M.V.F., Pinto, L.M.V.G. (1985) Application of Sensitivity Analsis of Load Supplying Capability to Interactive Transmission Expansion Planning. IEEE Transactions on Power apparatus and systems **PAS-104**, no. 2, 381–389
12. Romero, R., Monticelli, A. (1994) An Hierarchical Decomposition Approach for Transmission Network Expansion Planning. IEEE Transactions on Power systems **9**, no. 1, 373–380
13. Ruszczyński, A. (1997) Decomposition Methods in Stochastic Programming. Mathematical Programming **79**, 333–353
14. Ruszczyński, A. (1986) A Regularized Decomposition Method for Minimizing a Sum of Polyhedral Functions. Mathematical Programming **35**, 309–333

15. Ruszczyński, A., Świętanowski, A. (1996) On the Regularized Decomposition Method for the Two Stage Stochastic Linear Problems. IIASA, Working paper 96-014
16. Ruszczyński, A. (1993) Parallel Decomposition of Multistage Stochastic Programming Problems. Mathematical Programming **58**, 201–228
17. Tsamasphyrou, P., Renaud, A., Carpentier, P. (1998) Transmission Network Planning: an Efficient Benders Decomposition Scheme. Submitted to the 13th Power Systems Computetation Conference.
18. Van Slyke, R.M., Wets, R.J.B. (1969) L-Shaped Linear Programs with Applications to Optimal Control and Stochastic Programming. SIAM Journal on Applied Mathematics **17**, 638–663.

On Some Recent Advances and Applications of D.C. Optimization

Hoang Tuy

Institute of Mathematics
P.O. Box 631, Bo Ho, Hanoi, Vietnam
e-mail: htuy@hn.vnn.vn

Abstract. We review some recent advances of d.c. optimization methods in the analysis and solution of specially structured nonconvex optimization problems, including problems from continuous location, nonconvex quadratic programming and monotonic optimization.

Key words: D.C. Optimization, continuous location, distance geometry, nonconvex quadratic programming, semi-definite programming, branch and bound algorithm, tight bounding, normal branching, variables decoupling, SDP relaxations

1   Introduction

D.C. Optimization is a new field of deterministic global optimization concerned with the analysis and solution of nonlinear mathematical programs in which the objective function and/or the constraint left hand side functions can be expressed as differences of convex functions (d.c. functions). Applications include problems in economics (economies of scale), finances (bond portfolio optimization), computer science (VLSI chip design, data bases), control theory (robust control), physics (nuclear design, microcluster phenomena in themodynamics), chemistry (phase and chemical reaction equlibrium, molecular conformation), ecology (design and cost allocation for waste treatment systems), etc.

In the earlier period of developement of this field, most researches focussed on general methods and algorithms designed to solve nonconvex problems with no particular structure. With the spectacular advances of computer technology, many progresses have been achieved in this direction. However, due to the inherent difficulty of global optimization, general purpose methods and algorithms currently available are able to solve only problem instances of limited size.

On the other hand, many large scale problems encountered in the applications exhibit a particular mathematical structure making them relatively tractable, despite their large dimensionality. This has motivated in recent years an increasing interest to specially structured nonconvex optimization problems. A great deal of work has been done and remarkable advances have been achieved in the study of two classes of nonconvex optimization problems: low rank nonconvex problems, i.e., roughly speaking, problems whose dimensionality can be significantly reduced by means of suitable transformations, and nonconvex quadratic problems, i.e. problems described by means of quadratic, possibly indefinite, functions. It is on these problems whose special mathematical structure can be successfully exploited for numerical purposes that deterministic global optimization methods can most clearly demonstrate their superiority over nondeterministic approaches.

Aside from fast algorithms for multiplicative programming [16] (see also [17]), and decomposition methods for monotonic optimization problems, new methods for indefinite quadratic programming have been developed in relation with applications to combinatorial optimization, control theory, distance geometry, etc.

The primary aim of this paper is to review and discuss some of these advances. In a sense this is a complement to an earlier review by the author on d.c. optimization [30]. Aside from published results, we intend to present several new materials and also point out the challenging difficulties ahead, despite the progresses already achieved.

The paper consists of 5 sections. After this brief introduction, we state in Section 2 the prototype d.c. optimization problem and discuss a generic branch and bound algorithm for solving it. In the next Sections 3 and 4 we show the efficiency of this branch and bound method as applied to continuous location problems and quadratic indefinite programming. Section 5 is devoted to some challenging problems arising from most recent applications. A new outer approximation method for monotonic problems will be discussed which seems to outperform more classical approaches when applied to these problems. Although enormous difficulties still remain ahead, these results allow us to be optimistic about the future development of d.c. optimization methods.

2   The Prototype D.C. Optimization Problem

The general problem of global optimization is

$$\text{(GOP)} \qquad \min\{f(x) \mid x \in \Omega \subset R^n\}$$

where $f(x)$ and Ω are *not both convex*, such that a local minimizer may not be a global one (multiextremality). Usually it is assumed that $f(x)$ is continuous on some open set containing $\text{cl}\Omega$, while Ω is a closed, bounded and robust set, i.e. $\Omega = \text{cl}(\text{int}\Omega)$. In that case (GOP) is a continuous global optimization problem.

A *typical* (GOP) which is also one of the most extensively studied problem of global optimization [15] is the *concave minimization* problem, which corresponds to the case when $f(x)$ is concave and Ω is convex closed. It is well known that this problem is NP-hard even when Ω is a polytope.

Obviously the class of continuous global optimization problems is very large. Despite their great variety, however, most global optimization problems share a common structure rooted in the following rather surprising fact [26].

Theorem 1. *For any closed set $S \subset R^n$ there exists a closed convex function $g_S : R^n \to R$ such that*

$$S = \{x| \; g_S(x) - \|x\|^2 \leq 0\}.$$

A function $f(x)$ which can be represented as a difference of two convex functions is called a *d.c. function* (more precisely, a d.c. function on Ω, if these convex functions are defined on a convex set Ω). An inequality of the form $f(x) \leq 0$ or $f(x) \geq 0$ where $f(x)$ is a d.c. function is called a *d.c. inequality*. If $g(x)$ is a convex function then both the convex inequality $g(x) \leq 0$ and the reverse convex inequality $g(x) \geq 0$ are d.c. inequalities. Theorem 1 says that any closed set in R^n is the solution set of a d.c. inequality.

Since (GOP) can be rewritten as $\min\{t| \; (x,t) \in S\}$ where $S := \{(x,t)| \; f(x) - t \leq 0, \; x \in \Omega\}$, it follows from Theorem 1 that any continuous global optimization problem can, in principle, be converted to the form

$$\min\{c^T x \mid g(x) - h(x) \leq 0\} \tag{1}$$

where g, h are convex functions. Also, a set of boolean constraints of the form $x_i \in \{0,1\}$, $i \in I$, is equivalent to the set of d.c. constraints $0 \leq x_i \leq 1$, $\sum_{i \in I} x_i(x_i - 1) \geq 0$. Thus, the family of d.c. functions is sufficiently rich to be used for describing virtually every continuous or discrete global optimization problem of interest.

Practically, however, the conversion of a given continuous global optimization problem to the form (1) may be a hard task which may also lead to a problem (1) with very complicated functions $g(x), h(x)$ difficult to handle. Therefore, as **prototype d.c. optimization problem** it is often more convenient to consider the following:

(P) $\qquad \min\{f_0(x) \mid f_i(x) \leq 0 \; (i = 1, \ldots, m), \quad x \in X\}$

where X is a closed convex set in R^n and each function f_i, $i = 0, 1, \ldots, m$ is a d.c. function: $f_i(x) = g_i(x) - h_i(x)$, g_i, h_i convex on X. For solving this problem, a generic branch and bound (BB) method can be developed which has proven to perform well on problems of small dimension and is sufficiently flexible to allow adaptations to solve larger but specially structured problems. Earlier variants of this method have been proposed for separable concave minimization [15], and more generally, nonconvex optimization problems with separated nonconvex variables ([29], [32], and also [28]).

As is well known, a BB algorithm is characterized by two basic operations: *branching*, i.e. partitioning the space into polyhedral domains of the same kind (simplices, rectangles, or cones), and *bounding*, i.e. estimating, for each partition set M, a lower bound $\beta(M)$ for the values of the objective function over the feasible points in M.

2.1   Bounding

For definiteness, assume that the algorithm uses simplicial partitioning of the space, so that each partition set M is an n-simplex with vertex set $V(M)$. For such a partition set M consider the subproblem:

$$(\text{P}(M)) \qquad \min\{f_0(x) \mid f_i(x) \le 0 \ (i = 1, \dots, m), \ x \in X \cap M\}$$

To estimate a lower bound $\beta(M)$ for min $\text{P}(M)$ a common method is to relax $\text{P}(M)$ to a convex problem

$$(\text{RP}(M)) \qquad \min\{\varphi_0^M(x) \mid \varphi_i^M(x) \le 0 \ (i = 1, \dots, m), \ x \in X \cap M\}$$

such that $\varphi_i^M(x)$ is a *convex minorant* of $f_i(x)$ on M $(i = 0, 1, \dots, m)$, i.e. a convex function satisfying $\varphi_i^M(x) \le f_i(x) \ \forall x \in M$. In view of the latter inequalities, the optimal value in $\text{RP}(M)$ obviously gives a lower bound $\beta(M)$ for min $\text{P}(M)$.

A convex minorant $\varphi_i^M(x)$ of $f_i(x)$ on M is said to be *tight* (more precisely, *vertex-tight*) if $\varphi_M(v) = f_i(v) \ \forall v \in V(M)$. The bounding (or the relaxation $\text{RP}(M)$) is said to be *normal* if it is such that

(*) Any vertex of M which is feasible to $\text{RP}(M)$ is also feasible to $\text{QP}(M)$ and for any nested sequence $\{M_\nu\}$ of partition sets if a vertex x^* of $M_* = \cap_{\nu=1}^{+\infty} M_\nu$ is the limit of a sequence x^ν such that x^ν is feasible to $\text{RP}(M_\nu)$, then x^* is feasible to QP.

It can easily be proved that

Proposition 1. *If $\varphi_k^M(x), k = 0, 1, \dots, m$, is a tight convex minorant of $f_k(x)$ on the partition set M and for any nested sequence of partition sets M_ν we have $\varphi_k^{M_\nu}(x^\nu) - \varphi_k^{M_\nu}(v^\nu) \to 0$ whenever $x^\nu - v^\nu \to 0$ then the bounding $M \mapsto \beta(M)$ is normal, where*

$$\beta(M) = \min\{\varphi_0^M(x) \mid \varphi_i^M(x) \le 0 \ (i = 1, \dots, m), \ x \in X \cap M\}. \qquad (2)$$

For example, if $f_i(x) = g_i(x) - h_i(x)$ with g_i, h_i convex, then a tight convex minorant of $f_i(x)$ on M is

$$\varphi_i^M(x) = g_i(x) - \psi_i^M(x)$$

where $\psi_i^M(x)$ is the affine function that matches $h_i(x)$ at every $v \in V(M)$. Hence, a normal convex relaxation of P(M) is

$$\min g_0(x) - \sum_{j=1}^{n+1} \lambda_j h_0(v^j)$$

(RP(M))
$$\text{s.t. } g_i(x) - \sum_{j=1}^{n+1} \lambda_j h_i(v^j) \quad i = 1, \ldots, m$$

$$x = \sum_{j=1}^{n+1} \lambda_j v^j \in X, \ \sum_{j=1}^{n+1} \lambda_j = 1, \ \lambda_j \geq 0.$$

Remark 1. Often one may prefer to use linear rather than convex relaxation. For each $i = 0, 1, \ldots, m$ and $v^j \in V(M)$ let $p_{ij} \in \partial g_i(v^j)$. Then a tight convex minorant of $f_i(x)$ on M is the convex piecewise linear function

$$\varphi_i^M(x) = \max_{j=1,\ldots,n+1} [\langle p_{ij}, \ x - v^j \rangle + g_i(v^j)] - \psi_i^M(x)$$

so that $\beta(M)$ can be taken to be the optimal value in the linear program:

$$\min t_0 - \sum_{j=1}^{n+1} \lambda_j h_0(v^j)$$

(LR(M))
$$\text{s.t. } t_i - \sum_{j=1}^{n+1} \lambda_j h_i(v^j) \leq 0 \qquad i = 1, \ldots, m;$$

$$\langle p_{ij}, \ x - v^j \rangle + g_i(v^j) \leq t_i \quad i = 0, 1, \ldots, m; j = 1, \ldots, n+1$$

$$x = \sum_{j=1}^{n+1} \lambda_j v^j \in X, \quad \sum_{j=1}^{n+1} \lambda_j = 1, \ \lambda_j \geq 0.$$

Note that by means of simple manipulations it is always possible to replace a set of d.c. inequalities by a single one, thus reducing the number of d.c. constraints in the problem. To avoid having too numerous linear constraints in $LR(M)$ the above linear relaxation can be applied to the resulting equivalent problem. Sometimes, to get a tighter bound, one can take a point $v^0 \in M$, e.g. $v^0 = (v^1 + \ldots + v^{n+1})/(n+1)$, and add to LR($M$) the linear constraints

$$\langle p_{i0}, x - v^0 \rangle + g_i(v^0) \leq t_i \quad i = 0, 1, \ldots, m$$

where $p_{i0} \in \partial g_i(v^0)$. Also, when the functions $f_i(x)$ are quadratic, there are several different ways to construct linear or convex relaxations, including semidefinite programming relaxations (see e.g. [28]).

2.2 Branching

At each iteration, a partition set M with lowest $\beta(M)$ is further subdivided. A common subdivision rule is bisection, in which M is partitioned into two subsimplices of equal volume by a hyperplane through the midpoint of the longest edge of M or an edge chosen in such a way to ensure *exhaustiveness* of the subdivision process. The latter means that diam $M_{k_\nu} \to 0$ for any generated infinite nested sequence of simplices $\{M_{k_\nu}, \nu = 1, 2, \dots\}$. If the bounding is such that, as $\nu \to +\infty$, $\omega(M_{k_\nu})$ tends to a feasible solution of (P) then with an exhaustive branching rule the BB algorithm will be guaranteed to converge. However the convergence achieved this way may be very slow, as it may take too many iterations for diam M_{k_ν} to become sufficiently small.

When the bounding is normal, the following subdivision rule is more efficient.

Normal Simplicial Subdivision: Select an infinite sequence $\Delta \subset \{1, 2, \dots\}$. Set $\sigma(M) = 1$ for every initial simplex M and $\sigma(M') = \sigma(M) + 1$ whenever M' is a child of M (so $\sigma(M)$ is the generation index of M). If $\sigma(M) \in \Delta$ then bisect M otherwise divide M into subsimplices via $\omega(M)$.

In this subdivision rule the choice of Δ is up to the user and can be made adaptively, from iteration to iteration, to enhance the efficiency of the algorithm.

2.3   Algorithm

Once the bounding and branching rules have been specified, a BB algorithm can be defined according to a standard scheme (see e.g. [15]. Note that at any iteration k of this scheme:

- If for some partition set M the optimal solution $\omega(M)$ of the bounding subproblem (RP(M) or LR(M)) coincides with a vertex of M then, since the bound is normal, $\omega(M)$ is a feasible solution and can thus be used to update the incumbent (current best solution) $\bar{x}^k$.

- The candidate for branching is $M_k \in \operatorname{argmin}\{\beta(M)|\ M \in \mathcal{R}_k\}$, where $\mathcal{R}_k$ denotes the collection of all partition sets still of interest at this iteration.

- The algorithm terminates when $\mathcal{R}_k = \emptyset$ (then $\bar{x}^k$ is a global optimal solution), or when $f(\bar{x}^k) - \beta(M_k) \le \varepsilon$, if a tolerance $\varepsilon > 0$ is given (then $\bar{x}^k$ is an ε-optimal solution).

We shall refer to a BB algorithm with normal bounding and normal subdivision rule as a *normal BB algorithm*.

Theorem 2. *A normal BB algorithm terminates after finitely many iterations, yielding an ε-optimal solution.*

Proof. If the algorithm is infinite, it generates at least an infinite nested sequence $\{M_{k_\nu}\}$. By compactness, we may assume that $\bar{x}^{k_\nu} \to \bar{x}$. Furthermore, by normality of the subdivision, we may assume, see e.g. [28], that $\omega(M_{k_\nu}) \to x^*$, such that x^* is a vertex of the simplex $M_\infty = \cap_{k \in K} M_{k_\nu}$. It then follows from property (*) that x^* is a feasible solution of (P). Since $\beta(M_{k_\nu}) \le \beta(M)$ for all partition sets M still of interest at iteration k_ν it follows that $f(\omega(M_{k_\nu})) \leq f(x)$ for all feasible solutions x of (P) and hence, by continuity, $f(x^*) \leq f(x)$ for all feasible solutions x of (P). The feasibility of x^* then implies its optimality, and hence also the optimality of $\bar{x}$ since $f(\omega(M_{k_\nu})) \leq \bar{x}^{k_\nu}$ for every ν. We thus have $f(\bar{x}^{k_\nu}) - \beta(M_{k_\nu}) \to 0$, and therefore $f(\bar{x}^{k_\nu}) - \beta(M_{k_\nu}) \leq \varepsilon$ for sufficiently large ν. □

Remark 2. If rectangular subdivision is used, so that every partition set M is a rectangle $M = [p, q] = \{x \mid p \leq x \leq q\}$, then a tight convex minorant of a function $f(x)$ on M is a convex function $\varphi^M(x)$ satisfying $\varphi^M(x) \leq f(x) \ \forall x \in M$ and matching $f(x)$ at every corner (vertex) of M. The concept of normal bounding is defined as in (*) while a normal subdivision is defined as follows.

Normal Rectangular Subdivision: Let $M_k = [p_k, q_k]$ be the rectangle to be partitioned, $\omega^k = \omega(M_k)$ be an optimal solution of the bounding problem RP(M_k). Define

$$\eta_i^k = \min\{\omega_i^k - p_i^k, q_i^k - \omega_i^k\}, \quad i_k \in \operatorname{argmax}\{\eta_i^k \mid i = 1, \ldots, n\}$$

and divide M_k into two subrectangles

$$M_{k,1} = M_k \cap \{x \mid x_{i_k} \geq \omega_{i_k}^k\}, \quad M_{k,2} = M_k \cap \{x \mid x_{i_k} \leq \omega_{i_k}^k\}.$$

A normal BB algorithm is one with normal bounding and normal subdivision rule. Theorem 2 is still valid for rectangular algorihms.

Thus, given a problem (P) every normal bounding method along with a normal branching rule always define a convergent algorithm. Two issues may arise when applying the d.c. approach to a nonconvex optimization problem:

1) Problem formulation: although virtually every nonconvex optimization problem theoretically belongs to the realm of d.c. optimization, in practice finding an approriate d.c. formulation for a given problem is often far from being a trivial task.

2) "Curse of dimensionality": the computational difficulty increases rapidly with the number of nonconvex variables. It is therefore of utmost importance to choose a formulation involving as less nonconvex variables as possible.

In the next sections we shall see how these issues have been resolved successfully in a number of important cases.

3 Continuous Location Problems

The general problem of continuous location is to find the unknown locations of p points in R^n (most often $n = 2$ or 3) satisfying a set of given constraints and minimizing a cost (or maximizing an utility), which is a function of the mutual distances between these points and the distances from them to N

given fixed points. If $x^1, \ldots, x^p$ are the unknown points, $a^1, \ldots, a^N$ the fixed points, and $d(x, x') = \|x - x'\|$ then the problem is to minimize a given function $F(d(X))$ subject to $X \in \mathcal{X}$, where $X = (x^1, \ldots, x^p) \in R^{pn}, \mathcal{X}$ is a given subset of R^{pn} and $d(X)$ is a vector of components $d(a^i, x^j), d(x^l, x^j), i = 1, \ldots, N, j, l = 1, \ldots, p, l \leq j$.

It turns out that in many cases such a problem can be reformulated as a d.c. optimization problem by making use of the following facts from d.c. analysis (the proofs of which can be found e.g. in [17] or [28]) :

Proposition 2. *The pointwise maximum or pointwise minimum of a finite family of d.c. functions is also a d.c. function. In particular, if $f_i(x), i = 1, \ldots, m$ are convex, then*

$$\min_{i=1,\ldots,m} f_i(x) = \sum_{i=1}^{m} f_i(x) - \max_{j=1,\ldots,m} \sum_{i \neq j} f_i(x).$$

Proposition 3. *Let $u(x), v(x)$ be convex positive valued functions on a compact convex set $M \subset R^m$ such that $u(x) - v(x) \geq 0 \ \forall x \in M$. If $q : R_+ \to R$ is a convex nonincreasing function such that $q'_+(0) > -\infty$ then for any $K \geq |q'_+(0)|$ the function $g(x) := q(u(x) - v(x)) + K[u(x) + v(x)]$ is convex; in other words:*

$$q(u(x) - v(x)) = g(x) - K[u(x) + v(x)]$$

where both $g(x), K[u(x) + v(x)]$ are convex.

Similarly, with $u(x), v(x)$ as above, if $q : R_+ \to R$ is a concave nondecreasing function such that $q'_+(0) < +\infty$ then for any $K \geq |q'_+(0)|$ we can write

$$q(u(x) - v(x)) = K[u(x) + v(x)] - g(x).$$

3.1   Generalized Weber's Problem

An important continuous location problem is the generalized Weber's problem which traditionally is formulated as follows ([7]):

Suppose that p facilities (providing the same service) are designed to serve N users located at points $a^1, \ldots, a^N$ in the plane. Each user has an amount $r_j > 0$ of goods to send out to the facilities. The cost of sending t unit from a to x is $td(a, x)$. The problem is to determine the locations of the facilities so as to minimize the total cost, i.e.

$$\min_{x,w} \sum_{i=1}^{p} \sum_{j=1}^{N} w_{ij} d(x^i, a^j)$$

$$\text{s.t.} \quad \sum_{i=1}^{p} w_{ij} = r_j, \ j = 1, \ldots, N, \tag{3}$$

$$w_{ij} \geq 0, \ i = 1, \ldots, p, \ j = 1, \ldots, N,$$

$$x^i \in S, \ i = 1, \ldots, p,$$

where $x^1, \ldots, x^p$ are the unknown locations of facilities (factories); w_{ij} is an unknown weight representing the flow from facility i to the fixed point (warehouse) a^j and S is a rectangular domain in R^2 where the facilities must be located.

A serious disadvantage of this traditional formulation is that it involves a very large number of variables (pN variables w_{ij} and $2p$ variables x^i_1, x^i_2) and a highly nonconvex objective function. This makes the problem very difficult to handle, even for small values of p and N.

Using the d.c. approach it is possible to significantly reduce the number of variables, and express the objective function in a much simpler form. Noticing that in an optimal solution each user must be served by the closest facility and setting

$$h_j(x^1, \ldots, x^p) = \min_{i=1,\ldots,p} \|x^i - a^j\| \tag{4}$$

we can rewrite the problem as:

$$\min \sum_{j=1}^{N} r_j h_j(x^1, \ldots, x^p) \quad \text{s.t.} \quad x^i \in S \subset R^2, \ i = 1, \ldots, p. \tag{5}$$

Since by Proposition 2

$$h_j(x) = \sum_{i=1}^{p} \|x^i - a^j\| - \max_k \sum_{i \neq k} \|x^i - a^j\| \tag{6}$$

where the functions $\sum_{i=1}^{p} \|x^i - a^j\|$, $\max_k \sum_{i \neq k} \|x^i - a^j\|$ are convex it follows that (5) is a *d.c. optimization problem*.

An obvious advantage of this d.c. formulation over the standard one is that it uses only $2p$ variables : $x^i_1, x^i_2, i = 1, \ldots, p$, instead of $pN+2p$ variables. Furthermore, since it is a d.c. optimization problem of the form (P), it can be practically solved by the normal BB algorithm described in Section 3. Results of solving large-scale problems with $N = 100,000, p = 1$; $N = 10,000, p = 2$; $N = 1,000, p = 3$ by this approach have been reported in [32], [1], [2]. It is worth noticing that by the traditional approach, when $N = 500, p = 2$, the formulation (3) would lead to a formidable programming problem: globally minimizing a highly nonlinear function of more than 1000 variables!

3.2 Further Extensions

The efficiency of the d.c. approach is even more apparent when the model is extended to deal with complicated but more realistic objective functions.

FACILITY WITH ATTRACTION AND REPULSION

In realistic models, the attraction of facility i to user j at distance t away is measured by a convex decreasing function $q_j(t)$. Furthermore, for some users the attraction effect is positive, for others it is negative (which amounts to a repulsion). Let J_+ be the set of attraction points (attractors), J_- the set of repulsion points (repellers). Then the problem is to find the locations with maximal total effect, i.e. such that

$$\max \sum_{j \in J_+} q_j(h_j(X)) - \sum_{j \in J_-} q_j(h_j(X)) \quad \text{s.t.} \quad X = (x^1, \dots, x^p) \in R^{2p}$$

where $h_j(X)$ is given by (6). Using Propostion 3, each function $q_j(t)$ can easily be written as a difference of two convex functions, hence the above problem is still a d.c. optimization problem in R^{2p}. The normal BB algorithm performs efficiently on this problem, provided p is not too large (see [1], [2]).

MAXIMIN LOCATION

In other circumstances, rather than maximizing the total attraction one may wish to maximize the minimum distance from a user to the closest facility. This occurs for example when the facilities are obnoxious (such as nuclear plants, garbage dumps, sewage plants, etc). With $h_j(x^1, \dots, x^p)$ defined by (4) the mathematical formulation of the maximin location problem is then:

$$\max\{\min_{j=1,\dots,N} h_j(x^1, \dots, x^p) \mid x^i \in S, \ i = 1, \dots, p\}. \tag{7}$$

Analogously, in the case of emergency facilities, such as fire stations, hospitals, patrol car centers for a security guard company, etc., the problem is to minimize the maximum distance from a user to its nearest facility. With the same $h_j(x^1, \dots, x^p)$ as above, this problem, often referred to as the *p-center problem*, can be formulated as

$$\min\{\max_{j=1,\dots,N} h_j(x^1, \dots, x^p) \mid x^i \in S, i = 1, \dots, p\}. \tag{8}$$

Using Proposition 2, a d.c. representation of each of the functions

$$\min_{j=1,\dots,N} h_j(x^1, \dots, x^p), \qquad \max_{j=1,\dots,N} h_j(x^1, \dots, x^p)$$

can easily be derived from formula (6). Therefore, (7) as well as (8) are again d.c. optimization problems solvable by the normal BB algorithm (see [1]) for the case $p = 1$.

3.3 Constrained Location

A further advantage of the d.c. approach is that it can handle equally well constrained location problems. In fact, the latter can easily be reformulated as d.c. optimization problems under d.c. constraints.

To simplify the notation, we consider only single facility problems but the reader should be aware that the same approach applies to multisource and multifacility problems ([1]).

LOCATION ON UNION OF CONVEX SETS

In most real world problems, the facility can be located only in one of several given convex regions, i.e. the feasible domain is a union of several convex sets $C_i = \{x : c_i(x) \leq 0\}, i = 1 \ldots, k$ where $c_i(x)$ are convex functions ([14]). The constraint

$$x \in \cup_{i=1}^k C_i \tag{9}$$

can be expressed as

$$\min_{i=1,\ldots,k} c_i(x) \leq 0$$

or else $p(x) - q(x) \leq 0$, with $p(x) = \sum_{i=1}^k x_i(x), \ q(x) = \max_{j=1,\ldots,k} \sum_{i \neq j} c_i(x).$

Therefore, with the objective function as discussed above, and constraint (9), the problem can be written as:

$$\max\{G(x) - H(x) : \ x \in M, \ p(x) - q(x) \leq 0\} \tag{10}$$

where $G(x), H(x), p(x), q(x)$ are convex functions on R^2, M is a convex polygon in R^2.

The problem studied in [14] and [10] is a particular case when each C_i is a polygon, i.e. each $c_i(x)$ is a polyhedral function:

$$c_i(x) = \sup_{\nu \in N_i} l_{i\nu}(x),$$

with $|N_i| < +\infty$ and $l_{i\nu}(x)$ being affine functions.

LOCATION ON AREA WITH FORBIDDEN REGIONS

Another kind of realistic constraint is that the facility should not be located in certain forbidden regions. When these are open convex sets $C_i^\circ = \{x : c_i(x) < 0\}$, where $c_i(x)$ are convex functions (see e.g. [5]), the constraint $x \notin \cup_{i=1}^k C_i^\circ$ is equivalent to $\min_{i=1,\ldots,k} c_i(x) \geq 0$, which allows the problem to be written as

$$\max\{G(x) - H(x) : x \in M, \ p(x) - q(x) \geq 0\} \tag{11}$$

with $G(x), H(x), M, p(x), q(x)$ having the same meaning as previously. This is a problem of the same type as (10), but with the roles of $p(x)$ and $q(x)$ interchanged.

GENERAL CONSTRAINED LOCATION PROBLEM

The most general situation occurs when the constraint set is a closed, not necessarily convex, set. For a single source, we can formulate the problem:

$$\max\{G(x) - H(x)|\ x \in M,\ x \in D\}, \tag{12}$$

where $G(x), H(x), M$ are as previously and D is a closed subset of R^2. By introducing an additional variable t we can rewrite the problem as

$$\min\{H(x) - t|\ x \in M,\ G(x) \le t,\ x \in D\}. \tag{13}$$

When an explicit d.c. reformulation of the constraints is not readily available, we can solve this problem by an adaptation of an earlier outer approximation method by Tuy-Thuong, called the visible point method ([17], [28]).

4   Nonconvex Quadratic Optimization

An essential feature of location problems discussed in the previous section is their low rank of nonconvexity, expressed in the fact that by suitable transformations they can be reduced to d.c. optimization problems in a relatively low-dimensional space. We now turn to d.c. optimization problems with low degree of nonconvexity, namely quadratic optimization problems, whose general formulation is

(QP) $\min\{f_0(x)|\ f_i(x) \le 0\ (i = 1, \ldots, m),\quad x \in X\}$
where X is a nonempty polyhedron in R^n and $f_i(x) = \langle x, Q_i x \rangle + 2(c_i)^T x + d_i$ with Q_i being symmetric $n \times n$ matrices, $c_i \in R^n, d_i \in R$.

Our interest in this problem is motivated by various applications in many fields of economics and engineering. In a sense quadratic models are the most natural, because any twice differentiable function can be approximated in the neighbourhood of a given point by a quadratic function. Furthermore, from the mathematical point of view, quadratic functions are the simplest smooth functions whose derivatives are readily available and easy to manipulate. This gives the hope that, despite NP-hardness, quadratic programing problems should be relatively more tractable than other nonconvex optimization problems.

In fact, since (QP) is a special case of (P), it can be solved by a normal BB algorithm as described in Section 2. Branching is most conveniently performed by using rectangular subdivision, so the central issue in this method is *normal bounding*, i.e.

Given a rectangle $M = [p, q]$, construct a **normal** relaxation of

QP(M) $\left|\begin{array}{l} \min\ f_0(x)\quad \text{subject to} \\ f_k(x) \le 0, k = 1, \ldots, m,\ x \in X \cap M \end{array}\right.$

4.1   Linear Matrix Inequalities

It turns out that many different normal relaxations for $\mathrm{QP}(M)$ are possible. These relaxations often involve convex constraints of a special type called linear matrix inequalities. Denote by $\mathcal{S}_n$ the set of symmetric $n \times n$ matrices and for any $Q \in \mathcal{S}_n$ write $Q \succeq 0$ ($Q \succ 0$, resp.) to mean that Q is a semidefinite positive matrix (definite positive matrix, resp.). A linear matrix inequality (LMI) is an inequality of the form:

$$Q(x) := Q_0 + \sum_{j=1}^{n} x_j Q_j \succeq 0 \ (\preceq 0)$$

where $Q_0, Q_1, \ldots, Q_n \in \mathcal{S}_n$. Clearly, this is a *convex inequality* because

$$\{x|\ Q(x) \succeq 0\} = \cap_{y \in R^n} \{x|\ \langle y, Q(x)y \rangle \geq 0\}$$

where each $H_y := \{x|\ \langle y, Q(x)y \rangle \geq 0\}$ is a halfspace. A finite system of LMIs $Q_1(x) \succeq 0, \ldots, Q_h(x) \succeq 0$ can be expressed as a single LMI $\mathrm{diag}(Q_1(x), \ldots, Q_h(x) \succeq 0$.

A convex program of the form

(SDP) $\min\{L(x)|\ Q(x) \succeq 0\}$

where $L(x)$ is a linear function, is called a *semidefinite program* (SDP). Any linear or convex quadratic program can be cast into a SDP. A review of semidefinite programming and its applications can be found in [35]. For our purpose the usefulness of SDPs stems from the fact that these convex programs can be efficiently solved by recently developed interior point methods ([20]; [35]). Furthermore, using SDP many feasibility problems can be converted into d.c. optimization problems. For example, the well known BMI feasibility problem in control theory:

$$F_0 + \sum_{i=1}^{m} x_i F_i + \sum_{j,k=1}^{m} x_j x_k G_{jk} \prec 0$$

$(F_i, G_{jk} \in \mathcal{S}_n)$ can be shown [6] to be equivalent to minimizing the concave function $\mathrm{Tr}(W - x^T x)$ under the LMI constraints

$$F_0 + \sum_{i} x_i F_i + \sum_{j,k} w_{jk} G_{jk} \prec 0$$

$$\begin{bmatrix} W & x^T \\ x & 1 \end{bmatrix} \succeq 0.$$

4.2   Relaxation by Tight Convex Minorants

If $\rho(Q)$ denotes the absolute value of the smallest eigenvalue of a matrix $Q \in \mathcal{S}_n$ then it is well known that for any $r \geq \rho(Q)$ the matrix $Q + rI$ is semidefinite positive, i.e. the function $x \mapsto \langle x, (Q+rI)x \rangle$ is convex. Therefore, if r is any positive number satisfying

$$r \geq \rho := \max_{k=0,1,\dots,m} \rho(Q_k) \tag{14}$$

then the functions $g_k(x) = f_k(x) + r \sum_{j=1}^{n} (x_j - p_j)(x_j - q_j)$, $k = 0, 1, \dots, m$, are convex. Since $f_k(x) = g_k(x) - r\|x\|^2$, the problem QP can then be reformulated as the d.c. optimization problem

$$
\begin{aligned}
\min\ & g_0(x) - r\|x\|^2 \\
\text{s.t.}\ & g_k(x) - r\|x\|^2 \quad k = 1, \dots, m \\
& x \in X \cap M
\end{aligned} \tag{15}
$$

Note that the number ρ in (14) can be computed by solvig the SDP

$$\min\{t|\ Q_k + tI \succeq 0\}. \tag{16}$$

Proposition 4. *Given any rectangle $M = [p, q]$ a normal relaxation of $QP(M)$ is the convex program*

$$R_0 QP(M) \qquad \left|
\begin{aligned}
& \min\ f_0(x) + r \sum_{i=1}^{n} (x_i - p_i)(x_i - q_i) \quad s.t. \\
& f_k(x) + r \sum_{i=1}^{n} (x_i - p_i)(x_i - q_i) \leq 0, k = 1, \dots, m \\
& x \in X \cap M
\end{aligned}
\right.$$

Proof. The convex envelope of $-x_j^2$ on the segment $[p_j, q_j]$ is the affine function $p_j q_j - (p_j + q_j)x_j$ matching it at the endpoints of this segment. Hence the convex envelope of $-\|x\|^2$ on $M = [p, q]$ is $\sum_{j=1}^{n}[p_j q_j - (p_j + q_j)x_j]$ and a convex minorant of $f_k(x) = g_k(x) - r\|x\|^2$ on M is

$$\varphi_k(x) := (f_k(x) + r\|x\|^2) - r \sum_{j=1}^{n}[p_j q_j - (p_j + q_j)x_j]$$

$$= f_k(x) - r \sum_{j=1}^{n}(x_j - p_j)(x_j - q_j)$$

Clearly this convex minorant is tight, and condition (*) in the definition of a normal relaxation is satisfied. $\square$

The above relaxation was used in the context of a branch and bound algorithm in ([4]). Since this relaxation is normal, as has just been proved, a normal subdivision rule could be more efficient than the standard bisection rule in the latter paper.

4.3 Lagrange relaxation

Consider the Lagrangian for problem $\text{QP}(M)$

$$L(x, u) = f_0(x) + \sum_{k=1}^{m} u_k f_k(x). \tag{17}$$

It is well known that

$$\sup_{u \geq 0} \inf_{x \in X \cap M} L(x, u) \leq \min \text{QP}(X \cap M). \tag{18}$$

However, for every $u \geq 0$ such that the quadratic function $x \mapsto L(x, u)$ is nonconvex, the problem

$$\inf_{x \in X \cap M} L(x, u) \tag{19}$$

is difficult to solve. One way out of this difficulty is to take the infimum over only those values of u for which $L(x, u)$ is a convex function of x [25]. Since this will of course degrade the lower bound, a better way to get round the difficulty and obtain a normal relaxation is the following.

Suppose that $X = R^n$. Since $p_j \leq x_j \leq q_j$ if and only if $(x_j - p_j)(x_j - q_j) \leq 0$, we can replace the linear constraints $p \leq x \leq q$ by the set of n quadratic constraints

$$f_{k+j}(x) := (x_j - p_j)(x_j - q_j) \leq 0, \ j = 1, \ldots, n.$$

Thus, in the case $X = R^n$ we can rewrite $\text{QP}(M)$ as an all-quadratic program

$$\min\{f_0(x) |\ f_k(x) \leq 0, \ k = 1, \ldots, N\} \tag{20}$$

where $N = m + n$ and every function $f_k, k = 0, 1, \ldots, N$ is quadratic. For the Lagrangian of this problem

$$L(x, u) = f_0 + \sum_{k=1}^{m} u_k f_k(x) + \sum_{j=1}^{n} u_{m+j}(x_j - p_j)(x_j - q_j) \tag{21}$$

the constraint $x \in X \cap M$ in (18) should be replaced by $x \in R^n$, i.e.

$$\sup_{u \geq 0} \inf_{x \in R^n} L(x, u) \leq \min \text{QP}(M). \tag{22}$$

But now we have

$$\psi(u) := \inf_{x \in R^n} L(x, u) = -\infty \text{ if } L(x, u) \text{ is nonconvex,}$$

so we can write

$$\sup\{\psi(u) |\ u \geq 0\} = \sup\{\psi(u) |\ L(x, u) \text{ is convex in } x\}. \tag{23}$$

Proposition 5. *If at least one feasible point of* $QP(M)$ *does not coincide with a vertex of* M, *the lower bound for* $QP(M)$ *given by (23) is at least as tight as the lower bound given by* $R_0 QP(M)$.

Proof. By taking $u_{m+j} = \rho/(m+1)$, where ρ is defined by (16), we have the lower bound:

$$\sup_{u \in R_+^m} \inf_{x \in R^n} \{f_0(x) + \rho \sum_{j=1}^n (x_j - p_j)(x_j - q_j) + \sum_{k=1}^m u_k[f_k(x) + \rho \sum_{j=1}^n (x_j - p_j)$$
$$(x_j - q_j)]\} \tag{24}$$

But since there exists at least one feasible point x of $QP(M)$ satisfying $\sum_{j=1}^n (x_j - p_j)(x_j - q_j) < 0$, i.e. a point x such that

$$f_k(x) + \rho \sum_{j=1}^n (x_j - p_j)(x_j - q_j) < 0 \qquad k = 1, \ldots, m$$

the regularity condition is satisfied for the convex program $R_0 QP(M)$. It then follows from the duality theorem for convex programs that the optimal value of $R_0 QP(M)$ is equal to the optimal value of (24) .   □

Note that $L(x, u) = \langle x, Q(u)x \rangle + 2\langle c(u), x \rangle + d(u)$ where

$$Q(u) = Q_0 + \sum_{k=1}^N u_k Q_k, \quad c(u) = c_0 + \sum_{k=1}^N u_k c_k, \quad d(u) = d_0 + \sum_{k=1}^N u_k d_k,$$

and $Q_{m+j}, j = 1, \ldots, n$, is an $n \times n$ matrix with 1 at the entry (j, j) and 0 everywhere else. From (23) a lower bound for $QP(M)$ is $\sup\{\psi(u)| \ u \geq 0\} = \sup\{t| \ \psi(u) - t \geq 0\}$, but obviously $\psi(u) \geq t$, i.e. $\psi(u) > -\infty$ if and only if $L(x, u) \geq 0 \ \forall x \in R^n$, so

$$\psi(u) - t \geq 0 \Leftrightarrow \begin{bmatrix} x \\ 1 \end{bmatrix}^T \begin{bmatrix} Q(u) & c(u) \\ c(u)^T & d(u) - t \end{bmatrix} \begin{bmatrix} x \\ 1 \end{bmatrix} \succeq 0.$$

Therefore a lower bound for $QP(M)$ is

$$\max\{t| \ \begin{bmatrix} Q_0 & c_0 \\ c_0^T & d_0 - t \end{bmatrix} + u_1 \begin{bmatrix} Q_1 & c_1 \\ c_1^T & d_1 \end{bmatrix} + \ldots + u_N \begin{bmatrix} Q_N & c_N \\ c_N^T & d_N \end{bmatrix} \succeq 0, \ u \geq 0\}. \tag{25}$$

It can be checked that the dual to this SDP is the SDP

$$\min \ \mathrm{Tr} \ WQ_0 + c_0^T x + d_0 \quad \text{subject to}$$
$$\mathrm{Tr} \ WQ_k + c_k^T x + d_k \leq 0 \quad k = 1, \ldots, N,$$
$$\begin{bmatrix} W & x \\ x^T & 1 \end{bmatrix} \succeq 0.$$

which is a normal relaxation of $QP(M)$ (see subsection 4.5)

4.4 Decoupling Relaxation

The nonconvexity of $QP(M)$ is due to the presence of products $x_i x_j$ in which the variables are coupled. By replacing every product $x_i x_j$ in the expanded form of each quadratic function $f_k(x)$ with a new variable w_{ij}, these quadratic functions of x become affine functions of (x, w), namely:

$$\tilde{f}_k(x, w) = \sum_{i \leq j} a_{ij}^{(k)} w_{ij} + \sum_j c_{kj} x_j + b_k.$$

The problem $QP(M)$ then becomes

$$\tilde{QP}(M) \quad \left| \begin{array}{l} \min \ \tilde{f}_0(x, w) \quad \text{subject to} \\ \tilde{f}_k(x, w) \leq 0, \quad k = 1, \dots, m, \\ x \in X \cap M \\ w_{ij} = x_i x_j \ \forall i, j \end{array} \right.$$

where all the nonconvexity is now concentrated on the coupling constraints

$$w_{ij} = x_i x_j \quad \forall i, j.$$

Consider the set $E(M)$ of all (x, w) satisfying

$$x \in X \cap M \tag{26}$$

$$w_{ij} = x_i x_j \quad \forall i, j \tag{27}$$

A convex set $C(M) \supset E(M)$ is said to be a normal relaxation of $E(M)$ if

(**) Any corner of M which belongs to $C(M)$ also belongs to $E(M)$ and for any nested sequence $\{M_\nu\}$, whenever a corner x^* of $M_* = \cap_\nu M_\nu$ is the limit of a sequence $\{x^\nu\}$ of points $x^\nu \in C(M_\nu)$ then $x^* \in E(M_*)$.

Clearly by relaxing $E(M)$ to such a set $C(M)$ the resulting convex problem

$$\tilde{RP}(M) \quad \left| \begin{array}{l} \min \ \tilde{f}_0(x, w) \quad \text{subject to} \\ \tilde{f}_k(x, w) \leq 0, \quad k = 1, \dots, m, \\ (x, w) \in C(M) \end{array} \right.$$

will be a normal relaxation of $QP(M)$. Thus, for different sets $C(M)$ satisfying condition (**) we will have different normal boundings.

LINEAR RELAXATION
Let the polytope $X \cap M$ be described by the linear inequalities

$$\langle a^i, x \rangle \leq \alpha_i, \quad i = 1, \dots, l. \tag{28}$$

where $l \geq 2n$ and the inequalities $\langle a^i, x \rangle \leq \alpha_i, i = 1, \dots, 2n$, correspond to $p_j \leq x_j \leq q_j, j = 1, \dots, n$. It is easily seen that

$$x \in X \cap M \ \Leftrightarrow \ \left| \begin{array}{l} (\langle a^i, x \rangle - \alpha_i)(x_j - p_j) \leq 0; \\ (\langle a^i, x \rangle - \alpha_i)(q_j - x_j) \leq 0 \\ (i = 1, \dots, l; \ j = 1, \dots, n) \end{array} \right. \tag{29}$$

Now following [23], for any given quadratic function $f(x)$ denote by $[f(x)]_\ell$ the affine function of (x, w) obtained by replacing every product $x_i x_j$ in the expanded form of $f(x)$ with w_{ij}. Then by the above equivalence (29) $E(M)$ is contained in the polyhedron

$$[(\langle a^i, x \rangle - \alpha_i)(x_j - p_j)]_\ell \leq 0; \quad [(\langle a^i, x \rangle - \alpha_i)(q_j - x_j)]_\ell \leq 0 \quad \forall i, j. \tag{30}$$

Furthermore,

Proposition 6. *If (x, w) satisfies (30) and x is a corner of $M = [p, q]$ then (x, w) satisfies (27), i.e. x is a feasible solution to $QP(M)$.*

Proof. The system (30) includes all the inequalities

$$[(x_i - p_i)(p_j - x_j)]_\ell \leq 0, \quad [(x_i - p_i)(x_j - q_j)]_\ell \leq 0 \quad 1 \leq i \leq j \leq n \tag{31}$$

i.e.

$$g_{ij}(x_i, x_j) := \max\{p_i x_j + p_j x_i - p_i p_j, \; q_i x_j + q_j x_i - q_i q_j\} \leq w_{ij} \\ w_{ij} \leq \min\{q_i x_j + p_j x_i - p_j q_i, \; p_i x_j + q_j x_i - p_i q_j\} := h_{ij}(x_i, x_j). \quad \forall i \leq j. \tag{32}$$

But if (x, w) satisfies the latter system and x is a corner of $M = [p, q]$ then $w_{ij} = x_i x_j \; \forall i, j.$ □

As a consequence, the polyhedron (30) is a normal relaxation of $E(M)$. Hence a normal relaxation of $QP(M)$ is the linear program:

$$LR(M) \quad \min \; \tilde{f}_0(x, w) \quad \text{s.t.} \tag{33}$$
$$\tilde{f}_k(x, w) \leq 0 \quad k = 1, \ldots, m \tag{34}$$
$$\begin{aligned}[(\langle a^i, x \rangle - \alpha_i)(x_j - p_j)]_\ell \leq 0 \\ [(\langle a^i, x \rangle - \alpha_i)(q_j - x_j)]_\ell \leq 0\end{aligned} \quad i = 1, \ldots, l; \; j = 1, \ldots, n. \tag{35}$$

This relaxation is similar to the Reformulation-Linearization (RL) relaxation primarily proposed for linearly constrained quadratic programs in [23] (see also [24]). The RL relaxation involves many more linear constraints, namely all constraints of the form

$$[(\langle a^i, x \rangle - \alpha_i)(\langle a^j, x \rangle - \alpha_j)]_\ell \geq 0 \quad 2n + 1 \leq i \leq j \leq l.$$

It is not known to which extent these additional constraints may improve the bound. When $X = R^n$ the system (35) coincides with (32), see a related relaxation in [3].

4.5   SDP Relaxation

Denote by W the symmetric $n \times n$ matrix with elements w_{ij}. Condition $w_{ij} = x_i x_j \; \forall i, j$ can then be written as

$$W = xx^T. \tag{36}$$

which is equivalent to the inequality system [6]:

$$W \succeq xx^T \tag{37}$$

$$\text{Tr}(W - xx^T) \le 0. \tag{38}$$

Here (37) is a convex inequality because

$$W \succeq xx^T \Leftrightarrow \begin{bmatrix} W & x \\ x^T & 1 \end{bmatrix} \succeq 0,$$

while (38) is a reverse convex inequality because $\text{Tr}(W - xx^T) = \sum_{i=1}^{n}(w_{ii} - x_i^2)$ is a concave function of (x, w).

Noting that $\tilde{f}_k(x, w) = \text{Tr}WQ_k + c_k^T x + d_k$ we can thus reformulate QP(M) as a *convex program with an additional reverse convex constraint*, namely:

$$\min \text{Tr}WQ_0 + c_0^T \quad \text{subject to}$$
$$\text{Tr}WQ_k + c_k^T + d_k \le 0 \quad k = 1, \dots, m,$$
$$x \in X \cap M$$
$$\begin{bmatrix} W & x \\ x^T & 1 \end{bmatrix} \succeq 0$$
$$\text{Tr}(W - xx^T) \le 0$$

Lemma 1. *The convex envelope of the function* $\text{Tr}(W - xx^T)$ *on the set* $\{(x, w) | \; p \le x \le q\}$ *is*

$$\sum_{i=1}^{n} [w_{ii} - (p_i + q_i)x_i + p_i q_i]. \tag{39}$$

Proposition 7. *A normal relaxation of QP(M) is*

$$\text{DR}(M) \qquad \min \text{Tr}WQ_0 + c_0^T \quad \text{subject to} \tag{40}$$
$$\text{Tr}WQ_k + c_k^T + d_k \le 0 \quad k = 1, \dots, m, \tag{41}$$
$$\begin{bmatrix} W & x \\ x^T & 1 \end{bmatrix} \succeq 0. \tag{42}$$
$$\sum_{i=1}^{n} [w_{ii} - (p_i + q_i)x_i + p_i q_i] \le 0. \tag{43}$$

Proof. The Lemma follows from the fact that $\text{Tr}(W - xx^T) = \sum_{i=1}^{n}(w_{ii} - x_i^2)$ is a concave separable function of (x, w). To prove the Proposition , it suffices to check that if (x, w) satisfies DR(M) and x is a corner of M then x is a feasible solution of QP(M). But this is immediate because if (x, w) satisfies (43) and x is a corner of M, i.e. $x_i \in \{p_i, q_i\}$ $\forall i$, then $(p_i + q_i)x_i - p_i q_i = x_i^2$ $\forall i$, hence $\text{Tr}(W - xx^T) \leq 0$.   □

In several currently known SDP relaxations the equality (36) is relaxed to (37) or even simply to $W \succeq 0$ which is implied by (37) (see e.g. [12] and references therein).

Remark 3. We have thus obtained a full range of normal relaxations. While by Theorem 2 any one of these relaxations can be incorporated into a normal BB algorithm with guaranteed convergence, the choice of a proper relaxation for solving a given problem (QP) must be decided on the basis of a trade-off between the necessary computational effort and the desired quality of the bounds.

5   Challenging Problems: Monotonic Optimization

Many functions arising from applications are monotonic in the following sense: there exists a convex cone $K \subset R^n$, inducing a partial ordering $\succeq_K$ on R^n such that

$$x' \succeq_K x \Rightarrow f(x') \geq f(x).$$

This property is so frequent that solving optimization problems involving monotonic functions has become a fascinating research subject. From a computational viewpoint, the most attractive feature of this property is that in the search for a minimizer of a K-monotonic function $f(x)$ over a feasible set, once a feasible point x is known then the whole feasible region in the cone $x + K$ can be ignored. In particular, when the lineality space L of K is nontrivial then the problem is reduced to a search over the quotient space R^n/L, which is a lower dimensional space. An important amount of research work has been devoted in recent years to exploiting this feature for the decomposition of so called low rank nonconvex problems [17] in which monotonicity is combined with convexity or reverse convexity in some nice way.

When $K = R_+^n$, as it is often the case, K-monotonic functions are called *increasing functions*. Given two continuous increasing functions $f, g : R_+^n \to R_+$, a natural optimization problem is to

$$\text{maximize} \quad f(x) \quad \text{subject to} \quad g(x) \leq 1, \ x \in R_+^n. \tag{44}$$

Heuristically, this problem can be interpreted as maximizing the utility under a budget constraint on the cost. Assume that the set $G = \{x \in R_+^n \,|\, g(x) \leq 1\}$

is nonempty and compact and write the problem as

(MP) $$\max\{f(x)|\ x \in G\}$$

It is easy to check that G has the following properties:

(i) G is a compact subset of R^n_+ such that $x' \in G$ whenever $0 \leq x' \leq x$ and $x \in G$.

(ii) For every $x \in R^n_+ \setminus G$ there exists a unique point $\pi(x)$ on the ray through x such that $\pi(x) \in G$ but $\lambda \pi(x) \notin G$ for any $\lambda > 1$. Furthermore, $\{x \in G|\ x > \pi(x)\} = \emptyset$. In other words, the cone $K_{\pi(x)} := \pi(x) + R^n_{++}$ strictly separates x from G.

Although the problem (MP), with $f(x)$ increasing on R^n_+ and G satisfying the assumptions (i)(ii), may be highly nonconvex, it turns out that it can be solved efficiently by a very simple algorithm described below [21], [33].

Assume that $G \subset [0, b] \subset R^n_+$ and for a given tolerance $\varepsilon > 0$ let $a = (\varepsilon, \dots, \varepsilon)$. POLYBLOCK OUTER APPROXIMATION ALGORITHM

0. Start with the initial box $P_1 = [0, b]$. Set $x^1 = b, V_k = \{x^1\}, k = 1$.

1. Let z^k be the point on the ray through x^k such that $g(z^k) = 1$, but $g(\alpha z^k) > 1$ for all $\alpha > 1$.

2. Compute the n vertices of the box $[z^k, x^k]$ that are adjacent to x^k :

$$x^{k,i} = x^k - (x^k - z^k)e^i \quad i = 1, \dots, n$$

where e^i is the i-unit of R^n. Let $I_k = \{i|\ x^{k,i} > a\}$.

3. Set $V_{k+1} = (V_k \setminus \{x^k\}) \cup \{x^{k,i}|\ i \in I_k\}$. Compute $x^{k+1} \in \operatorname{argmax}\{f(x)|\ x \in V_{k+1}\}$. Increase k by 1 and return to Step 1.

For a proof of convergence and a report of numerical results with this algorithm we refer the reader to the papers [21], [33].

For any finite set $V \subset R^n_+$ the set $P = \cup_{z \in V}[0, z]$ is called a *polyblock* of vertex set V. Thus the above algorithm generates a nested sequence of polyblocks $P_1 \supset P_2 \supset \dots \supset G \cap H$, such that the vertex V_{k+1} of P_{k+1} is obtained from V_k by replacing $x^k \in V_k$ with $\{x^{k,i}, i \in I_k\}$. Since the maximum of an increasing function over a polyblock is obviously achieved at one of its vertices, x^k is a maximizer of $f(x)$ over P_k (see Step 3) and it can be proved that

$$f(x^k) \searrow \max\{f(x)|\ x \in G,\ x \geq a\}.$$

The name "polyblock outer approximation algorithm" refers to the fact that the algorithm generates a nested sequence of polyblocks outer approximating the feasible set $G \cap H$ more and more closely.

It is worth pointing out two important classes of nonconvex optimization problems which can be reduced to the form (MP) and solved by the polyblock outer approximation algorithm.

Example 1.

$$\max\{f(g(y))|\ y \in D\} \tag{45}$$

where $D \subset R^m$ is a compact convex set, $g : D \to R^n_{++}$ is a continuous (component-wise) convex mapping and $f(x)$ is an increasing function on R^n_+. A typical case is the convex multiplicative maximization problem:

$$\max\{\prod_{i=1}^n g_i(y)|\ y \in D\}$$

which corresponds to $f(x) = \prod_{i=1}^n x_i$. Clearly we can rewrite (45) as

$$\max\{f(x)|\ y \in D, g(y) \geq x\}$$
$$= \max\{f(x)|\ x \in G\}.$$

where $f(x)$ is increasing on R^n_+ while $G = \{x \in R^n_+|\ x \leq g(y),\ y \in D\}$ satisfies conditions (i) and (ii). Note that here both f and G are nonconvex.

Example 2.

$$\min\{f(g(y)|\ y \in D\} \tag{46}$$

where D, g, f are as previously. A typical case is the ordinary convex multiplicative programming problem:

$$\min\{\prod_{i=1}^n g_i(y)|\ y \in D\}$$

which corresponds to $f(x) = \prod_{i=1}^n x_i$. Setting $z_i = 1/x_i, 1/z = (1/z_1, \ldots, 1/z_m)$ we can rewrite (46) as

$$\min\{f(x)|\ y \in D, g(y) \leq x\}$$
$$= \min\{f(1/z)|\ y \in D, g(y) \leq 1/z\}$$
$$= \max\{\frac{1}{f(1/z)}|\ y \in D, g(y) \leq 1/z\}$$
$$= \max\{F(z)|\ z \in G\}$$

where $F(z) = 1/f(1/z)$ is increasing while $G = \{z \in R^n_{++}|\ 1/z \geq g(y), y \in D\}$ satisfies conditions (i) and (ii).

Thus the polyblock outer approximation method can be applied to a wide class of monotonic optimization problems. We close this paper by mentioning two challenging problems arising from the applications.

Example 3. (Distributing many points on a sphere)

$$\max \prod_{i,j} \|x^i - x^j\| \quad \text{s.t.} \quad \|x^i\|^2 = 1 \quad i = 1, \dots, N \tag{47}$$

([22] and references therein). This problem can be rewritten as

$$\max \prod_{i<j} y_{ij} \quad \text{s.t.} \quad x^i \in S, \ 0 \le y_{ij} \le \|x^i - x^j\| \quad 1 \le i < j \le N. \tag{48}$$

where S is the unit ball. Setting $G = \{y = (y_{ij})| \ 0 \le y_{ij} \le \|x^i - x^j\|(1 \le i < j \le N), \ x^i \in S\}$ it is easily seen that G is a set satisfying conditions (i) and (ii) in the orthant $\{y_{ij} \ge 0, 1 \le i < j \le N\}$. Thus, the problem (47) reduces to one of the form (MP). Obviously, considerable difficulties still remain in view of the size of the problem (the number of variables y_{ij}) which very quickly increases with N. Nevertheless, it may be of interest to investigate the above approach to this problem for relatively small values of N.

Example 4. (Molecular Conformation)
Given a cluster of p atoms (in R^3), locate their centers $x^1, \dots, x^N$ so as to minimize the potential energy function

$$V_N(x^1, \dots, x^N) = \sum_{1 \le i < j \le N} v(\|x^i - x^j\|)$$

where $v(r)$ is the interatomic pair potential (see e.g. [18]).
   For the Lennard-Jones model

$$v(r) = \frac{1}{r^{12}} - \frac{2}{r^6}$$

it has been shown in [36] that the minimum is at most $-(N-1)$ and that an optimal solution must satisfy $\|x^i - x^j\| \ge 0.5 \ \forall i < j$. One can thus assume $r \ge 0.5$.
   For $r \ge 0.5$ the function $v_1(r) = 1/r^{12}$ is a decreasing convex function of r with the derivative at $r = 0.5$ equal to $v_1'(0.5) = -12/r^{13} > -\infty$, so for $L = 12/(0.5)^{13}$ the function $v_1(r) + Lr$ is convex increasing. Since $r_{ij} = \|x^i - x^j\|$ is a convex function of $(x^1, \dots, x^N)$, it follows that $v_1(\|x^i - x^j\|) + L\|x^i - x^j\|$ is a convex function of $(x^1, \dots, x^N)$. Similarly, since $L \ge |v_2'(0.5)| = 6/(0.5)^7$ the function $v_2(\|x^i - x^j\|) + L\|x^i - x^j\|$ is a convex function of $(x^1, \dots, x^N)$. Therefore, $V_N(x^1, \dots, x^N) = F_1(x^1, \dots, x^N) - F_2(x^1, \dots, x^N)$, where

$$F_1(x^1, \dots, x^N) = \sum_{1 \le i \le j \le N} \left(\frac{1}{\|x^i - x^j\|^{12}} + L\|x^i - x^j\| \right)$$

$$F_2(x^1, \dots, x^N) = \sum_{1 \le i \le j \le N} \left(\frac{1}{\|x^i - x^j\|^6} + L\|x^i - x^j\| \right)$$

are convex functions of $(x^1, \ldots, x^N)$. The problem thus reduces to finding the global minimum of a d.c. function of $(x^1, \ldots, x^N)$ under the constraints $\|x^i - x^j\| \geq 0.5$ $(1 \leq i \leq j \leq N)$. Unfortunately, again the size of this d.c. optimization problem is very large even for moderate values of N.

References

1. Al-Khayyal, F. A., Tuy, H., Zhou, F. (1997) Large-scale single facility continuous location by d.c. optimization. Preprint, School of Industrial and Systems Engineering, Georgia Institute of Technology, Atlanta
2. Al-Khayyal, F.A., Tuy, H., Zhou, F. (1997) D.C. optimization methods for multisource location problems. Preprint, School of Industrial and Systems Engineering, Georgia Institute of Technology, Atlanta
3. Al-Khayyal, F.A., Larsen, C., Van Voorhis, T. (1995) A relaxation method for nonconvex quadratically constrained quadratic program. Journal of Global Optimization **6**, 215–230
4. Androulakis, I.P., Maranas, C.D., Floudas, C.A. (1995) αBB: A global optimization method for general constrained nonconvex problems. Journal of Global Optimization **7**, 337–363
5. Aneja, Y.P., Parlar, M. (1994) Algorithms for Weber facility location in the presence of forbidden regions and/or barriers to travel. Transportation Science **28**, 70–216
6. Apkarian, P., Tuan, H.D. (1998) Robust control via concave optimization: local and global algorithms. Proceedings of the 37th Conference on Decision and Control
7. Brimberg, J., Love, R.F. (1994) A location problem with economies of scale. Studies in Location Analysis **7**, 9–19
8. Chen, R. (1983) Solution of minisum and minimax location-allocation problems with euclidean distances. Naval Research Logistics Quaterly **30**, 449–459
9. Chen, R. (1988) Conditional minisum and minimax location-allocation problems in Euclidean space. Transportation Science **22**, 157–160
10. Chen, P., Hansen, P., Jaumard, B., Tuy, H. (1992) Weber's problem with attraction and repulsion. Journal of Regional Science **32**, 467–409
11. Floudas, C., Visweswaran, V. (1995) Quadratic Optimization. In: Horst R., Pardalos P. (Eds.), Handbook of Global Optimization. Kluwer, 217–269
12. Fujie, T., Kojima, M. (1997) Semidefinite programming relaxation for nonconvex quadratic programs. Journal of Global Optimization **10**, 367–380
13. Hansen, P., Jaumard, B., Tuy, H. (1995) Global optimization, in location. In: Zvi Dresner (Ed.) Facility Location: a Survey of Applications and Methods. Springer, Heidelberg, Berlin, 43–67
14. Hansen, P., Peeters, D., Thisse, J.F. (1982) An algorithm for a constrained Weber problem. Management Science **28**, 1285–1295
15. Horst, R., Tuy, H. (1996) Global Optimization. Springer, Heidelberg, Berlin, 3rd edition
16. Konno, H., Kuno, T. (1995) Multiplicative Programming Problems. In: Horst R., Pardalos P. (Eds.) Handbook of Global Optimization. Kluwer , 369–406
17. Konno, H., Thach, P.T., Tuy, H. (1997) Optimization on low rank nonconvex structures, Kluwer

18. Leary, R.H. (1997) Global optima of Lennard-Jones clusters. Journal of Global Optimization 11, 35–53
19. Maranas, C.D., Floudas, C.A. (1993) A global Optimization method for Weber's problem with attraction and repulsion. In: Hager W.W., Heran D.W., Pardalos P.M. (Eds.) Large Scale Optimization: State of the Art. Kluwer, 1–12
20. Nesterov, J.E., Nemirovski, A.S. (1994) Interior Point Polynomial Methods in Convex Programming: Theory and Applications. SIAM, Philadelphia
21. Rubinov, A., Tuy, H., Mays, H. (1998) Algorithm for a Monotonic Global Optimization Problem, Preprint, University of Ballarat, Ballarat, Australia
22. Saff, E.B., Kuijlaars, A.B.J. (1997) Distributing many points on a sphere. Mathematical Intelligencer 10, 5–11
23. Sherali, H.D., Tuncbilek, C.H. (1995) A reformulation-convexification approach to solving nonconvex quadratic programming problems. Journal of Global Optimization 7, 1–31
24. Sherali, H.D., Tuncbilek, C.H. (1997) New reformulation/linearization/ convexification relaxations for univariate and multivariate polynomial programming problems. Operations Research Letters 21, 1–9
25. Shor, N.Z., Stetsenko, S.I. (1989) Quadratic extremal problems and nondifferentiable optimization (in Russian). Naukova Dumka, Kiev
26. Thach, P.T. (1993) D.C. sets, D.C. functions and nonlinear equations. Mathematical Programming 58, 415–428
27. Tuy, H. (1995) D.C. Optimization: Theory, Methods and Algorithms. In: Horst R., Pardalos P. (Eds.) Handbook of Global Optimization. Kluwer, 149–216
28. Tuy, H. (1998) Convex Analysis and Global Optimization. Kluwer
29. Tuy, H. (1992) On nonconvex optimization problems with separated nonconvex variables. Journal of Global Optimization 2, 133–144
30. Tuy, H. (1995) D.C. Optimization: Theory, Methods and Algorithms. In: Horst R., Pardalos P. (Eds.) Handbook on Global Optimization. Kluwer, 149–216
31. Tuy, H. (1996) A General D.C. Approach to Location Problems. In: Floudas C.A., Pardalos P.M. (Eds.) State of the Art in Global Optimization: Computational Methods and Applications. Kluwer, 413–432
32. Tuy, H., Alkhayyal, F.A., Zhou, F. (1995) D.C. optimization method for single facility location problem. Journal of Global Optimization 7, 209–227
33. Tuy, H. (1999) Normal sets, polyblocks and monotonic optimization. To appear in Vietnam Journal of Mathematics, vol. 27.
34. L. Vandenberge, L., Boyd, S. (1995) A primal-dual potential reduction method for problems involving matrix inequalities. Mathematical Programming 69, 205–206
35. Vandenberge, L., Boyd, S. (1996) Semidefinite Programming. SIAM Review 38, 49–95
36. Xue, G.L. (1997) Minimum inter-particle distance at global minimizers of Lennard-Jones clusters. Journal of Global Optimization 11, 83–90

Proceedings of the previous French–German Conferences on Optimization were published in the following volumes:

- First Conference (Oberwolfach 1980): Optimization and Optimal Control, edited by A. Auslender, W. Oettli and J. Stoer (Lecture Notes in Control and Information Sciences, 30) Springer–Verlag, Berlin and Heidelberg, 1981.
- Second Conference (Confolant, 1981): Optimization, edited by J.B. Hiriart-Urruty, W. Oettli and J. Stoer (Lecture Notes in Pure and Applied Mathematics, 86) Marcel Dekker, New York and Basel, 1983.
- Third Conference (Luminy, 1984): Third Franco–German Conference in Optimization, edited by C. Lemaréchal. Institut National de Recherche en Informatique et en Automatique, Rocquencourt, 1984.
- Fourth Conference (Irsee, 1986): Trends in Mathematical Optimization, edited by K.-H. Hoffmann, J.B. Hiriart–Urruty, C. Lemaréchal and J. Zowe (International Series of Numerical Mathematics, 84) Birkhäuser Verlag, Basel and Boston, 1988.
- Fifth Conference (Varetz, 1988): Optimization, edited by S. Dolecki (Lecture Notes in Mathematics, 1405) Springer–Verlag, Berlin and Heidelberg, 1989.
- Sixth Conference (Lambrecht 1991): Advances in Optimization, edited by W. Oettli and D. Pallaschke (Lecture Notes in Economics and Mathematical Systems, 382) Springer–Verlag, Berlin and Heidelberg, 1992.
- Seventh Conference (Dijon, 1994): Recent Developments in Optimization, edited by R. Durier and Ch. Michelot (Lecture Notes in Economics and Mathematical Systems, 429) Springer–Verlag, Berlin and Heidelberg, 1995.
- Eigth Conference (Trier, 1996): Recent Advances in Optimization, edited by P. Gritzmann, R. Horst, E. Sachs and R. Tichatschke (Lecture Notes in Economics and Mathematical Systems, 452) Springer–Verlag, Berlin and Heidelberg, 1997.

Lecture Notes in Economics
and Mathematical Systems

For information about Vols. 1–295
please contact your bookseller or Springer-Verlag

Vol. 434: M. W. J. Blok, Dynamic Models of the Firm. VII, 193 pages. 1996.

Vol. 435: L. Chen, Interest Rate Dynamics, Derivatives Pricing, and Risk Management. XII, 149 pages. 1996.

Vol. 436: M. Klemisch-Ahlert, Bargaining in Economic and Ethical Environments. IX, 155 pages. 1996.

Vol. 437: C. Jordan, Batching and Scheduling. IX, 178 pages. 1996.

Vol. 438: A. Villar, General Equilibrium with Increasing Returns. XIII, 164 pages. 1996.

Vol. 439: M. Zenner, Learning to Become Rational. VII, 201 pages. 1996.

Vol. 440: W. Ryll, Litigation and Settlement in a Game with Incomplete Information. VIII, 174 pages. 1996.

Vol. 441: H. Dawid, Adaptive Learning by Genetic Algorithms. IX, 166 pages.1996.

Vol. 442: L. Corchón, Theories of Imperfectly Competitive Markets. XIII, 163 pages. 1996.

Vol. 443: G. Lang, On Overlapping Generations Models with Productive Capital. X, 98 pages. 1996.

Vol. 444: S. Jørgensen, G. Zaccour (Eds.), Dynamic Competitive Analysis in Marketing. X, 285 pages. 1996.

Vol. 445: A. H. Christer, S. Osaki, L. C. Thomas (Eds.), Stochastic Modelling in Innovative Manufactoring. X, 361 pages. 1997.

Vol. 446: G. Dhaene, Encompassing. X, 160 pages. 1997.

Vol. 447: A. Artale, Rings in Auctions. X, 172 pages. 1997.

Vol. 448: G. Fandel, T. Gal (Eds.), Multiple Criteria Decision Making. XII, 678 pages. 1997.

Vol. 449: F. Fang, M. Sanglier (Eds.), Complexity and Self-Organization in Social and Economic Systems. IX, 317 pages, 1997.

Vol. 450: P. M. Pardalos, D. W. Hearn, W. W. Hager, (Eds.), Network Optimization. VIII, 485 pages, 1997.

Vol. 451: M. Salge, Rational Bubbles. Theoretical Basis, Economic Relevance, and Empirical Evidence with a Special Emphasis on the German Stock Market.IX, 265 pages. 1997.

Vol. 452: P. Gritzmann, R. Horst, E. Sachs, R. Tichatschke (Eds.), Recent Advances in Optimization. VIII, 379 pages. 1997.

Vol. 453: A. S. Tangian, J. Gruber (Eds.), Constructing Scalar-Valued Objective Functions. VIII, 298 pages. 1997.

Vol. 454: H.-M. Krolzig, Markov-Switching Vector Auto-regressions. XIV, 358 pages. 1997.

Vol. 455: R. Caballero, F. Ruiz, R. E. Steuer (Eds.), Advances in Multiple Objective and Goal Programming. VIII, 391 pages. 1997.

Vol. 456: R. Conte, R. Hegselmann, P. Terna (Eds.), Simu-lating Social Phenomena. VIII, 536 pages. 1997.

Vol. 457: C. Hsu, Volume and the Nonlinear Dynamics of Stock Returns. VIII, 133 pages. 1998.

Vol. 458: K. Marti, P. Kall (Eds.), Stochastic Programming Methods and Technical Applications. X, 437 pages. 1998.

Vol. 459: H. K. Ryu, D. J. Slottje, Measuring Trends in U.S. Income Inequality. XI, 195 pages. 1998.

Vol. 460: B. Fleischmann, J. A. E. E. van Nunen, M. G. Speranza, P. Stähly, Advances in Distribution Logistic. XI. 535 pages. 1998.

Vol. 461: U. Schmidt, Axiomatic Utility Theory under Risk. XV, 201 pages. 1998.

Vol. 462: L. von Auer, Dynamic Preferences, Choice Mechanisms, and Welfare. XII, 226 pages. 1998.

Vol. 463: G. Abraham-Frois (Ed.), Non-Linear Dynamics and Endogenous Cycles. VI, 204 pages. 1998.

Vol. 464: A. Aulin, The Impact of Science on Economic Growth and its Cycles. IX, 204 pages. 1998.

Vol. 465: T. J. Stewart, R. C. van den Honert (Eds.), Trends in Multicriteria Decision Making. X, 448 pages. 1998.

Vol. 466: A. Sadrieh, The Alternating Double Auction Market. VII, 350 pages. 1998.

Vol. 467: H. Hennig-Schmidt, Bargaining in a Video Ex-periment. Determinants of Boundedly Rational Behavior. XII, 221 pages. 1999.

Vol. 468: A. Ziegler, A Game Theory Analysis of Options. XIV, 145 pages. 1999.

Vol. 469: M. P. Vogel, Environmental Kuznets Curves. XIII, 197 pages. 1999.

Vol. 470: M. Ammann, Pricing Derivative Credit Risk. XII, 228 pages. 1999.

Vol. 471: N. H. M. Wilson (Ed.), Computer-Aided Transit Scheduling. XI, 444 pages. 1999.

Vol. 472: J.-R. Tyran, Money Illusion and Strategic Complementarity as Causes of Monetary Non-Neutrality. X, 228 pages. 1999.

Vol. 473: S. Helber, Performance Analysis of Flow Lines with Non-Linear Flow of Material. IX, 280 pages. 1999.

Vol. 474: U. Schwalbe, The Core of Economies with Asymmetric Information. IX, 141 pages. 1999.

Vol. 475: L. Kaas, Dynamic Macroelectronics with Imperfect Competition. XI, 155 pages. 1999.

Vol. 476: R. Demel, Fiscal Policy, Public Debt and the Term Structure of Interest Rates. X, 279 pages. 1999.

Vol. 477: M. Théra, R. Tichatschke (Eds.), Ill-posed Variational Problems and Regularization Techniques. VIII, 274 pages. 1999.

Vol. 478: S. Hartmann, Project Scheduling under Limited Resources. XII, 221 pages. 1999.

Vol. 479: L. v. Thadden, Money, Inflation, and Capital Formation. IX, 192 pages. 1999.

Vol. 480: M. Grazia Speranza, P. Stähly (Eds.), New Trends in Distribution Logistics. X, 336 pages. 1999.

Vol. 481: V. H. Nguyen, J. J. Strodiot, P. Tossings (Eds.). Optimations. IX, 498 pages. 2000.